Historical & Cultural Astronomy

The Historical & Cultural Astronomy series includes high-level monographs and edited volumes covering a broad range of subjects in the history of astronomy, including interdisciplinary contributions from historians, sociologists, horologists, archaeologists, and other humanities fields. The authors are distinguished specialists in their fields of expertise. Each title is carefully supervised and aims to provide an in-depth understanding by offering detailed research. Rather than focusing on the scientific findings alone, these volumes explain the context of astronomical and space science progress from the pre-modern world to the future. The interdisciplinary Historical & Cultural Astronomy series offers a home for books addressing astronomical progress from a humanities perspective, encompassing the influence of religion, politics, social movements, and more on the growth of astronomical knowledge over the centuries.

The Historical & Cultural Astronomy Series Editors are: Wayne Orchiston, Marc Rothenberg, and Cliff Cunningham.

Clifford J. Cunningham
Editor

Cosmic Events

The Three Comets of 1618

Springer

Editor
Clifford J. Cunningham
University of Southern Queensland, Toowoomba, Australia

ISSN 2509-310X ISSN 2509-3118 (electronic)
Historical & Cultural Astronomy
ISBN 978-3-031-86949-5 ISBN 978-3-031-86950-1 (eBook)
https://doi.org/10.1007/978-3-031-86950-1

This Springer imprint is published by the registered company Springer Nature Switzerland AG
The registered company address is: Gewerbestrasse 11, 6330 Cham, Switzerland

Preface

> Experience teaches us, that all such things as are wonderful and extraordinary are not seen every day, Comets don't appear but at the distance of some Ages, Monsters, Deluges, Eruptions of Vesuvius, and Earthquakes happen rarely, and this uncommonness of the Appearance gives a Lustre and Beauty to abundance of things which suddenly lose it when they become too frequent.

This quote from the French scholar Gabriel Naudé (1600–1653), published in 1639 with an English translation from 1711 (pg 70–71), places comets in two contexts. Being listed first, comets were considered the most prominent exemplars of wonderful and uncommon events. Furthermore, they are critical agents that impart luster and beauty to our senses. Naudé likely had the extraordinary sight of the Great Comet of 1618 on his mind when he wrote this, as he was an impressionable 18 when it appeared in the sky.

The three comets of 1618 were the first to be observed through a telescope. The third of these, the Great Comet, achieved posthumous importance accorded to few other comets, as it was widely regarded as the harbinger of the Thirty Years' War that caused eight million deaths in Europe. This resonated throughout the seventeenth century. Samuel Fritz, a cook in Erfurt, wrote a chronicle *Cosmographia* between 1665 and 1676. He was only eight years old when he saw the Great comet, but he vividly recalled it and its baleful consequences.

> In the year 1618, this comet stood as a broomstick. Thence followed the Bohemian Unrest [a revolt against the Hapsburg monarchy], for God swept Germany a lot with the fiery broomstick of His grim Wrath. (quoted in Kurihara, 2014: 155)

Here again is Naudé:

> ...this great Circle of the Universe, since it has first begun its Course, has not ceased to introduce and change Monarchies, Religions, Sects, Cities, Men, Beasts, Trees, Stones, and generally all that is comprehended in this vast Machine; nor are the Heavens themselves free from their Change and their Corruption. (Naudé, 1711:124)

Nothing signified change in the heavens more than a bright comet, with 1618 being a bumper year: in addition to the Great Comet, which appeared in November and lasted into January 1619, there were two other comets that year. This book, in ten chapters written by the world's leading scholars in cometary history, sets out to examine these comets both individually and holistically.

The view of comets in the early seventeenth century was aptly described by the English philosopher Francis Bacon (1561–1626); this edited book examines all the cometary aspects that concerned him:

> Comets, out of question, have likewise Power and Effect, under the Grosse and Masse of Things: But they are rather gazed upon, and waited upon in their journey, then wisely observed in their Effects; Specially in their Respective Effects; That is, what Kinde of Comet, for Magnitude, Colour, Version of the Beames, Placing in the Region of Heaven, of Lasting, produceth what Kinde of Effects. (Bacon 1625: 332)

Following an Introduction by Paul Gabor SJ, this book (which had its origin in 2020) is divided into three broad categories. In Section 1, there are four chapters that consider the print publications that flowed from the presses in Europe during and after 1618. Many of these books and pamphlets included imagery associated with the comets, which are covered in two chapters. Poetry was also an essential component of many engagements with the Great Comet; this book's longest chapter takes a detailed look at the English-language poetry, with a few others in Dutch and Latin, some of which have previously only appeared in manuscript.

In Section 2, there are four more chapters with an international look at how the three comets were observed and perceived in Germany, Spain, Portugal, India, and China. The Great Comet was also seen by Native Americans in the area of Massachusetts: they thought it had to be a message from a manitou, and "…expected some strange things to follow" (Lipman, 2024: 116).

In Section 3, there are two chapters on how comets were widely perceived as harbingers of doom. The role of the Great Comet of 1618 in the Thirty Years' War is explored, as well as its astrological influence. While King James I of England, whose poem on the Great Comet is studied in detail in this book, discounted astrology by adopting a scientific attitude towards its appearance, even he recognized the comet embodied potential danger to his kingdom.

Recent scholarship emphasizes the effect of the comets on seventeenth-century life.

> In the arts, "baroque" generally designates an esthetics emphasizing the play of illusion and "truth," image and reality, and the overall drama of subjective perception and its "objects." Anne-Laure De Meyer (2021) convincingly expands the significance of the term to include the awareness of instability and crisis in every realm of early modern life. She mentions the Reformation, the climatological fluctuation that produced what might be called the "mini-ice age," the drop in population, chronic warfare, and the appearance of three comets in 1618, as factors contributing to a general sense of bewilderment and insecurity. De Meyer's work complements that of Anthony J. Cascardi in showing that what we think of as modern epistemology and science grew in and from acute awareness of incomprehension and danger. (Riggs, 2022: 35)

In the early modern era of the seventeenth century, the distinction between science and pseudo-science was not as clearly drawn as it has been since the Enlightenment. How the work of Johannes Kepler (1571–1630) was misused is a case in point, as recently studied by Ivo Schneider, emeritus professor of the history of science at the Universität der Bundeswehr, Munich. Kepler produced an *Ephemeris* for 1618, "In which for September the ecliptical longitude of Mars and the ecliptical longitude of the moon as well were fixed at $3^0\,33'$." Johannes Faulhaber (1580–1635), a teacher of elementary mathematics in Ulm "…had against all reason and every scientific rule interpreted the double occurrence of this value as the double occurrence of the number 333, that is, as the Number of the Beast 666, and had construed this as a divine indication of the occurrence of a special event" (Schneider, 2024: 360). When he saw the second comet of 1618, Faulhaber (1619) trumpeted his vindication, even though it was seen by him in November, not September as he predicted! In a Latin booklet of 1619, the principal of the grammar school in Ulm where Faulhaber taught launched a broadside against him. Johann Baptist Hebenstreit (1619) "extensively attacked the possibility of predicting the appearance of comets on the basis of Cabbalistic number speculations." (Schneider, 2024: 360). This induced the Catholic Church to convene a closed colloquium in the autumn of 1619, to determine if divine inspiration was involved. "Faulhaber's testimony that he owed his knowledge about biblical numbers only to his zeal when studying arithmetic, and to prayer, saved him from further sanctions" (Schneider, 2024: 361). It was one of the several controversies engendered by the great comets of 1572 (observed by Thomas Harriot and Tycho Brahe) and 1618 and the recent supernovas of 1572 (studied by Tycho) and 1604 (studied by Kepler). These sightings "…signalled for most contemporaries the beginnings of the final acts of history and the culmination of a decisive battle between the forces of good and evil" (Gibson, 1997: 71). It prompted Lutheran Paul Nagel to pen three books in 1619 dealing with the comets of 1618, culminating in 1620 with his book *Prodromus Astronomiae Apocalypticae*, literally, *Forewarning of the Astronomical Apocalypse* (Penman, 2009).

This book concludes with an Appendix on a digital library about a major controversy surrounding Galileo, Orazio Grassi, and the three comets. This is still a major topic of study, as witnessed in a recent book by Javier Loira (2024). He delves into an aspect of the comets that is worth quoting. Part 1 of his book, Comets,

> starts off with a lecture by Jesuit mathematician Orazio Grassi delivered in the winter of 1619. Humans are so made, Grassi maintained, that they feel excitement at new and rare phenomena that seem to defy nature's usual course. Take comets, for instance. They alter the existence of a world that would be perhaps more orderly without them but undeniably less thrilling. One would say nature introduces them artificially merely to avoid appearing repetitive. As one of Grassi's contemporaries aptly put it, conceits were 'poetry's comets,' which likewise catered to the hunger for novelty and artifice that is congenital to humans.

Speaking of poetry, it is examined in this book not just for its own sake but as a way of imparting information to a wider readership. It had long been a method of disseminating new discoveries:

> Columbus' "letter" describing what he thought he had accomplished, first written in Spanish and printed in Barcelona about April 1, 1493, was translated into Latin as *De Insulis Inventis* and published in Rome before the end of that month…and by mid-June 1493 the Latin Letter had been translated into a 68-stanza poem and published in Tuscan, the dialect of Florence. (Boorstin, 1995:21)

The poet William Cullen Bryant, in *The Ages of 1822* (Godwin, 1883), wrote

> On, like the comet's way through infinite space,
> Stretches the long untraveled path of light,
> Into the depths of ages.

The Great Comet of 1618, in particular, has blazed its way into the depths of ages, as evidenced by the continuing scholarly interest in that fiery appearance just over four centuries ago. As I commend the great intellectual effort of my co-authors to those who will further explore cometary history, I recall the enlightened thought of Johann Gottfried Herder in the Preface to his 1784 book *Ideas for the Philosophy of the History of Mankind.*

> The author fancied himself in the circle of those who actually took an interest in what he was writing and from whom he wished to entice their sympathetic and nobler thoughts. This is the finest reward that writing can bestow. (Herder, 2024: 4)

I conclude the Preface with a poem especially commissioned for this book. It was written in October 2024 by Conlan Salgado, a recent graduate in philosophy at Roosevelt University in Chicago.

The Great Comet of 1618

Galileo said that it was merely a grand effect.
The mathematician grabbed it by its tail
And found nothing solid, but only a number.
The orator wrote, "I know now how God throws away
His excess miracles." The astrologer read
The heavenly rhetoric, and found in it grave threats,
Not wasted miracles. Is not peace a wasted miracle?
Wondered the soldier who was secretly a father,
Secretly a husband, secretly a brother.
In this September of centuries, the long trail of light
Moved across millions of miles of darkness. Rilke thought:
How better to measure distance between two people?

Toowoomba, Australia — Clifford J. Cunningham

References

Bacon, F. (1625). *The Essayes or counsels, Civill and Morall.* John Haviland.
Boorstin, D. J. (1995). *Cleopatra's nose: Essays on the unexpected.* Vintage Books.
Douglas, M. (1975). *Implicit meanings: Essays in anthropology.* Routledge & Kegan Paul.

Faulhaber, J. (1619). *Fama Siderea Nova*. Daniel Mögling.
Gibson, K. (1997). Apocalyptic and millenarian prophecy in early Stuart Europe: Philip Ziegler, Ludwig Friedrich Gifftheil and the fifth monarchy. In T. Thornton (Ed.), *Prophecy: The power of inspired language in history 1300–2000*. Sutton Publishing.
Godwin, P. (Ed.). (1883). *The poetical works of William Cullen Bryant, 2 vols*. New York.
Hebenstreit, J. (1619). *De cabala log-Arithmo-Geometro-Mantica*. Johann Meder.
Herder, J. G. (2024). In G. M. Moore (Ed.), *Ideas for the philosophy of the history of mankind*. Princeton, Princeton University Press.
Kurihara, K. (2014). *Celestial wonders in reformation Germany*. Pickering & Chatto.
Lipman, A. (2024). *Squanto: A native odyssey*. Yale University Press.
Loira, J. (2024). *The age of subtlety: Nature and rhetorical conceits in early modern Europe*. University of Delaware Press.
Meyer, Anne-Laure de. (2021). *Sir Kenelm Digby (1603–1665)*. Honore Champion.
Nagel, P. (1620). *Prodromus Astronomia Apocalyptica*. Martin Rode.
Naudé, G. (1711/1623). In W. King (Ed.), *Political Considerations upon the Master-Strokes of State as practised by the Ancients and Moderns*. H. Clements.
Penman, L. T. I. (2009). Repulsive blasphemies: Paul Nagel's appropriation of unprinted works of Jakob Bohme and Valentin Weigel in his *Prodromus astronomiae apocalypticae*. *Daphnis, 38*, 599–622.
Riggs, L. W. (2022). Review of book by de Meyer. *Seventeenth Century News, 80*, 1–2.
Schneider, I. (2024). Climbing the social ladder: Johannes Faulhaber's path from schoolmaster to fortification engineer. In P. Beeley & C. D. Hollings (Eds.), *Beyond the learned academy: The practice of mathematics 1600–1850* (pp. 339–364). Oxford University Press.
Shapin, S. (2010). *Never pure*. Johns Hopkins University Press.

Contents

Part II The International Dimension

Part III Portents, Omens and Astrology

Editor and Contributors

About the Editor

Clifford J. Cunningham earned his PhD in the history of astronomy from the University of Southern Queensland in Australia (2016), where he is now a Research Fellow in the Astrophysics Group. His undergraduate degrees in science and classical studies are from the University of Waterloo in Canada. He wrote *Introduction to Asteroids* (1988), a five-volume series for Springer on nineteenth century asteroid research, seven volumes to date in the *Collected Correspondence of Baron Franz von Zach*, and (as editor) *The Scientific Legacy of William Herschel* (Springer, 2018). Reaktion Press published his book *Asteroids* (2022) and a book he co-authored, *The Solar System* (2025). He is currently editing a volume in Bloomsbury's *Cultural History of the Universe*. He was appointed by Springer as a Series Editor of their Historical & Cultural Astronomy book series in 2019. He is also an Associate Editor of the *Journal of Astronomical History and Heritage* (JAHH), a contributor to *Encyclopedia Britannica*, and from 2001 to 2025 the history of astronomy columnist for *Mercury* magazine. His scientific research ranging from ancient astronomy to Milton's *Paradise Lost* has been published in the *Journal for the History of Astronomy*, *Renaissance & Reformation*, *JAHH*, and *Annals of Science*. Asteroid (4276) was named Clifford in his honor in 1990 by the International Astronomical Union based on the recommendation of its bureau, the Harvard-Smithsonian Center for Astrophysics.

Contributors

Andreas Bähr European University Viadrina Frankfurt (Oder), Frankfurt (Oder), Germany

Nicholas Campion University of Wales Trinity Saint David, Lampeter, UK

Luís Miguel Carolino ISCTE - Instituto Universitário de Lisboa, CIES-IUL, Lisbon, Portugal

Clifford J. Cunningham University of Southern Queensland, Toowoomba, Australia

Paul Gabor Vatican Observatory, Vatican, Vatican City State

Miguel Á. Granada University of Barcelona, Barcelona, Spain

Doris Gruber Institute for Habsburg and Balkan Studies, Austrian Academy of Sciences, Vienna, Austria

R. C. Kapoor Indian Institute of Astrophysics, Bengaluru, India

David W. Pankenier Lehigh University, Bethlehem, PA, USA

Sara J. Schechner Department of History of Science, Harvard University, Cambridge, MA, USA

Giancarlo Truffa Società Italiana degli Storici della Fisica e dell'Astronomia, Padova, Italy

Chapter 1
Introduction

Paul Gabor

1.1 Introduction

The three comets of 1618 were not just a diversion to admire with your family and friends at a star party. The modern designations of the three comets are C/1618 Q1, V1, and W1. The first two, Q1 and V1, were less bright. The third one, W1, qualifies as a Great Comet (see the three https entries in the *References* section). Their appearance played an important historical role. They stimulated a discussion that led to an unfortunate rift between Galileo (who theorized that comets were not physical bodies but rather atmospheric effects) and several key Jesuit scholars (Drake & O'Malley, 1960). The study of these comets had a more durable impact, however. They were the first to be ever observed by telescopes, and as such, they represent a major milestone in the history of observational astronomy and later astrophysics. They were also the first to be observed with the view of determining their orbits by Kepler (1619). Both of these lines of study are active to this day.

They also played a significant role in the history of thought. Aristotelian natural philosophy of comets was challenged by Tycho already in 1577 when he determined that the Great Comet of that year (Fig. 1.1) could not possibly have been closer than the Moon, contrary to Aristotle, who had placed comets in the sublunary realm (Aristotle, n.d.-b*Meteorologica,* I, 7). The third comet of 1618, i.e., the Great Winter Comet, was the first Great Comet since 1577, and, staying visible for more than 2 months, it provided a great opportunity to advance the study of comets.

Last but not least, the Great Comet understandably agitated the general public, stimulating people's imagination, sometimes in fatal ways. This had significant political consequences. In particular, the Europeans could hardly be blamed for perceiving it as an omen of misery. This perception remained present in the popular

P. Gabor (✉)
Vatican Observatory, Vatican, Vatican City State
e-mail: pgabor@specola.va

C. J. Cunningham (ed.), *Cosmic Events*, Historical & Cultural Astronomy,
https://doi.org/10.1007/978-3-031-86950-1_1

Fig. 1.1 The Great Comet of 1577 over Prague. Engraving by Jiří Jakub Dašický. From a report entitled "About a terrible and marvelous comet as it appeared in the sky on the Tuesday after St. Martin's Day this current year 1577 of our Lord" (1577 Nov 12) written in German by Peter Codicillus of Tulechov

imagination for decades to come. After all, the comets came at a time when the Thirty Years' War was getting underway, and just a few weeks before the unexpected death, without legitimate issue, of the Holy Roman Emperor Matthias at the age of 62. In China, reactions to the Great Comet contributed to the demise of one dynasty and the accession of another.

1.2 Astronomy and Natural Philosophy

When Galileo first observed astronomical bodies with a telescope in 1609, he inaugurated a new era of astronomical research. This event was not an outlier, however. This was a time of significant and lasting cultural and social change. It would be overly ambitious of me to attempt an overview of all the upheaval in the late sixteenth and early seventeenth centuries worldwide. Let us merely mention a few aspects of the role astronomy played in these events.

Galileo's report on the details of the Moon's surface, published in his *Sidereus Nuncius* (Galileo 1610), stirred imaginations far and wide. If the "optical tube" (the word telescope had not entered wide use yet) demonstrates the presence of mountains and plains on the Moon, showing that it is similar to the Earth, many felt free to conjecture that the Moon is like the Earth in all respects. A simple syllogism then leads to an inescapable conclusion: The Moon is like the Earth in all respects, and the Earth is inhabited, therefore the Moon is likewise inhabited. Many proposed travel to the Moon. Some even proposed colonizing it. After all, at this point in history, the Europeans had just effectively usurped the whole of the New World in just a few decades. Considering how unequal their technology was to such a titanic coup, was it too great a stretch to envisage voyaging to the Moon and subjugating its denizens?[1]

Galileo's discovery of the Moon as an Earth-like body was of great philosophical significance. The heavens and the earth, the celestial and the earthly, represent polar opposites in the symbolic perception of the world. Heavenly bodies were perfect and unchanging, while the Earth was the opposite. It was Nicholas of Cusas's 1440 book *De Docta Ignorantia* that proposed an alternative. What if there is no fundamental difference between the Heavens and the Earth?[2] Galileo confirmed what Cusanus conjectured.

Cardinal Nicholas may not have been the first to say this,[3] but he certainly is the one whose views influenced those who observed the comets of 1618.[4] His thought was based on some Platonist ideas mixed with some mathematics. If the material Universe is imperfect, then the putative shells carrying the planets and other celestial bodies cannot be perfectly spherical, and they cannot have a perfectly defined center where the Earth would be.[5] But, argued Cusanus, if the mathematical perfection implied in the very notion of a center is absurd in an imperfect Universe, then the Earth cannot be the locus of elemental earth. In fact, the Earth may not differ fundamentally from the Heavens, the earthly from the celestial. Heavenly bodies may be like the Earth, and the Earth may be a heavenly body. And that is what Galileo's observations of the Moon demonstrated.

---

[1] Just to give one example, Bishop John Wilkins (Oliver Cromwell's brother-in-law, and a founding member of the Royal Society) writes: "Keplar [sic] doubts not, but that as soone as the art of flying is found out, some of their Nation will make one of the first colonies that shall inhabite that other world." (Wilkins 1638: 210) While Kepler (1634) was joking, Wilkins was not.

[2] "Therefore, the Earth is a noble star (II, 12: 166) Hence, since that entire region is unknown to us, those inhabitants remain altogether unknown. [...] But we are able to know disproportionally less about the inhabitants of another [cosmic] region [than we do about distant regions on Earth]. We surmise that in the solar region there are inhabitants which are more solar, brilliant, illustrious, and intellectual—being even more spiritlike than [those] on the moon, where [the inhabitants] are more (lunatic) than [those] on the earth, [where the latter are] more material and (coarse)." (II, 12: 171).

[3] Namely, John Philoponus of Alexandria. Even if Nicholas of Cusa knew this work (unlikely), his argument is structured differently from John's analysis.

[4] Copernicus refers to Cusa in the first book of *De Revolutionibus* (Nicholas Copernicus 1543).

[5] According to Aristotle (*De Coelo* Aristotle n.d.-a), the Earth was at the center of the Universe not because it was *the* privileged location in a spherical Universe, but rather because elemental earth's "natural location" was the Center, and the Earth as a body was simply the result of elemental earth's natural aggregation around the Center.

Let us also briefly note why it took astronomers nearly a decade after 1609 to finally observe comets with a telescope. Early telescopes were not very good for discovering comets, or indeed any other new celestial bodies. Their field of view was too small, the optical aberrations too distracting, and the mounts were too primitive. As a consequence, astronomers had a much greater chance of discovering comets with the naked eye than with such telescopes. Comets sufficiently bright for such discoveries are very rare, however.

My grandmother was born when the Great January Comet of 1910 (C/1910 A1) was dazzling humanity. It was visible even during daytime and had a tail up to 50° long. Recall the comets you have seen during your own lifetime. Comet C/1995 O1 Hale-Bopp was visible to the unaided eye for more than a year in 1996–97, reaching magnitude–1. Comet C/2006 P1 McNaught was significantly brighter but only graced observers in the Earth's southern hemisphere. Neither quite achieved the awe-inspiring stature of true Great Comets.

There is no doubt that the three comets of 1618 caused general agitation. The first two were, at first, noted only by specialists, but they became portentous in the context of the Great Winter Comet. What did they look like? As an introduction to the times and paths of their appearance on the celestial sphere, Sara Schechner's chapter in this volume contains an excellent summary, followed by a wide range of depictions, Miguel Granada's chapter contains gorgeous illustrations and descriptions by Wilhelm Schickard, and Giancarlo Truffa's chapter contains excellent ones by Orazio Grassi. The depictions and descriptions in Granada's and Truffa's chapters are by astronomers. Doris Gruber's chapter provides a representative example of a pamphlet with an engraving by a popular author, Lukas Schultes, presenting the Great Winter Comet as a *Wunderzeichen,*[6] and Sara Schechner's chapter provides a wide-ranging typology of cometary images.

Generally speaking, very bright comets are once-in-a-lifetime spectacles. The *International Comet Quarterly* provides a statistical estimate: "a comet that is brighter than fourth magnitude [can be seen on average] once every two years, a comet brighter than second mag once every 5.5 years or so, a comet brighter than first magnitude every 10 years or so, and a comet brighter than mag 0 roughly once every 15 years."[7] Note that a 0th magnitude comet may be cute, but it will not be an unforgettable spectacle. Comet C/1973 E1 Kohoutek reached magnitude −3 (about 16 times brighter than 0th magnitude), and it still became synonymous with spectacular disappointment.

It is therefore understandable that no comets were discovered between 1609 and 1618. There were no comets bright enough to be discovered with the unaided eye, and the telescope was not ready to help yet. Once a comet was discovered with the naked eye, then telescopes were brought to bear on it, providing new research

[6] Etymologically "wonder-sign" but semantically closer to "awesome/awful sign."

[7] "Brightest comets seen since 1935." ICQ web, http://www.icq.eps.harvard.edu/brightest.html (consulted Sep 1, 2024).

opportunities, studying the structure of the tail, its connection to what was considered the head of the comet, etc.

In particular, the careful study of comets, their position in space with respect to the Sun, and in particular the orientation of their tails, provided insights into the question of whether comets had their own light (fire) or whether they were merely "reflecting" (scattering) sunlight? More on this can be found in the chapters by Luis Carolino and Miguel Granada.

The dominant technical debate, however, centered on the location of comets in the Universe and on cometary paths. As we have already noted, the Aristotelian system made a strict division between the heavens and the earth. The limit was the sphere of the Moon. The sublunary world was composed of earth, water, wind, and fire, while aether, the fifth element or quintessence, was the only physical substance present in the supralunary world. The very idea of a vacuum was deemed absurd by Aristotle. Therefore, the heavenly spheres, or, more accurately, spherical shells, spun within each other, in full physical contact, the friction generating the *music or harmony of the spheres.*

Naturally, there was no room or place for comets, and other transient phenomena, in the supralunary world. What did empirical evidence say? In 1572, Tycho Brahe determined that a *stella nova* in Cassiopeia (new star; its modern designation is supernova SN 1572) was beyond the lunar sphere. True to his chosen name,[8] Tycho was lucky to observe another rare phenomenon just a few years later: the Great Comet of 1577 (C/1577 V1) allowed him to conduct further triangulation or parallactic measurements, determining that the comet was likewise beyond the sphere of the Moon. Following these important observations, the year 1618, with its three comets, brought a new set of opportunities to study, and perhaps decide once and for all, an important question of natural philosophy.

Johannes Kepler published his first two laws of planetary motion in 1609 and the third one in 1619, the same year he published his treatise on comets. The observational data he used were on Halley's Comet (1P/Halley) observed in 1607 by the Imperial Mathematician[9] himself, as well as the three comets of 1618. He was the first to observe comets with a telescope, which, for lesser humans, would have been the achievement of a lifetime. For Kepler, it barely registers among all of his other discoveries. Although his conclusions may not have been entirely correct, in his 1619 book, Kepler developed mathematical techniques to determine the trajectories of comets (Fig. 1.2).

One might think that the place of the comets would have instantly become *the* clinching argument with the power to disprove Aristotelian natural philosophy as a whole. Instead, most thinkers chose to tweak the properties of aether (making it compatible with the passage of comets) and to distinguish two types of cometary phenomena: the fiery Aristotelian sublunary ones and the new aethereal supralunary ones. This view was popularized by the influential Jesuit authors Athanasius Kircher

[8]Tycho is derived from τυχη, the Greek for luck or lucky.

[9]This was the title Kepler cherished the most.

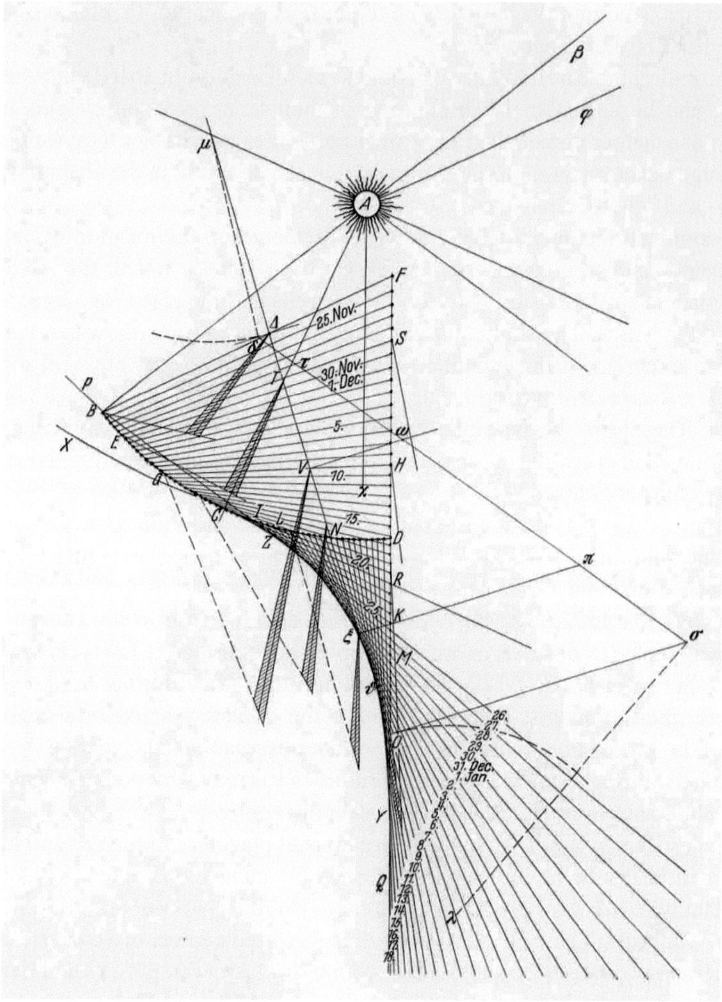

Fig. 1.2 Johannes Kepler: De Cometis libelli tres. Augsburg, 1619. Calculated path of the Great Winter Comet (i.p. 83 & 84)

(in his *Iter exstaticum*) and Kaspar Schott (who prepared the second, expanded edition of this popular textbook). It would be interesting to trace its origins. It seems that Wilhelm Schickard may have been a precursor of this strategy (see Andreas Bähr's chapter).

While the discovery of the land masses of the New World had made it clear that Aristotle's writings erred by omission regarding some specific "details" of our world, it would have been another matter entirely to reject the Aristotelian natural philosophy as a whole. What contributed to the intellectual inertia in this respect was the fact that there was no alternative system ready to provide a notional framework for natural philosophy and a basis for educational curricula. Change in the education system was (then as it is now) a slow process. Aristotelian natural philosophy would remain the backbone of school curricula until the eighteenth century. It was probably the need for a philosophical new system that made Cardinal de Bérulle tell Descartes to focus on philosophy,[10] and Rev. Pierre Gassendi (Gassendi, 1649) to suggest a revival of Epicureanism with its atomist natural philosophy.

I have mentioned two reasons why the comets of 1577 and 1618 did not cause an immediate rejection of Aristotelian natural philosophy: scholarly inertia and the ability to incorporate new facts into the Aristotelian framework. We need to remind ourselves of a third point that regarded astronomical phenomena in particular. The Ptolemaic system was, by and large, founded on Aristotelian natural philosophy. Ptolemy's mathematical model of planetary motions, however, was clearly incompatible with Aristotle's own writings.[11] The accepted solution was that the mathematical model was considered a *hypothesis*, i.e., a mere supposition used to obtain numerical results regarding the positions of the planets: a supposition that in no way described the physical reality of the inscrutable Universe. Astronomical observations were not considered on the same level as empirical evidence obtained by experimental means in other areas of natural philosophy, and the "unreasonable effectiveness of mathematics" (Wigner, 1960) in describing the physical world was only beginning to shine through.[12]

Finally, let us note that in our way of understanding the world, comets are natural phenomena produced by natural causes, and therefore possessing empirico-mathematical explanations as to these causes (*etiological* explanations to use Kepler's expression[13]). To those of us who are religiously inclined, a comet may also be a *sign*, i.e., it may be endowed with religious meaning, e.g., through Jungian synchronicity. The seventeenth century thinkers struggling with the appearance of the Great Winter Comet, however, did not see this distinction as clearly established as we do. Miguel Granada's chapter provides the example of Bartholomaeus Keckermann (1572–1609), a Calvinist theologian and an avid Aristotelian, who interpreted the Great Winter Comet not only as a sign with respect to its religious

---

[10] In November 1627, Pierre de Bérulle told René Descartes (then 31) that it was his duty of conscience to pursue philosophy, after which Descartes abandoned his life in Paris. His *Principia philosophiae* (Descartes 1644) includes a notable treatment on the place of the comets in the Universe.

[11] Aristotle's celestial spheres were homocentric spherical shells, each spinning at a constant rate around the one and the same Center of the Universe. Ptolemy needed to arrange the celestial spheres eccentrically, with the planets themselves carried not by the main spherical shells (*deferents*) but rather by epicyclic spheres, and the spin rates were not constant with respect to the Center but rather with respect to philosophically indefensible equant points.

[12] Galileo famously said, "Philosophy is written in this grand book, the universe, which stands continually open to our gaze. But the book cannot be understood unless one first learns to comprehend the language and read the letters in which it is composed. It is written in the language of mathematics, and its characters are triangles, circles, and other geometric figures without which it is humanly impossible to understand a single word of it; without these, one wanders about in a dark labyrinth." (*The Assayer*; Galileo (1623) incidentally, the book has a direct link to the comets of 1618.)

[13] From the title of his *Astronomia nova aitiologetos* (see also Gabor & Petrovićová, 2021). Galileo argues the point admirably in his *Letter to the Grand Duchess Christina* (Galileo 1615).

meaning but also as a miracle with respect to its etiology, i.e., a breach of natural order by God's direct intervention.

1.3 A Personal Digression

Comets of 1618 have been dear to me since 2004, when I attended the unveiling, by the Mayor of Kadan in Bohemia (Czech Republic), of a commemorative plaque to a notable son of the city, Father Wenceslaus Pantaleon Kirwitzer, S.J., launching a tradition of annual Kirwitzer Days in Kadan.

Kirwitzer received a part of his Jesuit training in Rome, where he also studied astronomy under Father Christopher Clavius, S.J. We could even imagine young Kirwitzer at the solemn occasion in 1611 when Galileo was received at the Roman College and honored for his 1609 astronomical discoveries, which the Jesuits were able to confirm.

Kirwitzer desired to become a missionary in China, and it was on his way there, while staying in Goa, that he was the first to observe a comet with a telescope in India, and indeed anywhere outside of Europe. While he sadly died of a tropical fever, never reaching China, his cometary account made it to Europe. Ramesh Kapoor's chapter contains a full report on the matter.

I made Dr. Kapoor's acquaintance at the 2018 General Assembly of the International Astronomical Union in Vienna, where he gave a talk mentioning Kirwitzer. He rekindled my interest in the comets of 1618, and pointed me to Kepler's book *De cometis libelli tres* (1619), where the illustrious Imperial Mathematician writes:

> *Prima Cometae fama literis mandata fuit Cassovia in Superiori Hungaria Viennam 1. Sept. [1618] scriptis, affirmatumque Cometam iam tunc ab octiduo visum. Itaque prima apparitio fuit die 25. Augusti St. Novo...* (p. 47).[14]

I was born in Cassovia[15] and so this detail understandably caught my attention. As we have already seen, this comet, first observed in my native city, is of great significance because of Kepler. He observed it with a telescope, and it so happened that it was the very first time that a comet was ever observed with a telescope.

Ever since 2018, I have been trying (unsuccessfully) to determine who in my native city could have discovered the comet. The discovery of C/1618 Q1 in Cassovia (Fig. 1.3), as described by Kepler, is not mentioned in other

[14] Here is a translation attempting to convey Kepler's concise style: "The first written report of the Comet was sent from Cassovia in Upper Hungary to Vienna on September 1, [1618,] claiming that the Comet had been observed for a week. Therefore, the first appearance was on August 25 New Style..." More literally, "First news of the Comet was sent in a written account from Cassovia in Upper Hungary to Vienna on September 1, claiming that the Comet had been observed for a week already by then."

[15] The city's Latin name is Cassovia, in German it is Kaschau, in Hungarian Kassa, and in Slovak Košice.

Fig. 1.3 "Cassovia: Superioris Hungariae Civitas Primaria (Upper Hungary's Principal City). Depictum ab Egidio Vander Rye Belga. Comm. Georgius Houfnaglius anno 1617." Published in G. Braun, F. Hogenberg, Civitates orbis terrarium, Cologne, 1617. The city's Latin name is Cassovia, in German it is Kaschau, in Hungarian Kassa, and in Slovak Košice

seventeenth-century cometary works I examined: Grassi (1619), Guiducci & Galileo (1619), Riccioli (1651), Lubieniecki (1667), Hevelius (1668).

Kepler mentions a report sent to Vienna. Perhaps the comet was discovered by the person who wrote the report? Therefore, I first asked myself, who in Cassovia would have been in the habit of sending reports to Vienna? The city council was Lutheran and generally not on the best of terms with the Habsburg capital. But the city was also the seat of the Royal military and administrative offices, which would have communicated with the King of Hungary and his government in Vienna. One possible option was the Captain General[16] (at the time Nagylucsei Dóczy András). Another was the Camera Scepusiensis (Zipser Kammer, Szepesi kamara, Spišská komora), which was, despite the name,[17] the tax office for the entire unoccupied part of Hungary.[18]

[16] Latin: capitaneus generalis; German: Kapitän-General; Hungarian: főkapitány; Spanish: capitán general. The title was probably of Spanish origin. Recall that the Habsburgs reigned in Spain as well as in the Holy Roman Empire, the Kingdoms of Hungary, Bohemia, and the Archduchy of Austria. It designated a military district and its commander. For most of the seventeenth century, Hungary had nine such districts.

[17] Scepusium (Zips, Szepes, Spiš) was a (then) affluent county northwest of Cassovia. Prior to 1526, there were four Royal Chambers in Hungary, administering taxes and some royal holdings. By mid-sixteenth century, only one Chamber remained. In 1567 Maximilian II moved it back to Cassovia (where a Royal Chamber had existed since the thirteenth century until 1536 when it had to move because of John Zapolya).

[18] Most of the Kingdom of Hungary was under Ottoman occupation, which started after the 1526 Battle of Mohacs. The rest was divided between the Principality of Transylvania (an Ottoman vassal) and "Royal Hungary."

The Archives of the Kingdom of Hungary remain in Vienna by mutual agreement of the respective governments of Austria and Hungary. *Ungarische Archivdelegation* kindly conducted a search following my request, responding that there is no such report in their care: a fact which was easy to determine because there are very few extant documents from the period.

Similarly, Cassovia's *Liber Civitatis* (kept in the City Archives) contains no record of astronomical observations, much less discoveries, at the time. Considering that the *Liber Civitatis* is primarily a book of the City Council's minutes naturally organized in chronological order, it would be hard to miss such an entry which would have had to be no more than a week or two after the discovery date itself, i.e., after August 25 (new style) or 15 (old style).

Johannes Bocatius,[19] the Rector of the city's grammar school, a notable poet and former Mayor, published a pamphlet about the Great Winter Comet, never once mentioning the other two comets at all, let alone their discovery. Considering his personality, it would have been impossible for him to discover a comet and not write about it. The pamphlet is simply entitled *Hornus cometa* (this year's comet), bearing roughly the same text in Latin, Hungarian, and German. The print recalls the idea that comets bring war, pestilence, and famine.

Bocatius refrains from the other frequent cometary trope, the death of kings. He does, however, mention the Emperor Matthias, "whose long life is our survival/welfare.[20]" The Captain General Nagylucsei Dóczy András was not fooled by this: everybody knew of Bocatius's sentiments towards the late Rudolph II and his then happily ruling brother. Dóczy immediately advised the Mayor and Council that it would be in the City's best interest to confiscate the pamphlet because of this oblique reference to the potential impending death of the Emperor Matthias, King of Hungary. On December 17, 1618, the Mayor and Council acted on this recommendation, bearing in mind that Master Bocatius's tactlessness could inflame His Imperial Majesty, and thus inconvenience the City.

The situation turned even more serious (as well as somewhat farcical)[21] when Matthias indeed died on March 20, 1619 (New Style). Then the printer, Johan Festus, petitioned the Mayor and Council to restitute the confiscated pamphlets to him (the Latin minutes in the *Liber Civitatis* call it "imago cometae") because, according to Festus, various merchants would like to sell it. The request was denied on March 14, 1619 (probably Old Style, i.e., March 24, 1619 New Style). Apparently, there is only one known surviving copy of the pamphlet, viz., in the *Leutschauer*

[19] Born in Vetschau, Germany (Lower Lusatia), in 1569, died in 1621. Staunch Lutheran. Emperor Rudolph II gave him the title *poeta laureatus caesareus*. Bocatius went on to serve the enemies of the Emperor (the Bocskai rebellion), which the latter took very personally. When Bocatius was captured, the Emperor held him in Prague's White Tower indefinitely. Bocatius escaped after 4 years. According to his own account, baking files in loaves and knotting sheets was involved (poetic license?).

[20] "mi meg maradasunk" (Hung.), "nostra est salus" (Lat.), "unsere Wo[h]lfahrt" (Ger.)

[21] The story is easy to follow thanks to the minutes in the *Liber Civitatis* (Kemény).

HORNVS COMETA 1618.

AZ Törvény tudoknak regulajok ez, hogy az intés elötte jarjon az büntetesnek. Példa beſedibenis vagyon; Az barátſágos embereknek tiſtek ez hogy kölczönben egy maſt inczék. Ez nyilvan való az mi hozzánk kegyelmes Iſtennek példajából, ki az embereknek bünért és törvény ſegéſit eleve való Attyai intés nélkül, nem ſokta meg vetni. Mert az mindenhato Vr Iſten, à mint az Propheta ſól, az alnokſágot büntetni nem akarja, minek elötte az Prophetak által meg nem jelentené. Tekints keresztyeni amaz réghi idöre, Ez világot viz özönel el akarvan veſzteni; Sodomat és Gomorrhat kénkeves tüzel; Ninivét fene vadaknak barlangjava akarvan tenni, jo ideien elöl jarok á'tal, jobb utra nem hitaé? Azont czelekedte az Ieruſalem veſzedelme elöt, ſoval és igyekel: Az többi közöt, illyenel (vizonſág erre az Ioſephus Hiſtoricus az Sidok hadakozáſaról való iraſaba az minemüvel moſt minket attyai képen gyaſzra int. Soha bizonyára, edes kereſztenem, ſoha büntetetlen az éghi legy fel nem tetzet. Minékünk kik Magyar kenyeret éſzünk, és hiſtoriakat olvaſunk, hog egyéb tartomanyokról ne ſzolyak, böven példa lehet, Ferdinand, Maximilian, Rudolphus és Matthias idejeben kinek hoſzſza élete mi meg maradaſunk, az uy Csillagok, avagy borzas üſtökös csillagok, mit hoztak magokal. Egy példa legyen ſok hellyet, Az 1596. eſztendei geriet üſtökös csillag, Egret tölünk el veſzté, és az Kereſztesi veſzedelmet hoza, mely meg moſtanis ſzégyen az kereſzteneknek. Ki mondhatatlan banatot kenſzeritetem, kereſzteny, Vytanom, mert valami haſonloval fényeget ha nagyobad nem, az igazſágos Iſtennek haragja, mellyet exſequalnijs akar moſt, ha az mi nehéz büneinktöl bucsut nem veſzünk. Vy Czéger ſzögeſztetet kü, nem tudom minemü bort ivoknak. Az Török az ö Portan költ levelét meg jegyzi, és Lofarkat törel küldven, gyülös hellyet hadi indulaſra köſzültet futat. Félek félek, hogy ezképen ne amaz mi Iſtenes Imperatorunk és viſzontag ſzemélyek változvan az réghi Comœdia ne játzottaſſék. Hád, Dög halal, Ehſég, magzati ez anyanak. Iobakra avagy lagyabakra, ha akarod kereſztény tanaczomat, ha akarjatok ö Magyar Orſzághnak lakoſi, kiké napkeletröl az fenyegetés. No térjetek meg, gyülyetek egybe, ſzeretetbe nem levö Nemzetſég, minek elötte ſzülyön ez az anya, mely az el múlt eſztendöben Scepesben egyben forrot kettös gyermeket, à honnét bodogtalan ſzaru nevekedet B. és az mit ez czoportos Ehgi csuda jelent, ſzöllye mondom, mint az port, az el múlo napot: Minek elötte reatok érkezzék az VR buſulaſanak haragja, minek elötte jöjön reatok az Vr haragjanak napja. Kereſſétek az Vrat ez földnek minden ſzelidi, kik az ö itiletit cselekedtétek. Kereſſétek az igazat, kereſſétek az engedelmeſt, hogy valamiképen el reytezheſſetek az Vr buſulaſanak napjan.

CAnon eſt Iuriſconſultorum: Admonitio præ-

DEs Rechts iſt eine Regel: Die Vermahnung

CASSOVIÆ, Typis Ioannis Festi.

Fig. 1.4 Johannes Bocatius, Hornus Cometa 1618. Printed by Johannes Festus. A photograph of the only remaining copy. Leutschauer Archive

Archive.[22] The text without the image was published in Bocatius's collected works (Fig. 1.4).

Another possibility would be clergymen. Lutheran and Calvinist clergy were unlikely to correspond with Vienna. That leaves the Catholic ones. It appears that there were very few Catholics in the City, at least few among the people "of substance." The City Council was Lutheran, as were most of the burghers. Some residents and passing nobles may have been Calvinist. The only Catholic individuals of rank residing in the city were there in an official capacity. Some of the Captain General's staff and soldiers and the *Camera Scepusiensis* officials were Catholic. Of course, there were likely also some simple folk who were Catholic. Thus, some servants and other employees in the households and businesses of the burghers and nobles were also Calvinist and Catholic. The Calvinists and the Catholics represented two vertices in a volatile triangle with the City's Lutheran establishment.

In any case, in late August 1618 (appearance of C/1618 Q1), there were few Catholic priests in the City. What follows is no more than a sketch of an idea. I know of only one Catholic priest who may have been in the vicinity. (Benevolent reader,

[22] Leutschau (Ger.) = Lőcse (Hung.) = Levoča (Slov.) = Leutsovia (Lat.) was another predominantly German-speaking city, linked to Cassovia in the *Pentapolitana* alliance with three other cities. In 1549, the alliance adopted *Confessio Pentapolitana* (akin to *Confessio Augustana*, i.e., Augsburg Confession).

if you know of another, please let me know.) It was St. Mark Crisinus.[23] He was a canon of the Metropolitan Chapter of *Strigonium* (Esztergom, Gran or Ostrihom), and he may have been in the area to oversee the Chapter's property formerly owned by the (then defunct) Benedictine Abbey of Széplak or Abaszéplak (now Krásna nad Hornádom). We know he was there the following summer, 1619. Perhaps he was there also in the summer of 1618. It would have been quite natural for him to correspond with Vienna. Additionally, if Crisinus was the one who made the discovery, it would not be surprising that the City Council (and the *Liber Civitatis*) would not be informed of the fact. In late August, his putative presence in Széplak may have been due to the harvest. The location is 7 km from Cassovia's center as the crow flies. Too far for a daily commute on horseback, perhaps, but close enough that Kepler could consider the report as sent from Cassovia, and, more importantly, Kepler could have considered the observations as conducted in Cassovia, and used the city's longitude in his calculations.[24]

Let us now return to the wider context of early seventeenth-century views on comets.

1.4 What Makes Comets Harbingers of Doom?

> *Calphurnia.* When beggars die there are no comets seen; The heavens themselves blaze forth the death of princes. (W. Shakespeare, *Julius Caesar*, II, 2).

In his chapter, Andreas Bähr provides a detailed exposition on the word *comet,* its true and false etymologies. He quotes Volkmar Happe, who explains that *cometa* comes from the Latin noun *comes* ("count" but originally "comrade")[25]: *comes et socius rerum fatalium* (an "ally and companion of fatal things") as the Bohemian prophet Wilhelm Neuheuser put it. Happe believes that the Thirty Years' War began with the Great Winter Comet, rather than with the Defenestration of Prague (May 23, 1618). Prague pastor Helwig Garth speaks of the Star of Bethlehem as a comet accompanying the first coming of Our Lord, while the Great Winter Comet heralds His second coming.

---

[23] He was martyred on September 7, 1619, by George Rakoczy's soldiers (Rakoczy was Calvinist) in Cassovia's Royal House together with the Jesuits Sts. Melchior Grodziecki and Stephen Pongracz, who had not arrived in Cassovia until 1619. Crisinus was not a Jesuit but he was an alumnus of Jesuit schools. The three martyrs were canonized in 1995.

[24] Kepler lists Cassovia's position in his *Catalogus locorum...*

[25] *Cometa* is a Greek word, derived from the adjective κομήτης ("long haired"). As a substantivized adjective it is used as a masculine because it is understood that the noun it goes with is the masculine *aster* ("star"): ἀστὴρ κομήτης. Thus in Latin, *cometa* is a masculine, and because of its morphology, it is of the first declension (most first declension words are feminine—making this one of the numerous exceptions) but with some Greek endings (if one desires to show off one's erudition), notably in the accusative singular *cometen* (rather than *cometam*).

Where does this perception originate? Claudius Ptolemy's *Tetrabiblos,* arguably the most classical astrological manual, provides concise instructions on how to interpret a comet's appearance:

> [Comets] naturally produce the effects peculiar to Mars and to Mercury — wars, hot weather, disturbed conditions, and the accompaniments of these; and they show, through the parts of the zodiac in which their heads appear and through the directions in which the shapes of their tails point, the regions upon which the misfortunes impend. Through the formations, as it were, of their heads they indicate the kind of the event and the class upon which the misfortune will take effect; through the time which they last, the duration of the events; and through their position relative to the sun likewise their beginning; for in general their appearance in the orient betokens rapidly approaching events and in the occident those that approach more slowly.[26]

Nicholas Campion provides a magisterial commentary in his chapter. Let us just note here that Ptolemy regards comets as omens or portents, i.e., as phenomena foretelling future events. It is not by accident that the word "ominous" is simply the adjective derived from the word "omen."

Why would anything as beautiful and captivating as a comet be a herald of misfortune? The reasons may go much deeper than Ptolemy's system itself. It is a clear example of a very old view that philosophers such as Plato and Aristotle merely attempted to systematize.

David Pankenier's chapter describes a case that could not have been influenced by Western Astrology in any way. The Great Winter Comet of 1618 significantly intervened in the history of the Chinese Empire: "Astral as well as other omens in 1618 played a significant role in the overthrow of the Ming dynasty (1368–1644) by the non-Chinese Manchus."

Let us develop a thought we have already introduced above. The heavens and the earth, the celestial and the earthly, represent polar opposites in the symbolic perception of the world. While earthly things are born, grow, wither, and die, always changing, always precarious, celestial things are constant, perfect, unchanging, unaffected by time; indeed, they are the closest to perfection and eternity at any phenomena perceivable by our perishable senses. When a comet appears, it is a disturbing paradox. It has the aspect of a celestial body, and yet it observes no laws of motion proper to the astronomical entities, thus disturbing the very order of the Cosmos. How deep this goes can be seen in Plato's *Timaeus*.[27] First, there is the famous distinction between that which always is and never becomes (changes), and that which always becomes and never (truly) is.

Change is clearly considered detrimental and incompatible with perfection and full existence. This may sound strange to us. We like to see change as a chance, as an opportunity. The ancients saw change as Chaos, destabilizing the Cosmos, the order which *was* the world. The Pythagorean Demiurge (Creator who in *Timaeus* is not omnipotent) makes the Cosmos a flawed reflection of the eternal ideas. It is the

[26] *Tetrabiblos*, II, 9. (English translation, pp. 90–91.)

[27] Let us note that the passages we shall peruse are devoted to a presentation of Pythagorean views. I would venture to add that these are likely exegesis of ancient myths and symbolism.

best he can do, considering that he is limited to using matter, the imperfect stuff of this physical world.

Thus, the unchanging motions of the celestial bodies are a moving image of eternity: uniform circular motion being the best achievable image of eternity's perfection. Comets appear and disappear, moving among the fixed stars in chaotic and unpredictable ways. They represent an anomaly (in the full meaning of the Greek word, i.e., lawlessness) perturbing the heavenly motions. In fact, it is on this basis alone that Aristotle teaches that comets cannot be truly celestial but rather must be atmospheric phenomena.

The great question of astrology in general applies to the comets, and perhaps even more so: Are the celestial phenomena causing terrestrial events, or are they merely correlated? Planetary conjunctions, ascendents, etc., repeat with mathematical regularity and thus are usually perceived as mere signs, but comets, perhaps piercing heavenly spheres, may be more readily seen as causes or instruments of cosmic upheaval.

Luis Carolino, in his chapter, speaks of the perception of the Great Comet of 1577 in the Iberian Peninsula. It was seen as heralding the death of King Sebastião (1578; aged 24) and the accession of Philip II of Spain to the throne of Portugal.

Comets were well established as omens, specifically of royal demise, the ascension of new monarchs, and even dynastic change, already at the time of Claudius's death on Oct 13, AD 54. The coincidence of Claudius's death with the appearance of a "long-tailed comet" was included by Suetonius in his best-seller *The Lives of Twelve Caesars* (n.d.) (29, 46). Claudius's successor, Nero, was thus perhaps understandably terrified when a prominent comet appeared. He sought to deflect Fate's blow from himself to others by slaughtering many prominent Romans.[28]

The conspicuous presence of a Great Comet in the heavens above presented a dilemma for monarchs. Nero may have overreacted, but even those kings who paid no heed to portents must have realized that a significant portion of their subjects would view such striking spectacles as signs from God. Clifford Cunningham's chapter shows how James I of England dealt with the problem through poetry.

1.5 Concluding Laudation

In this brief introduction, we have seen that the comets of 1618 are not a marginal or random matter, but rather that they merit our attention and study. This volume, "studded by landmark studies" (to quote a classic), offers many different approaches to the topic: the contemporary prints in the Holy Roman Empire

[28] Suetonius, *Twelve Caesars*, Nero, 36. The comet may have appeared in AD 60, 64 or 66. Nero died on June 9, AD 68, aged 30.

(Doris Gruber's chapter[29]), the typology of cometary images with a range of visual strategies (Sara Schechner's chapter), the comet's impact on the history of astrology (Nicholas Campion), a study of W. Schickard's noteworthy manuscript treatise (Miguel Granada), a detailed analysis of Orazio Grassi's charts (Giancarlo Truffa), overviews and evaluations of the writings on the comets in the context of the Thirty Years' War (Andreas Bähr), in Spain and Portugal (Luís Miguel Carolino), in China (David Pankenier), in India including the contribution of astronomically trained Jesuit missionaries (Ramesh Kapoor), and culminating (in my eyes) in the cometary poetry inspired, incited and induced by the comets (Clifford Cunningham). This volume's width and breadth provide a momentous reference for the educated public, while simultaneously stimulating specialists to further research.

References

Aristotle. (n.d.-a). *De Coelo.*

Aristotle. (n.d.-b). *Meteorologica.*

Bocatius, J. (1618). *Hornus cometa. Opera quae exstant omnia.* Published in three volumes (two volumes of poetry and one of prose) around 1990. https://mek.oszk.hu/20300/20373/

Descartes, R. (1644). *Principia philosophiae.*

Drake, S., (transl. & introduction), O'Malley, C. D., & (transl.). (1960). *The controversy on the comets of 1618. Galileo Galilei, Horatio Grassi, Mario Guiducci, Johann Kepler.* University of Pennsylvania Press.

Gabor, P., & Petrovićová, K. (2021). It's all Greek: Three of Kepler's book titles. *JAHH, 24*(3), 652–664.

Galileo, G. (1610). *Sidereus Nuncius.* Thomas Baglioni.

Galileo, G. (1615). *Letter to the Grand Duchess Christina.*

Galileo, G. (1623). *The Assayer.*

Gassendi, P. (1649). *De Vita et Moribus Epicuri (1647). Syntagma Philosophiae Epicuri.*

Grassi, O. (1619). *Disputatio astronomica de tribus cometis anni 1618.*

Gruber, D. (2020). *Frühneuzeitlicher Wissenswandel. Kometenerscheinungen in der Druckpublizistik des Heiligen Römischen Reiches.* Lumière.

Guiducci, M., & Galileo, G. (1619). *Discorso sulle comete... .*

Hevelius, J. (1668). *Cometographia.*

https://en.wikipedia.org/wiki/List_of_hyperbolic_comets

https://minorplanetcenter.net/db_search/show_object?object_id=C/1618+W1

https://ssd.jpl.nasa.gov/sbdb.cgi?sstr=C/1618+W1#content

John Philoponus of Alexandria (approx. AD 490 to 570), *On the eternity of the world against Aristotle*, (AD 530–534), and *On the creation of the world,* AD 546–549.

Kemény, L. Jr. (1892). *Magyar könyvszemle* (p. 380), and 1901 (p. 65).

Kepler, J. (1609). *Astronomia nova aitiologetos.*

[29] Fruit of her doctoral dissertation (Gruber, 2020).

Kepler, J. (1619). *De Cometis libelli tres.* The critical edition of Kepler's works: Kepler, Johannes; Bialas, Volker; Caspar, Max: *Gesammelte Werke.* Briefe 1612–1620. Beck, München, 1955 (*Johannes Kepler Gesammelte Werke*: 17). http://publikationen.badw.de/de/002334751

Kepler, J. (1634). *Somnium.*

Kepler, J. (n.d.). *Catalogus locorum Europae praecipue, set et africae asiaeque nonnulorum...* (Vol. 10). Gesammelte Werke.

Lubieniecki, S. (1667). *Historia omnium cometarum.*

Nicholas Copernicus. (1543). *De Revolutionibus orbium coelestium.*

Nicholas of Cusa. (1440). *De Docta Ignorantia.*

Ptolemy, C. (1940). English translation: Loeb Classical Library.

Riccioli, G. B. (1651). *Almagestum novum.*

Suetonius. (n.d.). *Twelve Caesars.*

Wigner, E. (1960). "The unreasonable effectiveness of mathematics in the natural sciences" a 1959 lecture by Eugene Wigner (delivered in New York on May 11, 1959). *Communication in Pure and Applied Mathematics, 13*(1), 1–4.

Wilkins, J. (1638). *The discovery of a world in the Moone.* (revised ed., 1640).

Pavel Gabor is the Vice Director of the Vatican Observatory, where he is responsible for its observational facilities in Arizona. His primary research interest lies in exoplanets. He also teaches the history of astronomy at the University of Arizona. As a Jesuit priest and an official of the Holy See, he further studies timekeeping, its history, and symbolism.

Part I
Publications, Poetry and Imagery

Chapter 2
The Comets of 1618 and the European Printing Press

Doris Gruber

2.1 Introduction

In 1618, three exceptionally bright comets appeared in rapid succession, all of them visible to the naked eye. The first (C/1618 Q1) was observed in August and the second (C/1618 V1) in early November. The third, known due to its outstanding size and brilliance as the Great Comet of 1618 (C1618/W1), was observed from November until January of the following year. News about these comets spread throughout Europe. Shortly after the initial sightings, the first publications on the topic appeared in print: John Bainbridge (1552–1643) published *An Astronomicall Description* in London, Bernat Mas (active 1618) published his *Pronostico natural* (*Natural Prediction*) in Barcelona, and Peter Crüger's (1580–1639) *Kurtzer Bericht* (*Short Relation*) appeared in at least two editions in Gdańsk,[1] while Provençal's (active 1618) *Les Prédictions des signes et prodiges* (*Predictions of Signs and Wonders*) was published in three editions in Paris, Bordeaux, and Lyon (Bainbridge, 1618; Crüger, 1618a and 1618b; Mas 1618; Provençal, 1618a, 1618b, and 1618c).

Curiosity but also, and more importantly, fear of the unusual phenomena witnessed and the will to find explanations for them unite these and many further publications. Previous research has already shown how fruitful the examination of these works can be. In particular, it has become clear by what diverse paths the knowledge of the time unfolded and that the appearance of comets not only inspired many leading thinkers of the time to suggest new explanations of the origin, nature, meaning,

[1] Volker Fritz Brüning lists an edition without an imprint; however, no copy of this could be found (Brüning, 2000: 735).

D. Gruber (✉)
Institute for Habsburg and Balkan Studies, Austrian Academy of Sciences, Vienna, Austria
e-mail: Doris.gruber@oeaw.ac.at

C. J. Cunningham (ed.), *Cosmic Events*, Historical & Cultural Astronomy,
https://doi.org/10.1007/978-3-031-86950-1_2

and effect of comets but also sparked a keen interest among the general population, whose perceptions helped shape and strengthen natural theories we consider true and right today (Drake et al., 1960; Hellman, 1971; Schechner Genuth, 1997; van Nouhuys, 1998; Gindhart, 2006; Meinel, 2009; Gruber, 2020; Jerratsch, 2020).

This chapter, therefore, does not focus on individual publications and the value of their content but discusses the significance of the printing press for the subject as a whole and outlines the possible benefits and challenges for future investigations. To this end, it provides the first comprehensive overview known to date of European publications in print concerning the comets of 1618.

2.2 The Printing Press and Comets

When Johannes Gutenberg (c. 1393–1406–1468) introduced movable-type printing to Europe, he drastically changed how people communicated. The printing press was a useful tool for disseminating information quickly, which not only enabled and accelerated the circulation of knowledge but also facilitated its permanence over a large geographical area. This affected both the literate and the illiterate, as printed information was not only sold but also passed on orally through discussion, singing, reading aloud, and preaching. Indeed, printed media intentionally targeted larger and more socially diverse audiences than most manuscripts or other media forms (McLuhan, 2011 [1962]: esp. 221; Chartier, 1987; Giesecke, 1991, 2007: 60–66; Luhmann, 1991 [1984]: esp. 600; Pettegree & Hall, 2004: esp. 785; Pettegree, 2014).

As a result, this new technology accelerated changes in what people accepted as knowledge. Operating with a socially determined concept of knowledge—as people at a given time, in a given place, and with a given sociocultural background continuously negotiate what they accept as true and right and thus as knowledge—Philipp Sarasin even calls for media in general to be viewed as "filters, which select, emphasize and suppress knowledges, change them and unite them with other knowledge stocks" (Sarasin, 2011: 168). If we follow this line of thought, media change takes on paramount importance in the process of knowledge change, even if it represents only one factor among many driving that process (Gugerli & Chassé, 2012: 95; Grunert & Syndikus, 2015: 27, 40 f.; Burke, 2016: 77 f., 123; Gruber, 2020: 21–26).

This acceleration affected discussions about comets and manifested itself in the proliferation of printed works on the subject. The number of such works peaked during and shortly after the apparitions of "great comets"—remarkably bright comets that were visible for exceptionally long periods (Brüning, 2000; Meinel, 2009: 122 f.; Stoyan, 2013; Jerratsch, 2019: 187; Jerratsch, 2020: 39–43). These comets, therefore, represented media events (Dayan & Katz, 1994; Carl et al., 2009; Meinel, 2009: 121 f.). The large numbers of surviving sources embody their contemporary significance and present myriad opinions on them. On the comets of 1618 alone, the present author identified over 350 sources published in the years 1618–81.[2] The

[2] As many of the publications deal with more than one of the three comets of 1618, no further distinction is made here between the individual apparitions. For the purpose of identifying publica-

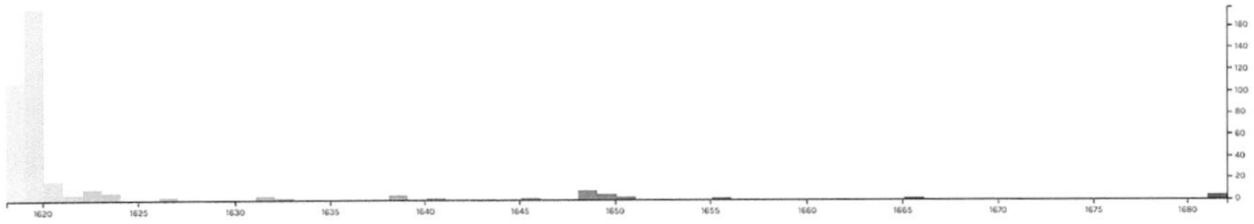

Fig. 2.1 Timeline of stand-alone publications concerning or referencing the comets of 1618 as collected by the author ($n = 364$)

great majority appeared simultaneously with or shortly after the appearance of the three comets. Later publications usually accompanied related political (see below) or natural events, such as the Great Comets of 1665 and 1680 (Fig. 2.1). New apparitions frequently inspired the reissuing of older works on the subject: Giovanni Ferrerio's (1502–1579) *De vera cometæ significatione* (*On the True Meaning of Comets*), for instance, first appeared in Paris in 1540, in Latin, and was reissued in Italian at the time of the Great Comets of 1577 and 1618; Nicolaus Mulerius' *Hemelsche Trompet* (*Celestial Trumpet*), meanwhile, was first published in 1618 and reappeared three times with the Great Comet of 1680 (Mulerius, 1680, 1681a, and 1681b).

As was the case with both previous and subsequent comets, many contemporaries interpreted the comets of 1618 as divine warning signs and astrological *omina*. This explains the peak in publications around 1648, as in retrospect, some believed that one or all of the three comets had heralded the beginning of the Thirty Years' War (Bähr, 2017; Böning, 2018: 13–22). The highly successful pamphlet *Von dem Dreißig-Jährigen Deutschen Kriege* (*Of the Thirty Years' German War*) testifies to this. The title of the work, published in at least twelve editions between 1648 and 1650, declares that "GOD has proclaimed the Thirty Years' War through the terrifyingly bright comet in 1618, which was seen over Europe for thirty days."[3]

An overview of the sources reveals some general trends. In terms of language, the majority (almost half) appeared in German. Relatively large shares of the total appeared in Latin, Spanish, Italian, French, Dutch, and English, as well as several

tions, the most important databases, above all the *Universal Short Title Catalogue* (USTC, 2025), as well as relevant bibliographies and literature, were searched (Harms et al., 1980–2018; Paas, 1985–2017; Schechner Genuth, 1997; Brüning, 2000; Gindhart, 2006; Portal Historischer Kalenderblätter 2016; Hamel, 2020; Jerratsch, 2020). Each edition and translation was counted individually. The complete bibliography is accessible via Zotero (Gruber, 2024; https://www.zotero.org/groups/5483973/bibliography_on_the_comets_of_1618) and includes "lost books," publications that are named in contemporary sources or more recent literature but of which no surviving copy is known to the present author. This, as well as the fact that it remains unclear how many sources still lie dormant in archives and libraries, must be taken into account in considering the following quantitative evaluations. The author thanks Cliff Cunningham for sharing his bibliography on the topic and warmly welcomes any additions, corrections, or recommendations regarding this collection.

[3] "*GOTT durch den 1618 erschrecklich leuchtenden Cometen/ welcher Dreißig Tage über Europen gesehen worden/ enen Dreißig-Jährigen Krieg verkündigt hätte*" (*Von dem Dreißig-Jährigen Deutschen Kriege*, 1648). This and all following translations are the present author's. For more details on this interpretation of the comet, see Bähr, 2017; Böning, 2018: 13–22.

publications each in Polish, Portuguese, Danish, Catalan, Swedish, and Czech (Fig. 2.2). The printing locations show a similar picture. Although publications appeared almost everywhere in Europe, the Holy Roman Empire presents the highest density (Fig. 2.3). A single publication was found to have been produced outside Europe: a pamphlet that describes the sighting of the comet in Paris and was printed in Lima (Nuevas II., 1619).

This reflects general trends in European print production in the early seventeenth century. In particular, we can observe West–East and North–South divides in the importance of printing. The market in the Holy Roman Empire was particularly strong, as were those in other countries in the West and South, including Spain, Italy, France, and the Netherlands, while in some parts of the North and East, manuscripts dominated public communication for much longer, for a range of reasons (Pettegree, 2014; Raymond & Moxham, 2016; Suire, 2020; Alazard-Fontbonne et al., 2021).

The supremacy of the Holy Roman Empire in this respect is, therefore, hardly surprising, but the extent of its dominance is. This was most likely rooted in several factors. First, the printing markets themselves varied widely. In France or England, print production was concentrated in locations such as Paris, Lyon, and London, whereas in the Holy Roman Empire, Spain, or Italy, a greater number of printing locations and publishers competed with one another. This competition promoted diversity (Blome & Böning, 2000; Bethencourt & Egmond, 2007; Pettegree & Weduwen since 2007; Wilke, 2008; Harms et al., 2013; Bellingradt et al., 2017; Rospocher et al., 2019; Suire, 2020). In this respect, it must be emphasized that in the early modern period in general, the printing business was highly risky: Printers and publishers across Europe regularly found themselves fighting for economic survival. As a result, the publications that appeared in print were mainly those expected to sell well or those in receipt of financial support (Weber, 1972: 46–50; Brednich,

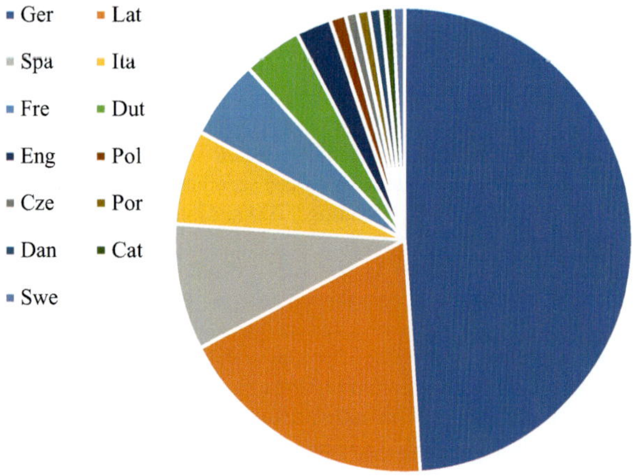

Fig. 2.2 Language shares of stand-alone publications concerning or referencing the comets of 1618 as collected by the author ($n = 364$)

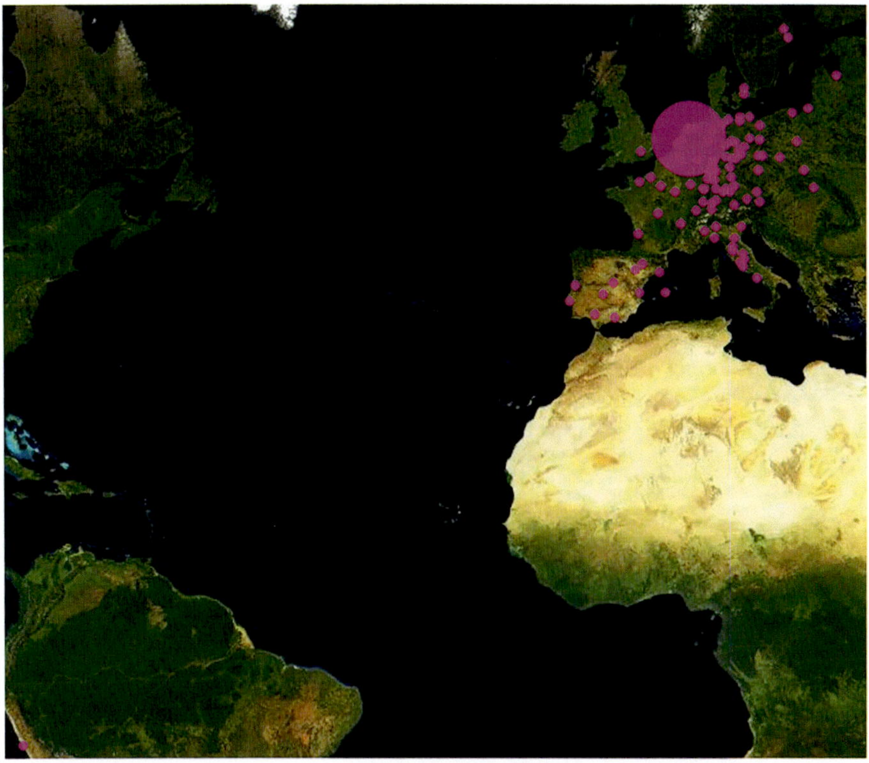

Fig. 2.3 Printing locations of stand-alone publications concerning or referencing the comets of 1618 as collected by the author. Points are sized according to quantity. If a publication names multiple locations, each is counted individually. Map created with Palladio by the author

1984: 286–287; Darnton, 1982, 2014; Schilling, 1990: 1–5, 16–17, 112; Bellingradt & Schilling, 2013: 282; Harms et al., 2013; Reske, 2015; Limbach, 2022).

From the perspective of content, the dominance of the Holy Roman Empire probably resulted from the belief in portents (*prodigia*), which seems to have been particularly strong in this region. In addition to comets, meanings were attributed to other extraordinary phenomena within and outside of nature, including earthquakes, conflagrations, and miraculous births. Mostly, people interpreted such events as divine warnings that called them to a penitent and blameless life (Warburg, 1980; Schenda, 1961 and 1997; Lever, 1993; Céard, 1996; Daston & Park, 1998; Burns, 2010; Carter, 2012; Barnes, 2016; Berns, 2016; Gruber, 2020; Jerratsch, 2020). German-language sources present this interpretation of the comets of 1618 particularly frequently. A publication that appeared in Christlingen, for instance, speculates as to the possible meaning of the Great Comet of 1618. Although the author acknowledges that it is not yet possible to know what exactly the phenomenon heralds ("*Was nun aber gegenwertigen Cometens deutung sein werde/kann man zur zeit noch nicht eytentlich wissen*"), he, she, or they assume that comets are generally followed by war, blood poisoning, famine, death, and similar events (*Kurtze gewisse Beschreibung*, 1618, Aiii r–v). The belief in *prodigia* was, however, widespread throughout Europe; it is represented in works in all of the languages mentioned here (Fig. 2.2). In France and the Netherlands, for example, publications repeatedly referred to comets as a "celestial trumpet," a literal mouthpiece of God (Mulerius, 1618; de Montl'hery, 1619; Fig. 2.4).

The dominance of the Holy Roman Empire in the present bibliography could also, to some extent, be a product of the current state of bibliographical work and the background of the present author. In Europe, access to sources varies depending on language, location, and materiality. For example, national bibliographies and source digitization are well advanced in the Netherlands and the English-and German-speaking world, whereas in southern and eastern Europe, much remains to be discovered. We therefore do not know how many relevant sources may await discovery in libraries, archives, or other public and private collections (Bruni & Pettegree, 2016).

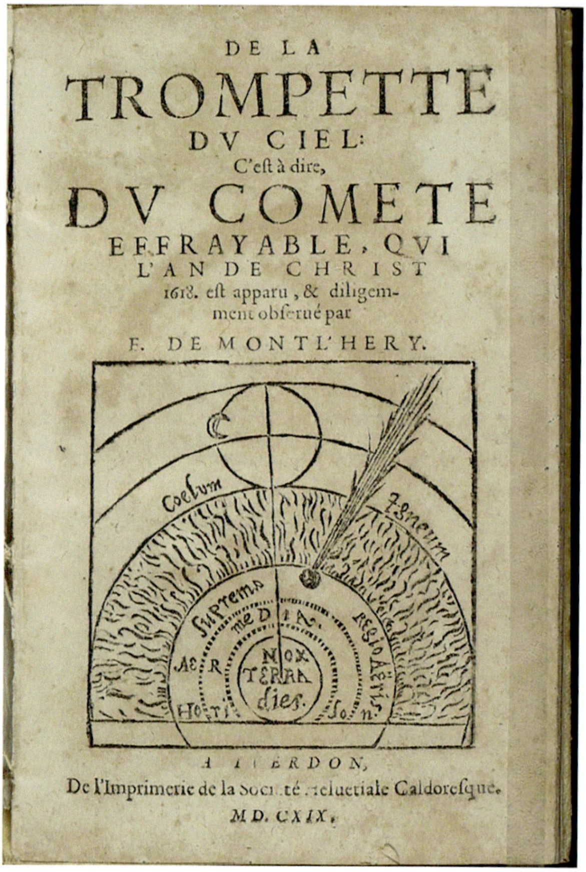
DE LA
TROMPETTE
DV CIEL:
C'est à dire,
DV COMETE
EFFRAYABLE, QVI
L'AN DE CHRIST
1618. est apparu, & diligem-
ment obserué par
F. DE MONTL'HERY.

A [illegible]ERDON,
De l'Imprimerie de la Soci[illegible]té [illegible]eluetiale Caldoresque.
MD. CXIX.

Fig. 2.4 E. de Montl'hery, *De la trompette dv ciel* […] (Yverdon-les-Bains: L'imprimerie de la Société belustiale caldoresque, 1619), not listed in the USTC; OCLC 732267938, title page. (*Source*: New Haven, United States, Yale University, Beinecke Rare Book & Manuscript Library [QB724 M65], https://collections.library.yale.edu/catalog/2041892)

Nonetheless, the present documentation of contemporary publications attests to a deeply interconnected European print market. News of the comets circulated across both linguistic and national borders, with many texts being translated from one language into the next or into multiple at once. Some works were published by the same printer in multiple languages, as was the case with Basilius Daniel z Deutschenberka's (1585–1628) *Saud Hwězdářský přirozený* (*Astronomical Natural Judgment*), which was published first in Czech and later in German by Joannes Strzibrsky (active 1615–1618) in Prague in (z Deutschenberka, 1618a and 1618b). Other works traveled further. Johann(es) Hoppener's (active 1618–1623) *Kurtz Erklärung* (*Brief Account*), for instance, first appeared in Rostock in German, was reprinted at an unknown location, and was subsequently translated into Danish and Swedish, reappearing in Helsingør as *Kort Forklaring* and twice in Stockholm as *Kort förklarning* (Hoppener, 1618a, 1618b, 1619a, 1619b and 1619c). The longest journey was made by the aforementioned publication from Lima, which reported news on the comet from Paris (Nuevas II., 1619; Fig. 2.5).

All Western printed publications, regardless of their language and place of printing, share several characteristics in their production and audience. Each was produced by a large number of people who shaped its content: In addition to the actual authors, this included printers, typesetters, editors, and artists, as well as authorities. The latter, indeed, are often named as dedicatees who supported or counteracted the circulation of information through financial support, the awarding of privileges, and censorship (Darnton, 1982, 2014; Blome et al., since 2000; Pettegree & Weduwen, since 2007; Testa, 2015; Limbach, 2022).

Depending on their scope and language, different publications were aimed at different audiences. The more pages a publication had, and the more numerous and complex pictorial elements it incorporated, the more expensive it was to produce and therefore also to purchase, and—as all over Europe, printed works of the time were mainly sold as goods—the smaller the number of people who could afford it. Language choice had a similar effect. Texts in Latin, the scholarly *lingua franca* of the time, targeted a different, usually more highly educated and financially better-endowed audience than those printed in the vernacular. This also affected the content, which usually reflected the taste and needs of its envisaged audience to at least some extent (Harms, 2008; Bellingradt, 2008: 77 f. and 2011: 15, note 31; Bellingradt & Schilling, 2013: 273; Gruber, 2020: 60–63).

2.3 Stand-Alone Publications

All of the sources mentioned so far are stand-alone publications, printed products that represented complete works in themselves and were not published periodically. These are also the print media forms that research on comets has examined most comprehensively to date (Schechner Genuth, 1997; van Nouhuys, 1998; Gindhart, 2006; Meinel, 2009; Gruber, 2020; Jerratsch, 2020). Stand-alone publications, however, took two different material forms: single-sheet and multisheet publications.

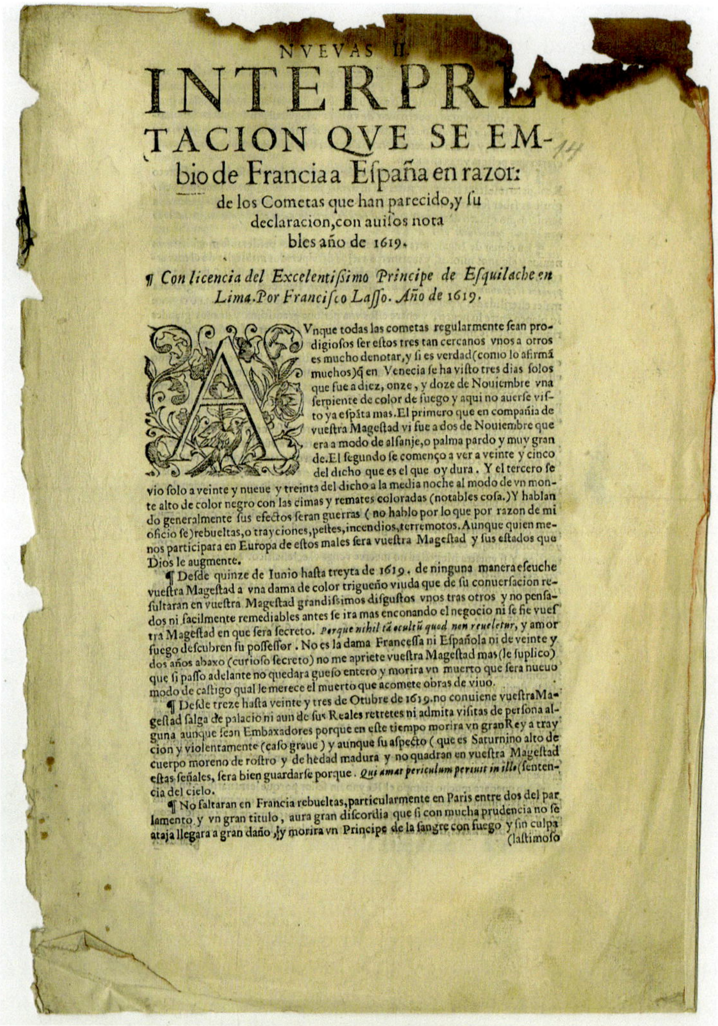
NVEVAS II.
INTERPRE-
TACION QVE SE EM-
bio de Francia a Eſpaña en razon
de los Cometas que han parecido,y ſu
declaracion,con auiſos nota
bles año de 1619.

¶ *Con licencia del Excelentiſsimo Principe de Eſquilache en Lima.Por Franciſco Laſſo. Año de 1619.*

AVnque todas las cometas regularmente ſean prodigioſos ſer eſtos tres tan cercanos vnos a otros es mucho denotar,y ſi es verdad(como lo afirmã muchos)q̃ en Venecia ſe ha viſto tres dias ſolos que fue a diez, onze, y doze de Nouiembre vna ſerpiente de color de fuego y aqui no auerſe viſto ya eſpãta mas.El primero que en compañia de vueſtra Mageſtad vi fue a dos de Nouiembre que era a modo de alfanje,o palma pardo y muy grande.El ſegundo ſe començo a ver a veinte y cinco del dicho que es el que oy dura. Y el tercero ſe vio ſolo a veinte y nueue y treinta del dicho a la media noche al modo de vn monte alto de color negro con las cimas y remates coloradas (notables coſa.)Y hablando generalmente ſus efectos ſeran guerras (no hablo por lo que por razon de mi oficio ſe)rebueltas,o trayciones,peſtes,incendios,terremotos.Aunque quien menos participara en Europa de eſtos males ſera vueſtra Mageſtad y ſus eſtados que Dios le augmente.

¶ Deſde quinze de Iunio haſta treyta de 1619. de ninguna manera eſcuche vueſtra Mageſtad a vna dama de color trigueño viuda que de ſu conuerſacion reſultaran en vueſtra Mageſtad grandiſsimos diſguſtos vnos tras otros y no penſados ni facilmente remediables antes ſe ira mas enconando el negocio ni ſe fie vueſtra Mageſtad en que ſera ſecreto. *Porque nihil tã ocultũ quod non reueletur*, y amor fuego deſcubren ſu poſſeſſor. No es la dama Franceſſa ni Eſpañola ni de veinte y dos años abaxo (curioſo ſecreto) no me apriete vueſtra Mageſtad mas(le ſuplico) que ſi paſſo adelante no quedara gueſo entero y morira vn muerto que ſera nueuo modo de caſtigo qual le merece el muerto que acomete obras de viuo.

¶ Deſde treze haſta veinte y tres de Otubre de 1619.no conuiene vueſtra Mageſtad ſalga de palacio ni aun de ſus Reales retretes ni admita viſitas de perſona alguna aunque ſean Embaxadores porque en eſte tiempo morira vn gran Rey a traycion y violentamente (caſo graue) y aunque ſu aſpecto (que es Saturnino alto de cuerpo moreno de roſtro y de hedad madura y no quadran en vueſtra Mageſtad eſtas ſeñales, ſera bien guardarſe porque. *Qui amat periculum peruit in illo*(ſentencia del cielo.

¶ No faltaran en Francia rebueltas,particularmente en Paris entre dos del parlamento y vn gran titulo, aura gran diſcordia que ſi con mucha prudencia no ſe ataja llegara a gran daño ,y morira vn Principe de la ſangre con fuego y ſin culpa
(laſtimoſo

Fig. 2.5 *Nuevas II. Interpretacion que se embio de Francia a España en razon de los Cometas que han parecido, y su declaracion, con auisos notables año de 1619* (Lima: Francisco Lasso, 1619), USTC 5033949, title page. (*Source*: New York, United States, courtesy of the Hispanic Museum & Library, [Nuevas II Interpretacion… 4 December 1618])

Single-sheet publications are usually referred to in English as broadsheets or broadsides. By 1618, they already had a comparatively long history as one of the first media forms that established themselves after the introduction of movable type. Broadsheets were the forerunners of today's newspapers, as they circulated news of all kinds to a socially diverse audience on an *ad hoc* basis, although this was not the case everywhere in Europe. In France, for instance, broadsheets seem to have been surprisingly absent in the early centuries of print. For presenting information on comets, broadsheets were particularly popular in German-speaking regions. In German, such publications are commonly called *Flugblätter* (Schilling, 1990; Bellingradt & Schilling, 2013: 273–279, 282–285; Pettegree, 2017; Bellingradt, 2020; Gruber, 2020: 60–83; Limbach, 2021; Broadsheet (broadside), in: Glossary, 2026).

One example is the *Warhaffte Newe Zeitung* (*True New Account*) published by Lukas Schultes (1593–1634) in Augsburg (Warhaffte Newe Zeitung 1619; Fig. 2.6). As was typical, the sheet consists of three parts: title, image, and text. The title reveals the content: the new comet that appeared over Augsburg and many other places in December 1618. The image depicts the comet above the city, bringing the moment of the event to its audience. The text is set in three columns and written as a ballad, underlining the point that it was intended to be sung. Here, the comet is interpreted as a divine omen: The text claims that both Luke's Gospel and scholars testified that war, bloodshed, famine, and starvation imminently followed comets.[4] It even includes a prophecy that the year 1622 would be particularly affected by misery and afflictions such as plague, robbery, murder, and crop failure. The portent (*Wunderzeichen*), it stresses, calls for repentance, which the text's numerous calls for mercy underline.

Both empirical observations and mathematical calculations are notably absent here, but they did appear, albeit less frequently than in multisheet publications, in broadsheets (Gruber, 2020: 193 f., 314–316; 428 f.). The scholar David Herlitz (1557–1636), for instance, first published his observations in the form of a broadsheet, his *Kurtzer vnd Warhafftiger Bericht* (*Short and True Account*, Herzlitz 1618a).

Multisheet stand-alone publications were similarly heterogeneous and closely related to broadsheets, sometimes even encompassing the same or similar texts and images as their single-sheet relatives. Such reuse was easier with images: While text had to be reset for each printing process, image stamps could be reused directly (Gaskell, 1972; Boghardt, 1977, 2008: 285–340). Three German editions of Elias Ehinger's (1573–1653) multisheet *Iudicium Astrologicum* (*Astrological Judgment*, 1618), for example, contain the same graphic as the broadsheet *Ein schönes neues Klaglied* (*A Beautiful New Lament*, 1619).[5]

As multiple sheets naturally offered more space for text and images than a single sheet, the former type of publication tended to include more elaborate reflections on comets. An expanded version of the text of the above-mentioned broadsheet by David Herlitz, for instance, surfaced in the form of a multisheet publication in numerous editions, newly entitled *Prodromus und Erster Vortrab, Oder Kurtze und einfältige Erklärung, Des Cometen* (*Preliminaries and First Presentation, or Short and Simple Explanation of the Comet*; Herlitz, 1618b).

---

[4] "There will be signs in the sun, moon, and stars. On the earth, nations will be in anguish and perplexity at the roaring and tossing of the sea" (Luke 21:25).

[5] For images of the title pages of two editions of Ehinger's publication and the broadsheet, see Gindhart, 2006: 274, figure 2, 280, figures 8a b.

Fig. 2.6 *Warhaffte Newe Zeitung/von dem newen Cometstern/so in dem 1618. Jahr den Decemb. Erschienen/welcher zu Augspurg vnd in vilen Landen ist gesehen worden/Gsangweiß gestellt sampt einem schönen geistlichen Lied/In dem Thon: Hilff Gott das mir gelinge/du Edler Schöpffer mein* (Augsburg: Lukas Schultes, 1619), USTC 2533529; USTC 2208994. (*Source*: Munich, Germany, Bavarian State Library [Einbl. II,16 m], urn:nbn:de:bvb:12-bsb00101876-4, © Public Domain 1.0)

Nonetheless, multisheet publications include a wide range of formats and scopes, ranging from two to several hundred pages, with or without pictorial elements. To address the diversity of their communicative functions, potential audiences, and prices, researchers have often suggested differentiating between specific forms of multisheet publications, such as pamphlets and books. Such proposals, however, often differ even within individual language areas, and in many such areas, a wide variety of terms has been established, from *chapbooks* and *canards* to *libels* and *pasquils*, *nouvelles* and *relations*, to name just a few. To complicate the prospect of differentiation even further, some of the terms developed, such as the German *Flugpublizistik*, refer to both single-and multisheet publications (Shepard, 1973: 13–36; Köhler, 1976: 41–50 and 1981; Schechner Genuth, 1997: 66 f.; Gruber, 2020: 60–63; Printed Book, in: Glossary, 2026). This makes it harder or, at times, impossible to compare research results at a European—not to mention a global—level. For this reason, no further differentiation is made here, but all multisheet publications are referred to in the following sections as books. It is, however, noted that the standardization of the descriptions of such publications is one of the fundamental research desiderata of book history globally.

2.4 Periodicals: Almanacs, Newspapers, and More

While the bibliography of stand-alone publications on the comets of 1618 continues to grow, much remains to be discovered in periodicals, most of all in almanacs and newspapers. Publications of this kind do not, for the most part, reference the comets they are describing in their titles and are, therefore, more difficult to find.

Almanacs are usually published annually and dedicated to a specific topic or theme. In the early modern period, they were mostly connected to calendars and contained observations of the seasons and the passage of time, holidays, as well as astronomical data and their interpretation. From the late fifteenth century onward, this material took on multiple printed formats, from single sheets, often intended to be pinned to a wall, to multisheet volumes meant for practical and repeated use. Subsequently, from the mid-sixteenth century onward, in many European areas, multisheet almanacs frequently included prognostications on the upcoming year, usually associated with astrological interpretations referencing outstanding events of the preceding year, such as apparitions of comets. Alongside this practical information, almanacs sometimes offered additional entertaining material such as histories, songs, poems, anecdotes, or images, which could also encompass the topic of comets (Bollème, 1969; Matthäus, 1969; Capp, 1979; Salman, 1999; Curth, 2007; Herbst, 2008; Arndt & Körber, 2010, 2020; Lüsebrink, 2013; Gruber, 2020: 87–97; Almanac, in: Glossary, 2026).

Almanacs are an excellent source on what the general population knew about a particular topic, such as comets. Thanks to their utilitarian character, large print runs, and comparatively low price, which was usually payable once a year, almanacs probably constituted the media form that—apart from the Bible and school books—reached the largest contemporary audience and consequently best reflects people's interests and needs. The form, however, has received little attention in many geographical areas, and historiography on comets has, for the most part, overlooked it. The great exceptions are the prognostic almanacs of Johannes Kepler (1571–1630), which have been studied frequently, albeit mostly due to the fame of their author and not the likely reach of the specific form of publication (e.g., Kepler, 1619a and 1619b; Gindhart, 2006: 163–181; Boner, 2013: 104–134), and the almanacs on the Great Comets of 1577, 1680, and 1743 that originated in the Holy Roman Empire (Gruber, 2018 and 2020). Nonetheless, much more material remains to be discovered. One example is Balthasar Schnurr's *Calendariolum Oeconomicum & perpetuum* (*Economic and Perpetual Calendar*), which offers a general statement on comets:

Of the comets/what they are.
A comet/which the common people generally call a peacock's tail/is/according to the scholars/an earthly/coarse/warm/dry/thick/dense/gluey inflation/direction and ascent of the vapor/raised from the earth by the force of the heavenly stars/ to the uppermost part of the air/where it is ignited by the elemental fire/and there it is moved by *primi mobilis*, the first movement: Like a star/it has its course/and in four and twenty hours it is carried from rising to setting;/it receives the form and secret power or direction to move into the lowest part/on earth/of those stars/ where it first rises and is seen.
The name and meaning of the comets.
Cometa has its name from the hair/Latin/a fiery and bright star.
Meaning.
It has the meaning of great winds/arid poisoned air/earthquake/pestilence/great lords' demise/and other terrifying coincidences. (Schnurr, 1619: 101 f.)[6]

This classic Aristotelian interpretation of comets as earthly vapors underlines the fact that theories about the celestial qualities of comets had not reached society as a whole but prevailed as a relatively specialist phenomenon in 1618. It remains to be

[6] "*Von den Cometen/ was sie seyen.*

Ein Comet/ welchen die gemeinen Leute gemeiniglich einen Pfauwenschwantz zu nennen pflegen/ ist/ nach der Gelehrten außsag/ eine irrdische/ grobe /warme/ truckene/ dicke/ zehe/ leimichte Auffblasung/ richtung vnd auffsteigung der Dünst/ durch die Krafft der himmlischen Gestirn/ auß der Erden erhaben/ biß zum obersten theil der Lufft/ allda durch das Elementische Feuwer angezündet/ vnd wirdt allda beweget durch hinrückung primi mobilis, *der ersten Bewegung: Wie ein Stern/ hat er seinen Lauff/ vnd wirdt in vier vnd zwantzig Stunden vom Auffgang zum Nidergang hingerissen/ bekompt die Gestallt vnd heimliche Krafft oder Wirckung in die vntersten Theil/ auff Erden zu wircken/ von denen Gestirnen/ da er zum ersten entspringt vnd gesehen wirdt.*

Vom Namen vnd Bedeutung der Cometen.

Cometa hat den Namen von dem Haar/ Lateinisch/ ein fewriger vnd härichter Stern.

Bedeutung.

Hat die Bedeutung grosser Winde/ dürrer vergiffter Lufft/ Erdbidem/ Pestilentz/ grosser Herren Vntergang/ vnnd anderer erschrecklicher Zufäll" (Schnurr, 1619: 101 f.).

seen, however, whether all European almanacs confirm this finding, or whether the answer differs depending on the geographical origin or language of a given almanac.

Similarly, newspapers have largely been excluded from research on the comets of 1618. At that time, newspapers were just becoming established and differed from their modern descendants. The term "newspaper" is itself a modern concept. In their early years, such publications often called themselves *Coranto*, *Gazette*, or *Relation*, and they appeared at first weekly, later twice a week, and finally, in some cases, daily. They were the work of post writers who collected news, organized it according to its place of origin, and usually reported it in a relatively objective way, without much interpretation. At first, newspapers were aimed at a relatively small, highly educated audience, which subsequently expanded rapidly. Although newspapers' content potentially included all current news, in their early days, they dealt mainly with political and military matters. The first newspaper known to date is the *Relation aller fürnemmen und gedenckwürdigen Historien* (*Relation of All Noble and Memorable Histories*), published by Johann Carolus (1575–1634) in Strasbourg from 1605 onward. In German-speaking lands, more titles soon followed, as did newspapers in other languages and countries (Blome, 2000; Böning, 2002a, 2002b; Schultheiß-Heinz, 2004; Welke & Wilke, 2008; Bauer & Böning, 2011; Pompe, 2012; der Weduwen, 2017; Newspaper, in: Glossary, 2026).

Few titles, however, can be identified that appeared at the time of or shortly after the appearance of the three comets, that is, between August 1618 and February 1619. Only eight, six in German and two in Dutch, with a total of 63 known surviving editions, can be confirmed.[7]

As Hölger Böning has shown, the comets are not mentioned in all of these newspaper titles. If the publications do mention them, they usually limit themselves to brief reports of comet sightings without interpreting the phenomenon (Böning, 2018: 13–22). This reflects the intention of the media form, which was to inform about rather than reflect on a subject. The boundaries between media forms, however, were sometimes blurred. In 1619, for instance, a multisheet volume titled *Beschreibung und kurtze Relation* (*Description and Short Relation*) appeared in Bremen. The first part of the publication is dedicated to the Great Comet of 1618 and lists numerous past comets and the negative events they supposedly foretold. This is followed by a number of news items not related to the comet, which are arranged according to their place of origin, that is, in a way that corresponds to the classic layout of a newspaper. Whether the *Beschreibung und kurtze Relation* is considered a book or a newspaper, this ambiguity highlights that the way in which information is presented—its material and medial form—does not always allow conclusions to be drawn about its content or audience.

[7] German: *Frankfurter Kayserl. Reichs-Ober-Post-Amts-Zeitung* (1615–1779; 26 editions); *Ordinari-Diensttags-Journal* (1618–1710; 30 editions); *Wochentliche Donnerstags Zeitung* (1618, 4 editions); *Zeitung, so sich hin vnnd wieder* (1619–1622; 8 editions); *Relation oder Zeitungen* (1619, 1 edition); *Wöchentliche Zeitung* (1619, 1 edition). See Historische Zeitungen (2013–2015). Dutch: *Courante uyt Italien, Duytslandt, &c.* (1618–1669; 3 editions); *Tijdinghen uyt verscheyde Quartieren* (1619–1671; 1 edition). See der Weduwen, 2017, vol. 2, 181–188, 260–265.

In addition to newspapers, several other periodicals appeared in print around 1618 that might offer further information on comets. In the German-speaking world, where print production diversified particularly early, this especially concerns *Messrelationen*. These emerged at the end of the sixteenth century, were usually published twice a year in connection with the book fairs in Frankfurt am Main and Leipzig, though occasionally also at New Year, and summarized the events of the preceding months. Similarly to the early newspapers, they initially targeted a financially potent, highly educated, and therefore comparatively small segment of society (Glüer, 2000; Rosseaux, 2004; Stöber, 2005: 53–61; Wilke, 2008: 30–33; Körber, 2016, 2018; Böning, 2018: 18–20; Gruber, 2020: 97 f.)

The belief in *prodigia* is also present in these works, although their interpretation of comets and similar events was not necessarily negative, but could be positive. An edition from New Year 1619 briefly summarizes empirical observations of the Great Comet of 1618 and subsequently asks that "God grant that this and similar portents will mean a happy and peaceful year for everyone" (Wintermonat, 1619: 24, 81, 88; Böning, 2018: 18–20; Körber, 2018: No. 982).[8]

We may draw from this that the portent-based interpretation of comets was not only widespread among the general population, usually taken to be the audience of broadsheets, but also circulated in the intellectually and financially elite circles targeted by newspapers and *Messrelationen*. Based on these findings, future studies might seek to shed light on the relationships between the individual reports and articles seen in these media in more detail. This could reveal previously unnoticed channels and agents of knowledge circulation.

2.5 Conclusions

This inquiry into European print production concerning the comets of 1618 brought to light an unexpectedly large number of publications. Many of these await more detailed examination. The overall overview, meanwhile, has revealed a deeply interconnected print market operating across Europe. Although the information presented in print tended to differ depending on the type of publication and the intended audience, the general belief in comets as divine demands for repentance was omnipresent across the continent. By far, the most publications on the comets, however, originated in the Holy Roman Empire. To some extent, certainly, this dominance was rooted in the operational nature of the print market at the time and the early diversification of printing in this region, but thematic preferences might also have favored this outstanding position.

Nevertheless, much remains to be discovered. More in-depth investigation of content might show whether location, social background, or other factors relating to

[8] "*Gott gebe daß diese vnd dergleichen Wunderzeichen ein fröliches vnd friedfertiges Jahr/ vber jedermans verhoffen bedeuten*" (Wintermonat, 1619).

the people involved in the production of these publications also affected that content—and if so, how. Moreover, the relationship between printed works and other material forms must be more clearly determined. This includes manuscripts, letters, diaries, chronicles, and other such forms, from both elite and common circles, which circulated at times in similar ways to printed publications, just as did medals or objects such as astrolabes or telescopes. The same need for investigation applies to information circulating outside Europe, which has rarely been compared with European sources (Schechner Genuth, 1997: esp. 106; Bentley, 2006; Ramerini, 2023). Ultimately, such analyses might help us understand why, when, and where people accepted which theories as true and right.

References

Sources

Bainbridge, J. (1618). *An Astronomicall description of the late comet from the 18. of Nouemb. 1618. To the 16. of December following. With Certaine Morall Progosticks or applications Drawne from the comets motion and irradiation amongst the Celestiall Hierglyphicks. By vigilant and diligent Obseruations of Iohn Bainbridge doctor of Physicke, and Louer of the Mathematicks*. Edward Griffin & Henry Featherstone. USTC 3008213.

Beschreibung und kurtze Relation. (1619). *Beschreibung und kurtze Relation, Deß Newen Cometen, welcher dieses abgeloffenen 1618 Jahrs, in Teudsch und Welschlanden gesehen worden: Was solcher bedeutet, oder künfftig bedeuten möchte. Sampt etlichen beygefügten Böheimischen und Wienerischen Zeytungen, so sich dieses 1619 Jahrs zugetragen*. Wilhelm Roß. USTC 2536475.

Courante uyt Italien, Duytslandt, &c. (1618–1669). s. n.

Crüger, P. (1618a). *M. Petri Crügeri Königl. Stadt Dantzigk Mathematici Kurtzer Bericht Von dem grossen noch zur zeit scheinenden Cometen: Prodromus subsecuturi Tractatus plenioris. concipiret den 18 Decemb. Anno 1618*. Andreas Hünefeld. USTC 2095925.

Crüger, P. (1618b). *M. Petri Crügeri Königl. Stadt Dantzigk Mathematici, Kurtzer Bericht Von dem grossen noch zur zeit scheinenden Cometen: Prodromus subsecuturi Tractatus plenioris: concipiret den 18. Decemb. Anno 1618*. Andreas Hünefeld. USTC 2534513.

de Montl'hery, E. (1619). *De la trompette dv ciel: C'est à dire, dv Comete effrayable, qvi l'an de Christ 1618. est apparu, & diligemment observé par E. de Montl'hery*. L'imprimerie de la Société belustiale caldoresque. Not listed in the USTC; OCLC 732267938.

Ehinger, E. (1618). *Iudicium Astrologicum Von dem Newen Cometa Welcher den 1. Decemb. 1618. am Morgen vor und nach 6. Uhren zu Augspurg und andern Orten von vielen Personen gesehen worden*. Hans Schultes the Elder. USTC 2043729. For further editions in German see: USTC 2069987, 2030731, 2157240. For the Italian translation see: USTC 4028952.

Ein schönes neues Klaglied. (1619). *Ein schönes neues Klaglied/von der letzten Zeit/was sich auch für Wunderzeichen alle Tag begeben/Wie sich auch zu Augspurg drey Engel haben sehen lassen Wie ihr solches in diesem Lied werdet vernemen den anden Tag Jenner/Anno 1619. [...] Von Dem Neuen Cometstern/so in dem 1618. Jahr/den 1. Decembr. erschienen/welcher zu Augspurg vnd in vielen Landen ist gesehen worden Gsangweis gestellt/[...]*. Jacob Senfft. Not listed in the USTC; OCLC 794011551: London, British Library [85/Tab.597.d.2.(13.)].

Ferrerio, G. (1540). *De vera cometæ significatione, contra astrologorum omnium vanitatem libellus*. Michel de Vascosan. USTC 147784.

Ferrerio, G. (1577). *La vera significatione della cometa, contro la oppenione di tutti gli astrologi. Di Giovanni Ferrerio piamontese*. Giorgio Marescotti. USTC 829277.

Ferrerio, G. (1618). *La vera significatione della cometa Contro la Openione di tutti gl'Astrologi apparsa l'Anno 1618*. Stefano Fantucci Tosi alle Scalee di Badia. USTC 4027833.

Frankfurter Kayserl. Reichs-Ober-Post-Amts-Zeitung. (1615–1779). [Thurn & Taxis] et al. Online edition. State and University Library Bremen. https://nbn-resolving.org/urn:nbn:de:gbv:46:1-2249.

Herlitz, D. (1618a), *Kurtzer vnd Warhafftiger Bericht, Welche sich begeben und zugetragen des Cometen und Geschwäntzten Sterns, so sich am vorgangenen Donnerstage den 19. November von etlichen allhier zu Stargard, des Morgens frü umb 4 uhr (biß bald an 7 uhr) ein langgeschwentzter Stern oder Comet, im Osten oder gegen auffgang der Sonnen, nach dem Südost zu, erstlich (meines wissens) gesehen worden, Gestellet durch D. Davidem Herlicium Medicum zu Stargard in Pommern*. Rhetische Druckerey. USTC 2532040.

Herlitz, D. (1618b). *Prodromus und Erster Vortrab/Oder Kurtze und einfeltige Erklerung/Deß Cometen/oder Geschwäntzten Sterns/so sich im November des 1618. Jahres hat sehen lassen*. Rhetische Druckerey. USTC 2082891. For further editions see: USTC 2004937; 2082969; 2069742; 2534684; 2536620; 2030657; 2133600; 2535970; 2070054; 2536105.

Hoppener, J. (1618a). *Kurtz Erklerung vnd Beschreibung DEs Newen Geschwantzten Cometens Oder vngewöhnlichen Sterns: Welcher im anfang des Novembris/dieses itzlauffenden Jahrs 1618. für der Sonnen auffgang int Südost ist erschienen vnnd gesehen worden [...]*. Joachim Fueß. USTC 2030716.

Hoppener, J. (1618b). *Kurtze Erklerung und Beschreibung Des Newen Geschwantzten Cometens Oder ungewöhnlichen Sterns: Welcher im anfang des Novembris/dieses jetztlauffenden Jahrs 1618.* In *für der Sonnen Auffgang int SüdOst ist erschienen und gesehen worden*. Joachim Fueß. USTC 2056943.

Hoppener, J. (1619a). *Kort Forklaring offuer den ny Comete med den lange Hale eller wsedvanlige Stiernis Beskriffuelse, hvilcken udi dette nærverendis Aar 1618 i Begyndelsen aff Novembris Maanet, for Solens Opgang i Sudost, haffuer ladit sig til siune stillet*. Hans Stockelman. USTC 270033.

Hoppener, J. (1619b). *Kort förklarning om then nyia cometen, medh then långe stierten och owanlige stiernes beskrifwelse*. Ignatius Meurer. USTC 252130.

Hoppener, J. (1619c). *Kort förklarning om then nyia cometen, medh then långe stierten och owanlige stiernes beskrifwelse*. Christoph Reusner. USTC 252131.

Kepler, J. (1619a). *Prognosticon, Von aller handt bedraulichen Vorbotten künfftigen Ubelstands/ in Regiments-und Kirchensachen/sonderlich von Cometen und Erdbidem/auff das 1618. und 1619. Jahr*. s. n. USTC 2070237.

Kepler, J. (1619b). *Prognosticum Auff das Jahr nach der Gnadenreichen Menschwerdung vnseres Herrn und Heylands JEsu Christi/nach gemainer bey der Occidentalischen Kirchen üblicher Rechnung M. DC. XX. Von Natürlicher Influencz deß Gestirns in diese Nidere Welt. So auch von Bedeuttungen der erschienenen Cometen/vergangenen oder künfftiger'n Finsternussen/ vnd anderer vngewohnlichen dinge. Zu Trewhertziger Warnung für mäniglich/auff den Merdidianum baider Ertzhertzogthümber Oesterreich ob der Ennß vnd Kärnthen/vnd der Stätte Lintz vnd Villach/gestellt vnd in Truck verfertiget*. Johann Blancke. Not listed in the USTC; OCLC 312579979.

Kurtze gewisse Beschreibung. (1618). *Kurtze gewisse Beschreibung deß Newen Cometsterns/ Welcher den 1. Decemb. diß 1618. Jar Am Morgen vor vnd nach 6. Vhren/von vielen Personen im Teutschland/hinvnd wider am Himmel gesehen worden*. s. n. Not listed in the USTC; OCLC 634706883.

Mas, B. (1618). *Pronostico natural y discurso del cometa y palma nubila, que han aprecido en el presente año de 1618*. Esteban Liberós. USTC 5038453.

Mulerius, N. (1618). *Hemelsche trompet morgenwecker, ofte Comeet met een langebaert erschenen anno 1618. in novembri ende decembri*. Hans Sas. USTC 1026093.

Mulerius, N. (1680). *Hemelsche trompet ende morgenwecker, ofte Comeet met een lange baert verschenen anno 1618. in novembri ende decembri*. Albertus Sas. USTC 1816460.

Mulerius, N. (1681a). *Hemelsche Trompet Ende Morgenwecker*. Rembertus Huysman. USTC 1568557.

Mulerius, N. (1681b). *Hemelsche trompet ende morgenwecker, ofte Comeet met een lange baert verschenen anno 1618. in novembri ende decembri*. Rembertus Huysman. USTC 1816887.

Nuevas II. (1619). *Nuevas II. Interpretacion que se embio de Francia a España en razon de los Cometas que han parecido, y su declaracion, con auisos notables año de 1619*. Francisco Lasso. USTC 5033949.

Ordinari-Diensttags-Journal. (1618–1710). s. n. Online edition. State and University Library Bremen. https://brema.suub.uni-bremen.de/zeitungen17/periodical/titleinfo/1053350

[Provençal]. (1618a). *Les prédictions des signes et prodiges qu'on a veu ceste présente année 1618, ensemble de la comete cheveluë sur ce florissant royaume de France*. Nicolas Rousset. USTC 6014387.

Provençal. (1618b). *Les Prédictions des signes et prodiges qu'on a veu ceste presente année 1618, ensemble la Comète chevelüe qui se voit depuis quinze jours sur ce florissant royaume de France descrites par le M., Provençal*. Pierre Roussin. USTC 6902449.

Provençal. (1618c). *Les Predictions des signes et prodiges qu'on a veu ceste presente année mil six cens dix-huit. Ensemble du comete chevelu qui s'est veu sur ce florissant royaume de France. Descrites par le M . Provençal*. Simon Millanges. USTC 6800440.

Relation oder Zeitungen. (1619). *Relation oder Zeitungen auß mehrerley Orten im Jahr [...] den vnd sonsten wochentlich von Nürnberg avisiert*. s. n. Online edition. State and University Library Bremen. https://brema.suub.uni-bremen.de/zeitungen17/periodical/titleinfo/1430852

Schnurr, B. (1619). *Calendariolum Oeconomicum & perpetuum: Das ist/Ein immerwährendes/ sehr nützliches/auch nötiges Calenderlein und Haußbuch: Allen und jeden Haußhältern/hohes und nidriges Stands Personen [...] auff alle Jar und zeit zu gebrauchen [...]. verfertiget/unnd zum erstenmal in Druck gegeben*, Edited by Hans Ludwig Gibelstatt Lochinger von und zu Archshoffen. Konrad Eifried. USTC 2070031.

Tijdinghen uyt verscheyde Quartieren. (1619–1671). s. n.

Von dem Dreißig-Jährigen Deutschen Kriege. (1648). *Von dem Dreißig-Jährigen Deutschen Kriege/ Welcher sich Anno 1618 angefangen und durch GOTTes Gnade Anno 1648 geendiget hat. Eben alß wann GOTT durch den 1618 erschrecklich leuchtenden Cometen/welcher Dreißig Tage über Europen gesehen worden/enen Dreißig-Jährigen Krieg verkündigt hätte. Ein Außzug./ Wie viel Schlachten diese Jahren über in Deutschland und dessen Nachbarschafft geschehen/ Wo und Wann Dann auch was in ieder Schlacht/Scharmützel/Beläg: und Eroberung der Städte und Festungen für Volck geblieben ist. Darbey auch Alter novnehmsten Helden welche sich in diesem Kriege tapffer gebraucht haben/Nahmen und Todt. Hierauff folget Eine Summa der Erschlagenen*. s. n. USTC 2024996. For further editions: USTC 2090187; 2513305; 2184832; 2103209; 2077154; 2090180; 2077325; 2090396; 2090361; 2090361; 2137200.

Warhaffte Newe Zeitung. (1619). *Warhaffte Newe Zeitung/von dem newen Cometstern/so in dem 1618. Jahr den Decemb. Erschienen/welcher zu Augspurg vnd in vilen Landen ist gesehen worden/Gsangweiß gestellt sampt einem schönen geistlichen Lied/In dem Thon: Hilff Gott das mir gelinge/du Edler Schöpffer mein*. Lukas Schultes. USTC 2533529; 2208994.

Wintermonat, G. (1619). *Continuatio XXX. Der Zehenjährigen Historischen Relation. Das ist: Warhafftige Beschreibung aller gedenckwürdigsten Historien, Handlungen und Geschichten, so seithero des nechstverschienenen Leipziger MichaelisMarckts, biß auff jetzige NewJahrsMesse [...]*. Abraham Lamberg & Kaspar Klosemann the Elder. USTC 2031087.

Wochentliche Donnerstags Zeitung. (1618). s. n. Online edition. State and University Library Hamburg. https://brema.suub.uni-bremen.de/zeitungen17/periodical/titleinfo/852676

Wöchentliche Zeitung. (1619). *Wöchentliche Zeitung aus mancherly Orten vom 15 Junij biß zum 2 Julij [...] gedruckt jm Jahr 1619*. s. n. Online edition. State and University Library Bremen. https://brema.suub.uni-bremen.de/zeitungen17/periodical/titleinfo/1451038

z Deutschenberka, B. D. (1618a). *Astronomisch gut düncken: Von dem schrecklichen Cometen/ welcher nach der Opposition der Sonnen vnd Saturni, den 28. Novembris in der Wag/dieses vnruhigen 1618. Jahrs erschienen ist: was er vor gestalt vnd natur an sich habe/vnd was er künfftig bedeute/zur warnung vnd anleitung zur Buß mit fleiß beschriben/von Daniele Basilio de Deutschenberg/der Rechten Doctore, vnd in der Vnicersit: zu Prag Astronomiae Professore: Auß dem Bömischen ins Deutsche verfertiget.* Joannes Strzibrsky. Not listed in the USTC; OCLC 245842576.

z Deutschenberka, B. D. (1618b). *Saud Hwězdářský přirozený: O strassliwé s ocasem Kométě kteráž se po welikém proti sobě patřenj Slunce s Hladolétem 28. dne Měsýce Listopadu na Znamenj Wáhy w Létu tomto baufliwém a zkormauceném 1618. wyskytla. Jakéhoby ta spůsobu a přjrozenj byla a co by budaucně wyznamenáwala pro wýstrahu a nabjdnuti ku pokánj, s bedliwosti sepsaný.* Joannes Strzibrsky. USTC 1783089.

Zeitung, so sich hin vnnd wieder. (1619–1622). *[...] Zeitung, so sich hin vnnd wieder, so wol im Römischen Reich, als andern Landen, täglich begiebet vnnd zuträgt, in diesem [...] Jahr.* [Joachim Gössel]. Online edition. State and University Library. https://nbn-resolving.org/urn:nbn:de:gbv:46:1-2802

Literature

Alazard-Fontbonne, J. et al. (Eds.). (2021). *Le monde de l'imprimé en Europe occidentale (vers 1470-vers 1680).* Bréal by Studyrama.

Arndt, J., & Körber, E.-B. (Eds.). (2010). *Das Mediensystem im Alten Reich der Frühen Neuzeit (1600–1750).* Vandenhoeck & Ruprecht.

Arndt, J., & Körber, E.-B. (2020). *Periodische Presse in der Frühaufklärung (1700–1750): Ein Vergleich zwischen Deutschland, Frankreich und den Niederlanden.* 2 vols. edition lumière.

Bähr, A. (2017). *Der grausame Komet. Himmelszeichen und Weltgeschehen im Dreißigjährigen Krieg.* Rowohlt.

Barnes, R. B. (2016). *Astrology and reformation.* Oxford University Press.

Bauer, V., & Böning, H. (2011). *Die Entstehung des Zeitungswesens im 17. Jahrhundert. Ein neues Medium und seine Folgen für das Kommunikationssystem der Frühen Neuzeit.* edition lumière.

Bellingradt, D. (2008). Die vergessenen Quellen des Alten Reiches. Ein Forschungsüberblick zu frühneuzeitlicher Flugpublizistik im Heiligen Römischen Reich deutscher Nation. In A. Blome & H. Böning (Eds.), *Presse und Geschichte. Leistungen und Perspektiven der historischen Presseforschung* (pp. 77–95). edition lumière.

Bellingradt, D. (2011). *Flugpublizistik und Öffentlichkeit um 1700. Dynamiken, Akteure und Strukturen im urbanen Raum des Alten Reiches.* Franz Steiner.

Bellingradt, D. (2020). Das Flugblatt im Medienverbund der Frühen Neuzeit: Bildtragendes Mediengut und Recycling-Produkt. *Daphnis, 48*(4), 516–538. https://doi.org/10.1163/18796583-04804001.

Bellingradt, D., & Schilling, M. (2013). Flugpublizistik. In N. Binczek, T. Dembeck, & J. Schäfer (Eds.), *Handbuch Medien Der Literatur* (pp. 273–289). De Gruyter.

Bellingradt, D., Nelles, P., & Salman, J. (Eds.). (2017). *Books in motion in early modern Europe. Beyond production, circulation and consumption.* Springer International Publishing.

Bentley, K. (2006). *"Her Hilf Uns". Der Komet von 1618. 'Zeitgenössische Wahrnehmung und Deutungen anhand von Selbstzeugnissen aus dem Süden des deutschsprachigen Raumes'* (Unpublished Master Thesis, University of Basel). https://edoc.unibas.ch/59940/.

Berns, J. J. (2016). Wunderzeichen am Himmel und auf der Erden. Der frühneuzeitliche Prodigiendiskurs und dessen medientechnische Bedingungen. In H. Jaumann & G. Stiening (Eds.), *Neue Diskurse der Gelehrtenkultur in der Frühen Neuzeit. Ein Handbuch* (pp. 99–162). De Gruyter.

Bethencourt, F., & Egmond, F. (Eds.). (2007). *Cultural exchange in early modern Europe. Vol 3: Correspondence and cultural exchange in Europe, 1400–1700.* Cambridge University Press.

Blome, A. (2000). *Zeitung, Zeitschrift, Intelligenzblatt und Kalender. Beiträge zur historischen Presseforschung*. edition lumière.

Blome, A., Böning, H., & Nagel, M. (Eds.). (since 2000). *Presse und Geschichte—Neue Beiträge*. Book series. edition lumière. https://www.editionlumiere.de/programm.html

Boghardt, M. (1977). *Analytische Druckforschung. Ein methodischer Beitrag zur Buchkunde und Textkritik*. Dr. Ernst Hauswedell & Co.

Boghardt, M. (2008). *Archäologie des gedruckten Buches*. Edited by P. Needham & J. Boghardt. Harrassowitz.

Bollème, G. (1969). *Les almanachs populaires aux XVIIe et XVIIIe siècles. Essai d'histoire sociale*. Mouton.

Boner, P. J. (2013). *Kepler's cosmological synthesis. Astrology, mechanism and the soul*. Brill.

Böning, H. (2002a). *Periodische Presse: Kommunikation und Aufklärung. Hamburg und Altona als Beispiel*. edition lumière.

Böning, H. (2002b). *Welteroberung durch ein neues Publikum: Die deutsche Presse und der Weg zur Aufklärung. Hamburg und Altona als Beispiel*. edition lumière.

Böning, H. (2018). *Dreißigjähriger Krieg und Öffentlichkeit. Zeitungsberichterstattung als Rohfassung der Geschichtsschreibung*. edition lumière.

Brednich, R. W. (1984). Flugblatt, Flugschrift. In K. Ranke & R. W. Brednich (Eds.), *Enzyklopädie des Märchens: Handwörterbuch zur historischen und vergleichenden Erzählforschung* (Vol. IV, pp. 1339–1358). De Gruyter.

Bruni, F., & Pettegree, A. (Eds.). (2016). *Lost books. Reconstructing the print world of pre-industrial Europe*. Brill.

Brüning, V. F. (2000). *Bibliographie der Kometenliteratur*. Hiersemann.

Burke, P. (2016). *What is the history of knowledge? What is history?* Polity Press.

Burns, W. E. (2010). *An age of wonders. Prodigies, politics and Providence in England. 1657–1727*. Manchester University Press.

Capp, B. S. (1979). *Astrology and the popular press: English almanacs 1500–1800*. Faber and Faber.

Carl, H., Schneider, H., & Vogel, C. (Eds.). (2009). *Medienereignisse im 18. und 19. Jahrhundert. Beiträge einer interdisziplinären Tagung aus Anlass des 65. Geburtstages von Rolf Reichardt*. R. Oldenbourg.

Carter, C. (2012). Meteors, prodigies, and signs. The interpretation of the unusual in sixteenth-century England. *Parergon, 29*, 107–134.

Céard, J. (1996). *La nature et les prodiges : l'insolite au XVIe siècle*. Droz.

Chartier, R. (Ed.). (1987). *Les usages de l'imprimé (XVe-XIXe siècle)*. Fayard.

Curth, L. H. (2007). *English almanacs, astrology and popular medicine, 1550–1700*. Manchester University Press.

Darnton, R. (1982). *The literary underground of the old regime*. Harvard University Press.

Darnton, R. (2014). *Censors at work: How states shaped literature*. Norton.

Daston, L., & Park, K. (1998). *Wonders and the order of nature. 1150–1750*. 2. Print. Zone Books.

Dayan, D., & Katz, E. (1994). *Media events. The live broadcasting of history*. Harvard University Press.

der Weduwen, A. (2017). *Dutch and Flemish newspapers of the seventeenth century, 1618–1700*. 2 vols. Brill.

Drake, S., et al. (1960). *The controversy on the comets of 1618: Galileo Galilei, Horatio Grassi, Mario Guiducci, Johann Kepler*. Translated by C. D. O'Malley. University of Pennsylvania Press.

Gaskell, P. (1972). *A new introduction to bibliography*. Clarendon Press.

Giesecke, M. (1991). *Der Buchdruck in der frühen Neuzeit: eine historische Fallstudie über die Durchsetzung neuer Informations-und Kommunikationstechnologien*. Suhrkamp.

Giesecke, M. (2007). *Die Entdeckung der kommunikativen Welt. Studien zur kulturvergleichenden Mediengeschichte*. Suhrkamp.

Gindhart, M. (2006). *Das Kometenjahr 1618. Antikes und zeitgenössisches Wissen in der frühneuzeitlichen Kometenliteratur des deutschsprachigen Raumes*. Dr. Ludwig Reichert.

Glossary. (2026). *Glossary of early modern popular print genres*, Edited by the project The European Dimensions of Popular Print Culture (EDPOP) led by J. Salman, Utrecht University. https://popularprint-glossary.sites.uu.nl/.
Glüer, J. (2000). *Meßrelationen um 1600—ein neues Medium zwischen aktueller Presse und Geschichtsschreibung: Eine textsortengeschichtliche Untersuchung*. Kümmerle.
Gruber, D. (2018). Der Komet von 1680 und die Kalenderpublizistik im Alten Reich. Eine exemplarische Annäherung. In W. Greiling & K.-D. Herbst (Eds.), *Schreibkalender und ihre Autoren in Mittel-, Ostund Ostmitteleuropa (1540–1850)* (pp. 77–96). edition lumière.
Gruber, D. (2020). *Frühneuzeitlicher Wissenswandel. Kometenerscheinungen in der Druckpublizistik des Heiligen Römischen Reiches*. edition lumière.
Gruber, D. (2024). Bibliography on the comets of 1618. In *Zotero*. Public Group Library. https://www.zotero.org/groups/5483973/bibliography_on_the_comets_of_1618.
Grunert, F., & Syndikus, A. (2015). Einleitung. In F. Grunert & A. Syndikus (Eds.), *Wissensspeicher der Frühen Neuzeit. Formen und Funktionen* (pp. 13–45). Akademie Verlag.
Gugerli, D., & Chassé, D. S. (2012). Wissensgeschichte. Eine Standortbestimmung. *Traverse. Zeitschrift Für Geschichte, 1*, 85–100.
Hamel, J. (2020). *Bibliographie der astronomischen Drucke bis 1700*. www.Astw.de/Astronomiegeschichte/Bibliographie_bis_1700/.
Harms, W. (2008). Lateinische Texte illustrierter Flugblätter. Der Gelehrte als möglicher Adressat eines breit wirksamen Mediums der Frühen Neuzeit. In W. Schilling & M. Schilling (Eds.), *Das illustrierte Flugblatt der frühen Neuzeit. Traditionen, Wirkungen, Kontexte* (pp. 123–134). Hierzel.
Harms, W., et al. (Eds.). (1980–2018). *Deutsche Illustrierte Flugblätter des 16. und 17. Jahrhunderts*. 7 vols. Niemeyer.
Harms, R., Raymond, J., & Salman, J. (Eds.). (2013). *Not dead things. The dissemination of popular print in England and Wales, Italy, and the low countries, 1500–1820*. Brill, La Vergne & MyiLibrary.
Hellman, C. D. (1971). *The comet of 1577. Its place in the history of astronomy*. Columbia University Press.
Herbst, K.-D. (2008). *Verzeichnis der Schreibkalender des 17. Jahrhunderts*. Verlag HKD.
Historische Zeitungen. (2013–2015). Edited by the State and University Library Bremen. https://brema.suub.uni-bremen.de/zeitungen17/.
Jerratsch, A. (2019). Celestial phenomena in early modernity: The integrated image of comets. In P. D. Omodeo & V. Wels (Eds.), *Natural knowledge and Aristotelianism at early modern Protestant Universities* (pp. 187–208). Harrassowitz.
Jerratsch, A. (2020). *Der frühneuzeitliche Kometendiskurs im Spiegel deutschsprachiger Flugschriften*. Franz Steiner.
Köhler, H.-J. (1976). Die Flugschriften. Versuch der Präzisierung eines geläufigen Begriffs. In H. Rabe et al. (Eds.), *Festgabe für Ernst Walter Zeeden zum 60. Geburtstag am 14. Mai 1976* (pp. 36–61). Aschendorff.
Köhler, H.-J. (Ed.). (1981). *Flugschriften als Massenmedium der Reformationszeit. Beiträge zum Tübinger Symposion 1980*. Klett-Cotta.
Körber, E.-B. (2016). *Messrelationen. Geschichte der deutsch-und lateinischsprachigen "messentlichen" Periodika von 1588 bis 1805*. edition lumière.
Körber, E.-B. (2018). *Messrelationen. Biobibliographie der deutsch-und lateinischsprachigen "messentlichen" Periodika von 1588 bis 1805*. 2 vols. edition lumière.
Lever, M. (1993). *Canards sanglants. Naissance du fait divers*. Fayard.
Luhmann, N. (1991). *Soziale Systeme. Grundriß einer allgemeinen Theorie*. Wissenschaftliche Buchgesellschaft.
Limbach, S. (2021). *Government use of print: Official publications in the holy Roman empire, 1500–1600*. Vittorio Klostermann.

Limbach, S. (2022). "Darzu mancher Mann sich viel zu schwach unnd zu wenig Befinden wu[e]rde". Buchdruckerinnen und ihre Tätigkeiten im Alten Reich, ca. 1550–1700. *Zeitschrift Für Historische Forschung, 49*(3), 399–440.

Lüsebrink, H.-J. (2013). *Französische Almanachkultur im deutschen Sprachraum: (1700–1815). Gattungsstrukturen, komparatistische Aspekte, Diskursformen*. V&R unipress et al.

Matthäus, K. (1969). Zur Geschichte des Nürnberger Kalenderwesens. Die Entwicklung der in Nürnberg gedruckten Jahreskalender in Buchform. *Archiv für Geschichte des Buchwesens, 9*, 965–1396.

McLuhan, M. (2011). *Die Gutenberg-Galaxis. Die Entstehung des typographischen Menschen*. Gingko Press.

Meinel, C. (2009). *Grenzgänger zwischen Himmel und Erde. Kometen in der frühen Neuzeit*. Universitätsverlag Regensburg.

Paas, J. R. (1985–2017). *The German political broadsheet. 1600–1700*. 14 vols. Harrassowitz.

Pettegree, A. (2014). *The invention of news. How the world came to know about itself*. Yale University Press.

Pettegree, A. (Ed.). (2017). *Broadsheets. Single-sheet publishing in the first age of print*. Brill.

Pettegree, A., & der Weduwen, A. (Eds.). (since 2007). *Library of the written word—The Handpress world*. Book series. Brill. https://brill.com/display/serial/LWWH.

Pettegree, A., & Hall, M. (2004). The reformation and the book. A reconsideration. *The Historical Journal, 47*(04), 785–808.

Pompe, H. (2012). *Famas Medium. Zur Theorie der Zeitung in Deutschland zwischen dem 17. und dem mittleren 19. Jahrhundert*. De Gruyter.

Portal Historischer Kalenderblätter. (2016). Edited by Friedrich-Schiller-Universität Jena, Thüringer Universitäts-und Landesbibliothek Jena & Universität Bremen, Institut Deutsche Presseforschung. http://zs.thulb.uni-jena.de/content/main/calendarList.xml#Z.

Ramerini, M. (2023). The three comets of 1618: A testimony from the spice islands, the Moluccas. *Colonial Voyage*. Retrieved March 27, 2024, from https://www.colonialvoyage.com/le-tre-comete-del-1618-una-testimonianza-dalle-isoledelle-spezie-le-molucche/.

Raymond, J., & Moxham, N. (Eds.). (2016). *News networks in early modern Europe*. Brill.

Reske, C. (2015). *Die Buchdrucker des 16. und 17. Jahrhunderts im deutschen Sprachgebiet: Auf der Grundlage des gleichnamigen Werkes von Josef Benzing*. Harrassowitz.

Rospocher, M., Salman, J., & Salmi, H. (Eds.). (2019). *Crossing borders, crossing cultures: Popular print in Europe (1450–1900)*. De Gruyter.

Rosseaux, U. (2004). Die Entstehung der Meßrelationen. Zur Entwicklung eines frühneuzeitlichen Nachrichtenmediums aus der Zeitgeschichtsschreibung des 16. Jahrhunderts. *Historisches Jahrbuch, 124*, 97–124.

Salman, J. (1999). *Populair drukwerk in de Gouden Eeuw: de almanak als lectuur en handelswaar*. Walburg Pers.

Sarasin, P. (2011). Was Ist Wissensgeschichte? *Internationales Archiv für Sozialgeschichte der deutschen Literatur (IASL), 36*(1), 159–172. https://doi.org/10.1515/iasl.2011.010.

Schechner Genuth, S. (1997). *Comets, popular culture, and the birth of modern cosmology*. Princeton University Press.

Schenda, R. (1961). *Die französische Prodigienliteratur in der zweiten Hälfte des 16. Jahrhunderts*. Max Hueber.

Schenda, R. (1997). Wunder-Zeichen. Die alten Prodigien in neuen Gewändern. Eine Studie zur Geschichte eines Denkmusters. *Fabula. Zeitschrift für Erzählforschung, 38*, 14–32.

Schilling, M. (1990). *Bildpublizistik der frühen Neuzeit. Aufgaben und Leistungen des illustrierten Flugblatts in Deutschland bis um 1700*. Niemeyer.

Schultheiß-Heinz, S. (2004). *Politik in der europäischen Publizistik. Eine historische Inhaltsanalyse von Zeitungen des 17. Jahrhunderts*. Franz Steiner.

Shepard, L. (1973). *The history of street literature. The story of broadside ballads, chapbooks, proclamations, news-sheets, election bills, tracts, pamphlets, cocks, catchpennies, and other ephemera*. David & Charles.
Stöber, R. (2005). *Deutsche Pressegeschichte. Von den Anfängen bis zur Gegenwart*. UVK-Verlags-Gesellschaft.
Stoyan, R. (2013). *Atlas Der Großen Kometen. Die 30 größten Kometen in Wissenschaft, Kultur Und Kunst*. Translated by Storm Dunlop. Oculum.
Suire, É. (Ed.). (2020). *Le monde de l'imprimé en Europe occidentale (vers 1470—vers 1680)*. Armand Colin.
Testa, S. (2015). *Italian academies and their networks, 1525–1700: From local to global*. Palgrave Macmillan.
USTC. (2025). *Universal Short Title Catalogue (USTC)*. https://www.ustc.ac.uk/
van Nouhuys, T. (1998). *The age of two-faced Janus. The comets of 1577 and 1618 and the decline of the Aristotelian world view in the Netherlands*. Brill.
Warburg, A. (1980). Heidnisch-Antike Weissagungen in Wort und Bild zu Luthers Zeiten (1920). In D. Wuttke (Ed.), *Aby Warburg. Ausgewählte Schriften und Würdigungen* (pp. 199–304). Koerner.
Weber, B. (Ed.). (1972). *Wunderzeichen und Winkeldrucker 1543–1586. Einblattdrucke aus der Sammlung Wikiana in der Zentralbibliothek Zürich*. Urs Graf.
Welke, M., & Wilke, J. (2008). *400 Jahre Zeitung. Die Entwicklung der Tagespresse im internationalen Kontext*. edition lumière.
Wilke, J. (2008). *Grundzüge der Medien-und Kommunikationsgeschichte*. Böhlau.

Doris Gruber heads the Research Unit Art History at the Austrian Academy of Sciences' Institute for Habsburg and Balkan Studies in Vienna. She studied history and art history at the universities of Graz, Vienna, and at Sciences Po, Paris. She has *authored Frühneuzeitlicher Wissenswandel. Kometenerscheinungen in der Druckpublizistik des Heiligen Römischen Reiches Deutscher Nation* (edition lumière 2020), as well as various articles on early modern travel writing, media and books, natural history, the history of knowledge, and digital humanities. She has received several awards for her research, including the *Jubiläumspreis des Böhlau Verlages Wien* and the *Francis Stephen Award* for her PhD thesis.

Chapter 3
The Visual Culture of the 1618 Comets

Sara J. Schechner

The first impression of a comet is visual—be it a fuzzy smudge or bright streak across a portion of the sky. It is not one of smell, taste, touch, or hearing, although acrid, noxious vapors, earthquakes, pestilence, monstrous births, thunderclaps, and wailing may be the alleged aftermath of its sighting or events coincident with or caused by its existence.[1]

About 60% of roughly 150 pamphlets and books published about the 1618 comets announced their purpose with an image of a comet (Fig. 3.1). Surprisingly, the other 40% did not, although they painted word pictures of the comets as to their shapes and significance, and where and when they were seen. The authors used descriptive words such as horrifying, treacherous, scary, prodigious, terrible, admonishing, warning, wonderous, marvelous, sword-shaped, pale, and transparent in reports they claimed were faithful, explanations they called diligent, and prognostications they promised would be meaningful. Such vivid language was not confined to authors of the non-illustrated pamphlets; the authors of the illustrated texts certainly deployed shocking phrases, but they let pictures do some of the figurative work.

The printed illustrations fall into several classes, which will be treated separately in this essay, along with their different functions. My typology of images includes: iconic or symbolic depictions; allegorical images and emblems; pictures of observed phenomena as seen by the naked eye or produced by an instrument; illustrations of a practical setup for observation or experiment; visual ways to map data (e.g., by cartography or graphs); diagrams of a supposed underlying structure of observed

[1] For a summary of cometary theories, historical beliefs about whether they are natural or supernatural, portents or causes of misfortune, and their relationship to prodigies, see Schechner (1997).

S. J. Schechner (✉)
Department of History of Science, Harvard University, Cambridge, MA, USA
e-mail: schechn@fas.harvard.edu

C. J. Cunningham (ed.), *Cosmic Events*, Historical & Cultural Astronomy,
https://doi.org/10.1007/978-3-031-86950-1_3

IVDICIVM ASTROLOGICVM
Von dem
Newen COMETA
Welcher den 1. Decemb. 1618. am Morgen vor
vnd nach 6. vhren zu Augſpurg von vilen Perſonen
geſehen worden.
Geſtelt von M. Elia Ehingero Profeſſore Gymnaſij Annæi
& Bibliothecario.
QVALES HOMINES, TALE COELVM.
Gedruckt zu Augſpurg/ bey Johann Schultes.

Fig. 3.1 The great comet of 1618 hovers ominously over the city of Augsburg on the title page of a typical pamphlet. Elias Ehinger, *Judicium astrologicum von dem newen Cometa welcher den 1. Decemb. 1618 am Morgen vor und nach 6. uhren zu Augspurg von vilen Personen gesehen worden* [Astrological judgment of the new comet which was seen by many people in Augsburg on December first, 1618 in the morning before and after 6 o'clock] (Augsburg, 1618a). Houghton Library, Harvard University, Offsite Storage 2023–24

phenomena (e.g., cosmological schemata); and mathematical diagrams illustrating geometrical relationships, methods, or proofs. The functional relationships of images to their texts were varied and overlapping. Principal purposes were to express an idea or tell a story; to prove or demonstrate a theory; to depict and share an observation; to organize information; to interpret that information; to train the "mind's eye" to see; to teach; and to make the book more aesthetically pleasing. Over and above these functions, writers attached images in order to convince readers of their authority, to serve as proof for what was witnessed, and by extension, to make the reader a witness too. And given the political and religious context, cometary images often served rhetorical and propagandist functions.

In recent years, there has been much discussion among historians about "epistemic" images in early modern texts that relate to the acquisition, production, and presentation of knowledge about the natural world and in mathematical disciplines.[2] Christoph Lüthy and Alexis Smets have taken an encompassing stance, calling an image epistemic if it was "made with the intention of expressing, demonstrating or illustrating a theory" (Lüthy & Smets, 2009: 399 n. 2). On the other hand, Lorraine Daston has posed a restrictive definition, saying that an epistemic image is made not only to depict an object of scientific inquiry but also to be a "stand-in for the too plentiful and too various objects of nature" in scientific discourse (Daston, 2015: 17–18). For my part, I lean towards Lüthy's and Smets's definition, but will not parse out if any particular images discussed in this essay are epistemic or otherwise. I am more interested in what the visual culture tells us about how the comets were experienced, analyzed, and discussed.

The visual imagery also offers insight into target audiences for the pamphlets and cultural predilections. There was a flow of information from educated writers (e.g., mathematicians, astronomers, physicians, and theologians) to their peers in some of the tracts and longer books, but more often the flow was to popular audiences. Indeed, not everyone who came in contact with one of these pamphlets or books would have necessarily been literate. Non-readers might have looked at the images over the shoulder of an owner who read aloud portions of the book, or summarized what evils the comet portended, or delivered a sermon on best behavior in light of the apparition. Therefore, the cometary visualizations link academic debates with popular culture.

A final word: In this essay, I will refer to the images as if the author played a key role in their design and execution and even the decision to have them on the title pages. However, we can presume that there were many exceptions, such as instances where the woodcutter or engraver did not understand the author's drawing or a publisher decided to insert an image for marketing reasons. To this end, we cannot say for sure what an author's intent was from the mere existence of the visual component attached to his text. Without further research into the steps of book production in each locale, we must be wary of this ambiguity.

---

[2] See for instance, Lüthy and Smets (2009); Daston (2015); Crowther and Barker (2013); Marr (2016); Marr and Heuer (2020); and Noyes (2023).

3.1 The Comets of 1618

Three bright comets appeared in the second half of 1618 and were widely seen. Before delving into an analysis of their visual culture, it is useful to erect a scaffolding to hold some common facts on what people saw and when, and what they generally believed about the nature of comets.

The first comet (C/1618 Q1) was discovered on August 25 by an observer in Hungary and was last seen on September 25. Johannes Kepler observed it from Linz, Austria, on August 27 before dawn. On September 6, he was the first person to look at a comet through a telescope and described it as resembling a cloud (Kronk, 1999: 333–335). Orazio Grassi, writing from Rome, did not see it but first heard news from Italian sources of the appearance in late August of a comet "licking the hind feet of the Great Bear." He gathered observations from Germany and Italy of its passage through Ursa Major saying that "in magnitude, it was equal to the stature of a man; its tail…verged southward wither the brilliance of the Sun clearly directed it" (Grassi, 1619: 5; Drake & O'Malley, 1960: 7–8).

The second comet of 1618 (C/1618 V1) was discovered on November 11 and last seen on December 9. It was a pre-dawn and early morning apparition in the southeast, first noticed by its long tail, which extended into Corvus from an unseen nucleus in Centaurus. The comet moved east to west, with the tail crossing into Crater and Hydra. Kepler reported that the tail was "a train of milky-white light," and Grassi that it extended 40° across the sky. Grassi described the comet as the "sword-shaped type called *xiphiae.*" But he acknowledged that few took notice of the comet. In spite of having a "long beam in the form of a sword unfolding itself toward the spirals of Hydra," the comet was "rather whitish, loose-textured, and of no considerable brightness" (Kronk, 1999: 335–338; Grassi, 1619: 4, 6; Drake & O'Malley, 1960: 6, 8).

The third comet of 1618 (C/1618 W1) was discovered on November 25 and last seen on January 22. Its overlap in timing with the second comet caused some confusion in early European reports. First seen with its head in the constellation of Libra, the comet then traveled across Boötes while its long tail swished Virgo's skirts and the hindquarters of Ursa Major. Observing from Ingolstadt in early December, Johannes Baptist Cysat noted that the head was brighter than third magnitude and the tail length as much as 70° (Kronk, 1999: 338–341). Grassi agreed that this great comet outshone and outlasted the two previous comets. He joked that the sky had realized its failure in getting attention from the first two comets, and "finally…it carried a very bright comet upward from the east, of such brightness that all eyes and minds were immediately turned toward it, and great throngs gathered on mountains and other very high places, with no thought for sleep and no fear of the cold north wind. Suddenly, men had no greater concern than that of observing the sky" (Grassi, 1619: 4; Drake & O'Malley, 1960: 6).

Grassi thought that the great comet had originated in the ecliptic at a spot in which the Sun and Mercury had been in conjunction in the sign of Scorpio on November 4. He called it an offspring of Mercury because of its mercurial countenance, going from super bright to pale in the course of its life. Mars was in Virgo at the time, but since Scorpio was the principal house of Mars, some observers were convinced that the comet had a Martian color in the beginning (Grassi, 1619: 7–8; Drake & O'Malley, 1960: 10).

On the physical nature of comets, most readers subscribed to the ancient theory of Aristotle that comets were sublunar meteors formed out of hot, dry, fatty exhalations rising from the Earth into the upper atmosphere, where, as condensed masses, they were ignited by the revolutions of the contiguous heavenly sphere. Sometimes a planet or stars played a role in gathering these exhalations together, but comets stayed below the Moon, and the celestial regions beyond the Moon remained unchanging and incorruptible. Comets might appear to move with the *primum mobile,* but their proper motion was due to them following their fuel.[3]

A minority of individuals, however, were reexamining the possibilities put forward in ancient times by the Pythagoreans, some Pre-Socratics, and Seneca that comets were mobile heavenly bodies of some eternal nature. The Senecan view was gaining more traction, especially among astronomers, since the nova of 1572 and the comet of 1577 were shown by Tycho Brahe's careful parallax measurements to be above the Moon (Brahe, 1913–1929: 4: 5–396; Hellman, 1967: 43–52; Schechner, 1997: 104–105, 261 n. 5).

Then there was the observation of Peter Apian and others after him that comet tails pointed away from the Sun. While the tail direction did not require comets to be supralunar, it did present problems for the notion of them being burning exhalations following a trail of fuel. Peter Apian, Gemma Frisius, Tycho Brahe, Johannes Kepler, and some authors writing about the 1618 comets thought that a comet could be made of some translucent material able to refract sunlight (Barker, 1993; Schechner, 1997: 105; van Nouhuys, 1998: 115–122).

Such observations poked holes in the Aristotelian cosmos, but did it do any damage to augury? The answer is no. Old and new theories about the nature of comets left a lot of room for interpretation and prognostication. If a scholar held fast to the Aristotelian theory that comets were fiery meteors, he expected the conditions that produced them to also produce other fiery meteors (such as shooting stars, auroras, fireballs, or lightning), parching of the air (leading to drought and wind storms), and phenomena caused by trapped terrestrial exhalations bursting from underground (for instance, earthquakes, stones tossed into the sky and raining down as meteorites, or tidal waves). Thus, apparitions of comets were taken as heralds of these correlated disasters. By late antiquity, comets were also seen in folklore as portents of the death of kings, war, insurrection, sedition, plague, and murder. No reason had to be given except that the gods willed it so, but a causal theory arose in the Middle Ages and was extremely popular in the 1618 cometary tracts. The arguments were that the hot, putrid vapors of comets parched and contaminated the air, thereby causing droughts, poor harvests, famine, sterility, and pestilence. Burning cometary

[3] See for example, António de Najera (1619); fols. B2v-B3r; or Discours (1618): 4–5; and Schechner (1997): 17–19.

exhalations inflamed the blood and provoked animosity between hot-tempered individuals, who were then disposed to uproot governments, overturn laws, and start wars. Princes were the first to die because their dissipated, and intoxicated lifestyles, consumption of delicate meats and lots of sex weakened their bodies and made them more prone to be harmed by comets than vulgar laborers. And as principals in wars, they were first in the line of fire (Schechner, 1997: Chap. 5).

On the other hand, if a scholar took the view that comets were celestial eternal bodies, it stood to reason that their elevated, starry, planetary nature would enable their powers to influence the Earth and its inhabitants. It also happened that some authors mixed and matched their explanations for why comets were reliable portents, taking, e.g., the position that comets were supralunar, translucent bodies, yet capable of inflaming spirits, inciting wars, and causing death with their noxious radiations (Schechner, 1997: 98–99).

But there was one thing everyone agreed on in 1618: Anything extraordinary—be it natural, preternatural, or miraculous—was a divine sign of impending judgment, and sinners would do well to repent and pray. Comets, therefore, kept company with mock suns (parhelia), eclipses, fire balls, apparitions of battling armies in the sky, rains of blood, stones, or frogs, two-headed calves, and other monsters in prodigy books like the *Augsburger Wunderzeichenbuch* (1545–1552) and illustrated works by sixteenth-century authors.[4] And like comets, monstrous births, ghostly visions, and wondrous celestial sights were each favored in their own illustrated, prognosticatory pamphlets and broadsides. Tycho Brahe put it well when he remarked, "Such unnatural births in the heavens…always have had something great to deliver to this lower world."[5] Comets were ciphers to be interpreted astrologically by their color, passage through the constellations and zodiacal signs, cardinal orientation, astrological houses, shapes and sizes, countries over which they hovered, and by historical induction (Schechner, 1997: Chap. 3).

3.2 Comet as Icon

Many title pages feature a woodcut vignette of the comet about which the book was written. However, the illustrations are highly stylized, rather crude, and far from naturalistic. The comet is shown as a pointed star with streams flowing behind it, and it is totally extracted from any sky or set among clouds.[6] (Fig. 3.2a–d) Although

[4] See the facsimile of the *Augsburger Wunderzeichenbuch*: Borchert and Waterman (2014). Also Lycosthenes (1557); Boaistuau (1561); Bateman (1581); Paré (1595); and Irenaeus (1566, 1595).

[5] See Tycho Brahe's German manuscript (1578), "De cometa anni 1577" in *Tychonis Brahe Dani opera omnia* (Dreyer, Raeder, and Nyström, eds.), 4: 379–396. It is translated in Christianson (1979): 110–140, see 137.

[6] Good examples showing the comet extracted from the sky include Velsius (1618); Dudith (1619); Grasser (1619); and Weber (1618). For the comet set among clouds, see: M. le Provençal (1618); Grasser (1618); Thurnman (1619).

Fig. 3.2 Iconic and stylized images of the 1618 comets. From left to right, top to bottom: (**a**) Andreas Dudith, *De cometarum significatione* (Breslau, 1619). Houghton Library, GEN *ZHC5. D8644.579db. (**b**) Johannes Velsius, *Prognosticatie van die nieuwe ende seer schrickelijcke comet* (Leeuwarden, 1618). KB, National Library of the Netherlands, KW Pflt 2802. (**c**) Johannes Thurnman, *Stella comans* (Erffurt, 1619). Houghton Library, GEN *GC6.Eh343.618c. (**d**) M. le Provençal [pseud.], *Les prédictions des signes et prodiges qu'on a veu ceste presente année 1618* (Paris, 1618). Houghton Library, GEN 21281.16*. (**e**) Niels Heldvad, *Resolution oc Forklaring paa den ny Comet oc wuanlige Stierne* (Copenhagen, 1618). Det Kongelige Bibliotek (Copenhagen), 19,-86. (**f**) António de Najera, *Discvrsos astrologicos sobre o cometa qve apareceo em 25. de nouembro de [1]618* (Lisbon, 1619). Bayerische Staatsbibliothek, Munich, Res/Astr.p.206. (**g**) Jean Belot, *Discours astrologique et description geographique et topographique du comette* (Paris, 1619). The Wellcome Library, London, 762/A

one might believe that the clouds were indicative of the author adopting the Aristotelian view that comets were sublunar meteorological phenomena, it appears to have been a convention in depicting the sky. Conversely, depictions of a comet against a starry backdrop must not be interpreted as an author's embrace of some sort of Senecan position or disbelief that the celestial regions were pure and immutable.[7] Similarly, a halo artistically drawn around a stellated comet nucleus should not be taken as proof that the author saw a coma of varying density.[8] The counterexamples are plentiful, and the images are too abstract and parsimonious for such correlations.

One might label such a simplified image as epistemic if one saw it as a stand-in for the too-varied instances of comets in the real world,[9] but I prefer to say they are iconic. They are road signs to draw the reader's eye and announce that "here be comets" without much effort to locate the "here."

Other vignettes do offer readers rudimentary landmarks (Fig. 3.2e, f), but just barely. For example, in a tract on the third comet, Niels Heldvad, a Danish Lutheran theologian and court mathematician to Christian IV of Denmark, positions the comet's head between the zodiacal symbol of Libra and a mythographic image of Scorpio (oddly mixing symbolic styles).[10] Similarly, António de Najera, a Portuguese cosmographer to Philip IV of Spain, puts the comet in a box placed between two shaded circles marked with 11 and 12. The left circle (11) has a small Sun and is partially eclipsed; the right circle (12) has a crescent Moon.[11] A new edition of Abraham Fleming's translation of bishop Friedrich Nausea's *Treatise of Blazing Starres* and a work by French chiromancer Jean Belot both locate the stylized comet inside a square bounded by cardinal directions.[12] Because they are so abstract, the interpretation of these images is difficult even after we have read the texts. At best, each author perhaps intended his image to be an *aide memoire* of a key aspect of the third comet of 1618—respectively, the zodiacal signs controlling its astrological significance, the role of a partial eclipse in generating the comet, and the southeast direction in which the comet pointed. Belot's woodcut has numbers arranged in a curious sequence to the borders of his square (Fig. 3.2g). Any knowledgeable astronomer or cartographer would have seen them as nonsense, so what could have been their purpose? Perhaps it was to communicate to less educated readers that the author's work was scientific or mathematical. I will come back to this sort of signaling when cometary charts are discussed.

[7] For example, Ehinger, *Iudicium astrologicum* (1618a); and also Avelar (1619). Avelar charts the second and third comets going through the constellations, but is committed to comets being meteorological according to Aristotle.

[8] Such a claim is made by Green in speaking about Thurnman, *Stella comans* (1619). See Green (2004): 94, 96.

[9] Definition according to Daston (2015).

[10] Heldvad (1618). On Heldvad's career, see Benediktsson (1980).

[11] Najera (1619). On Najera, see Domínguez-Castro et al. (2015): 21.

[12] Nausea (1618) was a new edition of the 1577 translation of *Super huis anni post Christum natum M.D.XXXI. & quolibet alio cometa exploratio*. Belot (1619).

3.3 Narrative and Allegorical Scenes

Many publishers chose to lead with an eye-catching image of the comet hovering over the heads of local observers. For example, the title page of Elia Ehinger's *Iudicium astrologicum* features a woodcut of Augsburg beneath the third comet of 1618 (Ehinger, 1618a) (Fig. 3.1). Now we are *somewhere*; now the celestial apparition is *grounded.* For Ehinger, a Lutheran theologian and Rector of the Augsburg Gymnasium, it bodes nothing good, but that interpretation is left inside the text. This is the same format used on comet broadsides—a blaring headline, a provocative image, and a text that warns of dire consequences presaged by the apparition.

Johann Baptista Hebenstreitt, Rector of the Ulm Gymnasium and "astrophilum," begins his 1618 tract with the same comet depicted more primitively over Ulm—along with a partial eclipse or conjunction of the Sun and Mercury (it is not clear what was intended) and symbols standing in for zodiacal signs and the planet Mars (Fig. 3.3a). Noteworthy are three individuals observing and discussing the meaning of the comet: a philosopher in an academic robe, an astronomer with an astrolabe, and a rich man in a cloak. A similar grouping plus a dog embellishes the pamphlet of Jacob Cats, a Dutch poet and politician writing at the time of the Synod of Dordrecht and a pause in the Dutch war with Spain (Fig. 3.3b).[13] Cats saw the comet of 1618 as a miraculous sign from God that peace would be restored if people returned to true religious faith. His text was in the form of a conversation between Wel-hem (a believer in cometary portents), Gheen-aert (a denier of astrology, who is consumed with material, terrestrial matters), and Reyn-hart (a bearded astronomer, who is cautious in using astrology to predict God's future actions). They conclude that comets are dire warnings of God's punishment, which might be averted through prayer and virtuous behavior (Cats, 1619). The vignette illustrates the conversation, and a Biblical quotation below it reinforces the interlocutors' conclusion: "This is what your Sovereign Lord says, your God, who defends his people: 'See, I have taken out of your hand the cup that made you stagger; from that cup, the goblet of my wrath, you will never drink again.'" (Isaiah 51:22).[14]

Another Dutch pamphlet reinforces the point that wonders like comets are to be collectively observed and discussed. The vignette invites readers to join other observers and count on the author, Balduinus Nicolai, to explain the apparition's meaning (Fig. 3.4). Nicolai's self-promotion was accomplished by an inset figure of him seated in his study—well-appointed with an armillary sphere—while drawing up a horoscope beside a window open to the stars. In spite of his reluctance to be called a "soothsayer," Nicolai forecast upheavals in Holland and Spain, and the continuation of slander and deceit by Remonstrants and Counter-Remonstrants.[15]

---

[13] Synod of Dordrecht (13 November 1618–9 May 1619) was a European meeting organized by the Dutch Reformed Church to settle controversies caused by a breakaway group of Arminians.

[14] Quotations from the Bible use the New International Version.

[15] Nicolai signs off as "beminder der loffelijcker matimatischer konsten" (lover of the praiseworthy mathematical arts) in Nicolai (1618). See also van Nouhuys (1998): 550–555.

Fig. 3.3 Narrative vignettes show people observing and discussing the comets. From top to bottom: (**a**) Johann Baptista Hebenstreitt, *Cometen Fragstück auß der reinen Philosophia* (Ulm, 1618). Herzog August Bibliothek, Wolfenbüttel, A: 49 Astron. (7). (**b**) Jacob Cats, *Aenmerckinghe op de tegenwoordige steert-sterre* (Middelburg, 1619). KB, National Library of the Netherlands, KW Pflt 3003

Verklaringhe
Om yegelijcken te adverteeren van
de dreyghementen Godts/ die wy aenmercken in't ae
schouwen vanden Comeet / gesien in't Jaer na de gheboorte ons
Heeren ende Salich-makers Jesu Christi 1618. (na de konst der Astronomie
van oude Doctooren daer op ghescreven/ te weten/ nae dat de vertoonin-
ghe geschiet/ in't Oosten ofte Westen/ oock nae de teeckenen des
Hemels ofte Zodiac/ daer de Comeet in gesien wort/ als
mede nae dat de steert lanck ofte kort is.
Ghepractiseert door Mr. Balduinum Nicolai, beminder der loffelijcker
Matimatischer konsten.
Gedruckt tot Amsterdam, by Broer Jansz. Woonende op de Nieuwe-zijds Kolc

Fig. 3.4 Balduinus Nicolai, *Verklaringhe om yegelijcken te adverteeren van de dreyghementen Godts, die wy aenmercken in 't aenschouwen vanden Comeet* (Amsterdam, 1618). KB, National Library of the Netherlands, KW Pflt 2800

Saxus Fontanus, a self-described "lover of the laudable art of astronomy," also chose to depict himself alongside the comet with the tools of the astronomical trade—a celestial globe and an armillary—in two tracts (Fontanus, 1618, 1619a). The second edition of his *De groote Practica oft Prognosticatie* (1619b) has cut away a corner of the frame of the portrait used in the first edition and inserted, rather clumsily, a large comet (Fig. 3.5). Like many others, Fontanus recounted cometary disasters since ancient times and concluded that the comet of 1577 and the nova of 1572 had heralded the horrible events in the Eighty Years War (also known as the Dutch Revolt) in the Habsburg Netherlands. These include the decimation of Zutphen and Naarden by the murderous Spanish army (1572), the Siege of Haarlem (1572–1573), the Battle of Moorkerheyde (1574), the Spanish Fury (the sacking of

Fig. 3.5 Portraits of Saxus Fontanus with the great comet of 1618, from title pages of three works by him. From left to right: *De groote Practica oft Prognosticatie*, first and second editions (Amsterdam, 1618, 1619b); and *Bediedenisse vande nievwe Comete ofte Sterre metten Steerte* (The Hague, 1619a). KB, National Library of the Netherlands, KW Pflt 2798, 3006, 2801

cities, 1572–1579), the Battle of Gembloux (1578), and so forth. The most recent 1618 comet, Fontanus warned, portended the death of a southwestern king (an allusion to Philip II, king of Spain), wars, seditions, treacherous councils, changes in government and religion, plus snow, ice, floods, winds, and fevers (Van Nouhuys, 1998: 550–555).

Jacob Marcus's engraved vignette shows us an astronomer observing with a cross staff, but the rest of the scene is allegorical and prognosticary (Fig. 3.6). In the foreground, Europa, personified as a woman, and Religion, symbolized by the overturned *globus cruciger* at her feet, are threatened by dangers approaching from behind—soldiers on foot and horseback, a city ablaze, and death depicted as a skeleton with a scythe (Marcus, 1619). And Johann Faulhaber, the Ulm mathematician, also cast his lot making apocalyptic predictions from the great comet of 1618 in his *Fama syderea nova*. The book illustrates the dire warnings, but significantly, it also shows him in the foreground with a cross staff, astrolabe, and a boy looking through a telescope (Faulhaber, 1618: fol. A4v) (Fig. 3.7).

Examples of allegorical images without reference to astronomers or their instruments include the title pages of David Herlitz (or Herlicius) and Caspar Hersbach. A German astronomer, physician, and poet, Herlitz interpreted comets as omens of war, famine, and death, as history had shown; he asserted that they would be among the signs of the end times. He reasoned that since the third comet of 1618 had appeared between the planets of Saturn and Mars, war would be the outcome. The book's woodcut (Fig. 3.8) shows a female figure precariously standing on a ball labeled Europa. She is the personification of Fortuna, and her fate does not look good. To the right, Saturn, personified as a man eating his child, stands on a cloud and reaches out to touch her left shoulder. On the left, Mars appears as a soldier in armor, holding a sword and a fiery brand (which touches her right shoulder). Above his head is a stylized comet. At his feet, a naked hag kneels and points a bow and arrow at Europa. In the text, Herlitz is vague in his forecast for Europe, and so it is in this image (Herlitz, 1619; van de Löcht, 2019). Caspar Hersbach, a mathematician who produced a celestial planisphere to be discussed below, exhorted readers to take notice of the awful apparition on the title page of his *Kurtzer Discurs*

Fig. 3.6 Europa and religion are threatened by war, conflagration, and death in this allegorical vignette from the title page of Jacob Marcus, *Corte beschrijvinghe ende affbeeldinghe des seer grooten ende schrickelijcken comeets* (Leiden, 1619). KB, National Library of the Netherlands, KW Pflt 3005

(Hersbach, 1619) (Fig. 3.9). In heavenly clouds, God points to a banner quoting Isaiah 47:3 ("I will take revenge"), while the comet blazes below carrying another warning ("I will visit their iniquities with a rod." Psalm 88:33). The pope on the left, a king on the right, and a wealthy merchant or gentleman between them each beseech God ("Spare, Lord, spare your people." Joel 2:17).

Emblems are worth noting here in the context of allegorical images. I found one on the title page of an English tract by John Bainbridge, the future Savilian Professor of Astronomy at Oxford (Fig. 3.10). The figure is circular and has a horizon line dividing the upper from the lower half. In the center is the Earth. At the top, the Hebrew name of God is inscribed in a cloud (it also heads each chapter in the book). Below this inscription appears the comet amidst a scattering of stars and the symbols of Mercury, Mars, and Saturn. The lower half has a Biblical quotation from Isaiah 40: "Lift up your eyes on hie, and behold who hath created these things" (Bainbridge, 1619: title page).

Visually these are story-telling images. They paint a quasi-realistic scene that draws spectators in and layers them with allegorical or symbolic figures to be

Fig. 3.7 Johannes Faulhaber (lower left) observes the first comet with an astrolabe and cross staff, while a boy looks at it with a telescope. Scientific observations aside, he then predicts apocalyptic events. Johannes Faulhaber, *Fama syderea nova* (Nuremberg, 1618). Houghton Library, GEN *GC6.F2735.618f

interpreted. To varying degrees, the images are also intended to shock audiences. The inclusion of astronomers' portraits and observers using scientific instruments was essential for the propagandist purpose of such pamphlets; they served to convince readers of the writer's expertise, his faithful description of the phenomena, and the worth of his analysis.

Fig. 3.8 Nothing good is predicted in this allegorical woodcut. David Herlitz, *Kurtzer Discurs vom Cometen, unnd dreyen Sonnen, so am Ende des 1618* (Szczecin, 1619). Bamberg, Staatsbibliothek, Astr.q.128(14)

As spectators of the twenty-first century, the narrative and allegorical images offer evidence of the people involved as observers and readers. The clothing depicted is indicative of social standing and profession. Astronomers are identified by their use of mathematical instruments. People observe in groups. Conversations (and shared reading matter) are paths to understanding the nature of the world. The occasional use of Latin phrases in the images as well as the texts, which are otherwise in the vernacular for the benefit of popular audiences, dress up the pamphlets. The flourishes signal the author's learning and sophistication.

3.4 Evidentiary Images

In addition to visual records of astronomical instruments being deployed to observe the comets of 1618, we have the astronomers' words as proof. Of the November–December comet, Swiss theologian and astronomer Elie Moléry says that "the sight of the ray in the shape of an elephant's tusk utterly excited and amazed me and bystanders because of its novelty and terrifying shape. But at once I judged that a comet had been formed…I took up my astronomical astrolabe quadrant and charts (*Instrumento Astronomico Quadrante Astrolabii* cum *chartis*)" to observe the comet (Moléry, 1619: 7, 19). In Lisbon, Manuel Bocarro Francês used a cross staff in observing both the second and third comets. He devoted a whole chapter to the

Kurtzer Discurs/
Darinnen die Natur/
Lauff vñ Bedröwung des Cometen/so Anno
1618. im DECEMBRI erschienen Physicè vnd
Astrologicè verzeichnet wirdt.
Durch M. Casparum Hersbach Agrippinatem Mathematicum.

Gedruckt zu Cölln/ Bey Peter von Brachel/ vnter Gülden Wagen
Im Spiegelberg/ Anno 1619.

Fig. 3.9 The Pope, a king, and wealthy merchant take heed of God's warning from the great comet of 1618. Caspar Hersbach, *Kurtzer Discurs, darinnen die Natur/ Lauff un[d] Bedröwung des Cometen/ so Anno 1618. im Decembri erschienen Physicè und Astrologicè verzeichnet wird* (Cologne, 1619). Bayerische Staatsbibliothek, Munich, Res/4 Astr.p. 517,18

method of parallax measurements and the computation of distances (Bocarro Francês, 1619: fols. 6r-8r). (Bocarro was a Portuguese *converso* physician, mathematician, and astronomer, who later took the name Jacob Rosales when he avowed his Jewish faith.[16]) Residing in London, John Bainbridge also used a cross staff to observe the third comet (Bainbridge, 1619: 3), and Johann Baptist Cysat, the Swiss Jesuit astronomer and professor of mathematics at the University of Ingolstadt, went all in on angle-measuring instruments to determine the third comet's position,

[16] On Bocarro's life and career, see Jacobs & Kayserling (1901–1906), 10: 470; and Carolino (2012).

Fig. 3.10 This emblem graces the title page of John Bainbridge's tract. *An Astronomicall Description of the Late Comet from the 18. of Nouemb. 1618. to the 16. of December Following: With Certaine Morall Prognosticks* (London, 1619). Linda Hall Library, QB724 .B34 1619

motion, and parallax. He used a wooden sextant of 5-foot radius, quadrants of 2- and 4-foot radii, and a 3-foot long cross staff (Cysat, 1619: 10, 72). And in Rome, Orazio Grassi, S.J. observed the second and third comets with angle-measuring instruments at the Collegio Romano, although he does not specify them (Grassi, 1619). The careful use of these instruments contributed to the production of charts of the comets' tracks by Moléry, Bainbridge, Cysat, and Grassi, and the conclusion by Bainbridge, Cysat, Grassi, and Bocarro based on parallax measurements that the comets traveled in celestial regions above the Moon. (More will be said about charts and diagrams below.)

The newest instrument to be pointed at any of the 1618 comets was the telescope (depicted in the lower left of Fig. 3.7). On 6 September 1618, Johannes Kepler led the way in observing the first comet with his "telescopium" (Kepler, 1619: 48). He was joined by Bainbridge and Cysat in studying the great third comet with a telescope (Kepler, 1619: 58). Bainbridge called the instrument a "Telescopium or Trunke-spectacle." He used the instrument to determine the separation of the comet from some faint stars (Bainbridge, 1619: 19). Cysat was perhaps the first Swiss individual to make a telescope (which he called "tubus opticus"), and built 6-foot and 9-foot long refractors to observe comets (Bolt, 2007). In *Mathemata astronomica*, he devoted a chapter to his observations of the comet's head with these telescopes during December, distinguishing the nucleus from the coma, and changes in their presentation (Cysat, 1619: 72–75) (Fig. 3.11). An anonymous French pamphlet also published two images of the comet's head, looking much like five-legged

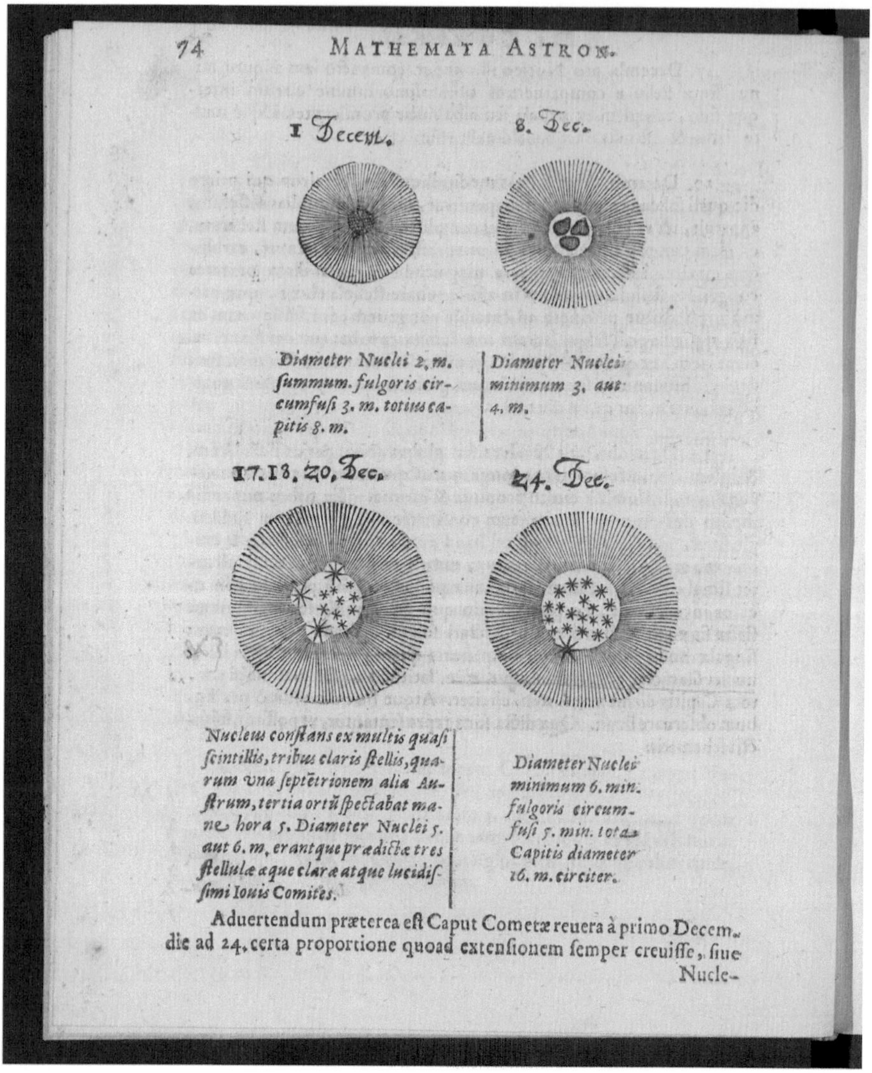
74 MATHEMATA ASTRON.

1 Decem. 8. Dec.

Diameter Nuclei 2. m. ſummum. fulgoris circumfuſi 3. m. totius capitis 8. m.

Diameter Nuclei minimum 3. aut 4. m.

17. 18. 20. Dec. 24. Dec.

Nucleus conſtans ex multis quaſi ſcintillis, tribus claris ſtellis, quarum vna ſeptētrionem alia Auſtrum, tertia ortũ ſpectabat mane hora 5. Diameter Nuclei 5. aut 6. m, erantque prædictæ tres ſtellulæ æque claræ atque lucidiſſimi Iouis Comites.

Diameter Nuclei minimum 6. min. fulgoris circumfuſi 5. min. totæ Capitis diameter 16. m. circiter.

Aduertendum prætereà eſt Caput Cometæ reuera à primo Decem. die ad 24. certa proportione quoad extenſionem ſemper creuiſſe, ſiue Nucle-

Fig. 3.11 Observations of the great comet's head through a telescope, showing the nucleus and coma over time. Jean Baptist Cysat, *Mathemata astronomica de loco, motu, magnitudine, et causis cometae* (Ingolstadt, 1619). ETH-Bibliothek Zürich, Rar 4254

starfish, presumably as seen through a telescope, but without explanation (Discours, 1618: 2) (Fig. 3.12).

Bocarro, a staunch critic of Aristotelian philosophy, argued that the heavens were as corruptible as the sublunar regions and made of the same stuff, but he retained a commonly held quasi-Aristotelian view that cometary material was gathered together by conjunctions of the superior planets (Bocarro Francês, 1619: fol. 5). He

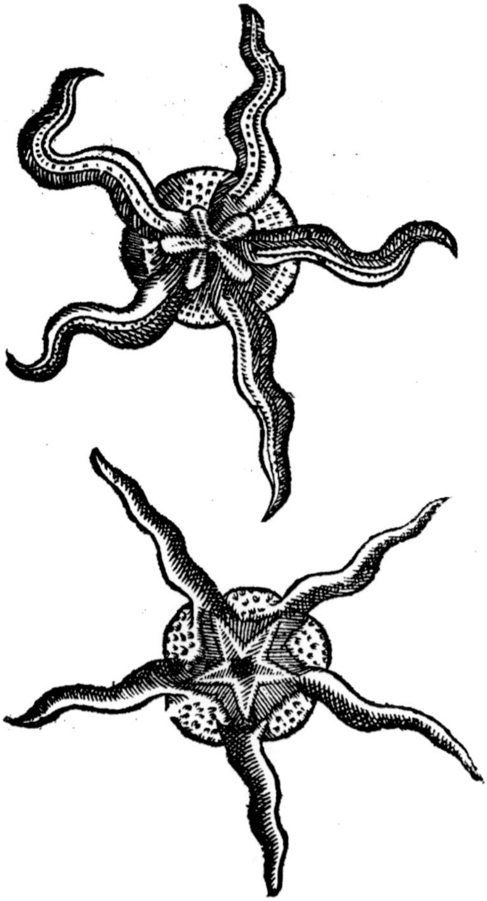

Fig. 3.12 Starfish-like image of the comet as seen through a telescope. *Discours sur la comète apparue sur la ville de Paris, le 29. et 30. Novembre et iours en suivant. 1618* (Paris, 1618). Bibliothèque municipale de Lyon, FC155-03

described the second comet, first seen on the ninth day of November, as "more than two spears long, of transparent light, white and silvery, the color of Saturn and Venus. I don't much specify because I saw stars through its body which was similar to a palm branch...and somewhat curved." Calling it "Cometa Palma," he published a drawing of it and concluded that a conjunction of Mars and Saturn had fashioned it in the region of Mars, where it caught fire and moved in its sphere (Fig. 3.13a). The third comet he described as "Caput Asconae," and found it to have the parallax of Mercury (Bocarro Francês (1619): fols. 8, 13v-14v, 18v; my translation). His depiction looks quite stylized, even though one assumes that he intended it to be as naturalistic a rendering of the observed phenomena as he had done for the earlier comet (Fig. 3.13b). Bocarro also drew other strange celestial apparitions witnessed in November 1618, including knives, a scimitar, a missile, triangles, and a helmut (Fig. 3.13c).

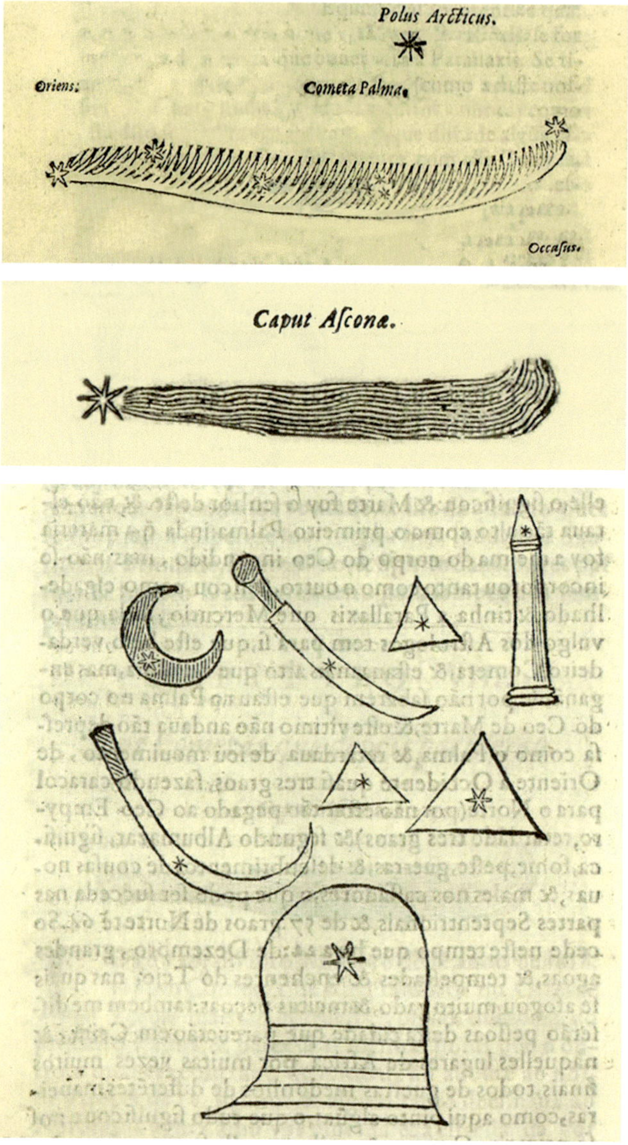

Fig. 3.13 Observations of the comets and other heavenly signs by Manuel Bocarro. From top to bottom: (**a**) The second comet, *Cometa palma*. (**b**) The third comet, *Caput Asconae*. (**c**) Celestial apparitions witnessed in November 1618. *Tratado dos cometas que apareceram em Novembro passado de 1618* (Lisbon, 1619). Biblioteca Nacional de Portugal, F. 1430

3.5 Maps, Charts, and Dynamical Representations

Many careful observers with astronomical instruments were inclined to produce a paper instrument—a celestial map or star chart[17]—to summarize wordy pages detailing the path of a given comet through the constellations and the changing length and orientation of its tail. Star charts, moreover, encoded observed locations, linked the data points into a smooth curve, and dynamically animated the comet's motion against the sphere of fixed stars. This made them easier to comprehend than data tables and wordy descriptions. They were more efficient and also helped to train the eyes of stargazers. The best charts were gridded with equatorial and ecliptic coordinates, which enabled precise placement of stars and interpolation of the comet's position between them. John Bainbridge (1619), for example, used a gnomonic projection of celestial longitude and latitude on his foldout chart (Fig. 3.14). From November 18 through December 16, 1618 old style (November 28–December 26 by the Gregorian calendar) he traced the third comet's daily path along a line divided into days on one side and declinations (10°S–0°–60°N) on the other. The map clearly shows the comet's head appearing in the constellation of Libra (and sign of Scorpio) with its tail sweeping across Virgo's skirt, and then tracks the comet up Boötes' right leg, torso, and left arm with its tail brushing Ursa Major's

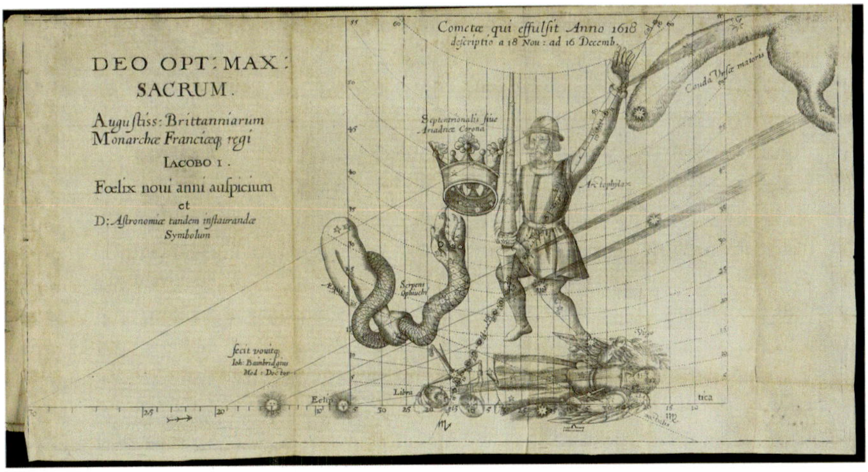

Fig. 3.14 Chart of the progression of the third comet of 1618 through the constellations. John Bainbridge, *An Astronomicall Description of the Late Comet from the 18. of Nouemb. 1618. to the 16. of December Following* (London, 1619). Linda Hall Library, QB724 .B34 1619

[17] Map and chart are used interchangeably in astronomy, although in some fields a map represents topography, landforms, and predetermined routes such as roads, whereas a chart represents detailed information and data (e.g., water depths, temperatures, or other conditions at a given spot) and gridlines that can be used to safely chart a course.

hindquarters. Throughout this passage, the chart also illustrates that the comet's tail remains anti-solar.

Grassi's *De tribus cometis anni 1618* had a foldout frontispiece that provided star charts of all three comets.[18] Luis de Avelar described himself as "baccalaureate of pontifical law, master of philosophy, student of mathematics, and alumnus of the Coimbra Academy." His chart (Fig. 3.15) combines the paths of the second and third comets on one sheet, but has few gridlines for reference. He thought that Mars and Mercury were instrumental in forming the comets, as was a conjunction of Saturn and Mars on November 4, but unlike Boccaro held to Aristotelian tradition that all the action was below the Moon. He is a good example of the disconnect that could exist between belief in comets being sublunar meteors and illustrating their paths through the stars (Avelar, 1619: 6). Another example is the anonymous author of the French pamphlet, *Discours sur la comète apparue sur la ville de Paris*, which has a pull-out star chart of the third comet's path (Fig. 3.16). This author acknowledges the theories of Anaxagoras, Democritus, and the Pythagoreans but dismisses them from experience and settles on Aristotle's (Discours, 1618: 8). Elie Moléry's book has a fine star chart (Fig. 3.17) (Moléry, 1619: after p. 20). Caspar Uttenhofer, who observed the comets from Nuremberg with his *circinus geometricus* (a "measuring circle" of his own devising), cross staff, quadrant, globe, put a star chart on his title page (Fig. 3.18).[19]

Caspar Hersbach published a broadside with an equatorial projection of the northern hemisphere (Fig. 3.19); it was a marketing ploy for his book (Hersbach broadside, 1618; 1619). According to the broadside's legend, "This comet first appeared on the sixth of December on the shinbone of Boötes over Africa and East India, then passed over this place of Cölln on the 22nd of December and finally disappeared over Sweden, Northern America and similar Northern countries, as can be read further in my little book."

The star charts by Bainbridge, Avelar, Moléry, and Grassi are geocentric projections—i.e., they show an "internal" perspective from inside the celestial sphere—with the stars mapped as seen by an Earthbound observer looking up at the sky (Figs. 3.14, 3.15 and 3.17). No doubt such maps were easiest to use by readers in the seventeenth century. Those by Uttenhoffer, Hersbach, and the French author are external projections as shown on celestial globes (Figs. 3.16, 3.18 and 3.19). They offer the view of a divine spectator whose perspective is external to the sphere of fixed stars with the Earth at its center. Such charts would have asked more of users, because they required them to do a mental flip of left and right, east and west, in order to relate the imagery to the reality of what they saw in the night sky. This would suggest that books with external-projection star charts were aimed at astronomical or mathematical users and not members of the general public.

---

[18] Grassi (1619): frontispiece. See the figure in Giancarlo Truffa's essay elsewhere in this volume.

[19] See Uttenhofer, 1619: title page; Uttenhofer (1626); and "Caspar Uttenhofer," Astronomie in Nürnberg, accessed 14 November 2024, https://www.astronomie-nuernberg.de/index.php?category=personen&page=uttenhofer-caspar

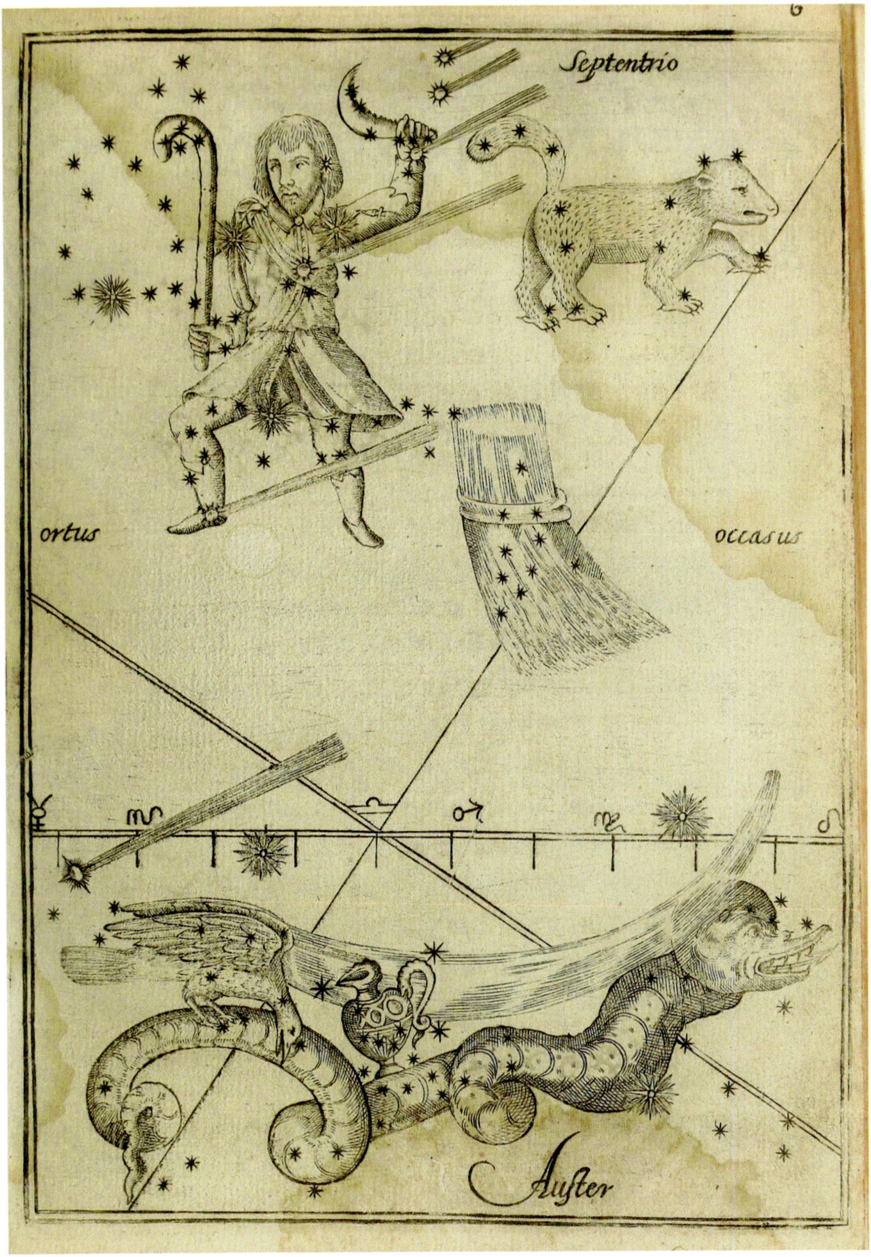

Fig. 3.15 Chart showing the paths of the second and third comets. Luis de Avelar, *Nox Attica* (Coimbra, 1619). John Carter Brown Library, Providence, RI, C619 .A949n

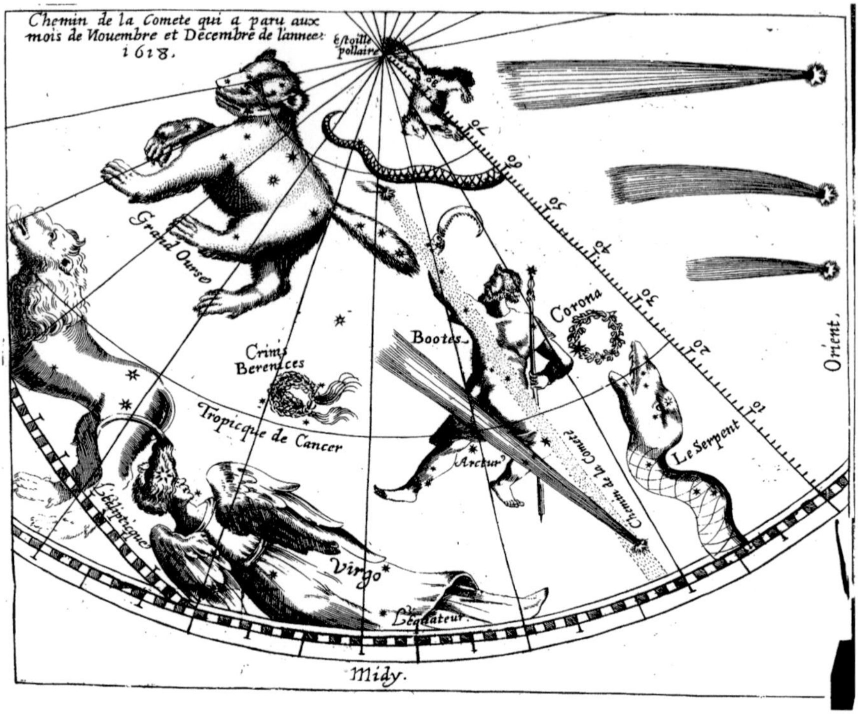

Fig. 3.16 The third comet's apparent path through the constellations. *Discours sur la comète apparue sur la ville de Paris* (Paris, 1618). Bibliothèque municipale de Lyon, FC155-03

Early modern cartography relied on the so-called Hipparchus rule.[20] This convention prescribes that human constellation figures should be depicted as facing the Earth's denizens. This means that celestial globes and charts with an external perspective ought to have figures illustrated from the back. On maps with an internal perspective, the figures were to be depicted from the front. Insofar as constellation figures served as finding aids and reference points for clear communication among observers, the convention was helpful—even necessary—so that a remark, such as "the comet went up the left arm of Boötes," would be consistent no matter which projection one was using. However, some cartographers have switched things up in the 1618–1619 pamphlets. Grassi, Moléry, and the French author depicted Boötes in reverse so that the comet went up his right arm. Consequently, these maps, although beautiful, would have presented difficulties to users wishing to compare reports or train the eye (Fig. 3.20).

Some cartographers eschewed the flat format and opted to give dimensionality to the celestial sphere by shading their two-dimensional representations of globe sections. Examples are found in books by Peter Crüger, professor of mathematics and

[20] For more on celestial maps, see Herlihy (2007).

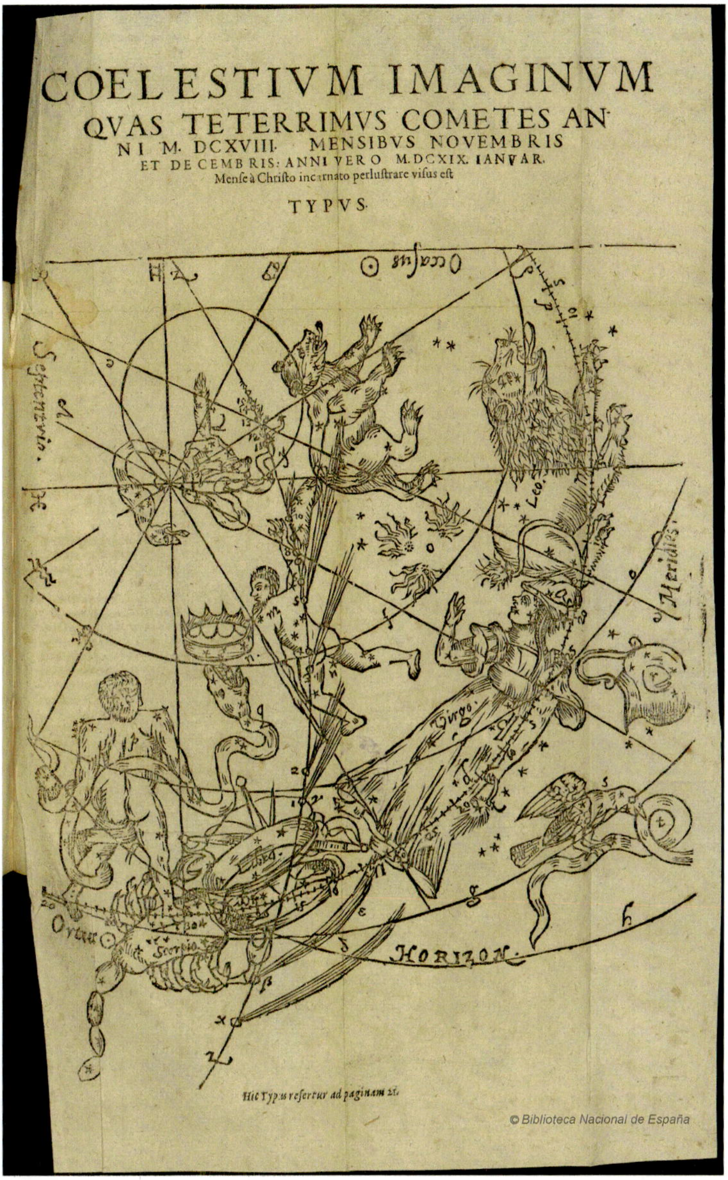

Fig. 3.17 Star chart with third comet. Elie Moléry, *De Tuba Coelesti, hoc est de Cometa Terrifico qui anno Christi 1618 apparuit et sedulo observatus est* (Embrun, 1619). Bibloteca Nacional de Espana, GMM/156(4)

Judicium de nupero Cometa Astro-
logo-Historicum.
Kurtzer Bericht und Erklärung/
Was von dem neuen Come-
ten/oder geschwentzten Stern/so sich die-
ses zu endlauffenden 1618. Jars/etliche zeit am Him-
mel in dem lüfftigen/warmen und feuchten Zeichen der Wag bey
den Bildern oder Sternen Boote und grossen Bären/der sonst Heer-
wagen genannt/sehen lassen/zuhalten/und desselben bedeu-
tung oder wirckung seyn möchte.
Auß Astrologischen gründen mit allem fleiß betrachtet/deß-
gleichen auß bewärten Historien und vielen Exempeln erwiesen/
unnd beschrieben/
Durch Casparum Uttenhofer zu Nürnberg/rc.

Non est CoMeta, qVI bonI qVID sVggerat.
Nürnberg in verlegung Simon Halbmayers Buchhändlers.
MDCXIX.

Fig. 3.18 Title page with star chart. Caspar Uttenhoffer, *Judicium de nupero cometa astrologo historicum* (Nuremberg, 1619). Bamberg, Staatsbibliothek, Astr.q.129

poetry at the Danzig Academy, Philipp Mueller, astronomer and physician in Leipzig, and Johann Caspar Odontius, an Altdorf mathematician who had worked with Kepler in Prague.[21] Crüger and Mueller used internal perspectives of a concave celestial hemisphere (which is hard to "see" because of the upward curving horizon line, Fig. 3.21a). Odontius deployed an external, globe-like cartography (Fig. 3.21b).

Instead of a "full-sky" view, we also find depictions of the comet's path across single constellations in a stop-motion format. Two approaches are represented by Cysat and Caspar Hersbach. Cysat presented a series of frames side by side (Cysat,

[21] See Crüger (1619): title page; Mueller (1618, 1619): title page; Odontius (1619), title page. On Odontius, see Astronomie in Nürnberg, accessed 14 October 2024, https://www.astronomie-nuernberg.de/index.php?category=personen&page=odontius-jc

Fig. 3.19 A fancy promotion for a book: an engraved broadside featuring an equatorial projection of the northern hemisphere with the comet track indicated. Caspar Hersbach, *Eigentliche verzeichnus, dieses im Jahr 1618. erscheinenden Cometen* (Cologne, 1618). Universitäts- und Landesbibliothek, Düsseldorf

1619: 63–69) (Fig. 3.22a). Hersbach's text included woodcuts of individual constellations—Scorpio, Virgo, Leo, Boötes, Ophiuchus, Ursa Major, and Draco—embedded in text devoted to their astrological interpretations. He overlaid a chopped path of the comet moving across Boötes, Ursa Major, and Draco (Fig. 3.22b).[22]

Highly abbreviated versions of charts, lacking the constellation figures or grid lines, were also produced. The *Assemblea celeste* charts (externally) the third comet across the sky with stars keyed to a legend (Danico il Riposato, 1619). Similarly, six representations of the comet move across the title page vignette of Eberhard

[22] Hersbach (1619). According to the Crawford Library Catalogue, the figures of figures of Boötes, Corona, Ursa Major, and Draco were printed with recycled blocks (comet added) used for Theodor Graminaeus, *Erklerung oder Außlegung eines Cometen* (Cologne, 1573); see Copeland (1890): 229.

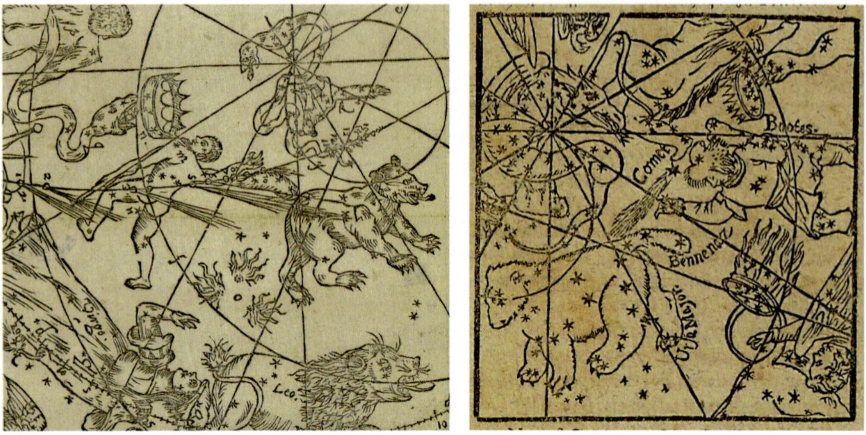

Fig. 3.20 On left: An internal perspective of the comet's path, but with Boötes reversed and not facing the Earth, detail from the chart by Elie Moléry, *De Tuba Coelesti*. On right: An external projection of the same portion of the sky, with Boötes correctly shown from the back side, Caspar Uttenhoffer, *Judicium de nupero cometa astrologo historicum*

Fig. 3.21 (**a**) Internal perspective of the great comet's path inside the concave dome of the heavens, but awkwardly drawn so that the stars appear to be projected onto a convex hemisphere. Peter Cruger, *Uranodromus cometicus* (Danzig, 1619). ETH-Bibliothek Zürich, Rar 4058. (**b**) Path of the great comet represented on a celestial globe using external projection. Johann Caspar Odontius, *Kometakribographia* (Nuremberg, 1619). Beinecke Rare Book and Manuscript Library, Yale University, 2008 2096

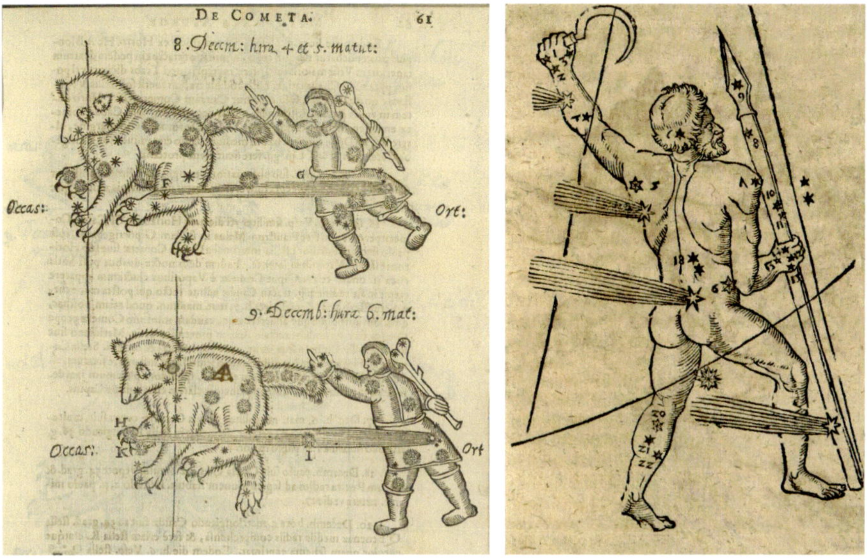

Fig. 3.22 Stop-motion depiction of the great comet going through the constellations Ursa Major and Boötes: (**a**) Jean Baptist Cysat, *Mathemata astronomica de loco, motu, magnitudine, et causis cometae* (Ingolstadt, 1619). ETH-Bibliothek Zürich, Rar 4254. (**b**) Caspar Hersbach, *Kurtzer Discurs, darinnen die Natur/ Lauff un[d] Bedröwung des Cometen/* (Cologne, 1619). Bayerische Staatsbibliothek, Munich, Res/4 Astr.p. 517,18

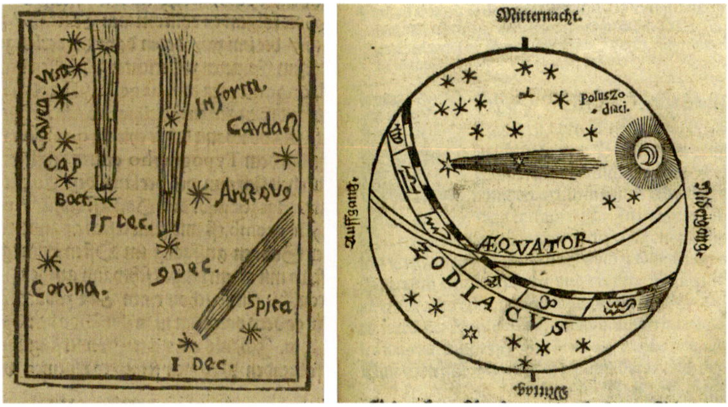

Fig. 3.23 Pseudo-scientific depictions of the comet on title pages: (**a**) Crüger's *Kurtzer Bericht von dem grossen noch zur zeit scheinenden Cometen* (Danzig, 1618). ETH-Bibliothek Zürich, Rar 2880. (**b**) Elias Ehinger, *Kometen Historia, das ist, Kurtze Beschreibung der fürnembsten Cometen* (Augsburg, 1618b) ETH-Bibliothek Zürich, Rar 4215: 3

Welper's tract (1619), with only stars of the Big Dipper, cardinal directions, and observed dates to locate them. These authors' intentions seem to have been faithful and informative. However, map-like figures were corrupted in Crüger's *Kurtzer Bericht* (Fig. 3.23a) and Ehinger's *Kometen Historia*, and August Rademan's

Grundliche to the degree that they are incomprehensible and even comically pseudoscientific.[23] Ehinger, for instance, has misplaced the equinoxes (Fig. 3.23b). What do we make of these? Perhaps it was a case of a confused woodcarver not following instructions, or a printer's addition (without the author's consent) to make the pamphlet more attractive to buyers who were not astronomers and would be impressed by an astronomical diagram and unaware that it was faulty. It may be significant that books and pamphlets with corrupted images were in the vernacular and not Latin.

3.6 Cosmological Schemes and Geometrical Diagrams

It is in the Latin texts that we would expect to see geometrical constructions and mathematical diagrams, and the books examined for this study do not disappoint.

Those who closely observed the comets and charted their passages through the constellations more often than not concluded that comets traveled in the heavenly regions above the Moon. There were a number of paths to reach this conclusion, but parallax measurements were empirical, and the concept was readily illustrated with a geometrical diagram (Fig. 3.24). A body close to the Earth, when viewed from different places on the planet, should appear against different background stars. The smaller the angular separation of those star markers—i.e., the smaller the parallax—the further away from the Earth the object must be. The Moon's parallax was a known quantity. A comet with little or no parallax had to be far "above" the Moon (Bainbridge, 1619: 16; Grassi, 1619: foldout plate; Cysat, 1619: 28, 37–42; Crüger, 1619: 35; Danico il Riposato, 1619: foldout plate). From there, astronomers might wish to observe any changes in parallax and position against known stars in order to calculate a comet's trajectory among the planets. This led to the insertion of

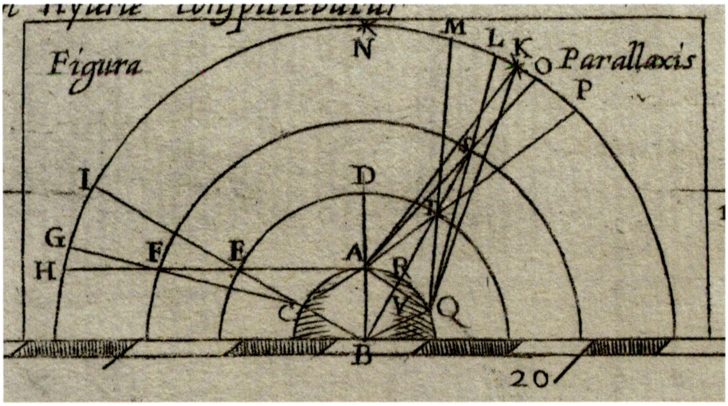

Fig. 3.24 Illustration of the use of parallax measurements to find the distance of the comet from the Earth. Orazio Grassi, *De tribus cometis* (Rome, 1619). Linda Hall Library, QB724 .G8 1619

[23] See the title pages of Crüger (1618), Ehinger (1618b), and Rademan (1619).

diagrams of spherical triangles (Fig. 3.25) and geometric constructions of the comet's track showing if it were curved or straight (Cysat, 1619: 12–17; Crüger, 1619). Perhaps best known of such diagrams is that of Kepler (Fig. 3.26) (Kepler, 1619: foldout plate after p. 8).

Accepting that comets moved through the heavenly realm raised a number of conundrums. What were they made of? Were they natural productions or miraculous signs? Did their existence mean that the heavens were mutable? The one

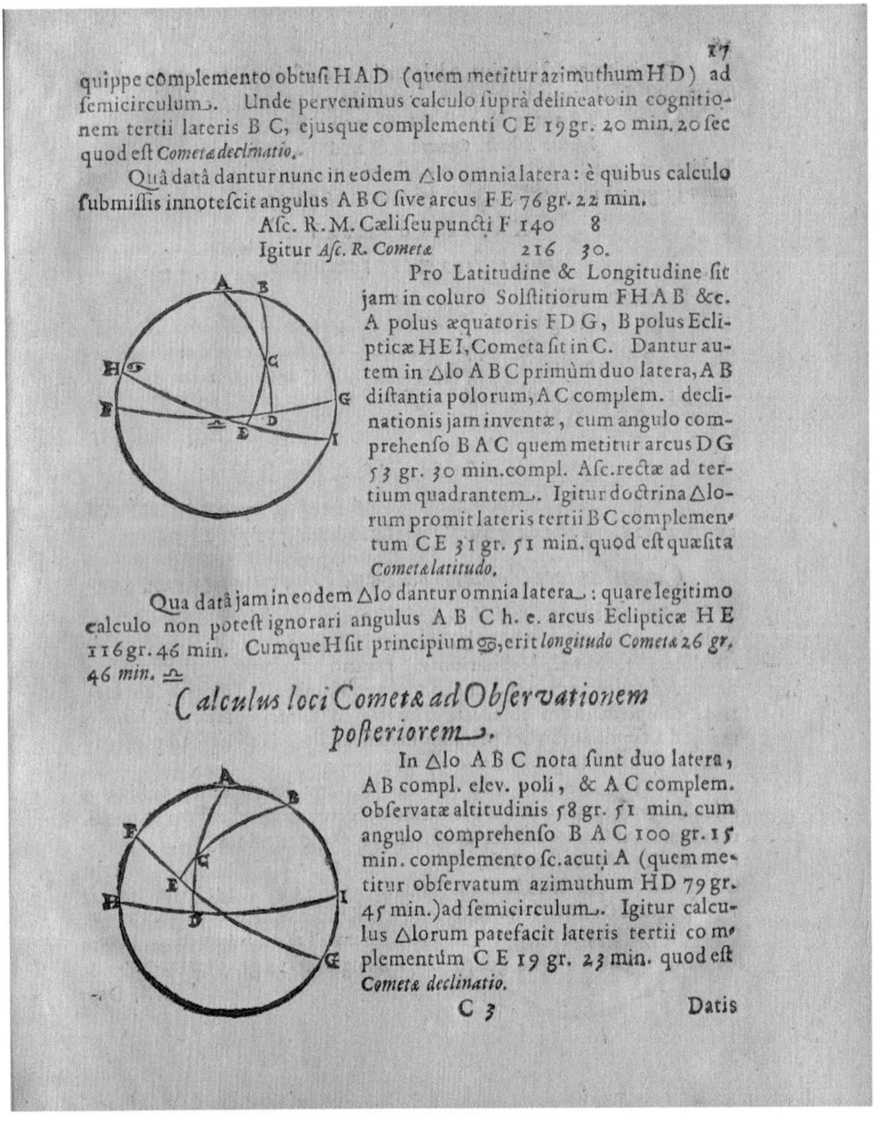
17

quippe complemento obtuſi HAD (quem metitur azimuthum HD) ad ſemicirculum. Unde pervenimus calculo ſuprà delineato in cognitionem tertii lateris B C, ejusque complementi C E 19 gr. 20 min. 20 ſec quod eſt *Cometæ declinatio.*

Quâ datâ dantur nunc in eodem △lo omnia latera: è quibus calculo ſubmiſſis innoteſcit angulus A B C ſive arcus F E 76 gr. 22 min.

Aſc. R. M. Cæli ſeu puncti F 140 8
Igitur *Aſc. R. Cometæ* 216 30.

Pro Latitudine & Longitudine ſit jam in coluro Solſtitiorum F H A B &c. A polus æquatoris F D G, B polus Eclipticæ H E I, Cometa ſit in C. Dantur autem in △lo A B C primùm duo latera, A B diſtantia polorum, A C complem. declinationis jam inventæ, cum angulo comprehenſo B A C quem metitur arcus D G 53 gr. 30 min. compl. Aſc. rectæ ad tertium quadrantem. Igitur doctrina △lorum promit lateris tertii B C complementum C E 31 gr. 51 min. quod eſt quæſita *Cometæ latitudo.*

Qua datâ jam in eodem △lo dantur omnia latera: quare legitimo calculo non poteſt ignorari angulus A B C h. e. arcus Eclipticæ H E 116 gr. 46 min. Cumque H ſit principium ♋, erit *longitudo Cometæ 26 gr. 46 min.* ♎

Calculus loci Cometæ ad Obſervationem poſteriorem.

In △lo A B C nota ſunt duo latera, A B compl. elev. poli, & A C complem. obſervatæ altitudinis 58 gr. 51 min. cum angulo comprehenſo B A C 100 gr. 15 min. complemento ſc. acuti A (quem metitur obſervatum azimuthum H D 79 gr. 45 min.) ad ſemicirculum. Igitur calculus △lorum patefacit lateris tertii complementum C E 19 gr. 23 min. quod eſt *Cometæ declinatio.*

C 3 Datis

Fig. 3.25 Diagrams show the usefulness of spherical triangles calculating a comet's trajectory. Peter Cruger, *Uranodromus cometicus* (Danzig, 1619). ETH-Bibliothek Zürich, Rar 4058

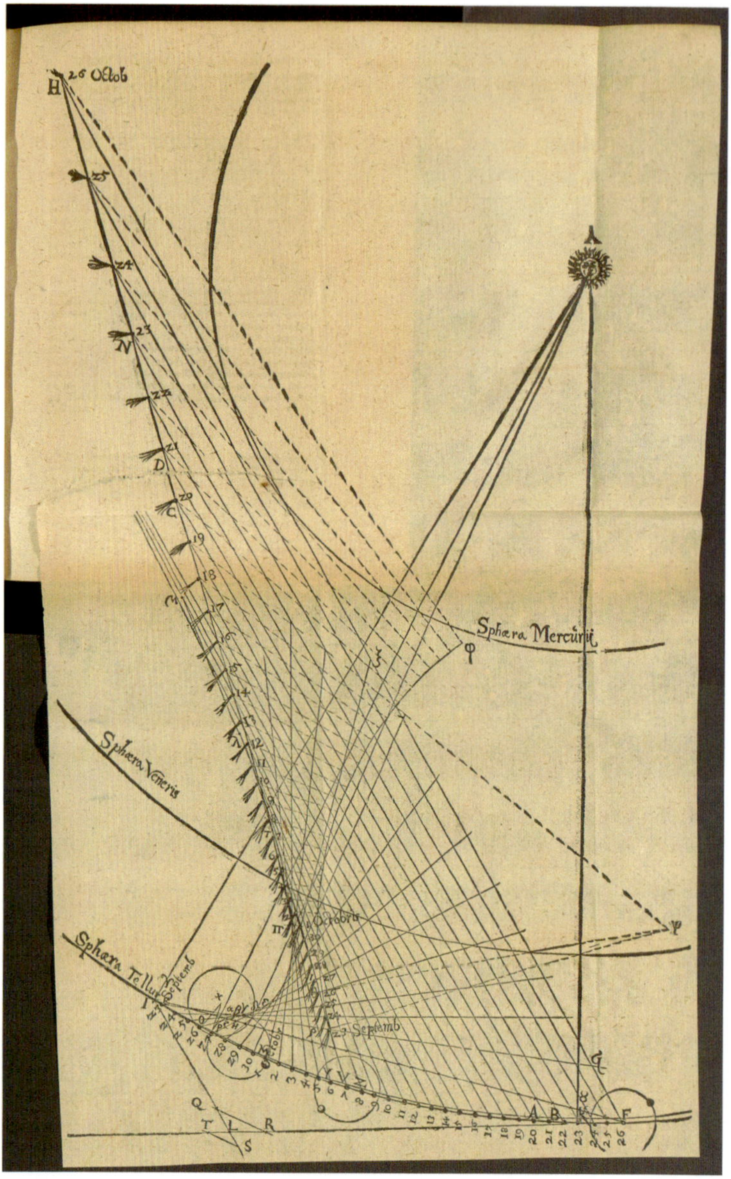

Fig. 3.26 Geometric construction of the trajectory of the comet. Johannes Kepler, *De cometis libelli tres* (Augsburg, 1619). ETH-Bibliothek Zürich, Rar 4389: 1

question whose answer was a resounding "yes" across the board in 1618 was: Did God send comets as signs of future events? Bainbridge, for instance, reasoned that it mattered not whether they were formed naturally or by God directly; in either case, comets were evidence that the heavens were mutable and without solid spheres, and divine providence could be counted on. In agreement on flaws in the

Aristotelian cosmology, Cysat adopted the Tychonic system, but wondered whether comets moved in straight lines or orbited the Sun. His book is illustrated with cosmological schemes of both possibilities (Cysat, 1619: 52–61) (Fig. 3.27). The Dutch physician and mathematician, Nicolaus Mulerius, was also convinced by his own observations that the great comet of 1618 was in a fixed orbit beyond the Moon, but was not prepared to throw out all of Aristotle, especially the part about the perfection and incorruptibility of the heavens. He argued for a Senecan view in which comets were a type of eternal body circulating in the heavens, only made visible by

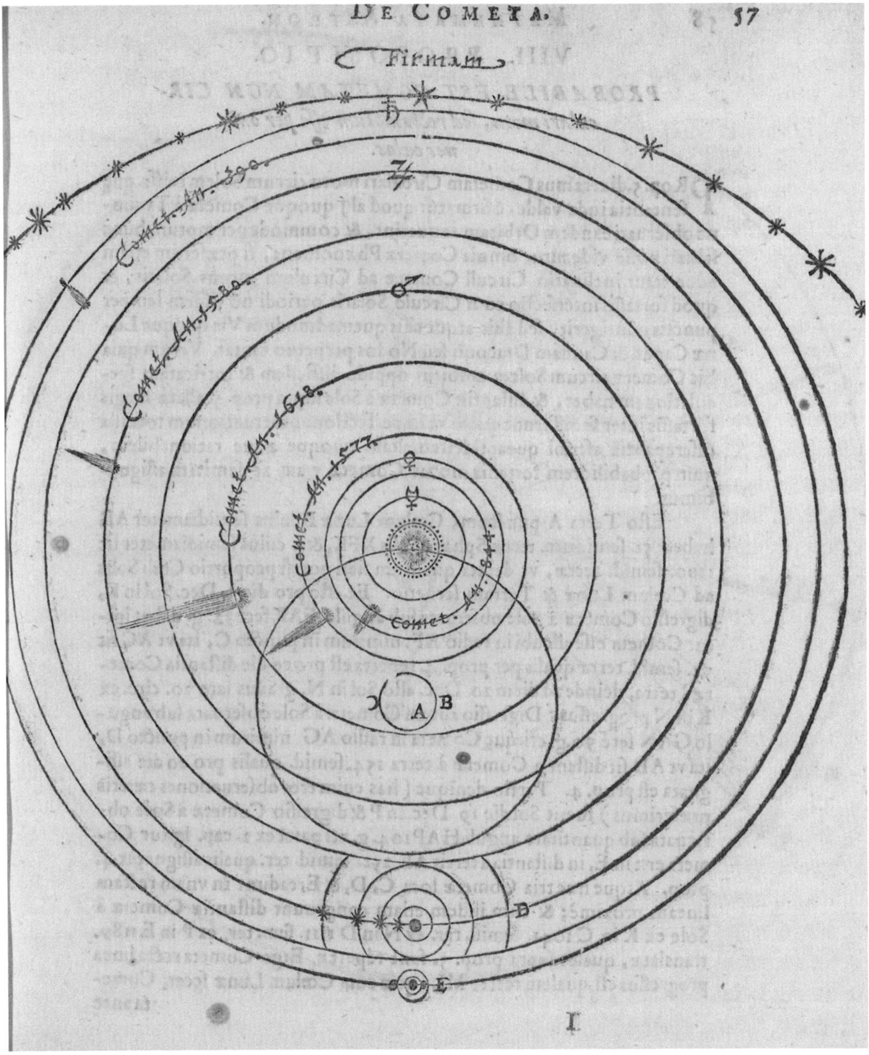

Fig. 3.27 A Tychonic cosmological diagram illustrating different comets orbiting the Sun. Jean Baptist Cysat, *Mathemata astronomica de loco, motu, magnitudine, et causis cometae* (Ingolstadt, 1619). ETH-Bibliothek Zürich, Rar 4254

God when he had an extraordinary message to deliver. Like planets, comets could powerfully influence life on Earth. He looked to the Dutch Revolt after the 1577 comet as proof, and warned that the 1618 comet portended worse and people ought to repent (Mulerius, 1618; van Nouhuys, 1998: 360–375, 539–542). He broadcast the gist of this message in his *Hemelsche Trompet Morgenwecker, ofte Comeet Met een Langebaert* (Heavenly Trumpet, Wake-Up Alarm Clock, or Comet with a Long Beard).

Mulerius's cosmological diagram (Fig. 3.28a) shows the comet in orbit beyond the Moon. This is not the conventional illustration of a "bearded star." That label traditionally referred to a comet moving away from the Sun with its tail before its head, like a man whose long, stiff beard precedes him. Mulerius, however, is playful in depicting the comet as a funny man with a long beard streaming sideways across the orbits.

In his book, Mulerius compares comets to comedians: Although great and important things (Sun, kings) should be studied, often novel, lesser things (comets, comedians) get more attention. But how are we to reconcile this with his point that the comet is a blaring trumpet, an alarm clock to wake up sinners? In addition, his funny diagram of the bearded star in orbit does not jibe with his belief that the tail is sunlight refracted by the comet head. This should warn us to be careful in how we read these figures.

Another cosmological scheme was depicted by Moléry, whose imagery may betray his indecision (Moléry, 1619: title page). In *De tubo coelesti* (another book entitled "Heavenly Trumpet"), *he* offered a quasi-Aristotelian theory that comets were thick, viscous terrestrial matter, stuck together and drawn up by planets or eclipses, and then set ablaze by Mars and Mercury, the motion of the fiery sphere, or heaven itself. The vignette on his title page shows the comet forming in the aery region just below the boundary with the fiery sphere, but its burning tail extends past the moon's sphere into the celestial regions (Fig. 3.28b).

Lastly, in the category of astronomical figures, I include horoscopes drawn up for the comets, because these are shorthand visualizations for the state of the celestial sphere and its planets at a given time. The horoscopic chart divides the sky into twelve astrological houses, starting below the eastern horizon with the first house, whose dividing line with the adjacent twelfth house is known as the cusp of the ascendant; this is the point of the Zodiac rising at the moment. The houses continue counterclockwise beneath the Earth to the lowest point at the cusp of the fourth house, which is the *Imum coeli*, the point where the meridian intersects the ecliptic in the north. From this point, the houses go up to the beginning of the seventh house, which is known as the cusp of the descendant. Then the houses rise above the western horizon to the cusp of the tenth house, the highest point known as the *Medium coeli*, which marks the intersection of the local meridian and the ecliptic facing south. And finally, they go down to the twelfth house, which touches the eastern horizon. These astrological houses are fixed for a given location, whereas the Sun, Moon, planets, fixed stars, and zodiacal signs rotate through them in the course of a day. The positions of the latter are recorded in the houses where they are found at a given date and time (such as a birth), and that time is recorded in the central square

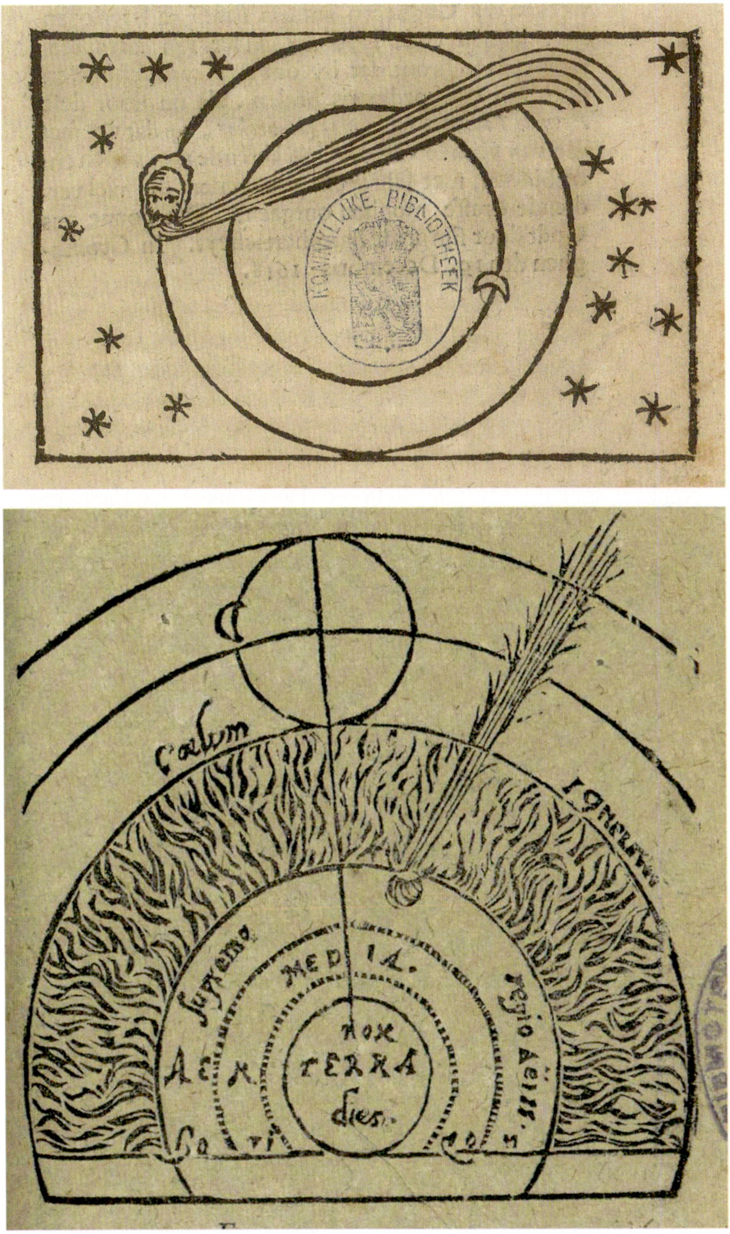

Fig. 3.28 Illustrations of two cosmological schemes: (**a**) A "bearded star" (comet) in orbit above the Moon. Nicolaus Mulerius, *Hemelsche Trompet Morgenwecker, Ofte Comeet Met een Langebaert Erschenen Anno 1618. in Novembri ende Decembri* (Groningen, 1618). KB, National Library of the Netherlands, KW Pflt 2799. (**b**) A comet burning in the upper atmosphere whose tail reaches beyond the Moon's sphere. Elie Moléry, *De Tuba Coelesti* (Embrun, 1619). Bibloteca Nacional de Espana, GMM/156(4)

along with the latitude. Each astrological house is thought to govern particular aspects of life, and its influence is reinforced by which celestial bodies or signs are within their boundaries. The astrologer uses this information along with the aspects of the planets (i.e., their geometrical configurations with each other, such as opposition, conjunction, trine, square, or sextile) to make astrological forecasts.

It was not uncommon for astrologers to draw up a horoscope for the "birth" of a comet (defined as when it was first seen) or for a particular date and time of its passage, as a tool to predict its significance and what it portended. Many horoscopes were prepared for the 1577 comet (Green, 2004: 94). Two that I found for the 1618 comets are by Moléry and Odontius, and Nicolai (1618) depicts himself in the act of drawing up a horoscope on the title page of his pamphlet (Fig. 3.4). Moléry's horoscope is titled "Dispositio coeli ad tempus, quo primum integer cometes anni 1618. in coelo fundere visus est [The disposition of the heavens at the time when the first whole comet of 1618 was seen to 'scatter' or 'pour' into the sky]" and is drawn up for 21 November 1618, 18 h, 23 min in the morning, at the northern latitude of 47°. The comet is shown in the twelfth house. He explains that the purpose of this horoscope is "astrological reasoning about the effects, power, and occult, yet sensible qualities of the comet" (Moléry, 1619: 55). Odontius's horoscope (Fig. 3.29) shows the comet in the tenth house at the latitude 49° 26′ on December 24 at 4

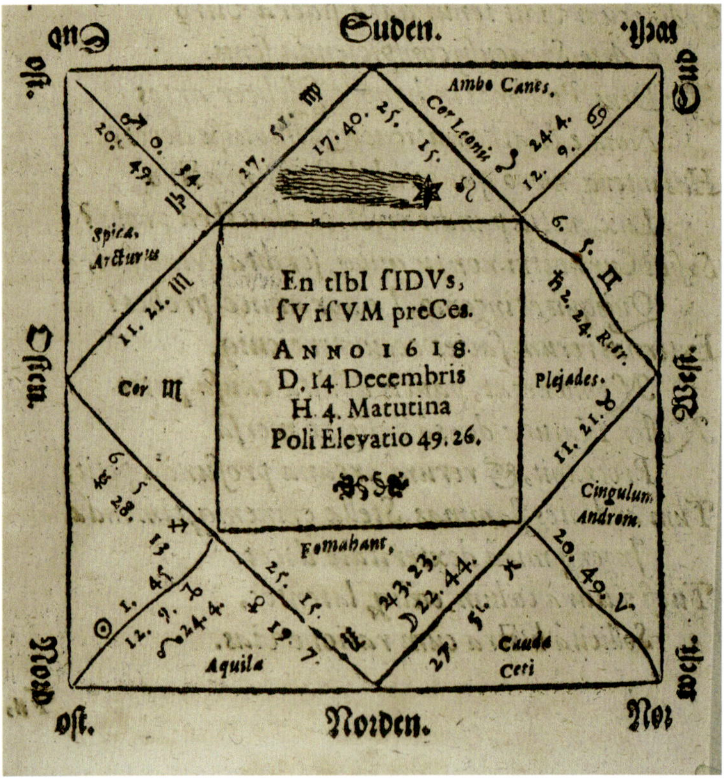

Fig. 3.29 Horoscope of the great comet of 1618. Johann Caspar Odontius, *Kometakribographia* (Nuremberg, 1619). Beinecke Rare Book and Manuscript Library, Yale University, 2008 2096

o'clock in the morning, when he says the comet was last seen in the sign of Virgo at 17° 40′. The horoscope shows the comet's relationship to the stars, planets, and houses "on which each student of astronomical matters may make an astrological judgment as he pleases" (Odontius, 1619: fol. Fiiir).

3.7 Conclusion

The visual culture of the comets of 1618 in Europe has been shown in this essay to be perhaps more diversified than expected, given the narrowness of the topic (comets of 1618) and the audience (generally popular). It included images that were iconic or symbolic; allegorical or emblematic; narrative, interpretative, or propagandistic; empirical and evidentiary; demonstrative of astronomical instruments in use; data visualizing and cartographic; dynamical representations of motion and change; cosmological schemes; and mathematical diagrams. These images had rhetorical, didactic, psychological, and commercial purposes. They aided the authors in illustrating, organizing, creating, and communicating knowledge. The variety of visual strategies shows us that one type of image does not fit all purposes.

The visual culture is informative to modern readers who are social or cultural historians of science. The clothing worn by individual observers suggests their social rank and learning, and perhaps the target audience for given tracts. Differences in the costume of Boötes the Herdsman vary wonderfully with the cold or warmer climes and the local cultures of the cartographers. However, scholars must be wary of being misled. The presence of a detailed chart of a comet's path through the constellations is not always indicative of the author's position concerning whether comets are sublunar or move through the spaces formerly reserved for planets.

References

Avelar, L. D. (1619). *Nox Attica: Hoc est. Dialogus de impressione metheorologica, et cometa anni domini 1618.*

Bainbridge, J. (1619). *An Astronomicall Description of the Late Comet from the 18. of Nouemb. 1618.* to the 16. *of December Following: With Certaine Morall Prognosticks or Applications Drawne from the Comets Motion and Irradiation amongst the Celestiall Hieroglyphicks.*

Barker, P. (1993). The optical theory of comets from Apian to Kepler. *Physis, 30*, 1–25.

Bateman, S. (1581). *The Doome warning all men to the Judgemente.*

Belot, J. (1619). *Instruction familiere et tres facile pour apprendre les sciences de chiromance et phisiognomie.... Plus un discours astrologique et description geographique et topographique du comette qui apparust sur notre hemisphere, l'année derniere 1618. Avec les predictions d'iceluy, dont les evenemens en seront formidables.*

Benediktsson, J. (1980). Niels Heldvad. In *Dansk biografisk leksikon* 16 vols. 3rd edn. Gyldendal. 1979–1984. 6: 624–625; Retrieved March 30, 2025, from https://biographicleksikon.lex.dk/Niels_Heldvad

Boaistuau, P. (1561). *Histoires prodigieuses.*

Bocarro Francês, M. (1619). *Tratado dos cometas que apareceram em Novembro passado de 1618.*

Bolt, M. (2007). Cysat, Johann Baptist. In T. Hockey et al. (Eds.), *Biographical encyclopedia of astronomers* 2 vols. Springer. 1: 267.

Borchert, T.-H., & Waterman, J. P. (Eds.). (2014). *The book of miracles: Das Wunderzeichenbuch—Le livre des miracles*. Taschen America.

Brahe, T. (1578). De cometa anni 1577. German manuscript, 1578. In J. L. E. Dreyer, J. Raeder, & E. Nyström (Eds.), *Tychonis Brahe Dani opera omnia* 15 vols. Gyldendal. 1913-1929. 4: 379–396.

Brahe, T. (1913–1929). *Tychonis Brahe Dani opera omnia* (J.L.E. Dreyer, J.Raeder, & E. Nyström, Eds.) 15 vols. Gyldendal.

Carolino, L. M. (2012). Manuel Bocarro Francês, the comet of 1618, and the impact of stoic cosmology in Portugal. In M. A. Granada (Ed.), *Novas y cometas entre 1572 y 1618: revolución cosmológica y renovación política y religiosa* (pp. 195–224). Universitat de Barcelona, Publicacions i Edicions.

Cats, J. (1619). *Aenmerckinghe op de tegenwoordige steert-sterre.*

Christianson, J. R. (1979). Tycho Brahe's German treatise on the Comet of 1577. *Isis, 70*, 110–140.

Copeland, R. (1890). *Catalogue of the Crawford Library of the Royal Observatory, Edinburgh.* Her Majesty's Government.

Crowther, K. M., & Barker, P. (2013). Training the intelligent eye: Understanding illustrations in early modern astronomy texts. *Isis, 104*(3), 429–470.

Crüger, P. (1618). *Kurtzer Bericht von dem grossen noch zur zeit scheinenden Cometen: Prodromus subsecuturi tractatus plenioris, concipiret den 18 Decemb. Anno 1618.*

Crüger, P. (1619). *Uranodromus cometicus: Ein aussführlicher Tractat vom grossen Cometen dess 1618 Jahrs ... neben eingefugtem Discurs von Cometen in gemein.*

Cysat, J. B. (1619). *Mathemata astronomica de loco, motu, magnitudine, et causis cometae qui sub finem anni 1618. et initium anni 1619. in coelo fulsit: ex assiduis legitimisq[ue] variorum phaenomenorum observationibus derivate.*

Danico il Riposato [pseudo]. (1619). *Assemblea celeste radunata novam[en]te in Parnasso sopra la nova cometa, Dall'Academico Danico Il Riposato, descritta in una lettera all'Academico Italico lo Spassionato.*

Daston, L. (2015). Epistemic images. In A. Payne (Ed.), *Vision and its instruments: Art, science, and technology in early modern Europe* (pp. 13–25). Pennsylvania State University Press.

Discours. (1618). *Discours sur la comète apparue sur la ville de Paris, le 29. et 30. Novembre et iours en suivant. 1618.*

Domínguez-Castro, F., García-Herrera, R., & Vaquero, J. M. (2015). An early weather diary from Iberia (Lisbon, 1631-1632). *Weather, 70*(1), 20–24.

Drake, S., & O'Malley, C. D. (Eds.). (1960). *The controversy on the comets of 1618.* University of Pennsylvania Press.

Dudith, A. (1619). *De cometarum significatione Cl. virorum Andreae Duditii Commentariolus & D. Thomae Erasti sententia/Elias Maior Vratislaviensis denuo edidit et adjecit.*

Ehinger, E. (1618a). *Iudicium astrologicum von dem newen Cometa welcher den 1. Decemb. 1618 am Morgen vor und nach 6. Uhren zu Augspurg von vilen Personen gesehen worden.*

Ehinger, E. (1618b). *Kometen Historia, das ist, Kurtze Beschreibung der fürnembsten Cometen, so von der Regierung an dess Römischen Kaysers Augusts, und der gnadenreichen Geburt Unsers Herrn und Heylands Jesu Christi, innerhalb 1618. Jahren gesehen worden.*

Faulhaber, J. (1618). *Fama syderea nova: Gemein offentliches Aussschreiben, des ehrnvesten, weitberühmbten und sinnreichen Herrn Johannis Faulhaberi ... Anlangend den newen vnd durch ein sonderbare Invention lang zuvor prognosticirten Cometstern, so den 6. Monats Tag Decembr. dess ablaufenden 1618 Jahrs, an alle Philosophos, Mathematicos, sonderlich Astronomos vnd Gelehrte dess gantzen Teutschlands, Authoris manu propria schriftlichen verfasst vnd abgangen.*

Fontanus, S. (1618). *De groote Practica oft Prognosticatie / op het Jaer... M. VI^{c}. XVIII.*

Fontanus, S. (1619a). *Bediedenisse vande nievwe Comete ofte Sterre metten Steerte, die int laetste vande maendt November, sich eerst aenden Hemel geopenbaert heft.*

Fontanus, S. (1619b). *De groote Practica oft Prognosticatie / op het Jaer... M. VI^c. XVIII.* Amsterdam.
Graminaeus, T. (1573). *Erklerung oder Außlegung eines Cometen.*
Grasser, J. J. (1618). *Christliches bedencken vber gegenwertigen Cometen: wie dann in gleichem vber andere Wunderzeichen, damit Gott der allmächtige, der vnbussfertigen bösen Welt, seine gerechte Gericht dräwet.*
Grasser, J. J. (1619). *Christliches bedencken vber gegenwertigen Cometen: wie dann in gleichem vber andere Wunderzeichen, damit Gott der allmächtige, der vnbussfertigen bösen Welt, seine gerechte Gericht dräwet.* 2nd ed.
Grassi, O. (1619). *De tribus cometis anni 1618.*
Green, D. W. E. (2004). *Assessment of early-modern observations of comets and supernovae: Focus on pre-telescopic European astrometric and physical data.* Durham University. Ph.D. thesis. Durham E-Theses Online: http://etheses.dur.ac.uk/3053
Hebenstreitt, J. B. (1618). *Cometen Fragstück auß der reinen Philosophia, Bey Anschawung deß in diesem 1618. Jahr in dem Obern Lufft schwebenden Cometen, erläutert.*
Heldvad, N. (1618). *Resolution oc Forklaring paa den ny Comet oc wuanlige Stierne, som bleff seet udi Novembri oc Decembri Maanet Aareffter Guds Byrd M.DC.XVIII.*
Hellman, C. D. (1967). The role of measurement in the downfall of a system: Some examples from sixteenth century comet and nova observations. *Vistas in Astronomy, 9*, 43–52.
Herlihy, A. F. (2007). Renaissance star charts. In D. Woodward (Ed.), *Cartography in the European renaissance* 2 vols. University of Chicago Press. 1: 99–122.
Herlitz, D. (1619). *Kurtzer Discurs vom Cometen, unnd dreyen Sonnen, so am Ende des 1618.*
Hersbach, C. (1618). [Broadside]. *Eigentliche Verzeichnus, dieses im Jahr 1618. erscheinenden Cometen, welcher Lauff oder Bewegung, durch die Zeichen des 8. Himmels, alhie vor augen gestellet wirdt.*
Hersbach, C. (1619). *Kurtzer Discurs, darinnen die Natur/ Lauff un[d] Bedröwung des Cometen/ so Anno 1618. im Decembri erschienen Physicè und Astrologicè verzeichnet wird.*
Irenaeus, C. (1566). *Wasserspiegel.*
Irenaeus, C. (1595). *Wasserspiegel.*
Jacobs, J., & Kayserling, M. (1901–1906). Rosales, Jacob Hebraeus (Immanuel Bocarro Frances Y Rosales). *Jewish encyclopedia.* 12 vols. Funk & Wagnalls.
Kepler, J. (1619). *De cometis libelli très.*
Kronk, G. W. (1999). *Cometography: A catalog of comets* (Vol. 1., Ancient-1799). Cambridge University Press.
le Provençal, M. [pseud.] (1618). *Les prédictions des signes et prodiges qu'on a veu ceste presente année 1618. Ensemble de la comete chevelue qui se voit depuis quinze jours sur ce florissant Royaume de France.*
Lüthy, C., & Smets, A. (2009). Words, lines, diagrams, images: Towards a history of scientific imagery. *Early Science and Medicine, 14*, 398–439.
Lycosthenes, C. (1557). *Prodigiorum ac ostentorum chronicon.*
Marcus, J. (1619). *Corte beschrijvinghe ende affbeeldinghe des seer grooten ende schrickelijcken comeets, de vvelcke den 27. November in't iaer 1618. ons verschenen is.*
Marr, A. (2016). Knowing images. *Renaissance Quarterly, 69*, 1000–1013.
Marr, A., & Heuer, C. P. (2020). Introduction: The uncertainty of epistemic images. *21: Inquiries into Art, History, and the Visual, 1*(2), 251–255.
Moléry, E. (1619). *De Tuba Coelesti, hoc est de Cometa Terrifico qui anno Christi 1618 apparuit et sedulo observatus est.*
Mueller, P. (1618). *De cometa anni M. DC. XVIII. Commentatis Physico Mathematica specialis & generalis.*
Mueller, P. (1619). *Hypotyposis cometae nuperrime visi.*
Mulerius, N. (1618). *Hemelsche Trompet Morgenwecker, Ofte Comeet Met een Langebaert Erschenen Anno 1618. in Novembri ende Decembri.*
Najera, A. D. (1619). *Discvrsos astrologicos sobre o cometa qve apareceo em 25. de nouembro de [1]618.*

Nausea, F. (1618). *A Treatise of Blazing Starres in Generall. As well supernaturall as naturall: To what Countries or people soever they appeare in the spacious world* (Abraham Fleming, trans.).

Nicolai, B. (1618). *Verklaringhe om yegelijcken te adverteeren van de dreyghementen Godts, die wy aenmercken in 't aenschouwen vanden Comeet.*

Noyes, R. S. (Ed.). (2023). *Reassessing epistemic images in the early modern world.* Amsterdam University Press.

Odontius, J. C. (1619). *Kometakribographia, das ist, Eygentliche, gruendliche Beschreibung dess im November vnd December erschienenen Cometen.*

Paré, A. (1595). *Des monstres et prodigies.* (5th ed.).

Rademan, A. (1619). *Gründliche vnd warhafftige Beschreibung des grossen vnd erschrecklichen Cometen, der mit gewisser Muthmassung in dem Monat Octobris, dieses 1618. Jahrs, seine Entzündung empfangen: worzu sonderliche Andeutung gegeben, die grosse, erschreckliche Conjunction des Saturni vnd Martis, den 10. Octobr. des obgemelten Jahrs: auch wie dieser Comet grosse Veränderung in dem gantzen H. Römischen Reich, mitbringen wird: zu einem Zeichen göttliches Zorns, vnd kunfftiger Strasse, Warnungsweise in Druck gegeben.*

Schechner, S. J. (1997). *Comets, popular culture, and the birth of modern cosmology*. Princeton University Press.

Thurnman, J. (1619). *Stella comans, oder, Gründliche vnd warhafftige Beschreibung, von dem newen Comet- vnd Wunderstern, auch Conjunctionen der obern Planeten: so im November vnd December des 1618 Jahres, für dem Auffgang so wol als Nidergang der Sonnen ... in der Lufft vnter dem Himmel erschienen vnd gesehen, sampt desselben Bedeutungen.*

Uttenhofer, C. (1619). *Judicium de nupero cometa astrologo historicum = Kurtzer Bericht vnd Erklärung, was von dem neuen Cometen, oder geschwentzten Stern.*

Uttenhofer, C. (1626). *Circinus Geometricus Zu Teutsch Meß-Circkel.*

Van de Löcht, J. (2019). 'Krieg/ groß Sterben/ vnd Hungers pein/l Bringt mit sich der Cometen schein': Publizistische Folgen der Kometen von 1607 und 1618/19 im Vergleich. *Daphnis, 47*(1–2), 85–112. https://doi.org/10.1163/18796583-04701011

Van Nouhuys, T. (1998). *The age of two-faced Janus: The comets of 1577 and 1618 and the decline of the Aristotelian world view in The Netherlands*. Brill.

Velsius, J. (1618). *Prognosticatie van die nieuwe ende seer schrickelijcke comeet ofte harige ster Anno 1618 verschenen.*

Weber, J. (1618). *Cometen Predigt: darin von Christi Geburt her, biss auff den jetzt Schwebenden, die vornembsten Cometen, vnd was darauff erfolget, kürtzlich erzehlet werden, mit angehefter, trewhertziger Vermahnung zu rechtschaffener Buss vnd Besserung des Lebens.*

Welper, E. (1619). *Observationes Astronomicae & Praedictiones Astrologicae Genommen auß dem Stand und Lauff des im nechstabgeloffenen 1618. Jahrs im Monat Novembri erschinenen grossen Cometens.*

Sara J. Schechner is the David P. Wheatland Curator Emerita of the Collection of Historical Scientific Instruments at Harvard University, retiring in 2024 after twenty-four years of service. She earned degrees from Harvard University and the University of Cambridge in physics and the history and philosophy of science. Her research, teaching, and exhibitions have earned her many prestigious international awards, including recognition as a Legacy Fellow of the American Astronomical Society (AAS) and the LeRoy E. Doggett Prize for Historical Astronomy. She is currently the president of the History of Astronomy Commission C3 of the International Astronomical Union. Her books include *Comets, Popular Culture, and the Birth of Modern Cosmology* (1997), *Tangible Things: Making History through Objects* (2015), and *Time of Our Lives: Sundials of the Adler Planetarium* (2019). Current research focuses on sundials, science, and social change; the representation of astronomers and their instruments in works of art; and scientific instrument making in America.

Chapter 4
The Great Comet of 1618: An Archaeopoetic Study

Clifford J. Cunningham

4.1 Introduction

The Roman rhetorician Quintilian (35–100) wrote that training of a man could not be complete "…if it stops short of music… nor again if he be ignorant of astronomy can he understand the poets." (Camargo, 2004: 95) Quintilian was stating not only that astronomy *could* be applied to understanding poetry, but that it *should* be applied to understanding poetry. Nowhere is such an understanding more appropriate than in a study of poetry relating to comets. However, even where comet poetry is quoted in a study of astronomical history, the reader is never compelled by the author to study it. As the poet T.S. Eliot (1888–1965) exhorted, "…the reader of a poem should take at least as much trouble as a barrister reading an important decision on a complicated case." (Atkins, 2012: 53) There is a permeable barrier between poetry and theology; a key question in this study is the meaning of 'signs.' In his poem *Gerontion*, Eliot (1920) wrote "Signs are taken for wonders," a phrase taken directly from a sermon preached before King James I by Lancelot Andrewes. The Great Comet of 1618 was a very recent memory the day of the sermon: Christmas Day 1618.

The role of astronomical poetry in the early seventeenth century was made clear by a charter that was drafted in 1605 by The Academy for Lynxes, a group Galileo was pleased to join in 1611.

> A Lynx must desire, in addition to knowledge of the things themselves, to give themself over to the pursuit of nature, particularly through mathematics. At the same time, they must not ignore poetry, nor the adornments of language, which decorate all of knowledge like a beautiful dress. (Fauber, 2019: 153)

C. J. Cunningham (✉)
University of Southern Queensland, Toowoomba, Australia
e-mail: Cliff.Cunningham@unisq.edu.au

C. J. Cunningham (ed.), *Cosmic Events*, Historical & Cultural Astronomy,
https://doi.org/10.1007/978-3-031-86950-1_4

There were three comets in 1618 (Nature, 1878), known by their modern designations as C/1618 Q1, C/1618 V1, and C/1618 W1. It is the third of these comets that is the "Great Comet" of 1618, the subject of this archaeopoetic study. As Bloomfield (2016: 6) explains. "Archaeopoetics is the poem's own formally and materially enacted exploration of what a poem can be and do as a form of historical inquiry." Aristotle taught that poetry is something more excellent and more philosophical than history. Thus, such an inquiry encompasses a philosophical dimension dealing in part with historical paradigms, which in the case of this comet study focuses on the nature of comets as understood in the early seventeenth century. Johannes Kepler (1571–1630), e.g., believed comets V1 and W1 were the same comet that had divided into two, and the sublunary nature of comets was still maintained by some. The astronomical heritage of the comet of 1618 extended far beyond the observational recording of its place and appearance in the sky. In addition to books that extended our knowledge of comets by both Galileo Galilei (1564–1642) and Johannes Kepler, more than a hundred pamphlets were published in Germany alone (Theibault, 1994; and Gruber in this book) and numerous medals were struck in Europe to commemorate its dramatic appearance (Walter, 1890).

The transient nature of poetry is another key element of archaeopoetics. Consider a modern example: In the mid-twentieth century, the French literary theorist Roland Barthes (1915–1980) said this in an interview about the poetry of his fellow Frenchman Paul Valery (1871–1945): "This poetry which had its glory days between 1920 and 1930, has gone completely out of fashion. It no longer has any kind of live, historical reality in the minds of those who work with literature today." (Wood, 2020: 20). If mere decades reek such change, how much more difficult, then, to breathe historical life into poetry of the seventeenth century, imbued with Epicurean and Stoic thoughts, for an audience of the twenty-first century? (Barbour, 1998). This is certainly a major reason why the astronomically-inspired poetry given in this study has been glossed over in cometary studies. But this ignores "poetry's power to preserve historical memory," as Virginia Closs (2020: 167) notes in a study of poetry revolving around the legacy of Nero. This study attempts to tap into that historical memory, but it is not made any easier if one considers the early seventeenth century as an age of wonder and superstition that embodies "a turning point where consciousness of a break with the past is bound up with the sense that memory has been torn...There are *lieux de mémoire*, sites of memory, because there are no longer *milieux de mémoire*, real environments of memory." (Nora, 1989: 7) For superstition, one need look no further than 1619, when three supposed witches in Lincolnshire were put on trial "for maleficent attempts against the Earl of Rutland's family." (Firth, 1911: 42) Despite the scientific attitude towards the comet evinced by England's King James I (1566–1625), it must be noted on the opposite scale of justice that 50 people were executed as witches during his reign (Tyson, 2011: 10) "As for them who think these witchcrafts to be but fantasies," he said in a speech of June 1591, "I remit them to be catechized and instructed in these most evident points…because I see the pride of these witches and their friends, which cannot be prevented but by mine own presence." (Tyson, 2011: 304). It was this sense of personal intervention that we find in the King's comet poem (Sect. 4.2.2). As is evident from the following passage, we are in an age before the founding of the Royal Society of London (1660) and the Académie française (1635). The portentous appearance of the Great Comet of 1618 in the skies of England is best expressed by the English historian and playwright Arthur Wilson (1595–1662):

> And now the Heavens declare the Glory of God; A mighty Blazing Comet appears in Libra, whose bearded Beams covered the Virgin Sign; it began on Wednesday morning, the 18 of November this year, and vanished away on Wednesday the 16 of December following...These Apparitions doe allwaies portend some horrid Events here below, and are Messengers of mischief to poor Mortals...What miserable Effects of War, Ruine, and Devastation, in most parts of the known World. (Wilson, 1653: 128–129)

The early seventeenth century was a time when poetry and astronomy were more than just complementary to one another——their relationship had been crystallized in the previous century by George Puttenham (1589: 53), who wrote that poets can be regarded as "...the first Astronomers and Philosophists." A recent study of the poetry of the English astronomer Jeremiah Horrocks (1618–1641), in his posthumously published 1662 book *Venus in sole visa* (Venus seen on the sun), also makes this link. R. Horrocks (2012: 119) writes "I enjoy their energy as examples of writing in a period when it was still permissible for a scientist to express his excitement, his sense of play, and his interest in philosophical and religious implications." The disregard in which seventeenth-century astronomical poetry is currently held was noted by Barton (2022) in his study of Horrocks. "Despite the long-recognized importance of Horrock's observations, his hexameter compositions have been largely ignored in later scholarship. In the latest translation of the *Venus in sole visa*, one poem—the longest and arguably the best—is omitted altogether." This lamentable state of affairs, which also permeates cometary studies, does little credit to the scholarly community.

"Until the late seventeenth century, humanists, natural philosophers, theologians, and poets operated within the same mental framework as far as the use of comets for prediction was concerned." (Jorink, 2010: 123). Historians of astronomy have written about comets as described by the first three of those groups, but the work of poets in cometary studies has been almost entirely ignored. To do so blunts our understanding of knowledge transmission; the important role of poetry in transmitting scientifically valid information has been studied by Goldstein (2017). Specifically for the comet of 1618, a case in point are rhymes put forth for German school children by two Protestant preachers of Basle, Switzerland—Johann Jacob Grasser (1579–1627) and Johann Georg Gross (1581–1630). They adopted a new style of preaching that "began to be heard in the second decade of the seventeenth century, a style that combined the methods of Ramism with the full doctrinal spectrum of Reformed Orthodoxy" (Burnet, 2006). Burnet refers here to Petrus Ramus (1515–1572), whose interaction with Johannes Kepler became "a central pillar of his new astronomy" (Cunningham, 2016a: 11). One Grasser and Gross rhyme reads:

> Eight things there be a Comet brings,
> When it on high doth horrid range;
> Wind, Famine, Plague, and Death to Kings,
> War, Earthquake, Floods and Direful Change.

This and other doggerel rhymes envisioned the comet of 1618 as a rod, such as

I am a Rod in God's right hand
Threatening the German and foreign land

were designed "to fasten the orthodox theory into the minds of school-children and peasants." (White, 1887: 132) This imagery was evoked at the time by everyday people; Hans Heberle (1597–1677), a cobbler by trade, began a diary in 1618 that, over the next 22 years, became a popular account of the Thirty Years' War (GHDI, 2023). His diary opened with this:

In 1618 a great comet appeared in the form of a great and terrible rod,
which was accorded us by and through God because of our sinful lives,
which we have richly earned and continue to earn daily.

The power of the Great Comet of 1618 was such that in 1650, 2 years after the Thirty Years' War ended (which many people believed began that conflict), a German medal was issued that once again evoked the rod of wrath: "God punished Germany for a long time with great fires, pestilence, war and famine. In that time there was a great comet star, a great rod to all." (Walter, 1890: 87) This was noted at the time by the English diarist John Eveyln, who wrote in 1624 "…the effects of that comet, 1618, still working in the prodigious revolutions now beginning in Europe, especially in Germany." (Evelyn, 1901: 4) While it may seem bizarre to us now, one can begin to understand this decades-long obsession with the Great Comet of 1618 in terms of a *cultural surface* (to use a concept initiated by Frade (2023: 7)) upon which the dread associated with comets is a manifestation of humanity's age-old fear of heavenly portents. The particular elements that were stimulated by the great comet on the cultural surface of seventeenth-century Europe will be explored in this chapter.

4.2 King James I of England (and King James VI of Scotland)

"James VI and I's poetry has received no historical and little critical examination." So declared Kevin Sharpe (1993: 127), and while that is not entirely true three decades later, the study of his comet poem in particular has never received a thorough study. This chapter examines it in five ways: rhetorical, astronomical, theological, satirical, and political. Before delving into the poetry, one must ask why the king turned to poetry to express his views. In the context of this study, James made it very clear in a poem he translated from du Bartas about Urania, the Muse of Astronomy [(Cunningham, 2024); I modernise the English here for ease of reading]. Poets, he wrote, were "Dame Nature's trunchmen, heavens interpreters true." (James, 1584) Trunchmen were often present at masques to offer glosses on the performance. By writing poetry, King James positioned himself as a trunchman: a true interpreter of messages from heaven. By 1618, King James had already established a long history of writing poems to convey his plans and beliefs to his subjects:

two volumes of his poetry were published in the sixteenth century, and seven more poems appeared between 1618 and 1623. He became King of Scotland when only 1 year old, and wrote his first poem at age 15 or 16. While still a young king, he styled himself as "the future patron of divine poetry in Britain." (Campbell, 1935: 40).

His early books of poetry, from 1584 and 1591, were widely read (Verweij, 2014) and, like the poems of Queen Elizabeth I (1533–1603), his poems were designed to "transmit royal policy." (Herman, 2001: 1501; Doran, 2024). James (while King of Scotland as James VI from 1567) also took to heart advice given to him by Elizabeth, who was Queen from 1558. "She reminded him of Isocrates' counsel to his emperor that royal words were the 'ensigns' of royal authority—badges or banners around which a ruler might gather force to counter rival or antagonistic ideologies and discourses." (Sharpe, 1993: 130) Such a deployment of words is in strong evidence in the King's comet poem, and (as a reader of Horace) he was certainly familiar with the injunction found in the 19th poem of Epistles 1: "He who trusts in himself will rule the hive." (Johnson, 1993: 90; Trinacty, 2012). In all the public buzz around the comet, he felt a mandate to rule. In addition to using poetry as a means of communicating royal policy, James wrote that he believed poetry might "lift up the mind" of fallen man to an understanding of God's natural order: "A breath divine in Poets breasts does blow." (quoted in Sharpe, 1993: 130).

John Koethe, Professor Emeritus of Philosophy (University of Wisconsin-Milwaukee), recently wrote about how or why poems engage us.

> …what makes certain linguistic expressions poetry is the way they engage us, rather than their character or content. Engage us in what ways though? Well, that is hard to say, though it's in ways we feel we can recognize, and that differ from the ways we're engaged by other forms of linguistic expression involved in science, philosophy, politics, or religion. (Koethe, 2022: 159).

King James not only captured but also harnessed concepts in all four of those realms of thought by a linguistic expression we call poetry. It is a study of his comet poem that is the centrepiece of this chapter. (See Note 1 for the publication history and archival locations of the poem.)

4.2.1 King James I of England

4.2.1.1 King James and Comets before 1618

Astronomers across Europe commented on the cosmic phenomenon, astrologers everywhere excitedly interpreted its meaning, and doomsayers and pessimists from Scotland to Sicily witnessed portents and prodigies in its fiery tail. Even King James I (of England) and VI (of Scotland) wrote about it. The Great Comet of 1618 certainly left its mark, but there was much to link King James with comets prior to 1618.

The royal penmanship of a poem to mark a celestial event has a long pedigree. Pliny the Elder (23–79) reports that the Emperor Titus (39–81) described in a poem what was termed *acontiae*. It was listed by Pliny as a type of comet, "like a dart with a very quick motion." It was "seen in his fifth consulship; and this was the last of these bodies which has been observed." (Pliny, 1855: 55–56).

It is quite likely James witnessed the Great Comet of 1577 (Stubbs, 2011) when only 11, and it is thought Tycho Brahe (1546–1601) personally gave James a copy of his manuscript on that comet when James visited Tycho on 20 March 1590. All discussions about James and his poem on the Great Comet of 1618 have overlooked the fact that it was not his first comet-related poem. The concluding two lines of a 14-line poem (undated but likely pre-1600) establish his purpose in giving a long list of things that exist (Westcott, 1911: 25):

All these for teaching man the Lord did frame
To honour him whose glory shines in them.

One of the things presented to man as a teaching tool is the appearance of a comet, here referred to as a 'bearded fire':

The sad and bearded fires, the monsters faire
The prodigies appearing in the air.

Even before he ascended to the English throne, he "acquired a reputation as the greatest royal poet in Christendom." (Rickard, 2015: 20) From the earliest years of his reign in England, which began in 1603 with the death of Elizabeth I, he associated his new position as King of both Scotland and England with a comet. His ceremonial entry into London in 1604 was designed to project to the populace his kingship as a successor to the first Roman Emperor, Augustus (Ronan, 1995). Comets were closely associated with Augustus (63 BCE–14 AD) and his uncle, Julius Caesar (Satterfield, 2016). A witness to the 1604 pageant described the cometary imagery erected in the central London street known as The Strand (quoted from Jonson, 1692: 311), featuring the mythological character Electra:

In the Strand, Electra, in the figure of a comet suspended between two 70-foot high Pyramids, had been provided with these words by Ben Jonson to conclude the pageantry:

Sun, moon, or stars shine more serenely fair.
This from that loud, blest oracle, I sing,
Who here, and first, pronounced thee Britain's king.
Long may'st thou live, and see me thus appear
As ominous a comet, from my sphere,
Unto thy reign; as that did auspicate
So lasting glory to Augustus' state

This required the audience to "literally see the meaning in the appearance, in the spectacle," to quote the German professor K. Ludwig Pfeiffer (1994: 20). By employing the mythological Electra (one of the Pleiades), Jonson used her appearance to bolster the abstract nature of comets with concrete cues, "to help out where science seems unable to touch base with ordinary perceptions of human reality." (Pfeiffer, 1994: 21) Electra was one of the famous seven sisters, but as she could not bear to witness the fall of Troy, she abandoned her sisters and translated into the heavens as a comet. The image of Electra in the pageant was therefore *emblematic* of a comet, and as it was linked to the glory of Augustus, it was thus a moral emblem, one of the four types of emblematic images (mathematical, theological, philosophical and moral) later codified by the French Jesuit Claude-Francois Ménestrier (1684: 5) "Moral figures are emblems," he wrote, "which are images intended to instruct men for the conduct of their life." (translation from Graham, 2016: 135) Thus, we see a comet visibly linked in the public mind to James as King of Britain; in his poem of 1618, James would personally link his kingdom with a comet. At the time Jonson welcomed the King to London in 1604, he had already established himself as a major playwright in the reign of Queen Elizabeth I. After receiving an annual pension from King James in 1616, he came to be regarded as England's first Poet Laureate [For more on Jonson see Sect. 4.6.1].

In 1617 (Anon, 1618), just a year before the great comet appeared, the Scottish poet William Drummond (1585–1649; Fig. 4.1) wrote a verse on the visit of King James to Scotland. Entitled *Forth Feasting*, it is quoted from Eyre-Todd (1895: 198). Here, the poet addresses the king as "Thou":

Thou sought'st to know this all's eternal source,
Of ever-turning heavens the restless course,
Their fixed eyes, their light which wand'ring run,

Fig. 4.1 William Drummond, a 1612 painting attributed to Abraham van Blijenberch. In the Scottish National Gallery. Courtesy of Wikimedia Commons, public domain. PD-1996

Whence moon her silver hath, his gold the sun.
If destine be or no, if planets can
By fierce aspects force the free-will of man.
The light and spiring fire, the liquid air,
The flaming dragons, comets with red hair,
Heaven's tilting lance, artillery, and bow,
Loud-sounding trumpets, darts of hail and snow.

Drummond writes with assentation that James, a true natural philosopher, actually did find the "causes, essence, might" of all these celestial happenings: the stars (fixed eyes of heaven), the planets (lights which wander), comets, aurora (the tilting lance, artillery, and bow). Also in 1617, we can get an idea of the position held by the king in the wider opinion of scholarly studies. Robert Fludd, supported in later years by the patronage of the king, dedicated the first volume of his *Utriusque cosmi* to James. In the book, James is accorded the Hermetic epithet "Ter maximus" (Thrice great), and saluted as "the most potent and wise prince in the world." (Pumfrey & Dawbarn, 2004: 167) This is a reference to Hermes Trismegistus, a combination of the Greek god Hermes (Roman god Mercury) and the Egyptian god Thoth, who was associated with wisdom and learning. This is an important point, as it offers a link between James and his view of the cosmos. King James believed he ruled by divine right: an ardent defender of this right was Edward Forset. In 1606, Forset published a pamphlet which "compared the commonwealth to the whole world and to the body of a man. Using neoplatonic and Hermetic references and quoting Trismegistus, he described a man as a giant stretching with his head and shoulders into the heavens while his feet firmly touched the earth." (Collins, 1989: 113) That man was King James (Fig. 4.2). In 1618, it was from his vantage point in the heavens that James encountered the Great Comet, which held no fear for him.

4.2.1.2 The Comet Poem by King James

Soon after the Great Comet of 1618 appeared on November 18, the King consulted scholars to ascertain its nature. He drew upon the knowledge of both Oxford and Cambridge, described as "lights of the Land" by the poet John Davies (1603: 11), who dedicated his book to the new King of England. One is these was Francis Mansell (1579–1665), Fellow of All Souls, Oxford. Mansell was well known to James, as he was the nephew of Sir Robert Mansell (1573–1656), Vice-Admiral of England, who had also appeared in 1605 in a masque written by Jonson. Francis was apparently unable to give James any answers about the comet, according to the London Puritan diarist Nehemiah Wallington (1598–1658; 1869: 11). John Bainbridge (1582–1643), who was writing a book on the comet at the time, told the king the comet was as far above the moon as the moon is above the earth, and not less than 2,300,000 miles away. (Chambers, 1858, vol. 1: 501). A resident of Cambridge University at the time, Sir Symonds D'Ewes, wrote "King James was enlightened [about the comet] by Dr John Gostlin (1569–1626), the Master of Caius College, at this time the Vice-Chancellor of the University, whose illustration of the

Fig. 4.2 King James, a 1605 portrait attributed to John de Critz. Museo de la Trinidad. Courtesy of Wikimedia Commons, public domain

subject gave the royal mind perfect satisfaction." (D'Ewes, 1851: 55). Sir Philip Mainwaring (1589–1661; who became an MP in 1625) stated on 22 November 1618, as a wonder, that the King "takes no more notice of the blazing starre than he has always done of the day-starre, nor will allow it be any other." (Chambers, 1830: 272) Thus, the king regarded the comet not as an omen, but as an astronomical object to be studied.

This is a critical point in the archaeopoetic analysis presented in this book: the king and a coterie of experts advising him together formed the sort of scientific spirit that the French philosopher of science Gaston Bachelard (1884–1962) identified as an epistemological rupture with everyday experience. "We believe," he wrote, "that scientific progress always shows itself in a rupture, in continuous ruptures, between ordinary knowledge and scientific knowledge." (Bachelard, 1953: 207). While it may not fully encompass the dynamics at play in the early seventeenth century, the epistemological rupture identified by Bachelard gives us a valuable hermeneutic with which to think about the impact the Great Comet had on both ordinary and scientific knowledge in and after 1618. I suggest this 'rupture' is ontologically coincident with the idea of Pierre Nora's 'torn memory.' (Nora, 1989) The memory of that year, more than any other, reverberated throughout the century. Two issues stand out in this regard, with a direct bearing on the two greatest physicists of that century.

1. "The awakening of René Descartes' interest in physics is often dated to late 1618, when Descartes first met Isaac Beeckman, an amateur scientist and mathematician who championed the new "mechanical" philosophy." (Slowik, 2021). That much has been established, but a brief consideration would conclude that it would have been nearly impossible for the two of them to have met without discussing the Great Comet that was blazing in the sky at that very time. Descartes wrote much about comets in the coming decade, but it was not published until after his death (Descartes, 1654).
2. It is often ignored that Newton's *Principia* of 1687 "is prefaced by a panegyric poem in which Edmond Halley—following Galileo whose study of the 1618 comet led him to declare that mathematics was the legible script of the Creation—proclaims": (Johnson, 2004: 419; Newton, 1687).

Now we know
The sharply veering ways of comets, once
A source of dread, nor longer do we quail
Beneath appearances of bearded stars.

The suggestion has been made that Halley's verses were beholden to the poem King James wrote about Tycho (Albury, 1978: 24). It was in this scientific frame of mind that the king wrote the following poem, *King James on the Blazeing Starr*, dated 9 December 1618 (I have modernised some words to make it more readable):

Yee men of Brittayne wherefore gaze ye so,
Upon an angry star? When as ye know
The Sun must turn to dark, the Moon to blood,
And then t'will be too late to turn to good.

O! bee so happy then whilst time doth last,
As to remember Doomesday is not past:
And misinterpret not with vain conceit
The character you see of Heaven's height:
Which though it bring the World some news from fate,

The letter is such as no man can translate:
And for to guess at God Almighties mind
Were such a thing might cozen all mankind:
Therefore I wish the curious man to Keep
His rash imaginations till he sleep

Then let him dream of famine plague and war
And think the match with Spain hath caus'd this star
Or let them think that I their Prince, and Minion
Will shortly change, or which is worse religion
And that he may have nothing else to fear,

Let him walk Paules, and meet the Devil there;
Or, if he bee a Puritan, and escapes,
Jesuites salute him in their proper shapes:
These jealousies I would not have be treason,
In him whose fancy overrules his reason;

Yet to be sure he did no hurt, 'twere fit
He should be bold to pray for no more wit,
But only to conceal his dreams—for there
Are they that would believe all he dares fear.

4.2.1.3 Rhetorical Analysis

While a rhetorical analysis may seem pedantic, that is far from the case. The German philosopher Moses Mendelssohn (1729–1786) referred to poetry and rhetoric as "beautiful sciences;" indeed, until the later seventeenth century rhetoric was affiliated "with 'scientific' pursuits, including the invention and discovery of new knowledge about the natural world." (Mann, 2012: 173). Thus, in a study of historical poetry, it cannot be divorced from the history of the science of astronomy (Mendelssohn, 1997: 179). Furthermore, in the era under study here, "rhetoric offered the only complete and integrated communications system." (Vickers, 1988: 341) James used several rhetorical figures of speech employed by Shakespeare, but he was well-versed in the art of rhetoric, deriving from ancient authors. First and foremost, the opening combines a personification with a rhetorical question, giving the poem a typically Baroque flavour (Harrison, 2023: 124).

One must first see the entire poem as a series of *couplets*: two lines of verse that rhyme, such as last/past; keep/sleep; fit/wit. In line 2, the 'angry star' is an example of *epitheton*, the act of naming something and adding a descriptive quality to it. Line 5 begins with the only exclamation of the poem; here James uses *ecphonisis*, from the Greek *ekphonein* (meaning 'to cry out, pronounce'), in this case bidding his subjects to be happy. Lines 7–10 are an *apothegm*—a brief, instructive saying, warning everyone against presuming to read the signs of Heaven. In lines 15–18 he deploys *metalepsis* by attributing a litany of dolorous (and merely anticipated) effects to the remote cause of a comet. He invokes the apoetic device of kinesthetic movement in line 20, by instructing his subjects to walk St. Pauls Cathedral. Beginning with 'And' in line 19, and continuing to 22, James establishes the punishment (encountering the Devil and Jesuits) that awaits those who don't keep their cometary beliefs born of imagination in check, but after the colon that marks a momentary pause after 'shapes', he balances this with a compensatory positive at 23–24, saying these jealousies (meaning the understandable guarding of a belief) are only a fancy, and do not rise to the level of treason against their Prince; this rhetorical form is an *antanagoge*. As for Jesuits, the King had a particular antipathy towards them: on 22 February 1604, he ordered all Jesuits to leave the country (McCoog, 2020). Overall, the poem embodies the form of a *paraenesis*: a moral exhortation in which his subjects are advised to abstain from superstition about comets. "Rhetorical, poetic conceptions of the unthinkable were a part of the early modern lexicon that natural philosophy relied on." (Killeen, 2023:1) The unknowable mind of God (line 11) is the key element in the first portion of the King's poem.

One aspect of the poem that is lost to "modern ears" is how the words sounded four centuries ago. This 'Original Pronunciation' of the words, O.P., has recently been reconstructed (Crystal, 2016). Andrew Lipman (Columbia University) offers an example of how this can be usefully applied here:

> The poem gives us a literal sense of the monarch's voice. One can 'hear' his Scots-inflected O.P. in the ways he rhymed 'bloode' with 'good' and 'conceyte' with 'heighte.' Both his verse and joke also captured James Stuart's distinctively crass and pompous personality. (Lipman, 2024: 117)

4.2.1.4 Astronomical and Astrological Analysis

The animating properties of *King James on the Blazeing Starr* are a fusion of politics, science, and religion. This would be a heady mix in the hands of any poet, but James was no ordinary poet. He was addressing this to the people of England and Scotland as Head of the Church of England (in its religious aspects) and as King (in its political aspects). An animating principle of his kingship was his belief in the divine right of kings, and he "saw himself as a sort of prophet king" (Doelman, 2000: 39), so it is important to understand that motivating factor when reading the comet poem. Even though Jonson derided the play, one reason *Pericles, Prince of Tyre* became a particular favourite of the Royal Court is its line "Kings are earth's gods." (While the play has been attributed to Shakespeare's late period, it is thought this line from the 1608 play was written by his collaborator George Wilkins). Whoever wrote it, "...a key unifying element in the late plays can be found in their engagement with topical concerns related to the person, politics and court of the King." (Kurland, 2013: 210). Thus, there was a synergistic relationship among poems, plays, and masques that the king read, wrote, or witnessed. Even though Court masques were invitation-only affairs seen by relatively few people, their relevance did not stop at the palace gate. "The Stuart court masques," writes Rob Browning (2005: 63), "…could achieve greater force by way of report and gossip than it ever could merely as a show for the spectators actually present."

The scientific aspect of King James' poem relating to comets comes from his position as a philosopher-king envisioned by Plato. It is part and parcel of what the Italian philosopher Italo Calvino (1923–1985) termed "The age of Shakespeare," which recognized "subtle forces connecting macrocosm and microcosm," including the Neo-Platonic firmament (Calvino, 1988). That his role as natural philosopher was no idle boast can be gleaned from the fact that Kepler (1619) dedicated his book *Harmonices Mundi* (Harmony of the World) not to any Continental monarch, but to King James, a personage he had never met but clearly admired. The copy presented by Kepler to James is at Oxford University (https://www.univ.ox.ac.uk/news/keplers-harmonices-mundi/). In his dedication, "Kepler hoped that this book, which dealt with the harmony of the heavenly bodies, would appeal to one who 'sought harmony and unity in the ecclesiastical and political spheres'." (Patterson, 1997: 126) *Harmonices Mundi* was one of the sources of Kepler's third law, which governed the orbits of comets and other solar system objects, but Kepler was just as adept at creating astrological forecasts.

Discussing the interplay between astrology and the dynamics prevalent in a Royal Court, Hilary Carey (1992: 19) states, "Astrology is a volatile material, but it needs the right chemical conditions to explode. It can exist as a stable constituent of the intellectual world of scholarship, or it can synthesise within the cauldron of the [royal] court into a potent and noxious compound." By writing this poem, James acted quickly to defuse the potentially explosive aspects of the comet's appearance, reminding his readers with the use of astronomical terms (line 3) that eclipses of the Sun (when it turns dark) and Moon (when its colour turns to blood red) are predictable—i.e., within the bounds of knowledge and reason, not astrology. This provides a certain answer to the question posed to James by Drummond the year before, whether or not fierce aspects of the planets can foretell destiny—as a serious and learned man, James came down firmly on the side of astronomy, not astrology. It must be noted, however, that "James's Scottish court had patronized several influential defenders of astrology." (Pumfrey & Dawbarn, 2004: 167) Instead of casting the comet as an astrological portent, he couches the argument in Biblical terms: Doomsday must be heralded by eclipses, not comets; his reference to the colour of lunar eclipses resembling blood comes directly from the Book of Revelations, but the entire opening of the poem derives directly from a passage in the King James version of the Bible, namely Matthew 24:29 (quoting from Isaiah 13:10 and 34:4).

> Immediately after the
> tribulation of those days shall
> the sun be darkened, and the
> moon shall not give her light,
> and the stars shall fall from
> heaven, and the powers of
> the heavens shall be shaken.

The link between Doomsday and eclipse was brilliantly evoked by Shakespeare in *Hamlet*, 1.1.120. In that passage (written in 1602, and thus well known to much of the populace), Horatio likens the ghosts' appearance to bad omens that predicated the assassination of Caesar: "...the moist star/ Upon whose influence Neptune's empire stands, /Was sick almost to doomsday with eclipse." Shakespeare must also be quoted here as describing what surely must have been the mindset of many. The Earl of Gloucester speaks in *King Lear*:

> These late eclipses in the sun and moon portend no good to us.
> Though the wisdom of nature can reason it thus and thus, yet nature
> finds itself scourg'd by the (sub)sequent effects.

Yes, astronomers can reason the true nature of eclipses. But the key word "yet" reveals that people still believe they must mean something, and that something portends no good. Even with reason on his side, King James was fighting an uphill battle.

4.2.1.5 Theological Analysis

By adopting an instructional tone throughout, James made this a didactic poem of the first order—an expression of his philosophical understanding of kingship; philosophy being one of the four realms of thought identified by Koethe that can (however imperfectly) be expressed poetically. As Deni Kasa (2024: 87) has written of the seventeenth century, "…poetry is essential for governing the passions of the people." As we saw in his witchcraft speech of 1591, James felt himself responsible for ensuring the instruction of his flock, giving impetus to his empowerment of poetry as an aid in ruling their passions.

Time is the key that leads to James's theological purpose in the comet poem. Temporal time is specifically called out in line 5. But the entire sweep of the first few lines presents a dizzying shift in temporal perspective. He moves quickly in lines 1 and 2 from the present continuous, with people gazing at the comet *right now* and in an ongoing manner, to the future in lines 3 and following. Lest people look to the past for comfort, James warns in line 6 that the past holds no solace. Rather than tremble with foreboding, he bids his subjects to be cheerful *for now*, while reminding them that the Last Judgment (Doom[e]sday) is yet to come and then it will be 'too late' to repent (line 4).

James was forced to issue this reminder because there was some theological dispute as to the timing of the Last Judgment, and this dispute continued in the years after the comet. Johannes Alsted, professor of Philosophy at Herborn (in Hesse, Germany), wrote that it would begin in 1694! His adoption of this view derived from the early disasters of the Thirty Years' War, which drove him into exile. The pro-James view was notably elucidated by the celebrated Cambridge theologian John Mede. He systematised all the prophecies and concluded that there was to be a thousand years' reign on earth, and that it was entirely in the future; it would be "in and during the day of judgement; which day of judgement should continue a thousand years, beginning with the ruin of the Antichrist." Many of the Protestant faith, such as Mede and Martin Luther, identified the Antichrist as the Pope in Rome. (Jue, 2006; Capp, 1972).

Quoting from Philippians 2:3, which admonishes "Do nothing out of selfish ambition or vain conceit," James in line 7 reinforces his edict with Scripture. Used in this sense, vain conceit is an incorrect sense of self. While eclipses may be understood, the signs of Heaven (particularly here the comet at "Heaven's height" in line 8) are not. By verbalizing these signs as "character," "letter," and "translate," James emphasizes the futility "of gazing, seeing, and guessing without understanding." (Rickard, 2007: 189) Here, James is following the semiotic system of Augustine, who distinguishes "between signs created by man (*signa propria*) and those bequeathed by God (*signa translata*)." (Dekoninck, 2016: 75) In this case, the sign bequeathed by God is the comet; however, translating that sign is not for mere mortals. The concept of 'heavenly writing' was already established: Agrippa von Nettesheim wrote of it 1533, and a manuscript by Rabbi Elcha ben David, called *De lectura per stellas* (*On Reading the Heavens*) "…presented a sort of heavenly alphabet made up of something like Hebraic letters." (Mulsow, 2022: 343).

But why did James invoke the Last Judgment in this theopoetic creation? The intersection of the Great Comet of 1618 with providential agitation became a flashpoint, evoking both a psychological and moral response from the king. Martin (2022) has shown that while much of historical scholarship ignores the subject, the providential project that was in full swing in the early seventeenth century was actually the well-spring of a multitude of actions (such as this poem) which only come into focus when we view them through the lens of the Apocalypse; Hackenbracht (2019: 6) explains its usage in the seventeenth century:

> As a heuristic for making sense of political events, the Last Judgment appealed to writers because it restructured existing communities according to the divine rule of reckoning…It offered, that is, a logic of political organisation backed by divine authority but enacted by a human writer, who could apply that logic to whomever he pleased.

In the hands of a writer who put great store by his own position as King by divine right, this passage of the poem (lines 3–6) was a very powerful one that leads into the political realm beginning in line 15. And at a time when Protestants were actively praying for the Last Judgment to happen soon, rather than fearing it at some distant time, it takes on added meaning. James was not the first to interpret comets in such a fashion. This longing embrace for the Last Judgment (a lucidity of want, I venture to call it) was also employed in the previous century by the German Lutheran theologian Jakob Andreae (1528–1590) who, like Grasser and Gross, refers to a comet as "a fiery rod" displayed as a dreadful sign by God to his children. Andreae writes that good and evil people regard comets in different ways. For the good, comets are a "joyful sign of Grace"; for evil ones, "God calls them to repentance and threatens them with his wrath through…signs in heaven." (Kurihara, 2016: 136) James renews the admonition against these signs, or characters, in heaven. Obviously, good people should *not* try to interpret these signs, as they are directed only at those who are evil. This relates to James's 1591 publication *News from Scotland*, which opens with these words:

> God by his omnipotent power, has at all times and daily does take such care, and is so vigilant, for the well-being and preservation of his own, that thereby he disappoints the wicked practices and evil intentions of all such as by any means whatsoever seek indirectly to conspire anything contrary to his holy will. (Tyson, 2011: 189)

The works of Andreae take on added significance when one considers that he was a major creator of the theological milieu in which Johannes Kepler was educated (Metheun, 1999). As a Lutheran, Andraea believed (as Martin Luther preached) that the Millennium was not in the future, but ended in the year 1000. Luther "maintained that the world of his day was in the predicted period of renewed persecution by the Antichrist immediately preceding the Last Judgment, which King James termed 'Doomesday'" in line 6. (Fix, 1991: 64) Of course, opinions differed. The astrologer Paul Nagel, spurred by the Great Comet, made a "bold prediction that the Second Coming of Christ and final age of the world would commence in 1624." (Jensen, 2006: 67).

Lines 9–12 are central to James' theological argument, and they derive ultimately from the work of the early Christian scholar Origen of Alexandria (184–253 CE). In book 23 of his *Philocalia*, Origen addresses the subject of astrological Fate. "The stars," he argues, "are signs," but "their messages are meant for the higher powers, not for humans." (Lewis, 2020: 570). Here, James accepts that Fate does exist, and that signs in the Heavens may contain a message, but it is written in a language "no man can translate." He especially warns in lines 11–12 against trying to "guess" the meaning, lest mankind be deceived ("cozen") and, by extension, fall into error. [see Sect. 4.4.1] Thus, he exhibits the very dichotomy adopted by Origen, who simultaneously excoriates those who use astrology while accepting "that constellations and planets do in fact have limited abilities to presage events." (Lewis, 2020: 570) In these lines, the King derives inspiration from Horace's *Odes* (Bk. 1.11), where the name of the addressee is Leuconoe, not Clarice:

> Don't ask, Clarice, we're not supposed to know
> What end the gods intend for us.
> Take my advice, don't gamble so
> On horoscopes of Babylon. For better, just
> Take what heaven might allot us.
> (McHugh, 2002: 43)

Line 9 is actually the linchpin of the poem, containing as it does the word 'fate.' While Origen may have been the ancient progenitor of the passage, its more immediate source was Heinrich Bullinger (1504–1575; Fig. 4.3). He was a key figure in the Swiss Reformation, and a close associate of John Calvin (1509–1564); this is crucial, as King James adhered to the Calvinist faith. Calvin was particularly concerned with divine justice, "part of a broader Reformed debate concerning the necessity of Christ's satisfaction for the redemption of sinful humanity." (Schendel (2022: xi), a theme expressed in the comet poetry with mention of Doomsday. Lines 7–12, admonishing overly speculative inquiry into heavenly writing, are directly inspired by the words of Calvin:

> Let us here remember that on the whole subject of religion one rule of modesty and soberness is to be observed, and it is this: in obscure matters not to speak or think, or even long to know more than the word of God has delivered. (quoted in Schendel, 2022: 38)

In the sixteenth century Bullinger created the most extensive correspondence of anyone in Reformation times; one of his correspondents was Queen Elizabeth

Fig. 4.3 Heinrich Bullinger, a 1550 painting by Hans Asper. Courtesy of Wikimedia Commons, public domain. PD-1996

I. Bullinger rejected "current beliefs in fate, fortune, and astrology. He also abhorred philosophical disputations about providence, which are contrary to the writings of the prophets and apostles." (Stephens, 2019: 113) And further to line 10, about the impossibility of understanding the character of the comet in "heaven's height," and also line 24 that evokes reason, Bullinger wrote that "reason itself teacheth us, that the mind of man is little or nothing moved, if the things themselves be not understood." (Harding, 1852: 291).

In addition to being a Calvinist, James was a Platonist. We have this on the highest authority. In his dedicatory letter to King James that appears in *Harmony of the World*, "Kepler portrayed King James as a Platonic ruler who also was the natural protector of the Protestant cause." (Pesic, 2022: 78) This is relevant as Plato divided the "divine frenzies" into four primary varieties, one of which is poetic. "The inclination of those discoursing on the remarkable powers of poetic frenzy was to point out its special affinity with the mystical and prophetic frenzies by commonly depicting poetry as a kind of versified prophecy." (Brann, 2002: 12) In his comet poem, James placed both mystical ideas and prophecy at the forefront by his mention of Doomsday, coupled with admonition and instruction to his flock.

This didactic exposition was a continuation of a policy King James (1598) had already established and publicly written while he was King of Scotland:

> For from the head, being the seat of judgement, proceedeth the care and
> foresight of guiding and preventing all evil that may come to the body…
> The head care for the body, so doth the king for his people.

On the subject of the evil of plague, e.g., he "championed" the biblical epic *Divine Weeks* by Salluste du Bartas (1544–1590), translated from French into English by Joshua Sylvester (1563–1618), as "an important example of a Christianized didactic presentation of plague." (Totaro, 2016:13). As in 2020, plague was indeed something to fear. The King's own entry into London to take the throne was delayed a year due to a plague in the capital that took 25,000 lives in 1603. (Totaro, 2016: 5). In his comet poem, the king is clearly not telling his subjects to discount fear of the plague, but is rather instructing them not to attribute any potential outbreak to the comet of 1618.

There would be another outbreak of plague in 1625, the year King James died as the longest-reigning monarch in Scottish history (57 years). The fact that his death was *not* accompanied by a comet was so startling it became the subject of an elegy by the ambassador to Venice, Sir Isaac Wake (1581–1632). Its opening lines are "Can a king die and we no Comet see?/Tell me, Astrologers, how can this be?" Wake, however, had a short memory: no comet heralded the death of Queen Elizabeth in 1603 either. The lack of a comet for her death was noted in a verse by Thomas Byng (Nichols, 1828, vol. 1: 9). An anonymous elegy similarly evoked wonder that the heavens did not herald James's demise (Heywood, 1641: 373)

> Can such a Monarch dye, and yet not have
> An earthquake for to open him a grave?
> Did there no Meteor fright the Universe?
> No comet hold a torch unto his hearse?

Such a forced linkage was not just an artifact of the early modern world. Nearly three centuries later, the appearance of a comet was connected to the death of a British monarch. The 9 May 1910 edition of the *Bermuda Observer* reported that unusual activity was observed in comet Halley when the king, Edward VII, died on May 6. (The Observatory, 1910: 226) For the use of the word 'meteor' in this era, see my study Cunningham (2023).

One cannot omit from this section a look at the employment of the term "blazing star" to headline the poem. King James had written several books prior to this, most notably in recent years his 1609 tract *Premonition to all Christian Monarchies, Free Princes and States*. It was notorious for accusing Pope Paul V of countenancing assassination, sorcery, and fornication. After the Venetian Inquisition read the book, it was banned, an act that was regarded as a great affront to the King. It was written that Catholics "…look upon his Majesties Bookes, as men look upon Blasing-Starres, with amazement, fearing that they portend some strange thing, and bring with them a certain influence to worke great change and alteration in the world." (quoted in Heilbron, 2021: 24) It is entirely likely the King was aware of this sentiment, so he may have deliberately chosen the 'blazing-star' title as a further poke in the eye to Catholics.

4.2.1.6 Satirical Analysis

Paul's Walk and Newsmongers

King James on the Blazeing Starr is not purely didactic, however. No scholar has heretofore recognised that it is also satirical, deploying both poetic forms in ways that mutually reinforce. (I use the word "mutual" in the seventeenth century meaning of "held or experienced in common," not the fourteenth and fifteenth century meaning of "reciprocally given and received.") As Gellar-Goad notes (2020: 166), "satiric humour depends on the audience's sense of safety from the aggression of the satirist." Here we see the King throws a safety net around his readership by assuring them nothing treasonable is afoot: widespread gossip and concern about his Spanish diplomacy is annoying but acceptable. The satire is evident when he suggests one might meet the Devil incarnate in a walk around St. Paul's Cathedral in London. By invoking the Devil as a metaphor for the newsmongers who haunt the environs of the Cathedral, he "acknowledges the extent of their ability to affect what people think," and is trying to position himself "to be the persuasive figure encountered by subjects curious for news." (Rickard, 2007: 191). This satirical swipe at newsmongers was made to measure for both the king and Jonson, who penned the astronomically-charged masque *News from the New World Discovered in the Moon* performed before the king on 6 January and 11 February 1621 (Jonson, 1692: 610–613). James is targeting printed news, rather than what remains in written form (Pritchard, 2025). Jonson expressed the royal view in his *Moon* Masque when the Factor of news tells the Printer that his untrue news does not offend him. Rather, "it is the Printing I am offended at, I would have no newes printed; for when they are printed they leave to bee newes [that is, no longer news]; while they are written, though they be false, they remaine newes still." Many people condemned the printing of books, maintaining that what was copied by hand "did not debase itself by being open to the common view." (Woudhuysen, 1996, 14–15). It was this 'common view' that the king was so much against.

Masques (which Jonson designated as poems) were typically dramatic performances with actors taking on different guises with the use of masks (Grantley, 1988); Court masques could be so elaborate that their equivalent cost in modern terms could reach nearly a million dollars.

In his 1599 play *Every Man Out of His Humour*, Jonson positioned Paul's Walk as "a site where meaning is contended, asserted, reified, and displaced." (Hoye, 2024: 102) He expanded on this in the *Moon* Masque. Like most masques, the *Moon* opens with a satiric antimasque (literally, a masque of antics). Here, a 'Printer' character states, "I'll give any thing for a good copy [story] now, be it true or false, so it be news." (Gifford, 1869a: 614) As today, one cannot be confident that what one reads in the press is true or false; in the kingdom of Scotland and England ruled by James, false information could be a very dangerous thing. The masque celebrates James as "the truth" and "the knowing king." This is juxtaposed with the anti-masque, which parodies the "irresponsible circulation of political gossip and news." (Perry, 2002: 207) The words of the Printer are astronomically inspired, as he surmises that news from the Moon has reached them through a telescope. It is a sterling example of how a masque can "encode and express a particular historical moment in a complex and often revealing way." (Lindley, 1984: 9).

The king's distaste for the newsmongers reached fever pitch in the following years. He appointed a licenser of the press to prevent the publication of anything that might upset Spain, and between 1621 and 1624, several authors and publishers "were fined and imprisoned and had their presses destroyed; others fled into exile." (Harris, 2014: 212) Paradoxically, for the king, the publication of thousands of newsletters weekly disseminated not only political gossip but also poetry, including his own. "It is quite common to find material of this kind recopied into the commonplace miscellanies of the period." (Perry, 2002: 207).

Dreams and Portents

James also raises a satiric smile on the lips of his more learned countrymen when he says many of them let their imagination confound their reason. A dream state bookends the last half of the poem, placing it in an otherworldly embrace. While the comet itself is very real, everything about it that relates to real-life events—"famine, plague, war" (line 15), and a bad omen on the proposed Spanish match—is nothing but a bad dream. The role of dreams in the mindset of James is diametrically opposed to the way the Gods of ancient Greece imparted knowledge. Hussey (1991:11) states that in Homer, the gods "Supply reliable information to the human race through the medium of dreams, omens, portents and oracles." For James, omens and portents in dreams are false information and must remain in the dream world. But we must look more closely at the metaphorical relationship between comets and "famine, plague, war." In modern metaphor theory, the source domain derives from real-life knowledge and experience of these three existential events. The more abstract target domain in this case is a comet. Once a metaphor becomes conventional (such as that between these events and the appearance of a comet), "the connections (or mappings) between the source and target domains do not tend to vary: the mode of mapping shifts from comparison to categorization." (Olive, 2021: 11) Thus it became lodged in the minds of many that comets are the harbingers of doom, a metaphorical connection elevated to a literal connection that has lasted for thousands of years, and one that King James was fighting against. The litany of disaster given by King James in line 15 is nearly identical to that in a couplet written about comets by Johannes Bocatius (1569–1621), the Imperial Poet Laureate of Hungary (see Csonka, 1990: 836). (Even though Bocatius wrote a book on the Great Comet of 1618, this Latin couplet does not appear in that publication). The only disaster enumerated by Bocatius that James prudently omitted was the death of the 'great head,' namely himself:

Mors magni capitis, bellum pestisque famesque,
  Quattuor haec semper (crede) cometa trahit.

The death of a great head/chief, war, and pestilence and famine,
  A comet (believe me) brings in its train these four things.

The King's stance of looking down on his subjects from a wise vantage point is very much in line with the position outlined by the Roman poet Lucretius (99 BCE–55 BCE) in *De Rerum Natura* (On the Nature of Things). In the proem of book 2 (lines 7–10), he writes

> But nothing is sweeter than to hold the wise men's self-defended precincts, lofty with placid teaching, wherefrom you could look down on the rest and see them all over, going astray and wandering as they seek life's way. (Gellar-Goad, 2020:145)

Seventeenth century philosophers paid much attention to the interpretation of plague in *De Rerum Natura*, and the king was a known reader of ancient satire (Rickard, 2007: 98). Thus, it is no surprise he specifically mentions plague, but he does so from the 'lofty precinct' of a natural philosopher, safe in the knowledge he has acquired from the 'placid teaching' of such eminent natural philosophers as Bainbridge and Gostlin. He is able to 'look down' on the masses, not only from his philosophic eyrie, but as the literal master of all he surveys as the King of England and Scotland. And like a true satirist, who "ridicule incorrect, inconsistent beliefs and behaviours" (Gellar-Goad (2020:32), King James recognises there is a high likelihood many believe everything portentous about the comet that another only fears. The poem actually concludes on a satirical note, as James says those whose fancy rules their minds "should pray for no more wit," where wit is used as a form of satirical speech to highlight folly in a person and society at large. However, he fuses this with a lesson of didactic instruction: he says no one should be so conceited that they can read the language (alphabet) in which fate is written, and nonsense about the comet should be "concealed" so as not to infect the minds of those who are mere vessels of autosuggestion. In this, he was in sync with other Early modern writers, who "agreed that the purpose of satire was moral instruction." (Ossa-Richardson, 2019: 220) Thus, the king is putting the onus of responsibility on those of his subjects who can read and understand his poem, appointing them to oversee the vast majority who are unable to read. Don't spread fear, the king wisely instructs. Here he pushes back on the 'ordinary knowledge' of the people, in the words of Bachelard, informed as he is with the 'scientific knowledge' of his natural philosophers.

"To put this in Jonson's terms," writes Perry (2002: 209–210) in reference to the *News from the Moon* masque, "the king needs to meddle with sublunary fantasies out of concern for unenlightened subjects…For all his insistence to the contrary, this poem records James's recognition that 'the knowing king' cannot remain safely above sublunary circulation…More broadly, as is suggested both by James's poem and Jonson's masque, the unauthorized expression of 'rash imaginations' was increasingly a matter of official concern." But the majority who could not read naturally responded to the power of appearance, which is why a comet provoked emotive reactions that mere words struggled to combat.

4.2.1.7 Political Analysis

King James on the Blazeing Starr has yet another layer, one that caused the greatest consternation in England for many years. The king had a favourite (believed by many to be his male lover), George Villiers, the first Duke of Buckingham (1592–1628; Fig. 4.4). Their closeness may be gauged by the fact James included his beloved George in this poem, referring to him as his "minion." (line 17) The

Fig. 4.4 George Villiers, the first Duke of Buckingham, a portrait from the 1620s by Peter Paul Rubens. In the Stirling Maxwell collection. Courtesy of Wikimedia Commons, public domain

English politician Sir Dudley Digges (1583–1639) likened the Duke to "a prodigious Comet" and a "Blazing Star," but not as a compliment. Even though he was comparing the duke to the supernova of 1572, his resemblance to the comet of 1618 was not lost on the readers of the day, especially as he used "star" and "comet" interchangeably in his text. "If this glorious sun," the king writes Digges:

> by his powerful beams of grace and favor, shall draw from the bowels of the earth an exhalation that shall take fire and burn and shine out like a star, it cannot be marveled at if the poor commons gaze and wonder at the comet and[,] if they feel the effects[,] impute them to the corruptible matter of it. (Bellany & Cogswell, 2015: 236)

The analogy that royal favour had drawn the comet's matter from the bowels of the earth as an exhalation harkens back to Aristotle's theory of comets as exhalations from the ground upwards (for more on James and Buckingham, see Russell (2025)).

Line 18, "Will shortly change…" signals the comet presaging the change from life to death for the King, Prince, and the Duke, or even worse than death, changing from Protestant to Catholic! The Duke outlived the King by 3 years; he was so widely loathed that his assassin was celebrated in the finest medium of the day to convey national sentiment: poetry. Protestants and Catholics are also mentioned more explicitly in the king's poem. Even though his comet tutor Bainbridge was a Puritan (a group of Protestants who were particularly repulsed by Catholic practices), James adopts a pejorative tone when referring to them here.

Line 23 employs the word treason, whose flip side is loyalty. The King enjoins his subjects to accept the king's truth, which is reflected in the micro-political context of Jonson's 1620 masque about the Moon. "Attentive and loyal subjects, in the masque, are imagined as being 'above the Moon far in speculation.' It is the centrality of such ideas to the ideology of James's court that makes his political, manuscript poetry seem at first glance so odd." (Perry, 2002: 208).

Two politically charged themes in King James's comet poem were revisited in the following years. In 1622, he wrote a poem as a riposte to an anonymous libel that dealt with "the undue prominence of the duke of Buckingham (the 'minion') and with his perceived opposition to anti-Spanish interests in the House of Commons." In 1623, the King turned to poetry once again to "advise his subjects not to worry about Prince Charles's marriage negotiations with the Infanta." (Perry, 2002: 211) Thus, the comet poem can be seen not as a one-off poem instigated by the comet's appearance, but as a *concerted poetic campaign* to inform the people of the realm of the King's views. Robert Bagg (1969: 413) offers a description of what such a connected suite of poems means in practice:

> Some poets write on the assumption that the readers knows nothing
> but the single poem before him; other poets, however, obviously
> make no such assumption. Their poems are full of echoes and
> cross references, as if the poet thought of his single developing
> community of poems as a counterpart to his life.

James certainly positioned his politically charged and didactic poems as a community of thought; in effect, they participate "in ideological debates about the nature of royal authority." (Perry, 2002: 211). In the absence of the internet, television, radio, or even daily newspapers [Note 2], poetry was the vehicle he chose to disseminate his authority, although as Perry (2002: 212) notes, "they may or may not have been literally produced by the hand and heart of the king." Ultimately, James's "descent into manuscript publication" of these poems shows that he realised the threat posed to the crown's authority was "extraordinary in nature." Perry (2002: 225). This may inadvertently have fed into the ideological origins of the Civil War of the 1640s, the greatest political upheaval in English history (Sharpe, 1993:117–138). Even in the nineteenth century, this link was being drawn. Commenting on the comet poem, Owain (1863: 324) writes

> The lines [in the poem] confirm in a remarkable manner the persistent folly of James in prosecuting a Spanish alliance for his unfortunate heir, and his studied contempt for the disapprobation and alarm of his subjects caused by his policy. It will be seen, too, that the royal poetaster was a true, if an involuntary, prophet: for (*mutatis mutandis*) the "dream of famine, plague and war," became a terrible reality to his family; whose misfortunes, it is not too much to say, sprung out of the proposed "match with Spayne."

As mentioned, the comet poem by King James is clearly instructional in tone, i.e., didactic. On this basis, one may ask what is its Stoff (express subject matter) and Thema (its true subject or theme) are, the two elements of didactic identified by Effe (1977). The comet is clearly its Stoff, but its Thema is very much rooted in the political furore caused by the intended marriage of his son. This has famously become known to history as 'The Spanish Match.' The match with Spain was the attempt to arrange nuptials between Charles, son of King James, and the King of Spain's very young sister Maria Anna (1606–1646). Charles (Fig. 4.5) later became King Charles I of England (1600–1649), while Maria Anna achieved the ultimate royal prize, becoming Empress of the Holy Roman Empire and Queen of Germany in 1637. The contentious negotiations about her betrothal to Charles dragged on for 5 years. The Oxford Platonist Thomas Jackson (1579–1640), "in his traditional anti-Catholic vein, discerned some great corruption within church and state, making dark allusions to the appearance of the comet of 1618, which had been interpreted by many as signifying God's disapproval of the Spanish Match." (Clarke, 2011: 39). That was in 1622, 4 years after James pushed back in his comet poem on the match, and the widespread concern he himself might "shortly change" his religion to Catholic. The following year, Prince Charles went on a trip to Spain to bring the match to fruition (Samson, 2006); his companion on the trip was Spencer Compton, son of the first Earl of Northampton. It was the first earl who wrote about the comet of 1618 (see Sect. 4.2.3). Spencer Compton became the second earl on the death of his father in 1630. The death of another courtier, Thomas Murray (1564–1623), coincided with Charles' trip to Spain. A poetic tribute to Murray, the provost of Eton and secretary to Charles, linked his death with the appearance of the comet. This is quoted from Doelman (2017):

Fig. 4.5 Prince (later King) Charles, a 1611 painting by Robert Peake. The Weiss Gallery, currently in The Royal Collection. Courtesy of Wikimedia Commons, public domain. PD-1996

In vaine we thinke the Commett did foretell
Greate men should dye & not the good as well:
Good men are starrs whose light does beautifie
The world; they're blazinge starres a Prodegy

A wonder of the world, whose fiery zeale
Whose horrid aspect frights both kinge & weale
Such as was thine thou starrelike shininge soule
Cheife of Charles Waine fixte to the Northen Pole;
On which all good like Adamant their eyes
Did cast mov'd by a virtuous sympathize.

Which madst our sunne his hastned course to staye,
And almoste quite forgott the Southerne way
Whilst by thy gentle rayes thou ruldest that light
Which god has made to rule; th*a*t Orbe so bright

Which other starrs doe borrow from even that
With thy coelestiall beames illuminate.
And how then could we be depriv'd of thee

But that our Sunne muste needs ecclipsed be?

This was indeed an age of wonder and prodigies (Park & Daston, 1998), as lines 4–5 evoke. It was not too many years before 1630 that heavenly knowledge was referred to as *Angells inwit* (Considine, 2022: 371); the angelic was never far away from thoughts of what appeared in the heavens. The poetics of wonder—derived from the late Greek tradition—came into vogue in England in the late sixteenth century (Biester, 1996).

Despite line 6 stating that even kings are frightened by comets, King James was not. The soul of Murray is described as star-like (line 7), and as linked to that of "Charles Waine" (equated here with Prince Charles). In the seventeenth century, "adamant" (line 9) was used as a synonym for diamond. The lines are saying all good people cast their eyes to Charles Waine in sympathy for the loss of Murray. The remainder of the poem suggests other stars borrow light from our Sun, but even that bright orb must be eclipsed as a result of Murray's death.

4.2.1.8 Edward Lapworth

Dr. Edward Lapworth (1574–1636) wrote a poem in response to the king's comet poem. He became associated with the reign of James I with his 1603 poem marking the accession of James as King of England; Lapworth also contributed a poem at the conclusion of the *Microcosmos* book by Davies (1603). In 1618 he was appointed first Sedleian reader in natural philosophy at the University of Oxford. His poem exists only in one manuscript, Folger V.a.345 (Marotti, 2021); the relevant portion is printed here for the first time.

Here in ye sky a blazing star doth shine
Whose bottom grosse seems itself to refine
Into far purer beames, which to th south
Shoot forth their luster, and their terror both,

The Sun was now in Sagittarius place't
Methought, forth of ye quiver he had cast
An army tipt with fire, or this seemed
Some usher to ye Sun ye would be deem'd
Great by his master's linary, but my man

Swears 'tis ye day star, which but lately ran
To meet ye Sun, and there hath burnt his wings
And gold with fire about ye walkin slings,
I smile but lik't his wit, that was content
Not to sift reasons deepe, least he repent
Like those star gazers, who doe reason for
Of Nature that they reason-les do grow.
Let them confine this star to Nature's laws
For matter, place, birth lasting, this but draw as
Men to confusion, Let them dare foretell
Death's Plagues, and threats, where with their
books so well
I say all may be false, this shall be true
In heaven and earth, nothing appeareth new
But his good will, men their use should make
And viewing heavenly threatenings should of shake
The cause of judgment sin, let our hearts blaze
And fire them salues with zeale, then none amaze
From heaven can dant us, than there shall arise
No fiery sword to fence in Paradise.

In line 1, the "blazing star" mirrors the "angry star" in line 2 of the king's poem. Line 5 gives us the astronomical fact that the Sun was in Sagittarius on Christmas Day, when the great comet was blazing. In line 8, "usher" meant 'to admit ceremoniously.'

The phrase "his masters linary" in line 9 is a complicated allusion to heraldry. Linary (which the OED considers an obsolete word used only in the 1500s) was equated with 'toad-flax,' a plant with golden flowers. In heraldry, a Couple-Close is defined as an "or [gold] linary, so termed from its enclosing the chevron by couples, being always borne in pairs." (Crabb, 1858: 113). The masters' linary, then, means two golden balls appeared on his shield or coat of arms. I suggest the "couple" is the comet and the real day-star (the Sun): two golden balls in the sky. Lapworth is saying "my man" (presumably a servant) was unable to distinguish between the Sun and the comet which "lately ran to meet" the Sun in Sagittarius. Thus, the comet was seen to approach the Sun, where it "burnt its wings" like Icarus (line 11). And both appeared to be golden: in line 10, one must compare "tis" with "his" in line 9, to discern that the comet ("ye day star") is being likened to the Sun (the real day star).

Regarding line 12: "walkin" is a variant of welkin, the vault of heaven. In this case, it is paired with "slings"; it appears the poet is giving us a vivid image of the great comet, manifest in the sky as golden, slinging out fiery (line 12) beams that allude to the words in line 4 as beams that "shoot forth their lustre." In lines 13 and 21, the poet intrudes himself into the poem with the first-person "I." In lines 13 and 14, Lapworth aligns himself with the stance taken by the king, who was "content not to sift reasons deep." In lines 15 and 16, he refers to scholars (astronomers) who lose their senses in a vain attempt at applying pure reason to the comet's appearance. In lines 17–21, he chides those who use the comet to foretell disaster. The Last Judgement is referenced in line 26, but the last 8 lines contain an uplifting message.

First, Lapworth writes (lines 23–24) that "nothing appeareth new but his good will," meaning that God is offering humanity something good in the comet's appearance. It admonishes people to exercise "zeale" (line 27) to save souls from damnation ("salues"); and arms us with the knowledge that if heaven cannot intimidate or dispirit ("dant") us because of our "hearts ablaze" (line 26) with such zeale and resolve, Paradise will not ultimately be denied us.

4.2.2 A Comet Poem by Prince Charles

In late January 1619, King James and his son Prince Charles (later King Charles I) attended a dinner held by Sir Nicholas Bacon in Culford, there to see Bacon's grandchild by his daughter Dorothy and her late husband Sir Bassingbourne Gawdey. Prince Charles, just 19, was so taken by Anne's appearance that he wrote a quatrain in her honour that invokes the great comet ("Heaven's wonder late") which had just disappeared from view a few days earlier. The finest letter-writer of the age, John Chamberlain (1553–1628), wrote of the encounter to Sir Dudley Carleton on 30 January 1619 (Nichols, 1828, vol. 3:525–526):

> The gentlewoman is marvellously commended both by the King, Prince, and Lord of Buckingham, and much made of by them all. The Prince is said to be so far in liking that these Verses I send you are fathered upon him, wherein she is compared to the late blazing star. But the chief grace and curiosity, they say, consists in new and gay, which is the anagram of her name:
>
> Heaven's wonder late, but now Earth's glorious ray,
> With wonder shines; that's gone, this new and gaye [Anne Gawdye]
> Still gazed on; in this is more than Heaven's light;-
> Day obscur'd that; this makes the day more bright.

4.2.3 The Death of Queen Anne

Anyone seeking portents of doom that winter in London did not have to look far. Immediately after the comet made its first appearance in November 1618, the Venetian Ambassador, Antonio Donato, wrote to the Doge and Senate of Venice about the wife of King James:

> Yesterday at sunrise a large and long comet appeared in the air towards the east. Everyone considers this unlucky, and a certain prognostication of ill. Accordingly they fear that the Queen will not live much longer. She is in a very bad condition. (quoted in Rickard, 2007: 189)

On Wednesday, November 21, 1618, Chamberlain wrote to his friend Sir Dudley Carleton (1573–1632), the King's ambassador at the Hague:

> On Wednesday, was the first day, we tooke notice here of the great blasing-star, though yt was observed at Oxford a full weeke before. Yt is now the only subject almost of our discourse, and not so much as little children, but as they go to schoole talke in the streets,

> that yt fore shewes the death of a king, or a queen, or some great warre towards. Upon which occasion (I thincke) yt was given out, all this towne over, that the Q. was dead on Thursday: but yesterday [Friday] I heard for certain that she is in a fayre way of amendment, and looked out of her window to see the hunting of a foxe. (McClure, 1939: 185)

This quote is important as it gives us a contemporary insight into the view of the comet by the people of London. More specifically, it takes us directly into "Paul's Walk," the newsmonger nave of St. Paul's Cathedral that so annoyed King James he felt compelled to mention it in his comet poem. One of the daily denizens of Paul's Walk was Chamberlain, who absorbed all the latest news and conveyed it to his friends.

The great Banqueting House, part of the Royal Palace of King James, burned to the ground on 12 January 1619. The blaze of the fire was compared to the comet's appearance as a blazing star by the first Earl of Northampton (1572–1630) in a letter of 27 January to the 21st Earl of Arundel (1585–1646): "I perceive that the fier of the Banqueting House will make a smoke in the Starre Chamber [see Sect. 4.4.3] which is almost as strange as the Blasing Starre." (Hervey, 1921:153). Just a few weeks later, on 2 March, England was plunged into mourning by the death of the Queen. Her death was quickly linked to the appearance of the comet. The Anglo-Welsh historian James Howell (1594–1666) wrote, "Queen Anne is lately dead of a

Fig. 4.6 Anne of Denmark, Queen of England, a 1605 painting by John de Critz. Royal Museums Greenwich. Courtesy of Wikimedia Commons, public domain

dropsy, which is held to be one of the fatal events that followed the last fearful comet."

After the passage quoted in the Introduction, (Wilson, 1653:130) turned from the general to the specific in his recitation of woe. In the following lines, he mentions the war in Bohemia (modern-day Czech Republic) that began the Thirty Years' War, which James was very much involved in, as his daughter briefly became Queen of Bohemia in 1619. She famously became known as the Winter Queen (Jeffrey, 2018). Between warfare and the accompanying plague, 20% of Europe's population perished in those 30 years. Wilson writes

> The first remarkable Accident that happened in England after this Prodigious Forerunner [the comet], was the death of Queen Anne...The Common People, who naturally admire their Princes, placing them in a Region above ordinary Mortals, thought this great Light in Heaven was sent as a Flambeau to her Funerall; their dark minds not discovering, while this Blaze was burning, the fire of War that broke out in Bohemia.

King James married Anne of Denmark (1574–1619; Fig. 4.6) in 1589; in her time as Queen, she was a great patron of masques, often asking Jonson to compose them, and sometimes even performing in them. When she died, the sad event was thus indelibly linked to the comet, as evinced by many poems. One in wide circulation (quoted in Gardiner, 1869: 240) was attributed to the King, who may have written it after he recovered from an illness in March and early April 1619. These verses reinterpret the comet as a divine invitation, depicting her demise in the most elevated terms as a "translation" of her very being to the heavens:

Thee to invite this guest God sent the star,
Whose friends and nearest kin good princes are;
Who, though they run the race of men, and die,
It serves but to refine their majesty.

So did the queen her court from hence remove,
And put off earth to be enthroned above.
She then is changed, not dead; no good prince dies,
But, like this day's sun, only sinks to rise!

Rickard suggests that the existence of at least 38 copies of this poem reflects a deliberate choice on the King's part to intervene in the public discourse of manuscript circulation. But Doelman (2021) writes, it "is just as plausible that the dynamic ran the other way: the attribution (whether accurate or not) to the King spurred widespread interest and circulation. Its popularity was also likely fuelled by its focus on the comet of late 1618, and the evolving way the King was supposed to have responded to that heavenly marker." Firth (1911) alleges these line "appear to be an altered version of the *Epicedium* sung at the Queen's funeral, and were not, seemingly, attributed to the King at the time." Firth refers here to a letter of 12 April 1619 from Rev. Thomas Lorkin to Sir Thomas Puckering (1592–1637); in that letter Lorkin sends his friend the full text of the poem just quoted, writing "The epicedium that is like to be sung upon the day of the funeral, I here send you enclosed."

(Williams, 1848: 150) There is no way to say for sure, but I am inclined to the view that James did not write the poem.

One is struck in this poem about the Queen by the similarity of its understanding of death with ancient Egyptian beliefs. The passage of the dead through the Underworld is mirrored by the passage of the Sun as it sinks, only to rise the next day (line 8).

The writer is saying in line 1 that "the star," namely the comet, is a friend and close kin to good princes, such as the King's wife. Here, the author closely echoes the lines of Shakespeare in his 1599 play *Julius Caesar*:

> When beggars die there are no comets seen;
> The heavens themselves blaze forth the death of princes.
> (Guthrie, 1964: 207)

An elegy, written on the occasion of the death of Queen Anne, was printed by John Tyler Hassam (1896: 43). Assuredly composed before 1631, it reads in part

> Usque adeo lugere juuat solatium inane,
> Quod pietate quidem faciunt humanitus omnes,
> Sed melior tua jam sors est, nec flebilis ulli,
> Hoccine erat modo quod tantum fulgere Cometam
> Vidimus, e terries et te poscebat Olympo?

> It is indeed a pleasure, though a vain solace, to mourn at length—and indeed it is natural to men that everyone does so out of piety (or reverence). But your lot is already a better one, and not to be lamented by anyone. For when we saw the Comet shining so brightly, wasn't it for this reason—because it was calling you from earth to Heaven

Hassam (1903: 40) states this poem "has long been erroneously ascribed to Ezekial Cheever, the famous master of the Boston Latin School." The authorship of the poem is undecided.

Another elegy upon the death of Queen Anne was written by Thomas Corbett. It concludes with these lines that, like the poem possibly attributed to the King himself, see the Queen's departure as a translation into Heaven. This is quoted from Bennett & Trevor-Roper (1955: 67):

> Her Father, Brother, Husband, Kinges;
> Royall relations: from her springes
> A Prince and Princesse; and from those
> Faire certaintyes, and rich hope growes.
>
> Here's Poetry shall be secure,
> While *Britaine*, *Denmarke*, *Rheine* endure
> Enough on Earth; what purchase higher,
> Save Heaven, to perfect her desire?
> And as a straying *Starr* intic't,
>
> And governd those *wise-men* to *Christ*:
> Ev's soe a Herauld-*Starr* this yeare
> Did Beckon to Her to appeare:
> A Starr which did not to our *Nation*
> Portend her *Death*, but her *Translation*:

Fig. 4.7 Dudley North, a 1615 painting. The National Trust. Courtesy of Wikimedia Commons, public domain

For when such Harbingers are seene,
God crownes a *Saint*, not kills a *Queene*.

This poem begins with family relationships: she was the daughter of King Frederick II of Denmark and the sister of King Christian IV; further, she is the mother of Prince Charles and Princess Elizabeth. Lines 9–10 evoke the Star of Bethlehem, thus linking her to Christ. Similarly, she did not actually die, but was translated to the sky or heaven (line 14). Putting a positive spin on the appearance of the comet, Corbett writes that they do not actually kill royalty, but herald their elevation to sainthood (line 16).

In 1645, Dudley North (1581–1666; Fig. 4.7), the third Baron North, wrote a 16-line elegy entitled *Upon the death of Anne of Denmark, Queen of Great Britain, and the blazing Starre appearing neere her death, taken for the stellified spirit of Prince Henry, dead not long before*. The concept of stellification (exemplfied by Julius Caesar's metamorphosis into a blazing star; James 2005) had a long history in English literature: Chaucer employed it in the sense of 'poetic ascent' in 1378 (Swinford, 2013). The poem title refers to the elder brother of Charles and thus the person who was groomed to be the next king; Prince Henry was born in 1594; his death at age 18 in 1612 elicited many literary responses (Streete, 2017). The first 4 lines of North's poem read

Brave soul! Thou hast prevail'd, God hath his own,
And we ill debtors were, nor paid the loan
Of such a jewel, bee thou Henries Star
Pointing thy mothers way and not our war.

The poet George Herbert (1593–1633) composed a 10-line Latin poem *On the Death of her most serene Majesty Queen Anne*. After praising her fame and saying he does not have the genius to render her justice, Herbert writes in lines 6–10 (Grosart, 1874: 170)

This joy alone to me remains,
That Sea and Sky excuse my pen's poor pains:
For Anne's praises are written in the open sky,
Our grief is written in the Ocean's face.

For Herbert, the comet's importance is expressed as the means by which praises of Anne can be written in the Heavens. The line regarding "Ocean's face" is clearly metaphorical, as writing on water was not possible until recently (Möller et al.,

Fig. 4.8 John Wilmot, the Earl of Rochester, a 1677 painting by Peter Lely. Victoria & Albert Museum. Courtesy of Wikimedia Commons, public domain

2023). While the idea of *writing* in Ocean's face seems to be a novel one, this was not the first time such a phrase was used: Gervase Markham (b. 1566) wrote this in his poem *The Most Honourable Tragedie of Sir Richard Grenville* (Markham, 1595: 79).

> But when the mornings dewie locks drunke up
> A mistie moisture from the Oceans face,
> Then might he see the source of sorrowes cup

Herbert may have been familiar with a book by Hugo Grotius (1609), where he described the sea as infinite like the cosmos (Fowler, 2023: 207), which would explain the metaphor mirroring the sky (the infinite repository of Anne) and the ocean (the infinite repository of human grief).

The aristocratic poet John Wilmot, the short-lived Earl of Rochester (1647–1680; Fig. 4.8), harkened back in 1675 to the death of Queen Anne in a composition entitled *A Very Heroical Epistle in Answer to Ephelia* (Veith, 1962: 114).

> You may as justly at the sun repine
> Because alike it does not always shine.
> No glorious thing was ever made to stay:
> My blazing star but visits, and away.
> As fatal, too, it shines as those i' th' skies:
> 'Tis never seen but some great lady dies.

There are clear semantic relations here with Rochester's most famous poem, *Upon Nothing*, published in 1679. As Brazgovskaya and Kholmanskikh (2014: 31) explain, Rochester was one of the "Cambridge neo-Platonists who believed that philosophy and poetry belong to the same sphere of rational and sensitive cognition." In the second line, he remarks on the absence of sunshine, and reinforces the 'lack of being' as a semantic allusion to 'nothing' by antonymically extending it to 'everything' glorious in line 3. He then makes particular reference to a comet in line 4, again emphasizing its ephemeral nature in line 6: when it goes away ('Tis never seen), there is nothing to look at. Finally, he relates all this nothingness or original state of emptiness to the passing of a great lady. As in his *Upon Nothing* poem, Wilmot here employs "the non-observable referent as mosaic," creating a lament based on the iconic image of the 1618 comet with metaphysics that "allows him access into the sphere of [the] transcendant" (Brazgovskaya and Kholmanskikh (2014: 33). The comet and Queen don't just leave us: they metaphysically depart together into an absent nothingness.

Several comet-related poems exist that were never published, but that does not mean they were never read. In a study of poetry manuscript circulation in Early Modern England, Marotti (2021: xviii) writes:

> In terms of poetic composition, the field of writing in the manuscript system was broader than that found in print culture, giving the impression that any literate individual was free to compose verse and effacing the boundary between literary producer and consumer.

A 74-line manuscript poem is entitled *On the death of Anne, queen of Britain, the Greek muse of Anne grieves, and Latin funeral rites complain, alas only the British sadness is silent.* It looks into the past, evoking the memory of Anne's first son Henry. The poet says the comet was sent by his spirit, and Anne's other 4 children who had died before her. Here are the first 17 lines that deal with the comet (from the online appendix to Doelman (2022)).

Nor can faire Isis hide her cristal streames
when muddy Tybur flowes pure Helicon
Nor English muse deny her choicest rimes
when Greekes and latins in a tragick tone
are seen her deer Pandorah's losse to moane
who would not gladly sacrifice his verse
when heaven it selfe sticks tapers on her hearse.

Unhallowed Hags, fel ministers of fate
was it you yt kindld yt prodigious fire
To blast ye beuty of our british state
'gainst it that wondring multitudes admire
Comets and furies needed not conspire
for they wch best deserve eternity
wthout such lights can see ye way to dy
No rather twas heroick Henryes ghosts spirit
And that quaternion of sweet infant ghosts
wch steed of earth, celestial thrones inherit

The opening of the poems alludes to the 'tears of Isis': the Egyptians believed the Nile flood was caused by the tears of the goddess. Here, the goddess sheds tears ("cristal streames," line 1) at the death of the Queen. The poet evokes two other rivers: the Tiber in Rome and Helicon, mentioned in Greek mythology. As a study in archaeopoetics, this poem shows that an understanding of classical times is essential for understanding comet-related poetry of the age. "The infusion and appropriation of classical history, its stories, figures and exempla, was widespread in seventeenth-century England…history and its lessons shaped sermons, poetry, drama, imagery and performances in early Stuart England." Gianoutsos (2021: 27).

The taper in line 7 is an allusion to the comet, which also possesses a flame. The 'unhallowed hags' (line 8) evokes the witches in Shakespeare's *Macbeth*, dated to 1603–1607; Shakespeare used as a prime source a book written in 1597 by King James himself: *Daemonologie*. The comet is referred to as a "prodigious fire" in line 9; the poet wonders if it was sent "to blast the beauty of the British state," namely Anne (line 10). The death of prince Henry was also linked to "A comet blazing terror to the east" in a poem by the Scotsman William Drummond (Johnson, 1810: 663), and Henry's chaplain Daniel Price (1613) expressed his death in astronomical terms: "Wee have yet the Sunne and the Moone [James and Anna], and stares of the Royal firmament; and though we have lost our morning Starre [Henry], yet we have Charles-waine [the pole star, Prince Charles] in our horizon." [see Sect. 4.4 for Charles-waine].

A 23-line poem that exists in only one manuscript (BL Harl. MS 3910) "rejects any prognostication or cause of her death apart from God's call." (Doelman, 2021: 98). It opens thus:

I chide no Blazing Starr yet did forgoe
ye Princess death: lett Hethen quarrel so
nor curse I the disease, nor ill aspect
of Planets, nor blame Physiks weak effect

because she's dead, I see a higher hand
beckoning her up, whom non can countermand.
He said that kings are Gods, but they shall dye
Like other men, his word doth verify
She that was rooted deep in Royalty:

& largely spread in Armes of Sov'rainty,
like to a goodly Cedar tree, from whom
such sprouts of growing majesty did come
as sett the world at gaze, & fixt each eye

with hope & wonder, on their Infancy.

In line 1, the poet does not rebuke (or chide) the comet for the death of the Queen, but rather understands that God, a "higher hand" (line 5), was responsible. The notion that kings are gods (line 7) fits neatly with James's concept of the divine right of kings. The poem suggests that Anne's children ('sprouts' in line 12) are "a better object for gazing wonder than the much remarked-upon blazing star." (Doelman, 2021: 100) The 'sprout' as part of a family tree plays on the idea in lines 9–11 that likened Anne to a Cedar, "rooted deep in Royalty."

Yet another poem, only in manuscript (BL Add. MS 25303, fol. 127 r.), evokes 'towering spirits.' Doelman (2021: 98) writes that it mocks those, including philosophers, who thought they had divined the comet's true portent.

Yow towringe spirits whose art-yrradiate eyne.
dare peirse the Conclaves of the sacred Time
Who, ere the fatall Crynite gane to'appere
durst calculate the fate of all the year

yet when y^{t} sin-fed flame a month had blazd,
true Samian Sophyes, silent stood and gaz'd,
Stand ever mute: heavens Oracle hath spoke
Shees dead: break yours our Jacobs staff is broke.
cease prodigies the supreame cause is knowne

Brittains bright Ann, above the sphears is flowne
Shott, like a starr, (brave Comet) fayre & right
th''ast [sic] hitt the harte of Albian, cleft hir whyte;
give me the Cataracts of Septemfluous Ryte
t'ore flowe each angle of this spatious Isle.

oh for a springe tyde, for a flood of verse
to beare that Arke up of ou[r] hopes, yr Hearse.
Dissolve you eternall ffountains of my head
to brynge Oceans, th'Oceans Queen is dead.

The word Crynite in line 3 means hairy-tailed star, which derives from Pliny's use of *crinitas* (haired); more often, the word was used as an adjective modifying 'comet.' In line 8, Jacob's staff (see also Sect. 4.4.1) was an instrument used to measure angles. Here, the staff represents Anne as the staff (support) of Jacob (a stand-in for King James); when she dies, the staff is broken. Her death has frustrated would-be astrologers (here obliquely referred to in line 6 as Samian Sophists; Pythagoras, Aesop, Epicurus, and Aristarchus are the most famous Samian philosophers) by the metaphorical breaking of their staffs. Line 12 refers to the White Cliffs of Dover, with Albion more generally England itself (the white land). Line 10 tells us Anne shot like a comet right into the heart of Albion (England). The pedantic word septemfluous (line 13) means divided or flowing in seven streams; it may refer to the seven kingdoms that formed the Heptarchy of Anglo-Saxon England, as the poet sees the streams flowing throughout "this spatious isle." In any case, it begins a watery trope that flows in a "flood of verse" (line 14) to the end of the poem, with the evocation of an ark (a la Noah's Ark) in the form of a hearse (line 16), fountains (line 17), and the ocean (line 18).

A 34-line elegy that bears the initials "w A" is discussed by Doelman (2021: 99) begins with *No further shalt thou choak me grief ympersoned.* He thinks the most likely author is William Alexander, a poet who often wrote about the royal family. Lines 10–30 are given here

When Englands Demi Goddess
her Queene gives place to Fate, and falls away
by those fell starres, she should rule not obey
when she's extinct, whom when I oft beheld
shyning in Gem*m*s (her minde all Gems exceld)

The Sunne, or rather in slow majestie
On Earth the Conglob'd starry company
And all Heaven mov'd methought, who not despares
to see Day more? or who not justly feares
Earths joynt dissolv'd, least all heaven reele a mayn

& tomble to ould Chaos once agayne.
will ye Sunne still be rising, & still sett?
Will day still follow night? Will Phoebe gett
repaire her lost Hornes, & recall anew
her former Raies, & Anne not hers renew?

But shee desrying rather for to smyte
all Heaven w^{t} blushes, darks y^{e} Sunnbeam quite
wth brighter Rayes, & giv's y^{e} world, now blynde
no more, $anoth^{r}$ eye; whether shee mynde
to shyne mongst starres, or els in some new sphere

becom fresh labor for th'Astronomere.

The "fell stares" in line 12 refers to the comet of 1618. "In its hyperbolic imagined construction of the universe, the poem places Anne (lines 16-17) as the central point" around which the stars and 'all Heaven mov'd'. The confluence of the comet and her death has "overthrown astronomical order," and the poet expresses fear that "all heaven" (line 19) could "tumble into Chaos" (line 20). This passage is clearly a "Petrarchan conceit"—a figure of speech quite popular with Renaissance poets, and named for Petrarch. One may often find a beautiful woman compared to the sun, e.g., but the poet here goes to the greatest possible extreme by linking Anne to the centre of the universe itself!

As noted by Oestreicher (2007) in his history of chaos theory, "Whether every effect can be precisely linked to a given cause or to a list of causes has been a matter of debate for centuries, particularly during the seventeenth century when astronomers became capable of predicting the trajectories of planets." This is particularly apt here as the poem is the only one in this study that specifically mentions astronomers (line 30). While this poem in linking the effect (death of Anne) with a specific cause (comet) does what all these poems do, this one is unique in further linking the effect with chaos on a cosmic scale. In this era, order versus chaos was a central conflict, and was even expressed in garden design with formal parterres and topiary (in France) and in England with the Oxford Physick Garden, established in 1621. The walls and gates of the garden, clearly evocative of order within holding back chaos without, were built by Nicholas Stone, master-mason to King James. [In the Netherlands, a survey of the wonders of divine Creation by explaining everything that can be seen in his garden was given in the poem *Hofwyck* in 1653 by Constantijn Huygens (1596–1687); see Sect. 4.9] "Indeed, the Stuart throne was celebrated as an ordered garden in the early years of the dynasty, as in Francis Bacon's *Masque of Flowers* performed in 1613. The players in this masque… begin as flowers and metamorphose into men in order to enact the tribute to King James I as their sun." (Svensson, 2019) This poem tells us Queen Anne now outshines sunbeams with even "brighter rayes" (lines 26–27), once again linking the Stuart monarchs to the sun. Lines 25–30 embody the concept of being "'translated" into the heavens. Anne triumphs over the blazing star, in effect replacing it as the object of the gazer's awe. This rhetorical strategy is similar to the James-ascribed poem ('Thee to invite'). He stresses the kinship between stars and princes, finding solace in the idea that she is now 'enthroned above.'" (Doelman, 2021: 99).

Sir Robert Aytoun (1570–1638) wrote 16 lines in Latin on the comet presaging the death of the Queen. It has not heretofore been included in scholarly studies of elegies devoted to her. (Roger, 1844:115–116)

Ad Jacobum aegrotantem cum cometa apparuit ante mortem reginae.

Vidit ut in caelo metuendum Regibus astrum
Anna, repentino corruit aegra metu.
Non tamen aegra sui formidine corruit, omnis
Pro charo fuerat cura metusque viro.
Tene inquit vir chare suo petit igne Cometes,
Nec fax placari vilius ista potest?
An nihil offensi Numinis iram,
Publica ni pereat te pereunte salus?
Di melius, liceat potius tua fata subire,
Proque viro conjunx victima laeta litet.

Alcestis potuit morituri fata mariti
Morte sua ad longos continuare dies.
Hanc laudem non sola feret. Vix dixerat, ecce
Lurida tabificus corripit ossa dolor,
Et moritur. Quid nunc metuas Rex maxime, plenum
Nonne piamentum sideris Anna fuit?

To James (VI) who was sick when a comet appeared before the death of the Queen:

When Anna saw in the sky a star to be dreaded by kings,
She fell sick and collapsed with a sudden dread.
But she didn't collapse in sickness because of terror on her own account,
Her every care and concern was for her dear husband.
"Dear husband," she said "does this comet come for you with its flame
And this torch not be placated any more cheaply?
Is it worth noting to the wrath of an offended deity
Unless the public good should be lost with your death?
Oh God, it would be better that [his] joyous wife be an acceptable sacrifice
in the place of [her] husband than to allow him to die
Alcestis was able to prolong by her own death
The life of her husband who was appointed to die."
She will not only bear this praise. For behold, scarcely had she spoken when
an all-consuming pain seized her pale bones.
And she died. Why now do you fear, great King? Wasn't Anna
a complete expiation for the star?

The introductory line (*aegrotantem* refers to a physical illness) makes it clear James was in fact infirm early in 1619: he suffered from kidney stones and was seriously crippled by gout. A more natural rendering of line 2 could be "a royally dreaded star." In line 3, the transferred epithet and word order of the Latin are interesting for its dramatic effect on the events narrated. Line 3 makes the dramatic statement that the Queen died of fright, being terrified of the comet! In line 5, personified and capitalized, the Latin word *cometes* is a reference to the father of Asterion, a Greek name that literally means "starry." The Asterion mentioned here was an Argonaut from Peirasia in Thessaly: he was the son of Cometes and Antigona, daughter of King Pheres of Pherae. Translation of "the Godhead" in line 7 depends on how one chooses to interpret Anne's theological outlook towards the end of her life. It may be rendered as God's will, or God. Line 10 literally means "than allow his to suffer your appointed fate for him." Lines 4–10 put forward the extraordinary hypothesis that Anna offered herself up as a sacrifice to save the life of her husband. In lines 11–12, Aytoun refers to the ancient legend of Alcestis, told in a tragedy by Euripides. To save her husband from death by the Fates, she sacrificed herself in his place. By elevating Anna to the same level as Alcestis, who was rescued from the underworld by Hercules, Aytoun is offering the reader solace that she is not really dead. But Aytoun takes a swipe at the King in the last two lines, putting him on the same level as many of his incredulous subjects who feared the comet. Despite this intemperance, he was buried in Westminster Abbey. Aytoun, a Scot, was a courtier as well as a poet: in 1612, he became secretary to Queen Anne. Thus, we have first-hand evidence that the Queen truly was terrified of the comet. His verses were often derived from contemporary events at court, including the death of the Queen.

The poet, composer, and physician Thomas Campion (1567–1620) wrote an epigram (*Ad Calliodorum*) about the comet (Vivian, 1909: 302):

Sollicitus ne sis signum fatale cometa
  Ut quid portendat, Calliodore, scias;
  Expectes cladem (domini natale propinquat)
  Non hominum, sed tu, Calliodore, boum.

Have no fear that the comet is a fatal sign,
  As you fathom it to foretell, Calliodorus;
  You may anticipate disaster (as the birthday of the Lord approaches),
  Not of men, Calliodorus, but of oxen.

"The comet referred to is the great comet of 1618, which began to be observed on Nov. 27, 1618 (N.S.). It created much concern, as the epigram suggests (Doelman, 2016); some thinking it to be a presage of the death of the Queen; some a warning against the Spanish Match, while others thought it to be connected with the fall of Barneveldt." (Vivian, 1909: 375) Calliodorus features in epigrams by Martial (Book 9) in which he became the owner of a farm, ploughing his fields; his name derives from *kallos*, "beauty." Campion's Latin lines are elegiac couplets: a line of dactylic hexameter followed by one of dactylic pentameter. "Campion, however, varies this, as many Latin poets did, especially those writing before Ovid. Campion often made his version of the elegiac couplet (also when he was writing in English) more complex, even seemingly closer to iambic meter, which he called 'licentiate iambic.'" (Engell, 2024; see also Note 3) In a poem from 1601, Campion alluded to signs in the sky. Writing of the life of an upright man, this is his fourth stanza:

Hee onely can behold
With unafrighted eyes
The horrors of the deepe,
And terrors of the Skies.

Campion (1601) is one of many poets in this era who also wrote verse in Latin (Oxford University, 1619; Lacrymae Cantabrigenses, 1619), as studied by Victoria Moul (2022). John Pyt wrote a Latin elegy that said the comet of 1618 "would show the high way to stars" for Anne.

Campion's epigram conforms precisely with the main features of an epigram as described by the famed Italian scholar Julius Caesar Scaliger (1484–1558); namely brevity and vividness (which here is the comet as a fatal sign). These two aspects combine in the epigram's witty concluding argument, which is achieved by comparing one thing with another; in this case, men and oxen (Riklius, 2023: 71). The effect of the comet on royalty was not limited to England. In Europe, the 1618 comet was linked to the deaths of the Habsburg Empress Anna, the Archduke Maximilian, and Emperor Matthias (Parker, 2013: 1647).

Fig. 4.9 George Wither. Frontispiece of the 1902 edition of *Wither's Poetry*. Courtesy of Wikimedia Commons, public domain. PD-1996

4.2.4 George Wither

The English poet George Wither (1588–1667; Fig. 4.9) wrote the lines below in 1625 in *History of the Pestilence* (which remained in a manuscript); he was describing the plague that killed King James. His words were incorporated in his widely read book of 1628, *Britain's Remembrancer*. Here (Wither, 1628: 275) clearly alludes to the comet of 1618, which was "but a while ago." Just as in the poem by King James in 1618, Wither lists plague, famine, and war. Are they the consequence of aurora ("prodigious lights" in line 1), a comet ("blazing star"), and "dreadful meteors"? For the King, these must be left to dreams; but for Wither, they are quite real, directly mediated by these fearful signs in the sky as a judgment from God. His woeful line about every nation being at war (lines 17–20) is a direct reference to what later became known as The Thirty Years' War.

I have already noted that King James considered himself to be at least partly a prophet. "With his first work of Charles' reign, *Britain's Remembrancer*, Wither

How often have we seene prodigious lights,
O'respread the face of heav'n in moonlesse nights?
How many dreadfull Meteors have there beene
In this our Climate, lately heard and seene?
Who knoweth nor that but a while agoe
A Blazing Star did threat, if not foreshow
Gods Judgements? In what age, tofore, did here
So many, who did Saints and Stars appeare,
Fall (as it were) from heav'n? Or who hath heard
Of greater Earth-quakes, then have lately scar'd
These quarters of the world? How oft, the touch
Of Famine have we had? But, when so much
Devoured by the Pestilence were we,
As in this present yeare our people be?
Of Wars, and martiall rumors, never more
Were heard within these confines heretofore;
When were all Kingdomes, and all Nations through
The world, so opposite as they are now?
We know no Country, whether nigh or far,
But is engag'd, or threatned with some War.

adopts a stance as national prophet. This lengthy work interprets the plague of 1625 as a warning to England of its sinful ways…Wither now sees himself as God's chosen messenger, but he still desires to be the elect of earthly king Charles as well." (Doelman, 2000: 52–53) The "remembrancer" of England is Wither himself. Wither was a well-known poet in the reign of King James, but he had evolved from being merely a lofty satirist who tentatively tried to advise the king. "There is a continuity between Wither's earlier works and *Britain's Remembrancer*, but in the latter work we find a greater recklessness, and a willingness to invoke a special status related to the will of God." (Doelman, 2000: 53) Jonson was clearly unimpressed by Wither, as he satirised him as a poetaster in the court masque *Time Vindicated* in 1623; it was said at the time that King James was quite angered by this negative representation of a person who was theologically important to him.

The reference in lines 10–11 to earthquakes is a bit curious, as there were no truly major earthquakes in the years immediately preceding 1618, but it is worth noting the 1580 quake felt strongly in London. It was one of the most powerful earthquakes recorded in the history of England. But why even mention earthquakes in a poem about a comet? Arthur Golding (1580), in a tract about the earthquake, neatly expressed the matter thus:

> God never powered out his grievous displeasure and wrath upon any Nation, Realm, City, Kingdom, State, or Country, but he gave some notable forewarning thereof by some dreadful wonder.

Comets and earthquakes were but two of the potential "dreadful wonders" by which God was supposed to have terrorised humanity with, but they are linked.

Comets are emblematic of famine, pestilence, and earthquakes, so much so that when these three terrors are mentioned, people at the time most likely conjured up the image of a comet even if it was not the subject under discussion (Kalff, 2017). We have already seen wonder specifically evoked in the poems quoted in Sections 4.2.1.5 and 4.2.2; and most importantly, the dismissal of the comet as a wonder by King James (Sect. 4.2.1). For a full treatment of how wonder has shaped the way we think and view the cosmos, see Park and Daston (1998) and De Cruz (2024).

Wither remarks on both meteors and a comet in his 1625 poem (lines 3 and 6) and the lines quoted below from a decade later. What he meant by these terms was based on the work of Aristotle, as he explained in poetic form in his most famous book, *A Collection of Emblemes,* from 1635. Emblems were widely used in this era, and each had an allegorical significance. This is the key to interpreting the comet poem by Wither. Like most poems that were composed to accompany images, this one establishes a visual scene (prodigious lights on moonless nights, lines 1–2, and the appearance of meteors, line 3). This sets up the remainder of the poem, with its elaboration of allegorical detail; allegory is at the heart of the widespread use of emblems (and their associated allegorical poems) in England from the 1580s to the 1670s (Freeman, 1978). (For a mid-seventeenth century comet emblem, see Daly & Silcox, 1989: 357–359) The following is a description of an emblem in Wither's book (listed as illustration XLV, Book 1), which does not depict a comet! Rather, it shows an infant resting on a skull; they are both surrounded by a snake, which signifies the circularity of creation.

The solid Stone, doth molder into Earth,
That Earth, e're long, to Water, rarifies;
That Water, gives an Airy Vapour birth,
And, thence, a Fiery-Comet doth arise:
That, moves, until it selfe it so impaire,
That from a burning-Meteor, backe again,
It sinketh downe, and thickens into Aire.

The verse completes the cycle, from air, to a cloud, then rain, which descends onto the ground that eventually hardens into a stone.

4.3 Sir Walter Raleigh

The following anonymous poem is quoted in Shawcross (2012:131–132):

On Sr Walter Raleigh, who was beheaded a little before the appearance of the Comett. 1618.	
I knew thee but by Fame, and they brave deeds, Those spoke thee loud; For where true worth exceeds. It cannot sleepe in Lethe. Who could but know Thee for the Muses Friend, and Spaines Arch-Foe?	1

Fig. 4.10 Sir Walter Raleigh, a 1588 portrait. National Portrait Gallery. Courtesy of Wikimedia Commons, public domain

Mee thinks the old Heroes weigh'd with Thee,
Homer was out, or they of meane degree;
Of Witt, and Valour, hee Two patternes sett;
In thee Both were, and both more strongly mett:
Thous sham'd his art, and spite of Rule and Fashion.
Mad'st Practice out goe Speculation.
And yet Thou hadst so much Mortality
To dy; though not without a Prodigy.
For Thou (o' Sunne) being sett, and darke Night come,
An upstart Starr would needs supply Thy roome,
And lend that light wee mist; yet 'twould not bee,
It shone bright, but not halfe so bright as thee:
It shone, but being outvied, itt streight was done,
As though a Meteor could out shine the Sunne.
Oh that I could tune out so full a straine,
As might become thy Ellegy. In vaine
I wishe itt. Englandes Muse Raleigh is deade
And blow spilt the balme of that rare heade.

Commenting on this poem, Shawcross explores the use of the comet as a metaphor:

> Of interest to the literary scholar is the use of a well-known astronomical event as metaphor, one that implies common belief in, or at least acceptance of, superstition. Whether the author, like Owen Glendower in Shakespeare's 1 *Henry IV*, really believed such a fiction or was only employing it as a metaphoric sign, we should recognize the apparent predictive viability of astronomic events for the people of the early seventeenth century.

In this poem, the comet is first mentioned as a "prodigy" in line 12, and then in line 14 as an "upstart starr." The author is directly implying in line 14 that the statesman and explorer Raleigh (1552–1618; Fig. 4.10) will be translated into the heavens as a hero of old. He writes that the comet, while bright, achieved less than half the brilliance of Raleigh. The comet could not compete with him, and was done in a short time (lines 18–19), thus suggesting the brilliance of Raleigh would remain undimmed for all time. In this, the poet was not wrong, as Raleigh remains an iconic figure of the age. As he writes, even Lethe (line 3), the river of hell whose waters induce forgetfulness, cannot wash away his memory.

Fig. 4.11 Richard Corbett, a painting done by Sylvester Harding between 1760 and 1809. British Museum. Courtesy of Wikimedia Commons, public domain. PD-1996

As the arch-foe of Spain (line 4), Raleigh also gets caught up in the imbroglio surrounding the Spanish Match. In this, D. Didacus Sarmiento de Acuna, Count of Gondomar (1567–1626), was a key player. James had sworn to Gondomar, the Spanish ambassador, that there would be no injury to the vassals of territories of the Spanish king, but Raleigh had attacked and destroyed the Spanish settlement of St. Tomé in Guayana. To maintain the Anglo-Spanish alliance, Raleigh became a human sacrifice to King Philip III of Spain (1578–1621). "Indeed, while Raleigh's execution in October 1618 allowed the King of England to demonstrate to the Spanish sovereign his continuing support for the dynastic union, it also contributed to the worsening of some of the political nation's opinion towards the marriage negotiations with the Spanish Habsburgs." (Caldari, 2015: 123). This precipitated King James to admonish those very people, and he chose his own poem about the comet as the vehicle to do it. Thus, we can regard this anonymous poem, with its cometary and Spanish elements, as a major impetus for the King to write his own comet poem.

4.4 Richard Corbett

Richard Corbett (1582–1635; Fig. 4.11) was appointed a royal chaplain by King James after an oration he delivered on the death of Prince Henry, the Prince of Wales, in 1612. Corbett became a Doctor of Divinity (1617), Bishop of Oxford (1628) and Norwich (1632). His verse epistle on the comet, dated 9 December 1618, was sent to Thomas Aylesbury (1576–1657), his friend and a patron of mathematical learning at Syon House (see Sect. 4.4.4).

The 60-line poem is the longest one in English dealing with the Great Comet of 1618. While Corbett does employ a few rhetorical flourishes (see below), this is—more than the poem by the King or anyone else—governed by dialectic, which is the reasoning part of the creative process studied and explained by the early modern humanist and poet Rudolph Agricola (1443–1485). For him, dialectic "was a super-science concerned with the whole of man's rational processes." (McNally, 1967: 393) Agricola (1479) stated that poetry (along with history and philosophy) is one of the three major types of exposition; Corbett's poem is nothing if not expository.

The famous American poet Wallace Stevens (1879–1955) was not thinking about a comet when he wrote "It is not every day that the world arranges itself into a poem," but when the great comet of 1618 was present as a catalyst, Corbett was inspired to express his worldview on an early December day in this extraordinary expository poem. In this endeavour, Corbett was expressing one of the chief purposes of a poet as expressed by Puttenham (1589: 6): "…to search after the celestiall courses and influences."

The poem was first published in Corbett (1647). For clarity of meaning, I have combined the texts of Bennett & Trevor-Roper (1955: 63-65) and Owain (1863); the latter copied it "from the holograph which is preserved in her Majesty's State Paper Office." I have removed the letter 'e' from several word endings to make the poem more readable. Note that in an early printing (Corbet (sic), 1672: 54), line 10 reads "First such a Friend to gaze on, next a Comet":

My Brother, and much more had'st thou been mine,
Had'st thou in one rich present with a line
Inclos'd Sir *Francis* (for of all thy Store
No Guift could cost thee *less* or bind mee more);
Had'st thou (*dear Churl*) imparted his return,
I should not with a tardy welcome burn,
But had let loose my joy at him long since,
Which now will seem but STUDIED NEGLIGENCE;
But, I forgive thee; two things keep thee from it;
First, such a *friend* to gaze on, then a *Comet*;
Which Comett we discern (though not so true
As you of *Sion*) as long-tayl'd as you.
We know already how will stand the case
With *Barnavelt*, and Universall Grace;
Though *Spayne* deserves the whole Starr, if the fall
Be true of *Lerma* Duke and Cardinall:
Marry, in *France* we fear no *bloud*, but *wine*;
Less danger's in her *Sword*, then in her *Vine*.
And thus we leave the Blazer coming over,
For our portents are wise, and end at *Dover*.
And though we use now forward censuring
Nor send our learned Proctors to the King;
Yet every morning, when the Starr doth rise,
There is no *black* for three howers in our Eyes:
But, like a *Puritan* Dreamer, to this light
All *eyes* turn *upward*, all are *Zeal* and *White*.
More: it is doubtfull this new Prodigy:
Will turn ten Schools to one *Astronomy*.
And this ANALYSIS wee justly fear,
Since every Art doth seek for rescue there:
Physitians, Lawyers, Glovers on the Stall,
The Shopp-keepers speak *Mathematiques*, all.
And though men read noe Gospell in these Signs,
Yet all Professions are turn'd Divines.
All measurers from the *Bodkin* to the *Pike*,
The Masons *Rule*, the Taylors *Yard* alike
Take *Altitudes*; and th'early Fidling Knaves
Of Flutes, and Hoe-Boyes, make them *Jacob-staves*.
Lastly, of fingers glasses we contrive,
And every Fist is made a Perspective.
Burton to *Gunter* writes, and Burton hears
From *Gunter*, and th' Exchange both tongue and ears
By carriage. Thus doth mired *Guy* complain
His waggon in their letters beares *Charles-waine*,

Charles-waine, to which they say the tayle doth reach;
And at this distance they both hear & teach.
Now, for the peace of God and men, advise,
Thou; who hast wherewithal to make us wise,
By thy rich Studyes, and deep Harriots Mind,
In which there is no dross, but all's refin'd.
O! tell us what to trust to; ere we wax
Al stiff and stupid with this Parallax.
Say, shall the old Philosophy be true,
Or doth He ride above the Moon, think you?
Is he a Meteor fired by the Sunn,
Or a FIRST BODY by Creation?
Hath this same *Starr* bin object of the wonder
Of our Fore-fathers? Shall the *same* come under
The sentence of our Nephewes? Write and send,
Or else this *Starr* a quarrel doth *portend.*

Upon the Same Starre

A Starr did late appear in Virgo's trayne,
The which from South to North did post amayne.
If England then be North, and South be Spayne,
Brave Charles sitt fast, and look upon thy Wayne!

We read in the State Papers (Green, 1858:601) that Corbett's poem was delivered to the Royal Court on Dec. 9: "Rhyming letter from Dr. Corbett to Mr. Alesbury, at Syon, ridiculing the general excitement caused by the appearance of the Court [sic, should read comet]." A contemporary witness to the comet, Sir Symonds D'Ewes, quoted from Corbett's poem in his diary, which was not published until 1851. Highlighting the line "of flute…," Symonds wrote "…the comet became the subject of universal interest. Whatever else it might portend, it portended the conversion of all orders of men into astronomers." (D'Ewes, 1851: 55). This alludes to lines 25–44, which relates that men of all stripes have suddenly turned to astronomy, thanks to the comet. It relates to a 1618 book by the German professor of medicine Daniel Sennert (1572–1637), where he writes "And who is there that dares reckon Smiths, Carpenters, Joyners, Weavers and such like Artificers, amongst Philosophers?" (Sennert, 1618: 7).

These poems are likely the ones referred to by John Castle in a letter he wrote to William Trumbull: "You will see by these verses what opinion our Courtiers have had of the prognostique part of the Comet. I heare the invention please the k. very well." Thus, it appears King James was quite familiar with Corbett's verse, and liked it. (Doelman, 2021) Indeed, we may well assume that Corbett wrote the poem with the full intention of having it circulated at Court, and since it was written before Dec. 9, it pre-dates the comet poem by the king. Perhaps Corbett's poem, along with the Raleigh-inspired poem, spurred James to write his own poem.

4.4.1 Rhetorical Analysis

While this analysis is not meant to be exhaustive, it clearly indicates Corbett was not just an idle versifier about the comet; while none of his poems were published during his lifetime, they did circulate in manuscript.

Lines 25–26 are an example of *Enjambement*, where one line of verse runs over at the end into the next one. Synathroesmus is an accumulation of individual items, leading to an explanation. This is used in lines 30–34, where professions are collectively referred to as the Divines. In lines 35 to 40, Corbett lists several 'weapons' that are being used to take "altitudes" or "Jacob-staves" (Jacob's staff) and finally that everyone is making his own telescope. This is an example of *Prolepsis*, where a general statement ("weapons" meant as tools of the trade) comes first, followed by a string of examples that are all astronomical observing instruments.

In lines 44 and 45, Corbett repeats "Charles-waine," an example of *Conduplication*—a duplication of words that gives the effect of emphasis. In *Permisio*, we encounter a figure in which we completely entrust a matter to another person and indicate that we are handing it over and resigning to the disposition of that person. Corbett does this twice when he turns over the investigation and explanation of the comet to Aylesbury. This occurs in lines 47 and 59–60. He further employs *Encomium* in lines 50–51, where he extols the intellect of the astronomer Thomas Harriot (1560–1621) with high praise. In his address to Aylesbury, which covers the concluding section of the poem (lines 47–60), Corbett employs an *Exclamatio* in line 51: "O! tell us what to trust." Also in the concluding section, lines 54, 56, and 60 all begin with the word "Or." This is the rhetorical figure *Conversio*, in which discourse is repeatedly connected by means of the same word. The entire passage from 47 to the end is an example of *Anacoenosis*, a figure where a speaker asks for an opinion about the point of a debate by appealing to someone absent. [Note 4] This is manifested by a series of four questions, which a writer in 1858 still found quite pertinent:

> Dr. Corbett comments on almost every subject of interest of that period, and concludes his amusing verses with a question, to which we should be very pleased, even now, to meet with a satisfactory answer. (Anon, 1858:570)

4.4.2 Astronomical Analysis

In 1618, Corbett was a 36-year-old student of Christ Church. What, asks Corbett, is the astronomical significance of this new star, and is it compatible with Aristotelian cosmology as in line 53, "Say, shall the old philosophy be true?" Everyone is discussing the comet, and he appealed to the Sion House experts to resolve the controversy. "By thy rich studies, and deep Harriot's minde." Corbett did not adhere to the Calvinist beliefs of the King; by rejecting the Calvinist doctrine of predestination, he was firmly in the camp of the Dutch Protestant theologian Jacobus Arminius

(1560–1609). That he was an Arminian sympathiser can be seen from his poem *The Distracted Puritane*, in the course of which he mocks the double predestinarian schema of the Puritan leader William Perkins (1558–1602); this nonsense was the idea that God chooses some to be saved AND also creates others who will be damned.

In lines 10–13, Corbett says everyone is able to *see* the comet, but those astute natural philosophers at Syon House are able to *understand* it. In lines 23–24, Corbett is saying that no one gets any sleep during the 3 hours normally devoted to that purpose, because they are outside looking up at the spectacle of the comet.

The entire section from 27 to 40 is predicated on analysis; its importance is emphasized by being in capital letters. The analysis here specifically refers to mathematics (line 32) as a method of discovery. While embraced by Galileo and Kepler, the application of mathematics to reveal the hidden aspects of Nature was feared (line 29) by many until the end of the eighteenth century. But this section is also a biting satire, where Corbett ridicules the "whimsical conduct of the *profanum vulgus*": everyone from Lawyers to shopkeepers, glovers, and knaves. (Owain, 1863: 323).

On lines 41–43, we read of Burton and Gunter, whose friendship was memorialised in this poem. The former was Robert Burton (1577–1640), of Christ Church, Oxford, author of the 1621 book *Anatomy of Melancholy*, one of the most famous books of the century. He was well read in astronomy and owned the book by Copernicus. On 3 December 1618, just a few days after the comet became visible, Burton was licensed to preach. Edmund Gunter (1581–1626), who added scales to a portable quadrant to solve problems in arithmetic (Jardine, 2024: 394), was the Gresham Professor of Astronomy in London from 1619 until his death (Apt, 1982). Burton mentions Gunter in his *Melancholy* book, and in 1623, Gunter gave Burton a copy of his book *The Description and Use of the Sector*. On the epistolary exchange mentioned in the poem, "Nothing is known to survive from this exchange of letters, and so the poem, and the books, provide vital testimony to their relationship and to Burton's interest in mathematics and geometry." (Lund, 2017:330) Lines 43–44 are more obscure, and refer to Josias Guy, who was known as a carrier in Oxford. Carriers were people who transported people and the mail from one locale to another. The line tells us that his horse-drawn waggon is so overloaded with letters that his waggon sticks in the muddy road (the mire). That some of these letters were of astronomical import is made clear in line 45, where we read the comet's tail reaches Charles-wain, which is effectively the hindquarters of Ursa Major. There is another classical connexion here which has never been noted before. The Roman poet Claudian wrote of the comet of 400 CE in the poem *The Gothic War*. He stated "…its errant tail dimmed the stars of the Getic Wain until at last its dying fires grew feeble and vanished." The Getic Wain is none other than Ursa Major, the same grouping of stars associated with the Great Comet of 1618. (Kronk, 2021: 690) Ben Jonson drew heavily on the works of Claudian, including a paraphrase of his lines in the 1604 pageant held for the entry of James I into London, the pageant that featured Electra. Furthermore, "In a collection of classical Latin verse extracts prepared for the future Charles I as a gift for his father, Claudian is the fourth most cited author." (Moul, 2021: 24) Thus, while Claudian is not overtly

Fig. 4.12 Henry Briggs by John Faber Jr., after Isaac Seeman mezzotint, 1738.

mentioned, Jonson and the royals were certainly aware of the astronomical link between Claudian and the comet of 1618.

The astronomer Thomas Harriot, mentioned in line 49, began as a client of the Earl of Northumberland's friend Sir Walter Raleigh. He remained resident at Northumberland's Syon House until he died in 1621. [see Sect. 4.4.4].

In lines 52–54, we encounter parallax and a question that dates back to Aristotle: were comets far-distant objects, or merely within the Earth's atmosphere? (Van Nouhuys, 1998) In the case of the great comet, this question was answered in November 1618, as we read in a 1 December letter from Lorkin to Puckering.

> Mr. Briggs conceives it to be a perfect comet, and therefore above the Moon (so mathematicians have demonstrated Aristotle's tenet in this point to be false), and thinks the comet to be a hundred thousand miles in length – I say, thinks, because, as yet, his observation hath found no ground for geometrical demonstration. (quoted in Williams, 1848: 110)

This was Henry Briggs (1561–1630; Fig. 4.12), a mathematician and mapmaker who was the first Gresham (and first Savilian) Professor of Geometry at Oxford (1596–1620). Being a mathematician, he likely used a measurement of parallax (line 52) to determine the comet's distance and length of its tail. He was certainly known at Court: he dedicated a book on logarithms to Prince Charles in 1624, and he was consulted by the Virginia Company, which James took a keen interest in. As one of the few scientists in England at the time who studied the Great Comet of 1618, his views must have reached the ears of the king, and they almost certainly met. His anti-astrological stance was also in tune with the thoughts of the king, even though theologically they were at odds.

More broadly, the mention of Briggs, Harriot, and all those anonymous people applying instruments and mathematics to the comet was clearly inspired by Francis Bacon. In his 1605 book *The Advancement of* Learning, Bacon "...elevates the method of experiment and the lowly mechanical arts into objects of the highest praise." (Steadman, 1987: 77). The New Science was thus expressed in this new form of poetry in the early seventeenth century:

> Since knowledge is power, the natural philosopher, the student of nature, may subject greater natural forces to his command than the strongest of classical worthies, accomplish nobler acts of benefit for the public good than the ancient benefactors, and achieve more admirable wonders than the heroes of epic and romance. (Steadman, 1987: 66).

It was Kepler himself who initiated "a poetic revolution," using his 1608 *Somnium* as a vehicle for discussing the New Astronomy (Aït-Touati, 2010). Following his lead, and acting as both poet and natural philosopher, Corbett muses about the origin of the comet in lines 55–56, wondering if it may have been expelled by the Sun, or a primordial body (the second option being correct). In lines 57–59, he wonders if the comet is periodic. The first comet to be proven periodic was Halley's comet, but Edmund Halley did not discover that the comet (which bears his name) was periodic until 1705. In a diary entry for 1 March 1665, Samuel Pepys also mused on the periodicity of the comet. Commenting on the bright comet of 1665, he writes that Robert Hooke (whose "very curious lecture about the late Comett" he had just attended at Gresham College) suggested it is "very probably that this is the very same Comett that appeared before in the year 1618, and that in such a time it will appear again—which is a very new opinion." (Wright, 2000:4.26) The mere suggestion the comet of 1618 may be periodic was thus quite an adventurous thought; alas, it has never been seen again.

4.4.3 Political Analysis

Several political figures are mentioned in the poem. In line 14, we encounter the Dutch statesman Sir John van Olden Barnavelt (1547–1619), a great negotiator and general, who was executed by the jealousy of Prince Maurice of Nassau on 3 May 1619. He was ambassador to England and France, and in command of the armies of the United Provinces. Massinger and Fletcher wrote a play about him in August 1619 that "attracted crowded audiences in London." (Firth, 1911: 47) The mention of Starr in line 15 is likely a veiled reference to the Starr Chamber, which was a popular law court in the reign of James I. This seems to be reinforced in line 13, with the word "case" meaning a law case. Here, in line 15, it suggests the full power of the Starr Chamber should be brought to bear against Spain.

Francisco de Sandoval, first Duke of Lerma (1552/1553–1625) was the favourite of King Philip III (line 16). On 26 March 1618, he was made a Cardinal by Pope Paul V. On 4 October 1618, the Duke retired to his paternal estate after having fallen due to both the intrigues of his son and an ungrateful monk whom he raised to be confessor of the king.

Corbett, in the 4-line addendum to his poem, is directly addressing Prince Charles in relation to the Spanish Match, advising him to stay in England (sit fast). Ultimately, Charles ignored this advice and set out for Spain in 1623 in an abortive attempt to secure his Spanish bride. These lines also incorporate a clever mention of Charles-wain, as discussed in 4.2; this was also an allusive link with the Roman Emperor Augustus, who was said to have moles on his body "…after the order and number in the Starres of Charles's-Wayne." (Browne, 1658: 58) It was not unusual for successors to King James to be linked to Charles-wain. Joshua Sylvester (1614) has a dedicatory poem to Thomas Murray and other members of Charles's court that addresses them as the "Grave *Guides* & *Guards* of Hopeful CHARLES *his Wayn.*" A wagon, or, its more elevated cousin the chariot, was used in connexion with Charles-wain and royalty by Jonson. He penned a Pindaric Ode in 1629, whose lines 59–60 read

When they shall read the Acts of Charles his reigne,
And see his Chariot triumph 'bove his waine.

This was written for James's son, King Charles, and directly relates to poetry. "It is the king's immortal person, present in the stars of Ursa Major (Charles's "Chariot" in Charles "waine"), whose patronage shall make possible the reformation of poesy for the new Carolingian Pleaides." (Brown, 2010: 46) The link was continued into the next generation with the fable of the birth-star prodigy of 29 May 1630, the date the future King Charles II was born (Brown, 2010: 48).

4.4.4 *Syon House and Thomas Harriot*

4.4.4.1 Corbett's Letter to Aylesbury

Syon (or Sion) House at Isleworth in Middlesex (near London) was a grand estate owned by Henry Percy, the ninth Earl of Northumberland. It was here that the first telescopic astronomer, Thomas Harriot, did much research (Chapman, 2008). But Harriot was not alone: he and his scientific companions became the "Syon House set." "Harriot discussed algebra, optics and astronomy with William Lower (1570–1615), keeper of Percy's Welsh Estates, and he made astronomical observations with Lower's successor John Protheroe (1582–1624). Through Protheroe,

Harriot also became acquainted with Thomas Aylesbury," (Stedall, 2003:4), who received the letter/poem from Corbett. At this time, Aylesbury was secretary to Buckingham, who was Lord Admiral of England. This is a measure of how significant the Syon group was regarded: Corbett sent the poem to them, not to the great Oxford University. Aylesbury was also quite important in the continuation of the Stuart monarchy: he was the great-grandfather of both Queen Mary II and Queen Anne.

We know that comets had been a topic of study at Syon House for several years. William Lower wrote Harriot a letter on 6 February 1610, where Lower remarks that Kepler's elliptical planetary orbits may help solve the problem of the "unknown walkes of comets." J. J. Roche (1985:183) writes, "This would seem to imply that Harriot's circle regarded comets as celestial objects, or perhaps even as planetary." Lower and Harriot had already observed Halley's comet in 1607, and Harriot would also see the Great Comet of 1618. The poem from Corbett to Aylesbury had the intended effect. Just a few weeks later, on 19 January 1619, Aylesbury wrote to Harriot that the keenest topic at the royal Court was the comet, even though it was no longer visible:

> The comet being spent, the talke of it still runnes current here. The King's Majesty before my coming spake with one of Cambridge called Olarentia (a name able to beget beleefe of some extraordinary qualities), but what satisfaction he gave I cannot yet learne; here are papers out of Spayne about it, yea and from Rome, which I will endeavour to get. (quoted in Halliwell-Phillipps, 1841: 44).

Cooper (1863) identified Olarentia as William Olerenshaw, "a native of Derbyshire; admitted a Fellow of St. John's College, Cambridge, April 6, 1609; and who was buried in the College chapel, Oct. 17, 1652."

4.4.4.2 Comet Portraits

Lest one think I am straying too far by discussing portraiture here, painting–in this era was widely regarded as a visual form of rhetoric (Riklius, 2023: 75). According to Olson and Pasachoff (1998: 21), the comet in the portrait of Thomas Hamilton (1563–1637) is "the second of that year (C/1618 W1)." While they are correct that the comet in the portrait is W1, they incorrectly state it was the second comet of 1618. "Because we do not know the motives for its inclusion in this portrait, we can only speculate on them. Did the sitter believe that the comet was a portent of his good fortune and/or his status, or did its appearance coincide with an occasion he deemed significant, or was he simply charmed by its astounding beauty." Moving past such idle speculation, the reason is likely based on his heredity. Thomas Hamilton was the son of a man of the same name, Thomas, and his wife, Elizabeth (c.1542–1612). She was the daughter of James Heriot, who died in 1580. It is no coincidence that the most notable astronomer living in England in 1618 was Thomas Harriot (sometimes spelled Heriot). While the ancestry of Thomas Harriot is unknown, there are significant gaps in the historical record that make it entirely possible that Thomas was a relative of Thomas Hamilton. To emphasize how much importance Hamilton attached to his Heriot ancestry, he took the title Earl of Haddington in 1626, in preference to the title Earl of Melrose that had already been granted him by King James in 1619. The Heriot family of the Earl Haddington's mother and grandfather was in fact based in Trabroun, Haddingtonshire, an area of Scotland immediately to the east of Edinburgh; the title he assumed was based on that location. The inclusion of the comet in the portrait can thus be regarded as an allusion or tribute to his Heriot ancestry, and a nod to his kinsman Thomas, who was studying it. (see Note 5).

The artist Pal van Somer, who settled in England in 1616, also painted the comet. "A respectable tradition gives to van Somer four head-size portraits of 'Elizabeth Countess of Exeter' and her three daughters, which bear either the date 1618 or a picture of the famous comet of that year." (Waterhouse, 1994: 53). One of these depicts Lady Diana Cecil (1596–1654), Countess of Oxford.

4.5 William Basse

Comets were quite often referred to as "starres." In the case of the Great Comet of 1618, various descriptors were employed. Thomas Jackson (1622: 26), a prebend of Canterbury Cathedral, in a sermon, called it a "prodigious starre," although "blazing star" was more often used. Here, the poet William Basse (1580–1653), uses the adjective "streaming." I have changed "u" to "v," and vice versa, to make this unpublished poem more readable.

Upon the late Starre.

This yeare there hath appear'd a streaming starre
Within our native Hemisphere or clime,
But whether it brings us newes of peace or warre,
Of plague, or famine, who i'st can divine?
Though some interprete it to change of State,
Hostile invasion, or some great man's end:
Rumors of warres here landed to us late,
Or like particulars that they intend:
But since the Character hath such a letter,
That none can understand but he that writ,
Let's feare the worst, our sins, and make us better,
And not to other ends interpret it,
For in the same there's matter underhil'd.
Which shall not to our knowledge be made plaine,
Till the portent and purpose bee fulfil'd:
For never came such messengers in vaine:
How ere with meekenesse, let us kisse the rod:
Hoping the best, yet leaving all to God.

The purpose of the poem by Basse is singular: the author hews to the instruction given by the king that humans cannot (and should not try to) interpret the characters ("such a letter" in line 9) that describe what the comet portends. Only "he that writ," namely God, can read it (line 10), but whatever the intent, the comet is there for a purpose (line 15). Ultimately, the message will not be in vain (line 16), but in the meantime, Basse holds out the promise of improving mankind ("make us better" in line 11). The poem is thus unique in putting a positive gloss on the comet's appearance, although that is predicated on humans not trying to "interpret it" (line 12) otherwise. The "rod" in line 17 was first encountered in this chapter in Sect. 4.1.

4.6 London's Playwrights

4.6.1 Ben Jonson

"Ben Jonson (Fig. 4.13) was one of the thousands and tens of thousands who, night after night through that month, gazed up, in different parts of Scotland, to the strange celestial visitant, and all his life afterwards he must have remembered that it was in

Fig. 4.13 Ben Jonson, a nineteenth-century copy of a 1617 portrait by Abraham Blyenberch. National Portrait Gallery. PD-1996

Scotland that he had seen the great comet of the year 1618." (Masson, 1893). Jonson stayed with William Drummond during his visit to Edinburgh. A 1632 poem by Ben Jonson in the second edition of the Plays of Shakespeare, entitled *To the Memory of My Beloved, the Author, Mr. William Shakespeare, And what he hath left Us* reads:

> I see thee in the hemisphere
> Advanc'd, and made a constellation there :-
> Shine forth, thou Star of Poets; and with rage,
> Or influence, chide, or cheer, the drooping stage.

The amateur astronomer Capel Lofft (1751–1824) wrote of the "star of Poets" line: "A Comet of 1618, very conspicuous, perhaps contributed to suggest this imagery." (Lofft, 1812: xxxiii) Was Jonson really thinking about the comet of 1618 when he was writing these words to honour his friend in 1632? I find it plausible, but only tenuously so.

The 6 January 1620 masque mentioned in 2.1.6, *News from the New World Discovr'd in the Moon*, was more than merely a cautionary tale of what might be newsworthy. "The masque's moon journey," writes Knowles (2015), "is combined with recycled Protestant apocalyptic interpretations of the 1618 comet as a harbinger of the Palatine crisis," which was a "key moment in the fissuring of the culture of consensus and as another step in the emergence of a public news culture." As such, it perfectly exemplified what Jonson himself wrote a masque should try to attain: "though their *voice* be taught to sound to present occasions, their *sense* should always lay hold on more removed mysteries." (quote is from the preface to his 1606 masque *Hymenaei*.) On the Continent, apocalyptic commentary on the Great Comet of 1618 was rife, an example being the pamphlet by the Leipzig physician Paul Nagel. "His works attracted growing interest and caused some furore." (Berg, 2011:103).

Jonson's fable of the Moon is related to the diplomatic initiatives arising from the assumption of the crown of Bohemia by James's son-in-law Frederick V of the Palatinate (1596–1632; King of Bohemia from 26 August 1619 to 8 November 1620), and to James's desire to discourage speculation about significant military intervention by England on the Continent. (Sellin, 1986) This is the so-called Palatine crisis just mentioned. "The general point is that in this masque, the false truths of religious fanaticism are replaced with royal Truth. So, it's like a royal antidote to the kind of masque of truth danced by the Protestant polemicists." (Knowles, 2023, priv. Comm.)

There is an astronomical link as well: Frederick and his wife Elizabeth (known to posterity as The Winter Queen; 1596–1662) supported Tycho's work at Uraniborg. The masque, which saw Prince Charles himself performing to his father, articulated a view that "excitable and ill-informed militancy must be restrained and subordinated to a judicious and far-seeing policy which had peace as its leading priority." (Butler, 1993:167).

In addition to his *Moon* masque, Jonson wrote another one whose ideas originate in the comet of 1618. Performed in 1622, *The Masque of Augurs* begins with an antimasque that makes reference to catoptrics—the use of mirrors for various visual effects. Here it is used very much in the modern sense of a telescope aimed at terrestrial objects, which can "multiply the vision" so that one can see objects up to five miles away. (Gifford, 1869b:632). Catoptrics later became the basis for the first Newtonian telescope. One must note the highly unusual inclusion of the science of optics in this antimasque, which is made particularly pointed as it is spoken by a Dutchman—it was well known that the telescope had originated in the Low Countries. Indeed, Harriot made his first observations with such a telescope in 1609 (Chapman, 2008). The reference to optics was almost certainly incorporated in the antimasque to please His Majesty, whose interest in matters of astronomy was well known to Jonson. Of the augurs in the masque proper, Jonson deploys another astronomical allusion. He tells us the augurs are ensconced in a temple, "each one guided by a star," where "star" can signify a comet, amongst other *ex caelo* objects or circumstances such as eclipses and meteors (shooting stars). That a particular comet is certainly being alluded to here is signalled by the line in the masque, "While the world about him [the king] is at odds." This is a reference to the ongoing conflict that would become known as the Thirty Years' War; it began in 1618. By 1648, it was written that the comet had been visible for just 30 days, each one corresponding to a year of war. (Thiebault, 1994: 445).

Further to these allusions I have identified, the famed Jonson scholar Anthony W. Johnson (2007) has discerned further links between the masque and the great comet. In this masque, as in ancient Roman religion, augurs were supposed to divine things such as the future course of action for the state. Noting that King James had invited Kepler to London, Johnson (2007: 176) states, "Kepler (and with him, Bainbridge) had led the way towards the idea that certain types of divinatory practice need not be incommensurable with Protestant thought." Despite his largely scientific approach to the comet of 1618, "Bainbridge was not shy to affirm that the five comets which had appeared over the ten years in which Luther was preaching were signs of God's approval." (Johnson, 2007: 176). He divined that the Great Comet of 1618 was actually a fortunate sign for King James; in this he was following the use of comets as a sign of "a new golden age" in the reign of Augustus (Satterfield, 2016: 338; Beacham, 1999:115), and it was meant in this sense by Jonson in his 1604 pageant as he writes "All Comets were not fatal, some were fortunately ominous, as this to which we allude." (Jonson, 1604). Here we can plainly see in action the motive force behind Jonson's works—to educate his rulers in virtue and deter them from vice. The link with Augustus is explained by Worden (1993: 71). The words in quotes are Jonson's own:

> In the alliance between royalty and poetry which he sought, poets, being of "far rare birth than kings," would not be junior partners. They should be ready with the "free and wholesome sharpness" which, in Jonson's account, Augustus had welcomed from Horace, Jonson's model.

This was reinforced by having the king's son, Prince Charles, take part in the masque as one of the augurs. The concurrent rebuilding of the Banqueting House with the presentation of the masque in 1622 was thus seen as "part of an inaugural act, aimed at completing the first part of a plan to promote diplomacy, negotiation and marital alliance." (Johnson, 2007: 176). The promotion of the Spanish Match at *The Masque of Augurs* is certain, as the Spanish ambassador Gondomar was present, along with King James. The ambassador was linked to the comet by the English geographer and chaplain to King James, Thomas Scott (1580–1626). Scott writes of the ambassador's presence in England

> ...where he had long been to the grief of many good men, hanging in their eyes, like a prodigious Comet, and threatening worse effects to the Church and State, then this other Comet [ie the real one] could effect or prefigure, if it had been effectual, as some maintain, or merely signall, as others judge. Scott (1620: 3).

The king himself felt the baleful influence of Gondomar (Doelman, 2011). The year 1620 saw a "joint Spanish-Imperialist invasion of the Palatinate [south-western Germany], and he felt that he had been deceived in this matter by Gondomar, with whom he had earlier collaborated to preserve peace." (Manning, 2018:172) Gondomar cast a long shadow: the 1624 play *A Game at Chess* by Thomas Middleton, which was wildly successful, cast Gondomar as an evil Black Knight (Fuchs, 2013: 58). One of the counsellors of state in Spain, Don Mendosa, turned the tables by comparing Prince Charles to the comet, but this time with admiration, not contempt. Scott (1624: 26), in a further report on the machinations of the Spanish match, reports Medosa said "...such a Prince as Charles of Wales, and only son to the King of Great Britain, upon whom and whose action's as a bright blazing Comet Europe begins to fix her eye."

A key element of King James's comet poem is an injunction about trying to read signs from heaven. This is also incorporated by Jonson in *Augurs*. "...in line with James's Calvinism [see Sect. 4.8 of this chapter], no presumption is made that humans can directly read the fates. For as Jove puts it, 'Till *Jove* ratifie, It is no *Augurie*,/Though utter'd by the mouth of *Destinie*.'" (Johnson, 2009: 190–191).

4.6.2 *John Fletcher and Philip Massinger*

Philip Massinger (1583–1640; Fig. 4.14) and John Fletcher (1579–1625) were responsible for several plays that incorporated the Great Comet of 1618, a testimony to its pervasive influence on the English theatrical world in 1619. Massinger succeeded Fletcher as company dramatist of the King's Men upon the latter's death. *The Laws of Candy*, *The Tragedy of Sir John van Olden Barnavelt*, *The Virgin Martyr*, *The Unnatural Combat*, *The Queen of Corinth*, and *The Island Princess* (together with a seventh play often attributed to Massinger, *The City Madam*) form a "comet suite" that is unprecedented in English theatre. (Boyle, 1886) The theatrical device of a "blazing star" had been used in English plays for several decades (Di Salvo, 2023), but it certainly took on more direct reference to the Great Comet of 1618 in these plays.

Fig. 4.14 Philip Massinger, from an engraving by Charles Grignion in 1750. Courtesy of Wikimedia Commons, public domain

4.6.2.1 The Laws of Candy

The Laws of Candy, which was not published until 1647, was likely written in 1619 or 1620. It invoked the fresh memory of the comet.

> Cassilane, a general of Candy (an archaic name for Crete), states in Act 2 (lines 202-210):
> If to bring home
> Long peace, long victory, even to your Capitol,
> If to secure your kingdom, wives and children,
> Your lives and liberties, if to renown
> Your honours through the world, to fix your names,
> Like blazing stars, admired and fear'd by all
> That have but heard of Candy or a Cretan,
> Be to deserve the approvement of my manhood,
> Then thus much have I done. (Chambers, 1908: 483)

Later in the play, Act 3 (lines 92–97), Cassilane again invokes the celestial visitation:

> Applause
> Is, Gaspero-puff-nothing.-Why, young lord,

Would you so much be sequester'd from those
That are the blazing comets of the time,
To live a solitary life with me,
A man forsaken? (Chambers, 1908: 508)

The editor of the play, Edmund Kerchever Chambers (1866–1954; 1908: 468), states in his analysis of the play that there was "no doubt the comet of 1618 made a sensation."

4.6.2.2 The Tragedy of sir John Van Olden Barnavelt

Another play that is definitely by Fletcher (Fig. 4.15) and Massinger, *The Tragedy of Sir John van Olden Barnavelt*, premiered in August 1619. It takes an overtly political stance in Scene 4 by mentioning the "master peace" that King James was trying to broker on the Continent. This is linked directly to the great comet, which was still visible telescopically in early 1619.

The Master peace is now a foot, which if it speed
And take but that sure hold I aim it at,
I make no doubt but once more, like a Comet,

Fig. 4.15 John Fletcher, a painting of unknown date. National Portrait Gallery. PD-Art

To shine out faire and blaze prodigiously
Even to the ruin of those men that hate me.
(Bullen, 1884: 45)

James had adopted *Beati Pacifici* (Blessed are the Peacemakers) as his personal motto. In reality, he had little choice as England had no standing army, and he was told in 1619 that his treasury was empty. Thus, he resorted to the infamous Spanish Match (so prominent in his comet poem) as a diplomatic solution to war in the Palatinate. This failed in September 1620 as Philip III invaded the Palatinate, and thus Europe began its long descent into the Thirty Years' War. Despite this, his plans for the Spanish Match continued until 1624! (see also 6.2.5).

The lines quoted above are spoken by Barnavelt himself; in real life, the Dutch politician Barnavelt (1547–1619) was indeed assailed by men who hated him. On 13 May 1619, he was executed: King James probably lost little sleep over this action by his ally The Netherlands, as he despised Barnavelt. The playwrights are turning the tables on Barnavelt: for daring to invoke the comet to save his life and ruin his enemies, he brings ruin upon himself.

4.6.2.3 The Virgin Martyr

This tragedy opens by directly linking the appearance of a great comet to Diocletian, the third-century Emperor of Rome. "Tis wondrous strange! The marches of great princes, Like to the motions of prodigious meteors, Are step by step observed." While Massinger wrote with one eye on the Emperor of Rome, he kept the other firmly fixed on his own monarch, King James—his very own "great prince." Perhaps it is no coincidence that James was the first English monarch to portray himself on his coinage as a Roman emperor. A darker allusion to the comet appears in Act II, where Massinger writes, "From such a star/ Blazing with fires of hate, what can be look'd for/But tragical events?" *The Virgin Martyr* was performed in October 1620 and was first published in 1622, later reprinted as Massinger (1844).

4.6.2.4 The Unnatural Combat

While the tragedy *The Unnatural Combat* by Massinger was not published until 1639, "From the comet allusion *The Unnatural Combat* seems to have been written about the beginning of 1619." (Boyle, 1886: 234) The play deals with the Barbary pirates and court corruption exemplified by the King's favourite, the Duke of Buckingham. The old warrior Malefort, Admiral of Marseilles, recalls a time "when my sword/Advanced thus, to my enemies appear'd/A hairy comet, threatening death and ruin." According to Gifford (1813, vol. 1: 138), this passage was one John Milton had in mind when he wrote the comet section of *Paradise Lost* [see Sect. 4.7].

4.6.2.5 The Island Princess

Three years after the appearance of the comet, a play by Fletcher entitled *The Island Princess* was first staged in December 1621. It conjured to mind the comet, and how it captured the attention of everyone on the island of England and Scotland:

> The fire I brought here with me shall do something
> Shall burst into material flames, and bright ones,
> That all the Island shall stand wondering at it,
> As if they had been stricken with a Comet.
> (quoted in Kafantaris, 2014:143)

Aside from the temporal link, this is clearly a reference to the comet of 1618, as the play by Fletcher deals with dynastic marriage. Just a few weeks earlier, in November 1621, King James made it clear to Parliament that he would not allow anything to jeopardise the dynastic treaty with Spain, which included the marriage of his son to a Spanish princess: the very subject of his comet poem of 1618. What so incensed many members of Parliament and the public was the possibility they might have to deal with the reality of a Catholic Queen, once Charles became king. Even though she would not be the ruling sovereign, it was an unacceptable scenario for many. In the event, the dreaded match never happened.

4.6.2.6 The Queen of Corinth

From its cometary reference, this play was performed in 1619; like other plays, it was first published in 1647. The play was set in the ancient Greek city of Corinth. In a conversation with a character named Conon in scene 4, the mean-spirited Crates speaks the following line: "I'll pledge you/ But for that glorious Comet lately fired."

4.6.2.7 The City Madam

Frederick Gard Fleay (1891: 227) states that a play often attributed to Massinger, *The City Madam*, "was surely not his," although it is still widely attributed to Massinger. The play, which Fleay dates to 1619, contains the text:

> You look, master Stargaze,
> As you had seen a strange comet, and had now foretold,
> The end of the world. (quoted in Gifford, 1813, vol. 4: 91)

This, writes Fleay (1891: 226), "cannot, surely, be long after the great comet of the 1618 winter." This is problematic, as modern sources cite the date of 1632 for the licensing of the play; it was not published until 1659.

4.6.3 George Chapman and Thomas Harriot

George Chapman (1559–1634), a dramatist whose life overlapped that of Shakespeare, was noted for his masques at the court of James I. Martin Butler has identified his *Masque of Twelve Months* as having been one of the two major Christmas shows at court in the winter of 1618/19. It was fittingly performed on Twelfth Night (early January 1619), and again on Shrovetide at Whitehall. Normally, the second show would have been performed at the Banqueting House, but that burned down on January 12; as that disaster was linked by many to the comet, its inclusion in the masque was especially ominous.

> Every night
> Make I attendance on this blessed bower,
> Where majesty and love are met in one,
> All harmful spirits frighting from his throne,
> And keeping watch that no ill-looking planet
> Fasten his beams here; all ill-looking comets,
> In all their influences so much feared,
> Converting into good and golden dews
> That peace and planet through the land diffuse. (32–40)

These lines are spoken early in the masque by a character named Madge Howlett. Her "words on the comet voiced an anxiety about the need to protect England's peace at this unsettling time…The comet's threat was dramatised later in the masque, when the Moons were drawn down close to Earth." (Butler, 2007: 381–382) Once the thirteen moons of the play are brought down toward Earth, a character named Beauty (the presenter of the masque in honour of a fairy king) asks, "What frolic person have we here?" Aglaia, an attendant of Beauty, responds:

> Prognostication, madam, that now enters,
> In prime of this new year, in all his honours
> Sought to for his predictions, and foreruns
> The Months, our masquers and new rising suns.

Of "Prognostication," Butler (2007: 398) has this to say: "Various predictions were made on the comet of 1618-19, including a set by the astronomer John Bainbridge." It was the honoured Bainbridge who, in the new year of 1619, was sought out for his predictions on what was happening in the sky; thus, he obliquely enters the masque. Bainbridge's treatise on the comet was addressed to the King, not the common people (Vermij & Vieth, 2022). The link between disaster and science was noted by Hugh Trevor-Roper (1965: 60):

> Ever since 1618 at least there had been talk of the dissolution of society, or of the world; and the undefined sense of gloom of which we are constantly aware in those years was justified sometimes by new interpretations of Scripture, sometimes by new phenomena in the skies. With the discovery of new stars, and particularly with the new comet of 1618, science seemed to support the prophets of disaster.

In 1621, the famous *Essex House Masque* (Raylor, 2000: 22) was performed at the riverside residence of James Hay, Viscount Doncaster. It also invokes the comet:

The fire of comets, such as do appear
More to the Artists; then the peoples fear,
Higher then Air hath place; met with the light
Snatch'd from the thunderbolts amazing flight.
The beauteous stares cool shooting; cunningly
Mingled, with sparks of Junoes Jealousy
Make up this lamp of life. (lines 168–174)

In his notes on the text, Raylor (2000: 35) states that lines 168–169 mean that comets are "so high in the firmament that they are invisible to all but an "artist"—scientist, or astrologer—through their optical equipment." Being placed in the same league as a true artist would surely have pleased the king, who was in the audience for the first and only performance of the *Essex House Masque* on 8 January 1621. Raylor concludes the "artist" in the masque was none other than Thomas Harriot. This takes on added relevance when one considers that the most likely author of the masque, George Chapman, was associated with Thomas Harriot (Fig. 4.16). It was

Fig. 4.16 Thomas Harriot, from a painting of 1602. Courtesy of Wikimedia Commons, public domain. PD-1996

Harriot who offered a critique of Chapman's work on Homer, immortalised in the 1816 sonnet by John Keats: *On First Looking into Chapman's Homer* (Rukeyser, 1971: 145). Furthermore, Viscount Doncaster (later first Earl of Carlisle; 1580–1636), who commissioned the Essex House Masque, was married to the ninth Earl of Northumberland's daughter. "After his marriage, Doncaster took up residence at Essex House, a Percy property, where Harriot also had rooms." (Raylor, 2000:150). Thus, the mention of comets in the masque can be linked to the observations of comets by Harriot, the most recent and prominent of which was the Great Comet of 1618. The final word of the comet poem by King James is "fear," the very word Chapman so crucially links with comets in the masques of 1619 and 1621. "The poet's denial that the comet would excite the people's fear was tactful: King James had attempted to allay public anxieties over the portentousness of the 1618 comet by announcing that it was 'nothing else but Venus with a fire-brand in her arse,' and by writing a poem instructing his subjects not to interpret it as an evil omen." (Raylor, 2000: 35–36) To round out our look at this astronomy-laden passage, line 170 puts the author in accord with the latest views on comets as being of celestial, rather than terrestrial, origin when he writes they are "higher then aire." Meteors, here termed "stares [stars] cool shooting," derive from the belief of Aristotle that they were caused by exhalations from earth, as explained in Sect. 4.2.4; Aristotle's view of the origin of comets is superseded in the masque by having them exist beyond the upper air. The masque author also very poetically relates meteors to Juno, Queen of the Gods, whose "sparks" of jealous rage directed towards humans she hated were legendary.

4.7 John Milton

John Milton (Fig. 4.17) was a careful reader of the plays of Massinger, and he was inspired by the comet passage of the play *The Unnatural Combat*; in Book 2 of *Paradise Lost*, he invoked a comet to describe Satan's appearance. The passage also linked fire (even the word Incensed, meaning "'fired," is etymological wit, as 'fires' two lines later confirms) with the northern sky, and war:

> Incensed with indignation Satan stood
> Unterrified, and like a Comet burned,
> That fires the length of Ophiuchus, huge
> In the Arctic Sky, and from his horrid hair
> Shakes pestilence and war. (2. 707–11)

As I showed in Cunningham (2016b), the passage has been misinterpreted by researchers for centuries. Two main issues arise: What comet was Milton referring to, and why did the comet appear "huge in the Arctic sky?"

Edward Gibbon (1788) was the first to suggest the comet of 1664 was the one meant by Milton, and by the early twentieth century, this view was widely accepted. However, Arthur Wilson Verity (1893: xv) correctly identified it as the Great Comet

Fig. 4.17 John Milton in a portrait of 1650. Christ's College, Cambridge. Courtesy of Wikimedia Commons, public domain. PD-1996

of 1618. Milton was 10 years of age in 1618; Orchard (1913: 257) writes, "the apparition of this imposing object must have made a deep and lasting impression upon his mind."

Both the Milton scholars Thomas Newton (in 1750) and Henry Todd (who quotes Newton in 1826) accept that the comet appeared in the constellation Ophiuchus "in the Arctic sky." Crucial questions arise here. Is that constellation, as Newton and others believed, "in the Arctic sky?" Did the comet really appear in Ophiuchus, as most Milton experts accept?

Ophiuchus is a far-reaching constellation that straddles the celestial equator, but most of it is south of the celestial equator. Thus, it is not regarded as an Arctic (far northern) constellation. This did not prevent the famed Milton scholar David Masson from glossing over the fact that most of it is in the south. He described Ophiuchus as "a large constellation in the northern heavens, stretching forty degrees."[76] Other researchers suggest that either Milton's knowledge of the constellations was not very precise by placing Ophiuchus in the Arctic sky, or he manipulated details to strengthen the relationship between Satan, the comet, and the serpent-bearer constellation. Regarding the former suggestion, the idea that a man of Milton's erudition could make a mistake of such magnitude is most unlikely.

Even the otherwise astute Orchard falters here when he says Milton "appears to have been unfamiliar" with the position of Ophiuchus. The latter suggestion can also be dismissed, as the author's premise was the now-discredited notion that *Paradise Lost* was a monument to scientific backwardness. Even more contrived seems the suggestion that the expression "arctic sky" refers not to the position of the comet, but to the position of the writer, meaning "as seen under the arctic or northern sky."

Noting that "Milton had a keen interest in astronomy," Sarkar (2012) traces the sources of Milton's knowledge of this comet to a treatise by John Bainbridge and in popular encyclopedias such as Swan's *Speculum Mundi*. Sarkar correctly states that Bainbridge explains the position of the comet with the help of a planisphere or sky-chart, "which shows it stretching from the equinoctial line to the north, the cometary line running parallel to Ophiuchus." But she goes too far in trying to support her thesis linking Satan and the serpent-bearer constellation Ophiuchus by repeatedly stating the comet of 1618 was actually *in* that constellation. Here is one instance:

> Within the context of seventeenth-century millennialism, the appearance of Satan as the comet and the new star in Ophiuchus enhances our perception of him as powerful and compelling. (Sarkar, 2012: 92)

To quote from my 2016 paper, "While it is true that the supernova ("new star") of 1604 was in Ophiuchus, it is *not* true that the Great Comet of 1618 was there too. Bainbridge never states the comet of 1618 was in that constellation: his book shows the comet running through Libra, Virgo, Bootes, and Ursa Major, adjacent to—*but not within*—Ophiuchus. Modern computer calculations agree with Bainbridge that it never passed through Ophiuchus. Eyewitnesses said the comet's tail was so long it extended into the northern reaches of the sky. Extensive tail lengths, up to 104 degrees in the sky, were reported for 1618 W1. In addition, the comet headed northwards until it reached an area near the Pole star, where it finally disappeared. Milton was saying that the comet extended across a length of sky *beside* Ophiuchus, but did *not* actually appear in that constellation. Crucially, he is saying the *comet* was huge in the Arctic sky, not that Ophiuchus was huge in the Arctic sky: "huge" modifies comet, not Ophiuchus, which justifies my intrusion of a comma between "Ophiuchus" and "huge" in 2.709. Thus, the tortured problem of trying to reconcile the comet of 1618 in a constellation that does *not* appear in the Arctic vanishes, solving a conundrum that has bedeviled students of Milton since at least 1750. As usual in his astronomical allusions, Milton was precise. The only confusion occurred in the minds of those who have under-analyzed what he wrote.

"Milton's use of a comet shows that he evoked fire, war, and the northern sky as emotive and celestial-geographical terms when describing both the aurora (meteor) and a comet. However, the imagery in 2.707–711 should be considered analogous to, but not identical with, that used by Milton in his description of the aurora, which was clearly delineated quite differently. This analysis also shows Milton had a full command of astronomical knowledge and sky lore, which he used as needed to advance the story of *Paradise Lost*. The careful deployment of shooting stars, meteors, and comets as *distinct* entities emanated from the mind of Milton. The conflation of these entities, and the confusion caused by their misinterpretation, is due entirely to other writers."

4.8 Jacob Cats and the Synod of Dordt

4.8.1 The Dutch Republic

The Great Comet of 1618 must be seen, in the case of The United Provinces of the Netherlands, as an important element of the Dutch Golden Age, "an age of capitalism, of painting and engraving, of literature, science, and technology, but above all, the Golden Age was famous because of its learning, with philology carrying the crown." (van Miert, 2018: 12) It lasted from roughly 1590 to 1670, and as the production of the King James version of the Bible in 1611 was nothing if not an exercise in philology, it is at once apparent that James felt a close affinity with the Dutch. There were many close links between the Dutch and English at this time. The most visible such link today is a cottage built by the London Dutch community and workers from the Low Countries on Canvey Island, close to the English town of Basildon; above its door is emblazoned the date 1618 (Joby, 2015: 38).

How comets were widely perceived in the Low Countries in the early seventeeth century may be gauged by how they were portrayed in a play seen by the populace in 1602. In his Senecan tragedy *Auriacus, sive libertas saucia* (written as a Latin poem), the scholar and playwright Daniel Heinsius (1580–1655) wrote, "We have seen a horrifying comet appear in the sky." It was said to portend things that ignored the petty concerns of the people.

> Let palaces and strong tyrants fear, 'the stars do not threaten
> private persons or the little shepherd, the sky does not know
> what is small, what is big, totters and prosperity opens up
> to a fall by its own weight.' (lines 1400–1405, quoted in Bloemendal, 2020: 242)

The collision between cometary history and religion in the Dutch Republic was forged in the dogma created by biblical philologists, who pushed biblical scholarship to the centre of historical studies in this period. Thus, a consideration of theological issues is essential to this study (Vermij, 2011).

4.8.2 The Synod of Dordt

The Synod of Dordt began on 13 November 1618 and concluded on 29 May 1619. An assembly of the Reformed Church of the Netherlands, it was held in Dordrecht and was attended by representatives from several countries. King James sent a high-level delegation: it consisted of Sir Dudley Carleton (who was very familiar with the comet and its reception in London, thanks to Chamberlain), George Carlton (Bishop of Llandaff), Dr. John Davenport (Professor of Divinity at Cambridge), Dr. Samuel Ward (translator of the English Bible), Dr. Thomas Goad (from St. Paul's Cathedral), and Walter Balcanqual (a Scottish Episcopalian). The source for this list is the Remonstrant Conrad Vorstius (1619: 75–76); it was his incendiary writings that led King James to intervene in the great theological debate at Dordt (Heilbron, 2021: 54–55). At the Synod, Vorstius was condemned as a heretic and banished due to the personal intervention of James; he directed Dudley Carleton to meet with the Prince of Orange in person to demand this harsh sentence. In the shadow of the great comet, James wrote *Meditations Upon the Lords Prayer* (1619), which he dedicated to Buckingham (the "minion" of the comet poem). To a large extent, *Meditations* is a polemic against Puritans (line 21 in the comet poem) and Vorstius. He states, "upon the foundation" laid by the Puritans "are also built up all these innumerable sects of new heresies, that now swarm in Amsterdam." (James, 1619: 15). One of the aims of the Synod of Dordt was to quash such heresies in the Netherlands.

King James "let it be known abroad that he was a passionate Calvinist in theology," and it was the Calvinists whose views were ultimately adopted when the Synod closed. (Colie, 1957: 17) The Dutch Republic was the "theological centre of international Calvinism," (Miert, 2018: 13), so it is not surprising to see an intervention by the King of England. Harkening back to James's poem and Doomsday, it is important to understand that for Calvinists, the Day of Judgment is a day of salvation and blessed resurrection. The judgment itself is not directed at individuals, "but of the judgment of Jesus Christ and His grace." (Quistorp, 1955: 12) Theologically, "The Calvinists made a clean sweep in 1619. After the Synod, the Remonstrants were in the position of traitors: in exile or in prison." (Colie, 1957: 18) A sermon preached in 1607 by Dr. Martin Fotherby, archdeacon of Canterbury, will surely not have been forgotten by the king. Fotherby denounced those who endeavoured by "a colourful pretence of reformation" to undermine the "blazing starre" of the early Stuart Church. (Walsham, 1999: 294) The theological link between his Church and the comet was embodied in the very title of the king's poem. (Goudriaan & van Lieburg, 2011: 398.)

John Bainbridge (1618: 40) mentions the Synod in his book on the comet of 1618: "I may not forget the coincidence of this celestiall messenger with the present Synode at Dort, divine providence actually with the rayes of this new Comet dispelling those foggie mists which began to over-shadow the glorious light of Evangelicall veritie." The linkage of the comet with providence is a manifestation of "special providence," and opposed to "general providence." Belief in the former declined in the latter half of the seventeenth century, although belief in the latter continued into the nineteenth in, e.g., the poetry of Joseph von Eichendorff (Avni, 1969). Isaac Newton himself believed in general providence, but his adherents "…did not believe that extraordinary natural occurrences were necessarily signs from God, because for them providence did not operate in nature by means of dramatic events like comets or earthquakes but rather through the imposition or order and harmony by natural law." (Fix, 1991: 8) Bainbridge employs many poetic quotations in his discussion of the comet, but they are all from ancient authors; none were written specifically about the events of 1618.

For the Dutch, the Synod was linked to the comet for many years. In Twisck, 1624, the Luther theologian Pieter Twisck (1565–1636) wrote a book about comets. It included a discussion of the "dread results of the 1618 comet (of which the disturbances caused by the Synod of Dort were for him the worst), and warned his countrymen of dismal fate awaiting them for their sins." (Tijdschrift, 1955) In 1682, confronted with another great comet, three Dutch authors were still linking the 1618 comet with the Synod: Jacobus Koelman, Gisbertus Cocq, and Johannes van Holst. "These three men were orthodox clergymen and their view of comets was clearly related to their view of seventeenth-century religious affairs—e.g., the comet of 1618 for them expressly foreshadowed the painful schism in the Dutch Church culminating in the Synod of Dort." (Colie, 1955: 203).

4.8.3 *The Comet Poem by Cats*

Heinsius' contemporary Jacob Cats (1577–1660; Fig. 4.18) hoped the Synod of Dordt would end civil and religious strife and was inclined to look benignly on the proceedings. Cats had close ties with England (he was knighted by King Charles I in 1627). He achieved fame through his emblem books, one of which was published

Fig. 4.18 Jacob Cats, from a 1634 portrait by Michiel van Mierevelt. Rijksmuseum. Courtesy of Wikimedia Commons, public domain

in 1618. Cats wrote a poem in which he appealed to man's ignorance and expressed his hope that the truth of Christ would drive out discord from the Low Countries. Importantly for this study, he framed it in terms of the Great Comet of 1618. The poem was first published by Cats (1619) and later reissued in Cats (1623: 230). To emphasise the link between the comet of 1618 and Christianity, Cats included in his pamphlet a comparison between the nova of 1572 with the star of Bethlehem. Both the nova and the comet were regarded as "new stars."

To preface the poem by Cats, written just after the "New Year" of 1619 began, Brandt (1721: 588) has this to say. "The poet (Cats) relates in these Verses, that at the birth of Christ there appeared a new Star [the Star of Bethlehem], which served for a guide to the Wise men: That there likewise appeared another new Star [supernova] at the St. Bartholomew-Massacre at Paris in 1572, when the blood of the Martyrs in France became the seed of the Reformed churches in that Country; and then he proceeds to give us his observations on the Times, which he thus expresses:"

Title of Poem: *Reflections on the meeting of the Modern Synod, occasioned by the present Comet.*

A Comet with strange Light, again adorns the Skies,
Strikes Fear in mortal Hearts, and draws our wondring Eyes;
As Thunder-struck we gaze, and, doubtful of th' event,
We wait to learn it's Course, and on what Errand sent.
True Son of God descend, the World's bright Sun appear !
Enlighten our dark Minds, and dissipate our Fear.
With Glory fill thy Church, and with thy Heav'nly Fire,
Cherish our frozen Hearts, and kindle new Desire
To serve thee this New Year. May we from hence presage
No fatal Crimes, but Vertues of a new-born Age.
True Prince of Peace appear ! Let Strife and Discord flee
Before thee, and give place to Godlike Unity.
Touch'd by thy gentle Spirit, may rocky Hearts relent,
And join'd by Friendship's Band, all future Jars prevent.

In the first four lines, Cats clearly positions the comet as a sign from heaven; he "defended the view that comets were supernatural beings miraculously created at God's command." (van Nouhuys, 1998: 550), and while its errand is as yet unknown, he holds out the expectation that it will become known. Cats exhorts his readers to come together in unity in a New Age of Peace; the catalyst for this is, of course, the comet. The use of the word "wait" in line 4 is also crucial here. In the seventeenth century, "wait" meant not just standing in readiness, but also placing hope in the divine.

4.8.3.1 Further Text by Cats (1619)

In the same book of 1619 (which is not paginated), there is a rhyming colloquy of three persons, *Dialogue about the present Comet.* Each of his three protagonists represented a point of view on comets. One, the Wel-hams, were convinced every human action was traceable to the stars; the Gheen-aerts were only interested in material things, and professed to be uninterested in the heavens; the Reyn-harts chose a mean between these extremes, advocating that celestial bodies needed to be studied. Cats used this triad to promote his own view that one should adopt the Reyn-harts position. Since van Nouhuys (1998) has written thoroughly of this, I will just give a couple of excerpts from the Wel-ham point of view to show he was writing the dialogue in the form of a poem.

If we view the strange Events
Which the daily Scene presents,
Rising Doubts we shall suppress,
And the Comet's Force confess
Changes both in Church and State
From this Star derive their Date.

Afterwards, taking notice of the influence of this Comet on other Countries, he says:

Let us take a farther View,
And the Comet's Course pursue;
Countries far remote from hence
Seem to feel it's Influence.
Spain itself, we understand,
Shares the Fate of Netherland.
Crimes of Great Ones veil'd in Night,
By this Star are brought to Light.

The reference to Spain likely alludes to the Spanish Match controversy; observationally, the great comet was observed in Spain by Juan Cedillo Diaz and his associates. "Apart from the observational data, in a manuscript written by a pupil of Cedillo, there is a discussion on the theory of the formation of comets, the character of signs and the causes of certain events." (Navarro-Brotons, 2007: 19). So, we can be sure the "influence" of the comet was as keenly felt in Spain as it was in the Netherlands, where there was "...hardly a Dutch philosopher, theologian or poet who was not concerned in some way with the [cometary] question of 'what this sign means.'" (Jorink, 2010: 110) Cats praised the works of the Dutch astronomer Philips Lansbergen (1561–1632) as contributions to religious piety. Lansbergen "...sanctioned the divine significance of such celestial phenomena through the Bible." (Jorink, 2010: 134).

In subsequent years, the High Dutch chronicles were full of dire links between the comet of 1618 and the war we know as the Thirty Years' War. Brinckmair (1638), at the 20-year mark of the conflict, collected these sources, which merit an entirely separate investigation.

4.9 The Comet as Seen from Russia

On 2 June 1618, an embassy left Gravesend, England, bound for Archangel, Russia. It was a diplomatic mission sent at the behest of King James to Tsar Mikhail, the first Romanov. The expedition was led by Sir Dudley Digges, and its chaplain was the 26-year-old Richard James, who had been elected probationary fellow of Corpus Christi College in 1615. The 18-member delegation only got as far as Kholmogory before Digges hastily returned to Archangel, and then back to England. The remaining group was instructed to proceed to Moscow. When Richard James reached Yaroslavl, some 270 km short of the capital city, he was inspired to write a Latin poem by a wondrous sight in the sky: the Great Comet of 1618. After spending seven months in Moscow, he eventually made his way back to England, laden with all the manuscripts he had written in Russia. These were eventually deposited in the Bodleian Library in Oxford; the comet poem being labelled Bodleian MS James 13, p. 239. After more than four centuries, it appears here in print for the first time (footnotes follow the references)

De Cometa quæ apparuit iuxta
vrsam maiorem ann*o* D*omini*
1618. per spatium mensis
in Decembri et Nouembri
quam vidimus in Varislauia Rutheno rum.

Tunc cum Sarmatici cœperunt foedera pacis
Frustrati Gallo milite ne caperent
Vir Michaele tibi steterint quæ moenia, fallar
In quibus haud tantum mobile vulgus erat.
Emicuit septem inter circumgestata triones
Promittens longam dira cometa facem.
Nos quod portentum flammæ quæsivimus acri
Suspicione o*mn*es. Quisq*ue* sibi, metuens.
Anglis ipse timens, ne forsitan invidiosa
Si foret occasu magne Jacobe tuo
Extimui et ne relligio vesana furensq*ue*
Armaret Batavos in sua fata duces.
Aut minor est forsan mundi iactura rubentis
Germani hoc sidus Cæsaris esse potest.
Plebeio at nunquam colescit sanguine regni
Jus dubium, regum est sanguinis istud opus.
Sarmaticosq*ue* Ruthenosq*ue* inter cælum interituram
Pacem quod bello miscuit inde rubet.

On the Comet that appeared near the Great Bear in the Year of Our Lord 1618 for the duration of a month in November and December, which we saw in [Yaroslavl][1] among the Ruthenians.

---

[1] Translation not certain, but the suggestion that Yaroslavl is meant (a suggestion made originally by Edward Taylor, while working on my research project as a post-doctoral student) seems plausible, given that we know the party was travelling between Archangel and Russia at this time.

At that time when the Sarmatians[2] began their negotiations for peace
Having been thwarted by a French soldier[3] from taking
Those walls which, o man Michael[4] stood firm for you, though I may be wrong[5]
In which there was a not so fickle mob.[6]
There flashed out, borne around among the seven Northern stars[7]
A dreadful comet, sending forth a great torch.
We wondered what this flame portended, all
Of us fearful and uncertain; and each afraid on behalf of his own people.[8]
I myself feared for the English, [thinking] perhaps it might be a hostile omen
[Presaging] your fall, great James.[9]
And I was terribly afraid that mad, raging religious fervour
Might prompt the Dutch[10] leaders to fight even to the death.
Or perhaps it was some lesser loss for the reddening[11] world:
This star could be the star of the German Caesar.
But never is the uncertain question of a kingdom's rule

[2] *Samartians*: i.e., the Polish-Lithuanian forces. The Poles and Russians signed the Truce of Deulino on December 11th.

[3] This appears to refer to the fact that just before the Polish–Lithuanian forces besieged Moscow in October 1618, two French engineers delivered the plans for the siege to the Russian side. You could translate either "by a French soldier" (though in fact I think there were two) or "the French soldiery."

[4] This seems to be a reference to the siege of Moscow in October 1618, the last major event of the Polish-Muscovite War. Moscow was defended by Michael Romanov, the "Michael" apparently addressed here, and the Polish-Lithuanian attempt to take the city failed. James apparently sympathised with the Russian side.

[5] I'm not sure what the Latin *fallar* is doing here. It may just be a sort of aside, amounting to something like "if I'm not mistaken."

[6] *mobile vulgus*: strictly speaking, borrowed from Seneca, *Hercules Furens*, 170; but so common a tag in Latin of this period that I wouldn't really call it an allusion as such. The point here seems to be that the people in this case did not change sides.

[7] The more usual *Septentriones* (one word) is here divided *septem. .. triones*. The word refers literally to the seven stars near the North Pole, though it is often used in Latin just to mean "the north." Here, however, it clearly refers to the constellation in which the comet appeared. The division of the word and the placement of *inter* (among) between the two parts is perhaps intended to emphasise the appearance of the comet "amid" the constellation.

[8] Here, the Latin *quisque sibi, metuens* means literally "each fears for himself." In the context, though, as he goes on to specify his personal fears for the English people, *sibi* probably carries the force of "his own people" as well as "himself" (as it quite often does in Latin).

[9] James I of England, here addressed directly, as if he were present. Though there's perhaps a hint of a pun here, too, since the author's name is also James.

[10] Apparently a reference to the conflict in the Netherlands between the Arminians and Calvinists, which came to a head in the summer of 1618.

[11] "Reddening" I think as in "with blood" as Europe began to slide into what would be the Thirty Years War. Since this poem seems to have been written roughly at the time of the events it depicts, it's an interesting depiction of a contemporary writer sensing this slide towards war—a bit like some of the famous poems from the summer of 1939, perhaps.

Settled[12] by the people's blood—that's a job for the blood[13] of kings.
The heaven is reddening rather because it has mixed with war[14]
The peace between the Poles and Russians that is about to collapse.

4.10 Conclusions

Why study poetry in publications inspired by the Great Comet of 1618? I could have posed this question at the outset, but the answer may have been taken as some form of special pleading. The contents of this chapter serve as an appropriate answer, but one must still explore why such a study has never been done before. To address it, I offer an analogy between comet poetry and the work of the Italian Giovanni Morelli (1816–1891; Fig. 4.19), who transformed art scholarship in the nineteenth century.

Fig. 4.19 Giovanni Morelli: detail from an 1886 portrait. Accademia Carrara. Courtesy of Wikimedia Commons, public domain. PD-1996

[12] The Latin *colescit* is an alternative spelling for the usual *coalescit*—to grow firm, cohere, unite. James is saying that civil war never solves anything, but also implying that a people will not rally around a non-royal leader (someone of "plebeian blood")

[13] Here, the "blood of kings" means, I think, not their death but the importance of royal lineage. The political view is conventional for the period.

[14] Here, the (slightly strained) metaphor is I think of how blood stains water (or perhaps milk or something white)—in any case, the idea is that the night sky is "reddening" with the comet because it presages war.

Here I quote from a book review by Jonathan Beckman (2022: 39), deputy editor of the Economist's *1843* magazine.

> Art criticism, he [Morelli] thought, suffered from a serious failing: no one was actually looking
> at the pictures. Academic art historians turned instead to archives for biographical materials or documents that would pin down authorship. Philosophers of art looked beyond the painting for the idea that preceded it, scarcely concerned with the subject depicted. And connoisseurs—critics, collectors, aesthetes—ascribed a painting to a particular artist based on its 'general impression', as though the essence of Botticelli rose off the surface like a scent.

The scores of comet publications that contain poetry did so for a reason, but no one has been looking at them. Historians of astronomy have looked at the mathematical content of the publications, philosophers have put the stir created by the Great Comet of 1618 into context by looking at what preceded it in the sixteenth century, and those who collect seventeenth-century pamphlets have lovingly catalogued them and kept them safe for future posterity. But no one was *actually reading* the poetry and trying to understand why the author either wrote or commissioned a poem related to the comet. Even otherwise excellent books on the cultural history of comets give poems little attention (and no serious study), despite Hegel's maxim that poetry is the "instructress of humanity." (quoted in Verene, 2021: 16). For example, Schechner Genuth (1997) includes a few poetic abstracts, but none on the comet of 1618; and a book on the comets of 1618 by Drake and O'Malley (1960) does not engage with poetry at all. A rare example of engagement is a study of seventeenth-century poetics in science by Claire Preston (2015), but while she does mention the comets of 1664–1665, the comets of 1618 are ignored. The Austrian philosopher Ludwig Wittgenstein best encapsulated it as Denk nicht, sondern schau! [Don't think, but look!). On the bright side, several scholars are now publishing studies of astronomical poetry from the sixteenth, seventeenth, and eighteenth centuries; these include Luis Miguel Carolino (2015), Per Pippin Aspaas (2016), Irina Tautschnig (2022), and Burgio (2023). Nonetheless, important poetry such as *Cometae* by Camillo Garulli (1777) and *Psychathanasia* by Henry More (1647) remains virtually unstudied as astronomical poetry. The difficulty in performing an anatomy of the poetry in this chapter is analogous to what Robert Burton wrote in 1621 (Note 6), which is certainly a key reason for the paucity of such studies.

Sigmund Freud saw in Morelli a kindred spirit, who divined "secret and concealed things from despised or unnoticed features, from the rubbish heap, as it were, of our observations." Nothing was more concealed than the "heavenly script" that King James warned against trying to read. This also had echoes in the nineteenth century, in the work of August Strindberg (1849–1912), whose celestographs contained "a kind of 'heavenly script,' an example of what he saw as nature's artistic striving form: an inherent cipher, hard to understand but still hopeful, like a glimpse of a great other, 'boundless' totality." (Feuk, 2001).

The extraordinary number of observations of the Great Comet of 1618, evidenced by the largest ever suite of publications about a comet which generated "too much factual information," altered the existing linguistic system of cometary poetry, to put the matter in a framework formulated by the nineteenth century French

Fig. 4.20 Constantijn Huygens in a 1641 portrait by Michiel van Mierevelt. Huygensmuseum Hofwijck, Voorburg. Courtesy of Wikimedia Commons, public domain

theoretical physicist and historian of science Pierre Duhem (Agassi, 1975). One can trace the change in the linguistic system of cometary poetry quite clearly. Aside from the specific details of the plays by Fletcher, the two masques by Jonson, and the poems by (or inspired by) King James, there had never before in England been such a comet-mediated hodology between these disparate forms of artistic expression. The fact that they are supercharged with great matters of state (the Spanish Match and the death of a Queen) collectively elevates them to a unique place in the history of astronomy. More amusingly, one is reminded of the sonnet *The Happy Encounter*, written by Walter de la Mare between 1901 and 1918. It depicts Science as a myopic hog, being "stroked and refreshed by Poetry until they are eventually able to peer into each other's eyes and rejoice in all they have in common." (Ford, 2022: 35).

On the Continent, even in late-seventeenth-century poetry, the effects of the Great Comet of 1618 reverberated. On 20 April 1681, the father of the famous Dutch astronomer Christiaan Huygens, Constantijn, completed a lengthy poem entitled *Cometen-werck* [finally published in Worp (1898: 259-262)]. In this 90-line poem, Constantijn Huygens (1596–1687; Fig. 4.20) questioned the traditional idea that comets were celestial portents or signs of divine anger. The poem was prompted by the appearance of the Great Comet of 1680–81, which made its appearance in November 1680. In December, Christiaan wrote to his father about the comet: “From some time one spoke of a comet... It appeared of surprising size, with a very long and conspicuous tail. I have never seen a comet of this strength, and you will be able to compare it to that of 1618.”

> It is remarkable that his father, old and retired to his Hofwijck, had concern for such phenomena. The son’s tone is of a man perfectly convinced that his father will as a matter of course observe the comet’s progress and be ready to offer such comparisons as he might between the Great Comet and that of 1618…it is something of a surprise that Constantijn Huygens should respond to the comet of 1618 just as he did, with *Cometen-werck*. (Colie, 1955: 196)

As the Great Comet of 1618 faded to oblivion in the early winter days of 1619, it left Europe with a sprinkling of cosmic dust in the form of poetry whose depths of meaning still have not been fully revealed four centuries later.

4.11 Notes

1. Maclean (1987) brought attention to a magazine article (Anon, 1858) that states the king’s poem had never before been in print. At the conclusion of the Maclean article, Howse (1987) writes: “The version of the poem given here is that in State Papers Domestic James I 14/104/16v. It is also published in Craigie, James (ed.), *The Poems of James VI of Scotland* II (unpublished and uncollected poems), uncollected poem no. x, p.173; and comments on pp. 255-256. Interestingly, another copy in British Library, MS Rawlinson Poet. 84,f.72 has the title *King James on the blazeing starr: octo: 28:1618*. There are also five copies in the Bodleian, discussed in the Bodleian Quarterly Record iii, no. 32 supp. 3, (1920-22). See also Notes & Queries Third Series, iii, 324 (1863).”

   Other known sources: *James VI and I* 2.172; Bodleian MS Ashmole 38, p. 45; Bodleian MS Eng. Poet. e.14, fol. 77v; Bodleian MS Eng. Poet. e.97, p. 11; Bodleian MS Malone 19, p. 39; Bodleian MS Smith 17, p. 141; Bodleian MS Tanner 465, fol. 81r; BL Add. MS 3910, fol. 29v; BL Add. MS 30982, fol. 21r; BL MS Harley 791, fol. 61r; BL MS Harley 1221, fol. 75r; BL MS Harley 6038, fol. 19r; CCRO MS CR 63/2/19, fol. 19v; Nottingham MS Portland PW V 37, p. 174; PRO SP 14/104/16; Folger MS V.a.162, fol. 31r; Folger MS V.a.345, p. 143 (see Marotti, 2016); Huntington MS HM 198, 1.29; Rosenbach MS 239/22, fol. 26v.

   For details on the manuscript poems of King James, see May (2008).
2. While there were no daily newspapers, the English printer John Wolfe (d. 1601) pioneered monthly publications that “can almost be regarded as England’s first printed newspapers.” (Dunthorne, 2013: 14) Begun in 1590, they contained reports from France, the Low Countries and Germany. They evolved into 24-page weekly newsbooks in the 1620s, much to the consternation of King James (Sharpe, 2006).

3. "Campion's meter is indeed elegiac couplets, and perfectly classical. The pentameter, the word we all use, is only five feet by virtue of addition, in reality the first two and a half feet of the dactylic hexameter (that meter being truly six feet), the last foot being catalectic, i.e., one element of the foot omitted, so just two syllables: –uu|–uu|–||–uu|–uu|–||. As with the first four feet of the dactylic hexameter, so the first two dactyls (but not the second two) may contract, giving two spondees, as in line 2 (– –| – – |–||) or one, as in line 4 (–uu|– –|–||). The position of the name Calliodorus, is the favored position for the vocative, as in Martial, the man's friend. Lines 1 and 3 are quite spondaic: da sp sp sp da (6) and sp sp da sp da (6) respectively." (Thomas, 2024)
4. Many of these rhetorical figures were well-known to seventeenth-century scholars, having been included in a popular text by the Jesuit scholar Cypriano Soarez (1562) that was printed 129 times between 1562 and 1700; translated into English in Moss and Wallace (2003). All the figures identified here were contemporaneous, and employed by Shakespeare (Pogue, 2009). My descriptions of each figure derive from these two sources.
5. The Heriots were also closely linked to the Royal Court: from another branch of the family, George Heriot (1563–1624) was goldsmith to King James I and Queen Anne. George's son James (d. 1634) succeeded as Court jeweller. After he died, his widow Elizabeth Jossey married David Cunningham of Robertland in 1637. David corresponded about the match with his cousin Sir David Cunningham of Auchenharvie, ancestor of the author of this chapter, who holds in his collection the original book of accounts of Prince (later King) Charles, written by Sir David in 1637.
6. "Doubt not but in the end you will say with [Burton], that to anatomize this [poetry] aright…is as great a task as to reconcile those chronological errors in the Assyrian monarchy and find out the quadrature of the circle…as great a trouble as to perfect the motion of Mars and Mercury, which so crucifies our astronomers, or to rectify the Gregorian calendar." (Burton, 1621: 17).

Acknowledgments Thanks to Svorad Zavarsky, Roger Ceragioli, Paul Gabor, Lorenzo Smerillo, and Joachim Schloer, for Latin translation, and James Lequeux for French translation. Gratitude to James Doelman for the text of the Basse poem. Thanks also to James Knowles for his insights into the meaning of masques, and to Peter Usher for his help on transcribing the Lapworth poem. For the translation of the difficult Aytoun and Richard James poems, I must thank both Victoria Moul (who offered valuable comments on my entire chapter, and the footnotes on the James poem) and Bryan Whitchurch, who also provided the following textual commentary on Aytoun; and Patrick Boner for his translation of Latin lines by Hassam and Campion. And a poetic thanks to the Harvard professors James Engell and Richard Thomas for their detailed analysis of the epigram by Campion. Thanks to the British Library, which allowed me to handle Thomas Harriot's original letters. Finally, I must acknowledge the continuing support of Dr. Milan Matt Emanuele for the past 20 years, without whom this chapter, and this book, would have never been written.

References

Agassi, J. (1975). When should we ignore evidence in favour of a hypothesis? In *Science in flux. Boston studies in the philosophy of science* (Vol. 28, pp. 127–154). Springer.

Agricola, R. (1479). *Published in 1515: De inventione dialectica*. Theod. Martinus Alost.

Aït-Touati, F. (2010). Thinking the cosmos in the seventeenth century: Fiction, hypothesis and astronomy from Kepler to Huygens. *Annales. Histoire, Sciences Sociales, 65*(2), 325–344.

Albury, W. R. (1978). Halley's ode on the principia of Newton and the epicurean revival in England. *Journal of the History of Ideas, 39*(1), 24–43.

Anon. (1618). *The muses welcome to the high and Mightie Prince James by the grace of god king of Great Britain*. Thomas Finlason.

Anon. (1858). The two comets—1618, 1858. *Literary Gazette* 1 (New Series), 570. 30 October issue.

Apt, A. (1982). *The reception of Kepler's astronomy in England: 1596–1650*. D.Phil thesis, St. Catherine's College.

Aspaas, P. (2016). Astronomy, Latinity, enlightenment: Niels Krog Bredal's poems commemorating the transits of Venus, 1761 and 1769. *Symbolae Osloensis, 90*, 2-5-234.

Atkins, G. D. (2012). *Reading T.S. Eliot*. Palgrave Macmillan.

Avni, A. (1969). *The bible and romanticism*. Mouton.

Bachelard, G. (1953). *Le matérialisme rationnel*. PUF.

Bagg, R. (1969). The electromagnet and the shred of platinum. *Arion, 8*(3), 407–429.

Bainbridge, J. (1618). An astronomicall description of the late comet from the 18. of Novemb. 1618. To the 16. of December following. Henry Fetherstone.

Barbour, R. (1998). *English epicures and stoics: Ancient legacies in early Stuart culture*. University of Massachusetts Press.

Barton, W. (2022). The poetry of Jeremiah Horrocks's *Venus in sole visa* (1662): Astronomy, authority, and the 'new science'. *Perspectives on Science, 30*(6), 982–1004.

Beacham, R. (1999). *Spectacle entertainments of early imperial Rome*. Yale University Press.

Beckman, J. (2022). Just look at them. *London Review of Books, 44*(2), 39–40.

Bellany, A., & Cogswell, T. (2015). *The murder of King James I*. Yale University Press.

Bennett, J. A. W., & Trevor-Roper, H. R. (Eds.). (1955). *The poems of Richard Corbett*. Clarendon Press

Berg, H. (2011). *Military occupation under the eyes of the Lord: Studies in Erfurt during the thirty years war*. Vandenhoeck & Ruprecht.

Biester, J. (1996). Admirable wit: Deinotēs and the rise and fall of lyric wonder. *Rhetorica, 14*(3), 289–331.

Bloemendal, J. (2020). *Daniel Heinsius, Auriacus, sive Libertas saucis +(orange, or liberty wounded)*. Brill.

Bloomfield, M. (2016). *Archaeopoetics: Word, image*. University of Alabama Press.

Boyle, R. (1886). Beaumont, Fletcher and Massinger. In Eugen Kolbing (Ed.), *Englische Studien* (pp. 209–239). Gebr. Henninger.

Brandt, G. (1721). *History of the reformation* (Vol. 2). John Childe.

Brann, N. L. (2002). *The debate over the origin of genius during the Italian renaissance*. Brill.

Brazgovskaya, E., & Kholmanskikh, J. (2014). On the notion of nothing: J. Wilmot and Metaphysics in images. *European Journal of Science and Theology, 10*(6), 29–36.

Brinckmair, L. (1638). The warnings of Germany by wonderfull signes, and strange prodigies seene in divers parts of that countrey of Germany, betweene the yeare 1618. And 1638. In *Together with a briefe relation of the miserable events which ensued. All faithfully collected out of credible high Dutch chronicles, and other histories*.

Brown, M. A. (2010). *The Pleiadic Age of Stuart Poesie: Restoration Uranography, Dryden's Judicial Astrology, and the Fate of Anne Killigrew*. PhD thesis, Georgia State University.

Browne, T. (1658). *Pseudodoxia Epidemica,,, whereunto is added the garden of Cyrus*. Edward Dod.

Browning, R. (2005). "To serve my purpose": Interpretative Agency in George Wither's *a collection of Emblemes*. In Y. Bruce (Ed.), *Images of matter* (pp. 47–71). University of Delaware Press.
Bullen, A. H. (1884). *The tragedy of Sir John van olden Barnavelt*. Martinus Nijhoff.
Burgio, B. (2023). British physico-theological poetry and Newtonian physics: The use of *Principia Mathematica* (1687) in sir Richard Blackmore's *creation* (1712). *English Literature, 9*, 5–24.
Burton, R. (1621). *The anatomy of melancholy*. Henry Cripps.
Butler, M. (1993). Jonson's news from the New World, the "running masque," and the season on 1619–20. *Medieval & Renaissance Drama in England, 6*, 153–178.
Butler, M. (2007). George Chapman's 'masque of the twelve months' (1619). *English Literary Renaissance, 37*(3), 360–400.
Caldari, V. (2015). *The end of the Anglo-Spanish match in global context, 1617–1624*. PhD thesis, University of Kent.
Calvino, I. (1988). *Six memos for the next millennium*. Harvard University Press.
Camargo, M. (2004). Time as rhetorical Topos in Chaucer's poetry. In S. D. Troyan (Ed.), *Medieval rhetoric: A casebook* (pp. 91–108). Taylor & Francis. [Passage quoted is in Book 1.iv of *The Institutio Oratoria* of Quintilian].
Campbell, L. B. (1935). The Christian muse. *The Huntington Library Bulletin, 8*, 29–70.
Campion, T. (1601). *Booke of Ayres*. 'The man of life upright', the second song in Book 1. London.
Capp, B. (1972). *The fifth monarchy men: A study in seventeenth-century English millenarianism*. Faber & Faber.
Carey, H. M. (1992). *Courting disaster: Astrology at the English court and University in the Later Middle Ages*. Martin's Press/Springer.
Carolino, L. M. (2015). A poetical critique of Aristotle: The role of cosmological poetry in late-renaissance Portugal. *Portuguese Studies, 31*(1), 7–24.
Cats, J. (1619). *Aenmerckinghe op de tegenwoordige steert-sterre: Met aenwijsinghe vande rechte wetenschap om alle teykenen des Hemels, ende vreemde Gesternten wel ende loffelijck uyt te leggen. (Remarks on the present tailed star: with use of correct science to properly and excellently explain all celestial signs and strange constellations)*. no publisher given.
Cats, J. (1623). Bedenckinghe op de steert-sterre, gheisen int iaer ende gheduydet op de vergaderinghe van de synode, ghehouden tot Dordrecht. In *Zeeusche Nachtegael* (pp. 230–231).
Chambers, R. (1830). *The life of King James the first, in 2 volumes*. Constable and Co.
Chambers, R. (1858). *Domestic annals of Scotland: From the reformation to the revolution* (Vol. 2). W. & R. Chambers.
Chambers, E. K. (1908). The laws of candy. In A. H. Bullen (Ed.), *The works of Francis Beaumont and John Fletcher* (Vol. 3, pp. 465–544). George Bell and Sons.
Chapman, A. (2008). Thomas Harriot: The first telescopic astronomer. *Journal of the British Astronomical Association, 118*(6), 315–325.
Clarke, E. (2011). *Politics, religion and the song of songs in seventeenth-century England*. Palgrave Macmillan.
Closs, V. (2020). *While Rome burned*. University of Michigan Press.
Colie, R. L. (1955). Constantijn Huygens and the rationalist revolution. In *Tijdschrift voor Nederlandse Taal- en Letterkunde* (Vol. 193). Brill.
Colie, R. L. (1957). *Light and enlightenment: A study of the Cambridge Platonists and the Dutch Arminians*. Cambridge University Press.
Collins, S. L. (1989). *From divine cosmos to sovereign state: An intellectual history of consciousness and the idea of order in renaissance England*. Oxford University Press.
Considine, J. (2022). *Sixteenth-century English dictionaries*. Oxford University Press.
Cooper, T. (1863). King James I and the comet of 1618. *Notes and Queries, 3*, 417.
Corbet, R. (1647). *Certain elegant poems*. William Crook.
Corbet, R. (1672). Poems. In *Corrected and enlarged* (3rd ed.). William Crook.
Crabb, G. (1858). *The book of knowledge*. Leavitt & Allen.
Crystal, D. (2016). *The Oxford dictionary of original Shakespearean pronunciation*. Oxford University Press.

Csonka, F. (1990). *Opera quae exstant omnia: Poetica of Ioannes Bocatius*. Akadémiai Kladó.

Cunningham, C. J. (2016a). *Early investigations of Ceres and the discovery of Pallas*. Springer.

Cunningham, C. J. (2016b). Milton's paradise lost: Previously unrecognized allusions to the Aurora borealis, and a solution to the comet conundrum in book 2. *Renaissance and Reformation, 39*, 5–33.

Cunningham, C. J. (2023). The evolution of 'meteor' as an astronomical trope 1560-1760. In S. Gullberg & P. Robertson (Eds.), *Essays on astronomical history and heritage: A tribute to Wayne Orchiston on his 80th birthday* (pp. 341–366). Springer.

Cunningham, C. (2024). Tycho's Conversation with Urania, and other engagements with the Muse. *Journal of Astronomical History and Heritage 27*(1), 105–126.

D'Ewes, S. (1851). *College life in the times of James the first, as illustrated by an unpublished diary of Sir Symonds D'Ewes*. John W Parker and Son.

Daly, P., & Silcox, M. (1989). William Marshall's emblems (1650) rediscovered. *English Literary Renaissance, 19*(3), 346–374.

Davies, J. (1603). *Microcosmos*. Joseph Barnes.

De Cruz, H. (2024). *Wonderstruck: How wonder and awe shape the way we think*. Princeton University Press.

Dekoninck, R. (2016). The Jesuit *Ars* and *Scientia Symbolica*: From Richeome and Sandaeus to Masen and Ménestrier. In W. de Boer, K. Enenkel, & W. Melion (Eds.), *Jesuit image theory*. Brill.

Descartes, R. (1654). Chapter 9: On the origin and the course of the planets and comets in general; and of comets in particular. In *Le Monde de M. Descartes ou le Traite de la Lumiere (The World)*. Theodore Girad.

Di Salvo, G. (2023). The blazing star in renaissance drama. In D. Rosvally & D. Sherman (Eds.), *Early modern liveness* (pp. 196–216). The Arden Shakespeare.

Doelman, J. (2000). *King James I and the religious culture of England*. D.S. Brewer.

Doelman, J. (2011). Two poems on the death of the Duke of Lennox and Richmond. *Opuscula: Shorts Texts of the Middle Ages and Renaissance, 1*(6), 1–10.

Doelman, J. (2016). *The epigram in England, 1590–1640*. Manchester University Press.

Doelman, J. (2017). "The daring pen of sorrow": Elegies on the death of Thomas Murray (1564–1623). *The Seventeenth Century Journal, 34*(1), 27–63.

Doelman, J. (2021). *The daring muse of the early Stuart funeral elegy*. Manchester University Press.

Doran, S. (2024). *From Tudor to Stuart: The regime change from Elizabeth I to James I*. Oxford University Press.

Drake, S., & O'Malley, C. D. (1960). *The controversy on the comets of 1618*. University of Pennsylvania Press.

Dunthorne, H. (2013). *Britain and the Dutch Revolt, 1560–1700*. Cambridge University Press.

Effe, B. (1977). *Dichtung und Lehre: Untersuchungen zur Typologie des antiken Lehrgedichts [poetry and teaching: Studies on the typology of the ancient teaching poem]*. C.H. Beck.

Eliot, T. S. (1920). *Poems*. A. A. Knopf.

Engell, J. (2024). *Private communication*. June 20.

Evelyn, J. (1901). *The diary of John Evelyn* (Vol. 11). M. W. Dunne.

Eyre-Todd, G. (1895). Forth feasting. In *Scottish poetry of the seventeenth century* (Vol. 5, pp. 193–206). William Hodge.

Fauber, L. S. (2019). *Heaven on Earth*. Pegasus Books.

Feuk, D. (2001). The Celestographs of August Strindberg. *Cabinet* 3, summer issue.

Firth, C. H. (1911). The ballad history of the reign of James I. *Transactions of the Royal Historical Society, 5*, 21–61.

Fix, A. C. (1991). *Prophecy and reason: The Dutch Collegiants in the early enlightenment*. Princeton University Press.

Fleay, F. G. (1891). *A biographical Chronicle of the English Drama 1559–1642* (Vol. 1). Reeves and Turner.

Ford, M. (2022). Thee, thou, twixt. *London Review of Books, 44*(6), 35–38.

Fowler, A. (1970). *Triumphal forms: Structural patterns in Elizabethan poetry*. Cambridge University Press.
Fowler, C. (2023). Delineating the sea. In S. Whiteman (Ed.), *Landscape and authority in the early modern period* (pp. 197–214). University of Pennsylvania Press.
Frade, S. (2023). *Heracles and Athenian propaganda*. Bloomsbury.
Freeman, R. (1978). *English emblem books*. Octagon Books.
Fuchs, B. (2013). *The poetics of piracy: Emulating Spain in English literature*. University of Pennsylvania Press.
Gardiner, S. R. (1869). *Prince Charles and the Spanish marriage: 1617–1623* (Vol. 1). Hurst and Blackett.
Garulli, C. (1777). *Hypothesis Copernicana, Cometae et Elegiarum Monobiblos*. G. Salomoni.
Gellar-Goad, T. H. M. (2020). *Laughing atoms, laughing matter: Lucretius' De Rerum Natura and satire*. University of Michigan.
GHDI (German History in Documents and Images). (2023). https://germanhistorydocs.ghi-dc.org/sub_document.cfm?document_id=3709
Gianoutsos, J. A. (2021). *The rule of manhood: Tyranny, gender, and classical republicanism in England, 1603–1660*. Cambridge University Press.
Gibbon, E. (1788). *Decline and fall of the Roman Empire* (Vol. 4, p. 325). A. Strahan.
Gifford, W. (1813). *The plays of Philip Massinger, 4 vols*. G. and W. Nicol.
Gifford, W. (1869a). News from the new world discover'd in the Moone: A masque, as it was presented at court before King James. In *The works of Ben Jonson* (pp. 614–617). Routledge, Warne and Routledge.
Gifford, W. (1869b). The masque of augers. In *The works of Ben Jonson* (pp. 630–634). Routledge, Warne and Routledge.
Golding, A. (1580). *A discourse upon the earthquake*. Henry Binneman.
Goldstein, A. J. (2017). *Sweet science: Romantic materialism and the new logics of life*. University of Chicago Press.
Goudriaan, A., & van Lieburg, F. (Eds.). (2011). *Revisiting the Synod of Dordt (1618–1619)*. Brill.
Graham, D. (2016). Claude-Francois Ménestrier: The founder of 'Early Modern Grounded Theory'. In W. de Boer, K. Enenkel, & W. Melion (Eds.), *Jesuit image theory*. Brill.
Grantley, D. (1988). Masques and murderers: Dramatic method and ideology in revenge tragedy and the court masque. In C. Bloom (Ed.), *Jacobean poetry and prose* (pp. 194–212). St. Martin's Press.
Green, M. A. E. (1858). *Calendar of state papers, domestic series, of the reign of James I 1611–1618*. Longman, Brown, Green, Longmans and Roberts.
Grosart, A. (Ed.). (1874). *The complete works of George Herbert: Verse* (Vol. 2 of 3). Robson.
Grotius, H. (1609). Mare liberum. English translation. In D. Armitage (Ed.), *The free sea*. Liberty Fund.
Guthrie, W. G. (1964). The astronomy of Shakespeare. *Irish Astronomical Journal, 6*(6), 201–211.
Hackenbracht, R. (2019). *National Reckonings: The last judgment and literature in Milton's England*. Cornell University Press.
Halliwell-Phillipps, J. O. (1841). *A collection of letters on scientific subjects, 44*. Historical Society of Science, R. and J.E. Taylor.
Harding, T. (1852). *The decades of Henry Bullinger. The fifth decade*. The University Press.
Harris, T. (2014). *Rebellion: Britain's first Stuart kings, 1567–1642*. Oxford University Press.
Harrison, S. J. (2023). Maffeo Barberini's poems for the Farnese family in early baroque Rome. In J. Glomski, G. Manuwald, & A. Taylor (Eds.), *Baroque Latinity* (pp. 121–135). Bloomsbury.
Hassam, J. T. (1896). *The Hassam family*. Privately printed.
Hassam, J. T. (1903). Ezekial Cheever. *The New England Historical and Genealogical Register, 62*, 40–50.
Heilbron, J. L. (2021). *The ghost of Galileo*. Oxford University Press.

Herman, P. C. (2001). Authorship and the royal "I:" King James VI/I and the politics of monarchic verse. *Renaissance Quarterly, 54*(4), 1495–1530.

Hervey, M. F. S. (1921). *The life correspondence & collections of Thomas Howard earl of Arundel.* Cambridge University Press.

Heywood, T. (1641). *Life of Merlin.* J. Okes.

Horrocks, R. (2012). Jeremiah Horrocks, astronomer and poet. *Journal of the Royal Society of New Zealand, 42*(2), 113–120.

Howell, J. (1645). *Epistolae ho-Elianae; or, familiar letters, domestic and foreign.* Humphrey Moseley.

Howse, D. (1987). Comments appended to the article 'King James' poem on the great comet of 1618' by A. Maclean. *Journal of the British Astronomical Association, 97*(2), 74.

Hoye, J. M. (2024). *Sovereignty as a vocation in Hobbes's Leviathan.* Amsterdam University Press.

Hussey, E. (1991). The beginnings of epistemology: From Homer to Philolaus. In E. Hussey (Ed.), *Epistemology, companions to ancient thought* (Vol. 1). Cambridge University Press.

Jackson, T. (1622). *Six sermons on Jerem. 7.16.* I. Haviland.

James, V. I. K. (1584). *The Essayes of a Prentise, in the divine art of Poesie.* Thomas Vautroullier.

James, V. I. K. (1597). *Daemonology, in forme of a dialogue.* Robert Waldegrave. see also Tyson (2011).

James, V. I. K. (1598). *The true law of free monarchies.* Robert Waldegrave.

James, H. (2005). Ovid and the question of politics in early modern England, in Yvonne Brice (Ed.), *Images of matter* (pp. 92–122). University of Delaware Press.

Jardine, B. (2024). The life mathematick. In P. Beeley & C. Hollings (Eds.), *Beyond the learned academy* (pp. 387–407). Oxford University Press.

Jeffrey, R. (2018). *Princess Elisabeth of Bohemia.* Lexington Books.

Jensen, D. (2006). *The science of the stars in Danzig from Rheticus to Hevelius.* Doctoral dissertation, University of California, San Diego.

Joby, C. (2015). *The Dutch language in Britain (1550–1702).* Brill.

Johnson, S. (1810). *The works of the English poets* (Vol. 5 of 21). J. Johnson.

Johnson, W. R. (1993). *Horace and the dialectic of freedom: Readings in epistles 1.* Cornell University Press.

Johnson, C. (2004). "Periwigged heralds": Epistemology and intertextuality in early American Cometography. *Journal of the History of Ideas, 65*(3), 399–419.

Johnson, A. W. (2007). Four faces of Apollo: Divination, music, cosmology and healing in Ben Jonson's masque of augurs (1622). In P. Curry & A. Voss (Eds.), *Seeing with different eyes: Essays in astrology and divination* (pp. 173–208). Cambridge Scholars Publishing.

Johnson, A. W. (2009). Jonson's Eirenic community: The case of *the masque of augurs.* In R. D. Sell & A. W. Johnson (Eds.), *Writing and religion in England, 1558–1689* (pp. 169–193). Taylor & Francis.

Jonson, B. (1604). *B (Jon: His part of king James his Royall and magnificent entertainment through his honorable Citte of London, Thurseday the 15th.Of March.1603*). Edward Blount.

Jonson, B. (1692). *The works.* H. Herringman.

Jorink, E. (2010). *Reading the book of nature in the Dutch golden age, 1575–1715.* Brill.

Jue, J. (2006). *Joseph Mede (1586–1638) and the legacy of millenarianism.* Springer.

Kafantaris, M.A. (2014). *Royal marriage and the politics of transition in Stuart drama 1603–1630.* PhD Dissertation, Ohio State University.

Kalff, S. (2017). Comets–celestial objects in the emblem tradition of the late seventeenth and early eighteenth century. In K. Enenkel & P. Smith (Eds.), *Emblems in the natural world. Intersection* (Vol. 50, pp. 319–350). Brill.

Kasi, D. (2024). *The politics of grace in early modern literature.* Stanford University Press.

Kepler, J. (1619). *Harmonices mundi [Harmony of the worlds].* Godofredi Tampachi.

Killeen, K. (2023). *The unknowable in early modern thought: Natural philosophy and the poetics of the ineffable.* Stanford University Press.

Knowles, J. (2015). *Politics and political culture in the court masque*. Palgrave Macmillan/Springer.
Koethe, J. (2022). *Thought and poetry*. Bloomsbury.
Kronk, G. W. (2021). A new and improved orbit for comet C/400 F1. *Journal of Astronomical History & Heritage, 24*(3), 688–696.
Kurihara, K. (2016). *Celestial wonders in Reformation Germany*. Routledge.
Kurland, S. M. (2013). Shakespeare and James I: Personal rule and public responsibility. In A. J. Power & R. Loughnane (Eds.), *Late Shakespeare, 1608–1613* (pp. 209–224). Cambridge University Press.
Lacrymae Cantabrigensis: In obitum Serenissimae Reginae Annae. (1619). Cambridge University.
Lewis, N. D. (2020). Hellenistic astronomy in early Christianities. In A. C. Bowen & F. Rochberg (Eds.), *Hellenistic astronomy: The science in its contexts* (pp. 551–571). Brill.
Lindley, D. (Ed.). (1984). *The court masque*. Manchester University Press.
Lipman, A. (2024). *Squanto: A native odyssey*. Yale University Press.
Lofft, C. (1812). *Aphorisms from Shakespeare*. Longman, Hurst, Rees, Orme and Brown.
Lund, M. A. (2017). Reading. In J. Lee (Ed.), *A handbook of English renaissance literary studies*. John Wiley & Sons.
Maclean, A. (1987). King James' poem on the great comet of 1618. *Journal of the British Astronomical Association, 97*(2), 74.
Mann, J. (2012). *Outlaw rhetoric: Figuring vernacular eloquence in Shakespeare's England*. Cornell University Press.
Manning, R. (2018). The Jacobean peace: The irenic policy of James VI and I and its legacy. *Quidditas, 39*, 148–189.
Markham, G. (1595). *The most honourable tragedie of Sir Richard Grenville*. I. Roberts.
Marotti, A. (2016). Christ Church, Oxford, and beyond: Folger MS V.a.345 and its manuscript and print sources. *Studies in Philology* 113(4), 850-878.
Marotti, A. (2021). *The circulation of poetry in manuscript in early modern England*. Routledge.
Martin, J. J. (2022). *A beautiful ending: The apocalyptic imagination and the making of the modern world*. Yale University Press.
Massinger, P. (1844). *The virgin martyr*. James Burns.
Masson, D. (1893). Ben Jonson in Edinburgh. *Blackwood's Edinburgh Magazine, 154*(938), 790–804.
May, S. W. (2008). The circulation in manuscript of poems by King James VI and I. In J. Dutcher & A. L. Prescott (Eds.), *Renaissance historicisms* (pp. 206–224). University of Delaware Press.
McCLure, N. E. (1939). *The letters of John Chamberlain* (Vol. 2). American Philosophical Society.
McCoog, T. (2020). Converting a King: The Jesuit William Crichton and King James VI and I. *Journal of Jesuit Studies, 7*, 11–33.
McHugh, H. (2002). Translation of ode 11, book 1. In J. D. McClatchy (Ed.), *Horace: The odes, page 43*. Princeton University Press.
McNally, J. R. (1967). Rudolph Agricola's *De Inventione Dialectica Libri Tres*: A translation of selected chapters. *Speech Monographs, 36*, 393–422.
Mendelssohn, M. (1997). In D. Dahlstrom tr (Ed.), *Philosophical writings* (p. 179). Cambridge University Press.
Ménestrier, C.-F. (1684). *L'Art des emblêms*. Robert J.B. de la Caille.
Metheun, C. (1999). The teachers of Johannes Kepler: Theological impulses to the study of the heavens. In *Sciences et Religions: de Copernic à Galilée* (pp. 183–203). de l'Ecole Française de Rome.
Möller, N., Hecht, L., Niu, R., Liebchen, B., & Palberg, T. (2023). Writing into water. *Nano-micro. Small, 19*(49), 1.
Moss, J., & Wallace, W. (2003). *Rhetoric & dialectic in the time of Galileo*. The Catholic University of America Press.

Moul, V. (2021). England's Stilicho: Claudian's political poetry in early modern England. *International Journal of the Classical Tradition, 28*, 23–50.
Moul, V. (2022). *A literary history of Latin & English poetry: Bilingual verse culture in early modern England.* Cambridge University Press.
Mulsow, M. (2022). *Knowledge lost: A new view of early modern intellectual history* (Tr. H.C. Erik Midelfort). Princeton University Press.
Nature. (1878). *Our astronomical column: The comets of 1618* (pp. 247–248).
Navarro-Brotons, V. (2007). Astronomy and cosmology in Spain in the seventeenth century: The new practice of astronomy and the end of the Aristotelian-scholastic cosmos. *Cronos, 10*, 15–40.
Newton, I. (1687). *Philosophiæ Naturalis principia Mathematica*. The Royal Society.
Nichols, J. (1828). *The progresses of King James the first, 4 volumes*. J. B. Nichols.
Nora, P. (1989). Between memory and history: Les Lieux de Mémoire. *Representations, 26*, 7–24.
North, D. (1645). *A forest of varieties*. Richard Cotes. https://quod.lib.umich.edu/e/eebo/A52444.0001.001?rgn=main;view=fulltext
Oestreicher, C. (2007). A history of chaos theory. *Dialogues in Clinical Neuroscience, 9*(3), 279–289.
Olive, P. (2021). Red herrings and perceptual filters: Problems and opportunities for Aeschylus's Supplices. *Arethusa, 54*(1), 1–29.
Olson, R., & Pasachoff, J. (1998). *Fire in the sky: Comets and meteors, the decisive centuries, in British art and science*. Cambridge University Press.
Orchard, T. N. (1913). *Milton's astronomy: The astronomy of Milton's 'Paradise Lost'*. Longman's, Green, & Co.
Ossa-Richardson, A. (2019). *A history of ambiguity*. Princeton University Press.
Owain, G. (1863). King James I and the comet of 1618. *Notes and Queries, third series, 3*, 323–324.
Oxford University. (1619). *Academiae Oxoniensis Funebria sacra aeternae memoriae serenissimae Reginae Annae potentissimi monarchae Iacobi ... sponsae dicata*. Oxford University Press.
Park, K., & Daston, L. (1998). *Wonders and the order of nature: 1150–1750.* Princeton University Press.
Parker, G. (2013). *Global crisis: War, climate change, & catastrophe in the seventeenth century.* Yale University Press.
Patterson, W. B. (1997). *King James VI and I and the reunion of Christendom.* Cambridge University Press.
Perry, C. (2002). "If proclamations will not serve": The late manuscript poetry of James I and the culture of libel. In D. Fischlin & M. Fortier (Eds.), *Royal subjects: Essays on the writings of James VI and I* (pp. 205–232). Wayne State University Press.
Pesic, P. (2022). *Sounding bodies: Music and the making of biomedical science*. MIT Press.
Pfeiffer, K. L. (1994). Suggestiveness or interpretation: On the vitality of appearances. In W. Pape & F. Burwick (Eds.), *Reflecting senses* (pp. 15–32). Walter de Gruyter.
Pliny. (1855). *The natural history of Pliny* (J. Bostock & H.T. Riley, Eds.) (pp. 55–56). Henry G. Bohn.
Pogue, K. E. (2009). *Shakespeare's figures of speech*. iUniverse.
Preston, C. (2015). *The poetics of scientific investigation in seventeenth-century England.* Oxford University Press.
Price, D. (1613). *Spiritual odours to the memory of Prince Henry*. Oxford University Press.
Pritchard, T. (2025). News, Memory and the Culture of the Stuart Intervention into the Thirty Years' War, 1624–1630. Brill.
Pumfrey, S., & Dawbarn, F. (2004). Science and patronage in England, 1570–1625. *History of Science, 42*, 137–188.
Puttenham, G. (1589). *The Arte of English Poesie*. Richard Field.
Quistorp, H. (1955). *Calvin's doctrine of the last things*. John Know Press.
Raylor, T. (2000). *The Essex house masque of 1621*. Duquesne University Press.

Rickard, J. (2007). *Authorship and authority: The writings of James VI and I*. Manchester University Press.
Rickard, J. (2015). *Writing the monarch in Jacobean England*. Cambridge University Press.
Riklius, T. (2023). Innovation and fusion: Sarbiewski's theory of Baroque literary style. In J. Glomski, G. Manuwald, & A. Taylor (Eds.), *Baroque Latinity* (pp. 69–83). Bloomsbury.
Roche, J. J. (1985). Thomas Harriot's observations of Halley's comet in 1607. In J. D. North & J. J. Roche (Eds.), *The light of nature: Essays in the history and philosophy of science* (pp. 175–192). Martinus Nijhoff.
Roger, C. (1844). *The poems of Sir Robert Aytoun*. Adam & Charles Black.
Ronan, C. (1995). *"Antike Roman" power symbology and the Roman play in early modern England, 1585–1635*. University of Georgia Press.
Rukeyser, M. (1971). *The traces of Thomas Hariot*. Random House.
Russell, G. (2025). The Six Loves of James I. Atria.
Samson, A. (Ed.). (2006). *The Spanish match: Prince Charles's journey to Madrid, 1623*. Routledge.
Sarkar, M. (2012). *Cosmos and character in Paradise Lost*. Palgrave Macmillan.
Satterfield, S. (2016). The prodigies of 17 b.c.e. and the *Ludi Seaculares*. *Transactions of the American Philological Association, 146*(2), 325–348.
Schechner Genuth, S. (1997). *Comets, popular culture, and the birth of modern cosmology*. Princeton University Press.
Schendel, J. D. (2022). *The necessity of Christ's satisfaction*. Brill.
Scott, T. (1620). *Vox Populi, or Newes from Spayne*. no publisher given.
Scott, T. (1624). *The second part of Vox Populi*. Ashuerus Janss.
Sellin, P. R. (1986). The politics of Ben Jonson's Newes from the new world discover'd in the Moon. *Viator, 17*, 321–327.
Sennert, D. (1618). *Epitome naturalis scientaie*. Caspari Heiden.
Sharpe, K. (1993). The king's writ: Royal authors and Royal authority in early modern England. In K. Sharpe & P. Lake (Eds.), *Culture and politics in early Stuart England* (pp. 117–138). Stanford University Press.
Sharpe, K. (2006). Sacralization and demystification. The publicization of monarchy in early modern England. In J. Deploige & G. Deneckere (Eds.), *Mystifying the monarch* (pp. 99–115). Amsterdam University Press.
Shawcross, J. T. (2012). A contemporary view of Sir Walter Raleigh. *ANQ, 5*(3), 131–133.
Slowik, E. (2021). Descartes' physics. In E. N. Zalta (Ed.), *The Stanford encyclopedia of philosophy*. https://plato.stanford.edu/archives/win2021/entries/descartes-physics/
Soarez, C. (1562). *De arte rhetorica*. Ioannem Barrerium.
Steadman, J. M. (1987). *Milton and the paradoxes of Renaissance heroism*. Louisiana State University Press.
Stedall, J. A. (2003). *The Greate invention of algebra: Thomas Harriot's treatise on equations*. Oxford University Press.
Stephens, W. (2019). *The theology of Heinrich Bullinger*. Vandenhoeck & Ruprecht.
Streete, A. (2017). Elegy, prophecy, and politics: Literary responses to the death of Prince Henry Stuart, 1612-1614. *Renaissance Studies, 31*(1), 87–106.
Stubbs, C.E. (2011). *Abraham Fleming: Writer, cleric and preacher in Elizabethan and Jacobean London*. Doctoral thesis, Royal Holloway, University of London.
Svensson, A. (2019). 'And *Eden* from the *Chaos* rose': Utopian order and rebellion in the Oxford Physick garden. *Annals of Science, 76*, 157–183.
Swinford, D. (2013). Stellification and poetic ascent in the house of fame. *Modern Philology, 111*(1), 1–22.
Sylvester, J. (1614). *The parliament of vertues royal*. Humphrey Lownes.
Tautschnig, I. (2022). Constructing authority in the paratext: The poems to Johannes Hevelius' *Selenographia*. *Perspectives on Science, 30*(6), 1005–1041.
The Observatory. (1910). From the St James's Gazette for 16 April. *The Observatory, 33*, 226.

Theibault, J. (1994). Jeremiah in the village: Prophecy, preaching, pamphlets, and penance in the thirty years' war. *Central European History, 27*(4), 441–460.

Thomas, R. (2024). *Personal communication*. June 20.

Totaro, R. (2016). *The plague epic in early modern England*. Routledge.

Trevor-Roper, H. (1965). *Crisis in Europe, 1560–1660*. Routledge & Kegan Paul Ltd..

Trinacty, C. (2012). The fox and the bee: Horace's first book of epistles. *Arethusa, 45*, 57–77.

Twisck, P. (1624). *Comeet-Boecxken. Zijnde een corte Chronijcsche beschrijvinge van alle de grouwelijcke ende schrickelijcke Cometen die haer aen den Hemel vertoont hebben*. Zacharias Cornelisz.

Tyson, D. (2011). *The demonology of King James I*. Llewellyn.

van Miert, D. (2018). *The emancipation of biblical philology in the Dutch Republic, 1590–1670*. Oxford University Press.

Van Nouhuys, T. (1998). *The age of two-faced Janus: The comets of 1577 and 1618 and the decline of the Aristotelian world view in the Netherlands*. Brill.

Veith, D. M. (Ed.). (1962). *The complete poems of John Wilmot, Earl of Rochester*. Yale University Press.

Verene, D. P. (2021). *The rhetorical sense of philosophy*. Cornell University Press.

Verity, A. W. (1893). *Paradise Lost*. Cambridge University Press.

Vermij, R. (2011). Johannes Phocylides Holwarda and the interpretation of new stars in the Dutch Republic. In P. Boner (Ed.), *Change and continuity in early modern cosmology* (pp. 135–152). Springer.

Vermij, R., & Vieth, P. (2022). Confessionalization and comets. John Bainbridge on the comet of 1618. *Annals of Science, 79*(3), 275–291.

Verweij, S. (2014). "Booke, go thy ways": The publication, reading and reception of James VI/I's early poetic works. *Huntington Library Quarterly, 77*(2), 111–131.

Vickers, B. (1988). *Defense of rhetoric*. Oxford University Press.

Vivian, P. (1909). *Campion's works*. Clarendon Press.

Vorstius, C. (1619). *The judgement of the synode holden at Dort, concerning the five articles*. John Bill.

Wallington, N. (1869). *Historical notices of events chiefly in the reign of Charles I* (Vol. 1). Richard Bentley.

Walsham, A. (1999). *Providence in early modern England*. University of Oxford Press.

Walter, D. L. (1890). Medallic memorials of the great comets. VI. The comets of the year 1618. *American Journal of Numismatics, 24*, 84–87.

Waterhouse, E. (1994). *Painting in Britain 1530 to 1790*. Yale University Press.

Westcott, A. (1911). *New poems of James I of England*. PhD thesis. Columbia University, New York.

White, A. D. (1887). *A history of the doctrine of comets*. A paper read before the American Historical Association, Saratoga, 10 September 1885. In *Papers of the American Historical Association 2(2)*. G.P. Putnam's Sons.

Williams, R. (1848). *The court and times of James the first, 2 vols*. Henry Colburn.

Wilson, A. (1653). *The history of Great Britain, being the life and reign of King James the first*. Richard Lownds.

Wither, G. (1628). *Britain's Remembrancer*. John Grismond.

Wither, G. (1635). *A collection of emblemes, ancient and moderne*. Richard Royston. Reprinted in 2019. Good Press.

Wood, M. (2020). Bitten by a snake. *London Review of Books, 42*(10), 20–22.

Worden, B. (1993). Ben Jonson among the historians. In K. Sharpe & P. Lake (Eds.), *Culture and politics in early Stuart England* (pp. 67–90). Stanford University Press.

Worp, J. A. (1898). *De gedichten van Constantijn Huygens* (Vol. 8 of 9). Wolters.

Woudhuysen, H. R. (1996). *Sir Philip Sydney and the circulation of manuscripts 1558–1640*. Oxford University Press.

Wright, D. (2000). The astronomy in Pepys' diary. *A&G, 41*, 4.23–4.27.

Clifford J. Cunningham's Ph.D. in history of astronomy dealt largely with William Herschel and the asteroids. He is Research Fellow in the Astrophysics Group at the University of Southern Queensland (Australia). He has written or edited 18 books on history of astronomy including a 5-volume history of early asteroid studies published by Springer. He is a Series Editor of Springer's Historical & Cultural Astronomy books, and Associate editor of the Journal of Astronomical History and Heritage. Asteroid (4276) was named Clifford in his honour in 1990. His discoveries include identifying Charles Burney Jr. as the person who created the word 'asteroid' in 1802; showing Manilius (not Hipparchus) introduced the numerical six-magnitude system; and recognising a description of the aurora borealis in Milton's Paradise Lost which had eluded scholars for 350 years.

Chapter 5
The Astronomical Charts of the Jesuits: The Table of *De Tribus Cometis Years MDCXVIII Disputatio Astronomica*

Giancarlo Truffa

5.1 Introduction

One of the most important cartographic documents testifying the observations of the three comets appeared in 1618 is constituted by the table included in the *Disputatio Astronomica,* the short booklet published anonymously in February 1619 in Rome, but immediately attributed to the mathematics teacher of the Roman College, the Jesuit Orazio Grassi (1583–1654), a text which began the controversy that involved Galileo, his student Mario Guiducci and Grassi himself for several years. Ottavio Besomi and Mario Helbing published a detailed analysis of the table in their edition of Galileo/Guiducci's *Discourse on the Comets,*[1] which also included the publication of the Latin text of the *Disputatio astronomica* with the Italian translation, but it is useful to reconsider this table adding further details to better define its contents.

The table can be divided into five boxes (Fig. 5.1), which I have indicated with the letters A, B, C, D, E.

[1] O. Besomi and M. Helbing, Annotazioni alla "De tribus cometis disputatio", in Besomi and Helbing (2002), pp. 444–446.

G. Truffa (✉)
Società Italiana degli Storici della Fisica e dell'Astronomia, Padova, Italy

C. J. Cunningham (ed.), *Cosmic Events*, Historical & Cultural Astronomy,
https://doi.org/10.1007/978-3-031-86950-1_5

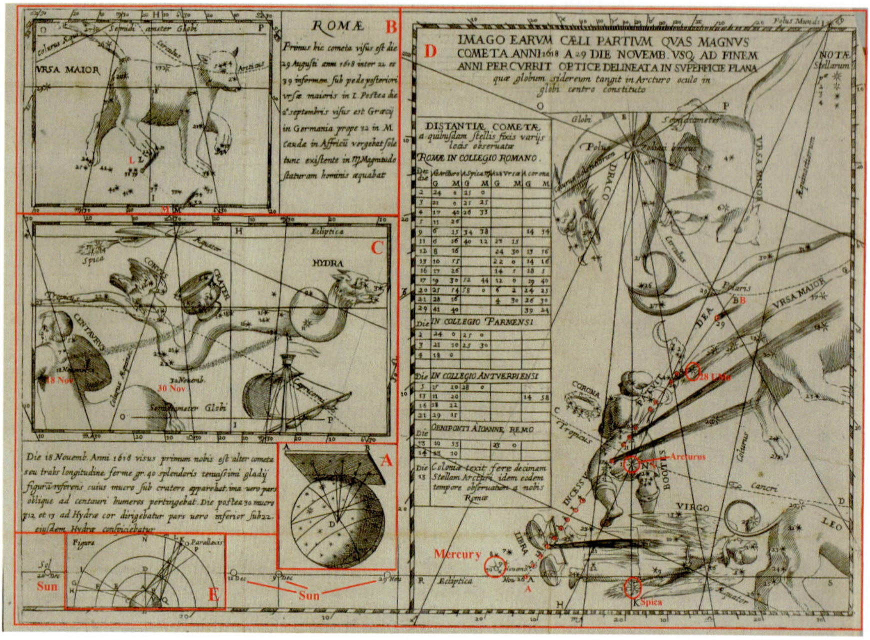

Fig. 5.1 The table in Orazio Grassi *Disputatio astronomica*, Rome, 1619 (public domain image)

5.2 The Projection Used

First of all, the gnomonic projection, used to construct the three celestial maps representing the places and motion of the three comets of 1618, although already known in antiquity, does not appear to have had great success among cartographers. As we will see, there are few examples that illustrate its use. Box A shows the scheme of the gnomonic projection. The principle used is to project the position of the stars on the surface of the celestial sphere starting from point D, center of the sphere of fixed stars, i.e., the center of the Earth, onto a plane AB tangential to the celestial sphere at point C. The main property of this projection is that the great circles of the celestial sphere, i.e., the Colures of the Solstices and Equinoxes, the Equator and the Ecliptic, are represented by straight lines. The other circles, the Tropics of Cancer and Capricorn and the Polar Circles, are portions of ellipses. Grassi owed the inspiration to use this type of cartographic projection to his teacher and colleague at the Collegio Romano, Christoph Grienberger (1561–1636), who used it for the celestial maps contained in the *Catalogus veteres affixarum longitudines, ac latitudines conferens* cum *nouis. Imaginum Caelestium prospectiua duplex. Altera rara ex polis mundi, in duobus hemisphaeriis aequinoctialibus* ..., Rome 1612, even if, as we will see, Grassi differs from it in various details.

The gnomonic projection is described, probably for the first time, by the Dutch mathematician, cartographer and instrument maker, Reiner Gemma Frisius (1508–1555) in his work *De astrolabio catholico*, printed in Antwerp posthumously in 1556, dedicated to the construction of a variant of the planispheric astrolabe which used the so-called universal Rojas projection. On pages 4r-5r the author provides the explanation of the gnomonic projection, accompanied by the illustration of its geometric scheme.[2] Recent research has determined that the first use of this projection in cartography is represented in the work of the German painter and engraver, Philipp Uffenbach (1566–1636), in his 1598 work *Bericht und Erklärung zweyer beygelegten künstlichen Kupfferstücken, oder Zeitweiser der Sonnen über die gantze Welt* [*Relation and explanation of two artificial pieces of copper or indicators of the time of the sun on the whole world*], explained the construction of a diptych sundial in which the two plates, in addition to the hour lines and other indications temporal, presented two geographical maps in gnomonic projection inserted in the publication as tables outside the text.[3] Subsequently, the Nuremberg mathematician, Franz Ritter (15??–1641), used the same projection[4] in a plate out of text of his work *Speculum solis, Das ist, Sonnenspiegel: Beschreibung und unterricht, derer in das Kupffer gestichenen Sonnenuhren, In welcher der gantze Lauff der Sonnen, unnd deß Mondten durch deroselben Schatten zu sehen ist* [*Speculum solis*, i.e. *solar mirror: description and instruction of sundials engraved in copper, in which the entire course of the sun and moon can be seen through the same shadow*].

In the astronomical field,[5] the first to have used the gnomonic projection seems to have been Johannes Kepler, to create the celestial map that represented the position of the stella nova which appeared in 1604 in the constellation of Ophiuchus, a table contained in *De stella nova in pede serpentarii et qui sub ejus exortum* de novo *iniit igneous trigono: libellus astronomicis, physicis, metaphysicis, meteorologicis & Astrologicos ...* [*Of the new star at the foot of the Serpentario and which, at its rising, entered a new fiery trine: a booklet full of astronomical discussions, physical, metaphysical, meteorological and astrological...*],[6] and immediately afterwards in the table representing the path of the comet that appeared in 1607, contained in the *Außführlicher Bericht Von dem diß 1607. Jahrs erschienenen Cometen* [*Detailed report on the comet that appeared in the year 1607*].[7] The Jesuit mathematician Christoph Grienberger was the first to produce a celestial atlas, i.e., a collection of celestial maps, using the gnomonic projection in the *Catalogus veteres*.

---

[2] Gemma Frisius (1556) Keuning, J. (1955), p. 7

[3] Folk, R. (2014)

[4] Snyder, J. P. (2007), pp. 379, 380

[5] Warner, D. J. (1979), pp. 99, 100, 135; Friedman (2007), p.117

[6] Kepler, J. (1606), table between pages 76 e 77

[7] Kepler, J. (1608), table outside text

5.3 The Maps

Box B shows the first comet, 1618 I,[8] in the constellation Ursa Major, depicted according to an image almost identical to the image in Grienberger's star atlas, second plate. Two observations, labelled L and M, are reported, as the caption next to the illustration reports:

> *ROME Primus hic cometa visus est die 29 augusti anni 1618 inter 22 et 39 informam sub pede posterior UMa in L. Postea die 2 septembris visus est Graecij in Germania prope 32 in M. Cauda in Affricum vergebat sole tunc existentente in Virgo. Magnitudo staturam hominis aequabat* [*This comet was seen for the first time in Rome on August 29 of the year 1618 in **L**, between the star 22 and the shapeless (star) 39 under the hind leg of the Big Dipper. Then on September 2 it was seen in Graz, Germany, near the star 22 in **M**. The tail was pointing towards Africo (the direction of the South West wind) and the Sun was in Virgo. The length is equal to the stature of a man*]

Box C shows comet 1618 III,[9] which was observed moving between the constellations Centaurus and Hydra. Two observations made on November 18 and 30 are drawn with the long tail of the comet, as the caption below reads:

> *Die 18 November anni 1618 visus primum nobis est alter cometa seu trabs longitudine ferme gr.40 splendoris tenuissimi gladii figuram referens cuius mucro sub cratere apparebat, ima vero pars oblique ad centauri humeros pertingebat. Die postea 30 mucro pars 12 et 13 ad Hydrae cor dirigebatur pars vero inferior sub 22 eiusdem Hydrae conspiciebatur* [*On November 18 of the year 1618 we saw for the first time another comet, or beam, of longitude close to 40 degrees, which it had the shape of a sword of faint luminosity, the tip of which appeared under the constellation of Vasus, while the lower part reached obliquely behind the Centaurus. Subsequently, on day 30 the tip, between (stars) 12 and 13, headed towards the heart of Hydra, while the lower part was visible under (star) 22 of the same Hydra*]

For a comparison with the representations of the constellations contained in this box, Centaurus, Raven, Vase, Hydra, and Ship, reference must be made to three different tables from Grienberger's star atlas (plates 23, 24, and 25). In this case we notice many differences in the illustrations, such as the heads of the Hydra, the Vase which has the shape of a crater in one case and of an amphora with two handles in the other, the Centaurus which is portrayed by Grienberger as a bearded man with long hair, while in Grassi he is beardless and with short hair, and the top of the mast of the Ship which in one case has the sail blown out by the wind and no sail in the other. Only the depiction of the Raven is similar in both illustrations.

Box D is the most detailed one. It presents the path of comet 1618 II[10] from November 26 to December 29 (even if the caption of the figure mentions a little different period, between November 29 to the end of the year) when the comet crossed the sky through the constellations of Libra, Bootes, Ursa Major and up to the Dragon. Again, the comparison of the drawing in Grassi with various plates in Grienberger (plates 1, 2, 4, 14, 15) demonstrates that most of the details of the

[8] Grassi, O. (1619), pp. 5–6; Besomi and Helbing (2002), pp. 258–261

[9] Grassi, O. (1619), p. 6; Besomi and Helbing (2002), pp. 260–261

[10] Grassi, O. (1619), pp. 6–7; Besomi and Helbing (2002), pp. 260–265

figures are similar but, for example, the figure of Bootes, which is seen from behind, looks in opposite directions and, in one case, he is dressed as a peasant with a cap on his head while in the other he appears to be dressed but bareheaded. The Coma of Berenice asterism, which appears in Grienberger's table of the constellation of Leo, and the constellation of the Hunting Dogs, which appears in the table of Bootes, are not present in Grassi's table.

The caption of the figure reads

> *Imago earum caeli partium quas magnus comet years 1618 to 29 November usque ad finem anni percurrit, optice outlined on the surface plana quae globum sidereum tangit in Arcturo oculo in globi centro constituto* [*Graphic representation of those parts of the sky that the great comet of the 1618 from November 29 to the end of the year; this graphic representation is obtained with optical projection on a plane tangential to the celestial sphere in the star Arcturus and placing the viewing point (the eye, the center of projection) in the center of the world*]

Compared to the other two boxes with the comets, this one has added a table reporting the observations of the angular distances in degrees and minutes of the comet from the stars: Arcturus (alpha Bootes), Spica (alpha Virgo), 28 UMa, 1 Corona Borealis, and made in different European places: the Roman College in Rome, the Jesuit College of Parma, the Jesuit College of Antwerp and by Remus, mathematician of Leopold of Austria, in Innsbruck. In the last row of the table the following indication is highlighted for the date 13 December 1618:

> *Die 13 Coloniae texit fere decimam stellam Arcturi; idem eodem tempore observatum a nobis Romae* [*The comet in Cologne almost completely obscured the tenth star of the constellation Bootes; the same day the same occultation was observed in Rome*]

Grassi underlined in the text how these two observations were clear evidence that the comet had no parallax and therefore must have been at a distance from the Earth much greater than that of the Moon.[11]

Close to the first observations of the comet at the end of November in the constellation of Libra, the position of Mercury appears for November 30th, in fact, the planet could be seen at dawn in the last week of November and in the first week of December of 1618. Kepler also remembered in *De cometis libelli tres* that he observed it on 21 November 1618 "*Mercurius apparuit in clara aurora clarissime*".[12] Some positions of the Sun also appear in boxes A and E (29 November, 9, 12, 20 December), which, however, are part of the representation of comet 1618 II. At the top right is the table for recognizing stellar magnitudes from the first to the fourth.

Grassi also used letters within the celestial map of this comet: A and B indicate the first and last day on which the comet was observed, so the segment AB represents the observed path of the comet, I the north pole of the celestial sphere, L the north pole of the ecliptic, N the star Arcturus (alpha Bootes), INH the meridian line passing through N, KNL the great circle of the ecliptic passing through N, CD the Tropic of Cancer, EG the Arctic Circle, RS the ecliptic. The equator line and the equinoctial colure line are also indicated. The choice of the gnomonic projection, in

[11] Grassi, O. (1619), p. 11; Besomi and Helbing (2002), pp. 274–277

[12] Kepler, J. (1619), in KGW, VIII, 183

which, as we said, the great circles are represented as straight lines, while the minor circles are represented as portions of ellipses, allowed Grassi to visualize very simply one of the arguments he intended to demonstrate, that is, that the trajectory of the comet was a great circle, in this case the segment AB, resulting from the conjunction of all the daily observations carried out at the Collegio Romano.[13]

5.4 The Schema for the Parallax

Box E illustrates the parallax calculation scheme, described by Grassi on pages 8–11.[14] The semicircle CAQ represents the northern hemisphere of the Earth, whose center is in point B; EDT is the semicircle representing the distance of 100 miles from the Earth's surface; FS the semicircle representing the distance to the Moon; HNP the semicircle of the fixed stars; the points C, A, Q represent three cities on the surface distant from each other, in particular A corresponds to Antwerp and C to Rome.

To explain how to determine the distance of a celestial body from the observer on Earth, Grassi first considers a body at point T on the EDT semicircle. Observed from A, the point T would appear on the sphere of the fixed stars in P, while observed from Q, it would appear at the point M. The parallax of the body T is given by the angular distance MP. Grassi then considers a body located at point S, at a distance from the Moon. In this case, S would appear in O if observed from A and in L if observed from Q, so the parallax of the body S is lower and, therefore, its distance from the Earth is greater.

5.5 Some Manuscript Pages

Besomi and Helbing (2002), in addition to the printed booklet, described one manuscript copy of the *Disputatio* in their edition of the text, with parallel Italian translation,[15] the manuscript Ges.458, ff.17r-23r, of the Central National Library in Rome. Grassi's text is bound with the other texts (five in total) read during some meetings in the Collegio Romano at the end of 1618 to discuss the nature of the comets. The texts are written by two different hands and Besomi–Helbing supposed they were assembled sometime later, taking Grassi's text from the print. The text also contains a copy of the table with the paths of the three comets. Apparently, this copy, made with a pen, is very similar to the printed version, but the editors

[13] Grassi, O. (1619), p. 13; Besomi and Helbing (2002), pp. 278–281

[14] idem, pp. 266–277

[15] Besomi and Helbing (2002), pp. 83–85

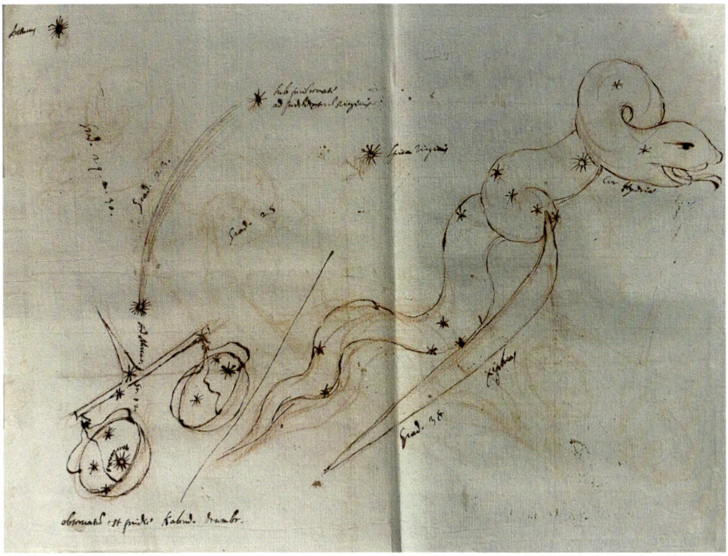

Fig. 5.2 The constellations Hydra and Libra, with some observation of C/1618/III. Sheet inserted in a copy of Grassi's *Disputatio astronomica* (Casanatense Library, Rome, Signature VOL MISC.2504 12)

identified several differences, the most important is the absence of the schema for the parallax in the handwritten copy.[16]

Unfortunately, there are no elements to establish whether this copy has been a preliminary copy made for the preparation of the print or if the copyist copied it from the print with some missing parts.

A very interesting copy of Grassi's work is preserved in another library in Rome, the Casanatense Library.[17] The *Disputatio* is bound in a miscellaneous volume which includes 15 short treatises on different subjects, a collection possibly assembled in the eighteenth century.[18]

Some handwritten pages have been inserted in the text of *Disputatio:*

- the drawings of the position of Comet 1618 III in the constellation of Hydra and of Comet 1618 II in the constellation of Libra on the same page (Fig. 5.2);
- the drawings of the positions and the tail of Comet 1618 II between November 29 and December 14 with some stars (Fig. 5.3);

[16] The other missed details are: the label "Circulus Arcticus" in box B; the label "Spica" in box C; several stars in box A; the table with star magnitudes, label "L" indicating the Zodiacal pole and number "29" for the position of the comet in box D.

[17] Signature VOL MISC.2504 12. A digital copy is available here https://books.google.it/books?id=3mgMb8OKHskC&hl

[18] A handwritten table of contents is on p. 705

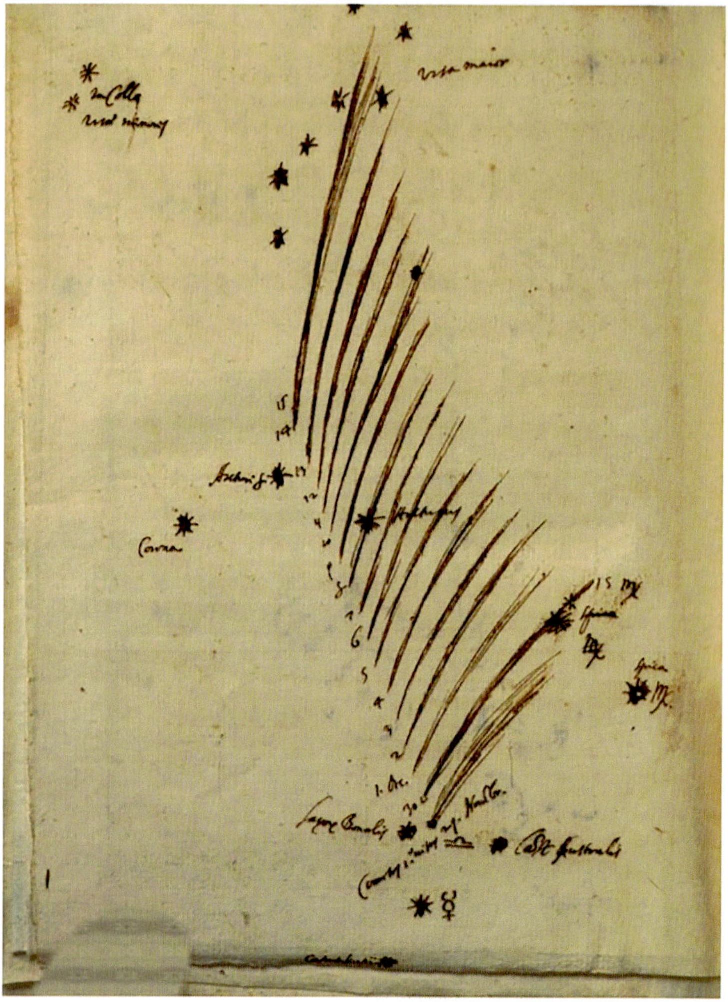

Fig. 5.3 The positions of the comet C/1618/II. Sheet inserted in a copy of Grassi's *Disputatio astronomica* (Casanatense Library, Rome, Signature VOL MISC.2504 12)

- the list of observed angular distances of Comet 1618 II from the stars Arcturus, Spica, Ursa Major and the Corona Borealis. This list, which covers the period from 30 November to 14 December, reports observations that would have been taken at the Collegio Romano (Figs. 5.4 and 5.5).

Unfortunately, this insert is anonymous and there is no indication to identify the author or the authors, since the hand seems to change between the drawings and the list of distances. The comparison with the printed table also reveals several differences.

Let us first consider Fig. 5.2, where the positions of both the 1618 II and 1618 III are drawn, divided by a transversal line, the comet 1618 II on the left, just above the constellation of Libra, the comet 1618 III, below the constellation of Hydra, on the

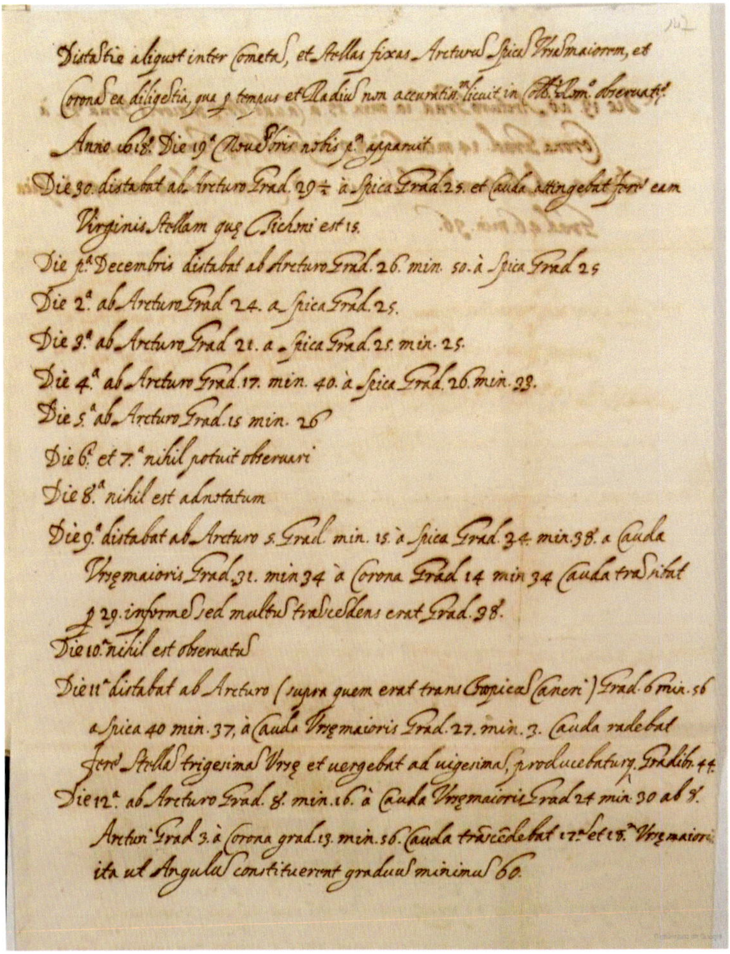

Fig. 5.4 List of the observations of comet C/1618/II in the Collegio Romano. Sheet inserted in a copy of Grassi's *Disputatio astronomica* (Casanatense Library, Rome, Signature VOL MISC.2504 12)

right. Several labels are present: in the bottom left, "observatio est primi Kalendae dicembr." could be the date of the observations for both the comets. Some stars are labelled: "Cor Hydrae," under the head of Hydra, "Spica Virginis," in the middle of the page, and "Arcturus," in the top left, "Mercurius" in the left scale plate of Libra. For comet 1618 II, the distances between the comet with Mercury "Ad Mercur. gr. 12," Spica, "grad. 25," and with Arcturus, "grad. 29 m. 30," are indicated; "grad. 23" is written along the tail, but it is not clear if it is referred to the length of the tail or to the distance of the comet with a star with label "sub perizomate ad?ermo destro virginis," i.e., the star 15 in the printed table and in Grienberger's star atlas. For comet 1618 III, the nature of the comet "xiphias" and the length of the tail "grad. 38" are transcribed. While the drawing of Libra is conventional, the head of Hydra is different from the print and in Grienberger's star atlas.

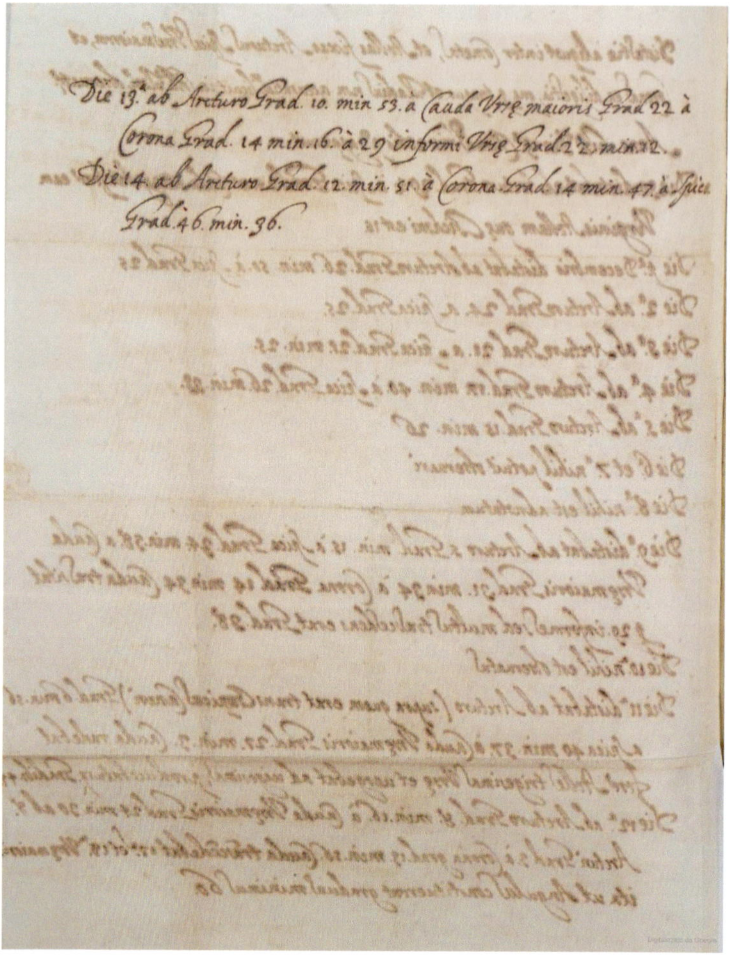
Die 13.ª ab Arcturo Grad. 10. min 53. à Cauda Ursæ maioris Grad 22 à
Corona Grad. 14 min. 16. à 29 informi Ursæ Grad. 22. min. 32.
Die 14. ab Arcturo Grad. 12. min. 51. à Corona Grad. 14 min. 47. à Spica
Grad. 46. min. 36.

Fig. 5.5 List of the observations of comet C/1618/II in the Collegio Romano. Sheet inserted in a copy of Grassi's *Disputatio astronomica* (Casanatense Library, Rome)

In Fig. 5.3 the positions and the tail of the comet 1618 II are drawn for each day from November 29 to December 14. The following stars are drawn and labelled: "Caput Scorpii," Mercury represented by its symbol, "Lans(?) Borealis," "Lans(?) Australis," "Spica" with the symbol of Virgo, "Spica" crossed by a line to delete it and substituted by "15" with the symbol of Virgo, "Arcturus," "Corona," "Aschenis," "Ursa Maior," "?Colla Ursa minoris," with the addition of the note "Cometa 1a visa 29 Novembris" at the beginning of the observations reported.

Figures 5.4 and 5.5 contain the list of observed angular distances from the relevant stars Arcturus (alfa BOO), Spica (alfa VIR), star 28 in UMa and Corona (alfa CRb). In Table 5.1 I reported the data from this list and the data from the table included in box D of the print. The two lists present several differences, both for the

Table 5.1 Comparison of the distances reported in the table in Box D of the print and the distances reported in the handwritten list of Figs. 5.4 and 5.5

Table in Box D									Data in ms									
Dec	Ab Arcturo		A Spica Vir		A 28 Ursae		A corona		Dec	Ab Arcturo		A Spica Vir		A 28 Ursae		A corona		Note
die	G	M	G	M	G	M	G	M	die	G	M	G	M	G	M	G	M	
																		Anno 1618 die 19 novembris nobis prima apparuit
									30 Nov	29	1/2	25						Cauda attingebat fere eam Virginis stellam quae Bichoni(?) est 15
									1 Dec	26	50	25						
2	24	0	25	0					2 Dec	24		25						
3	21	0	25	25					3 Dec	21		25	25					
4	17	40	26	33					4 Dec	17	40	26	33					
5	15	26							5 Dec	15	26							
9	6	15	34	88			14	34	9 Dec	5	15	34	38	32	34	14	34	Cauda transibat per 29 informe sed multus trascendens erat grad. 38
11	6	16	40	12	27	15			11 Dec	6	56	40	37	27	3			Cauda radebat fere stella trigesima Urse et vergebat ad vigesima, producebaturque gradibus 44

(continued)

Table 5.1 (continued)

Table in Box D									Data in ms									
Dec	Ab Arcturo		A Spica Vir		A 28 Ursae		A corona		Dec	Ab Arcturo		A Spica Vir		A 28 Ursae		A corona		Note
die	G	M	G	M	G	M	G	M	die	G	M	G	M	G	M	G	M	
12	8	16			24	30	13	56	12 Dec	8	16			24	30	13	56	Cauda trascendebat 17a et 18a Urse maioris ita ut Angulus constituerunt graduus minimus 60
13	10	55			22	0	14	16	13 Dec	10	53			22		14	16	
									14 Dec	12	51	46	36			14	57	
16	17	26			14	5	18	5										
17	19	30	52	44	12	0	19	40										
20	25	54	58	0	6	2	24	25										
21	28	16			4	30	26	30										
29	41	40					39	24										

days of the observations and for the data related to the same days, so it is clear the two lists are independent and the handwritten list represents a complement of the data in the printed list.

5.6 Conclusions

The Table included in Grassi's *Disputatio astronomica* is a good example of a very detailed cartographic representation of the three comets that appeared in 1618, but it attracted very little attention from the historians of astronomy and of celestial cartography due to the few existing copies of this small book. Galileo basically ignored this table in his reaction to Grassi's pamphlet, and later it was never mentioned, ending up forgotten.

References

Besomi & Helbing. (2002). Galileo Galilei e Mario Guiducci, *Discorso delle comete*, edizione critica e commento a cura di Ottavio Besomi e Mario Helbing. Antenore.

Folk, R. (2014). Uffenbach's "Zeitweiser" published 1598. *The Compendium. Journal of the North American Sundial Society, 21*, 4–13.

Friedman, A. F. (2007). Renaissance star charts. In D. Woodward (Ed.), *The history of cartography, volume 3.1, cartography in the European renaissance* (pp. 99–122). University of Chicago Press.

Gemma Frisius, R. (1556). *De astrolabio catholico …*, Anversa.

Grassi, O. (1619). (attrib.), *De tribus cometis anni MDCXVIII disputatio astronomica publice habita in collegio Romano Societatis Iesu.* ex Typ. Iacobi Mascardi. https://books.google.it/books?id=3mgMb8OKHskC&hl

Grienberger, C. (1612). *Catalogus veteres affixarum longitudines, ac latitudines conferens cum nouis. Imaginum Caelestium prospectiua duplex. Altera rara ex polis mundi, in duobus hemisphaeriis aequinoctialibus* .... B. Zanetti.

Hersbach, C. (1618). *Eigentliche verzeichnus dieses im jahr 1618 erscheinenden cometen, welcher lauss oder bewegung, duirch die zeichen des 8 himmels, alhie vor augen gestellet wirdt*, Colonia https://www.britishmuseum.org/collection/object/P_2000-0521-46

Kepler, J. (1606). *De stella nova in pede serpentarii et qui sub ejus exortum de novo iniit trigono igneo: libellus astronomicis, physicis, metaphysicis, meteorologicis & Astrologicis* ..., Pragae. Edited in *Johannes Kepler Gesammelte Werk* (Vol. I) Hrsg. von Max Caspar. C.H. Beck, 1938, pp. 149–356.

Kepler, J. (1608). *Außführlicher Bericht Von dem diß 1607. Jahrs erschienenen Cometen*, Hall in Sachsen. Edito in *Johannes Kepler Gesammelte Werk* (Vol. IV) Hrsg. von Max Caspar und Franz Hammer. C.H. Beck, 1941, pp. 57–76.

Kepler, J. (1619). *De cometis libellis tres*, Augustae Vindelicorum [Augsburg]: typis Andrae Apergeri; sumptibus Sebastiani Mylii, 1619. Edito in *Johannes Kepler Gesammelte Werk*, vol. VIII, Bearbeitet von Franz Hammer. C.H. Beck, pp.131–262.

Keuning, J. (1955). The history of geographical map projections until 1600. *Imago Mundi: The International Journal for the History of Cartography, 12*, 1–24.

Ritter, F. (1607). *Speculum solis, Das ist, Sonnenspiegel: Beschreibung und unterricht, derer in das Kupffer gestochenen Sonnenuhren, In welcher der gantze Lauff der Sonnen, unnd deß Mondten durch deroselben Schatten zu sehen ist*. Christoff Lochner https://www.e-rara.ch/zut/content/titleinfo/1951396

Snyder, J. P. (2007). Map projections in the renaissance. In D. Woodward (Ed.), *The history of cartography, volume 3.1, cartography in the European renaissance* (pp. 365–381). University of Chicago Press.

Uffenbach, P. (1598). *Bericht und Erklärung zweyer beygelegten künstlichen Kupfferstücken, oder Zeitweiser der Sonnen über die gantze Welt*. Paul Brachfeldt. Retrieved May 30, 2024, from https://books.google.it/books?id=SElbAAAAcAAJ

Warner, D. J. (1979). *The sky explored: Celestial cartography, 1500–1800*. A. R. Liss.

Giancarlo Truffa received a degree in Physics from the University of Pavia, after which he worked as software engineer and project manager. Upon retirement, he has dedicated his career to the history of astronomy, specifically the period between the Middle Ages and the Scientific Revolution, with focus on astronomical instruments, tables and star catalogues, celestial cartography and ancient horology. He is vice-president of HORA, the Italian Association of Ancient Horology.

Part II
The International Dimension

Chapter 6
The Comets of 1618 in Germany and Wilhelm Schickard's Manuscript Treatise

Miguel Á. Granada

6.1 Germany's Primacy in Renaissance and Early Modern Astronomy

Petrus Ramus (1515–1572)—the French reformer of the liberal arts and promoter in France of the mathematical disciplines, with special insistence on their usefulness—published in 1567 in Paris a *Prooemium mathematicum*, whose three books traced a history of the mathematical sciences, their development in ancient Greece, their decline in the dark ages and their restoration in Renaissance Europe. In 1569, these three books were reprinted in Basel as the first three of the *Scholarum mathematicarum libri unus et triginta*. There, in the second book, Ramus extolled the role of Germany in the restoration of the mathematical disciplines: "Ergo mathesis in Germania praecipue effloruit" (Ramus, 1569: 65). Ramus praised Regiomontanus (1436–1476) and the promotion of the mathematical arts in Nuremberg after his death[1]; he also extolled the appointment in Nuremberg and other cities of public teachers of mathematics who taught these disciplines not only to the learned but also to artisans who did not know the Latin language, thus making possible the

[1] Ramus, 1569: 65: "Ergo nobiles illi quondam in Graecia atque Aegypto artifices hodie nulli sunt: Regiomontani Noribergae supersunt. Senatus enim populusque Noribergensis operam dedit, ut perpetuos Regiomontanos haberet. Itaque Vernerus [Johannes Werner (1468–1522)] primum: deinde Sconeri pater [Johann Schöner (1477–1547)] et filius [Andreas Schöner, 1528–1590] Regiomontani animam deinceps excitarunt".

M. Á. Granada (✉)
University of Barcelona, Barcelona, Spain
e-mail: granada@ub.edu

C. J. Cunningham (ed.), *Cosmic Events*, Historical & Cultural Astronomy,
https://doi.org/10.1007/978-3-031-86950-1_6

emergence of "nobiles sine literis artifices."[2] Ramus also reviews the flowering of mathematics in German universities, giving pride of place to the work of Melanchthon at Wittenberg, which was continued by his disciples: Jacob Milichius, Erasmus Reinhold, Caspar Peucer, and others (Ramus, 1569: 66–67). Ramus found space to praise authors who are perhaps better known today for their astrological work, such as Siderocrates [Samuel Eisenmenger, 1534–1585][3] and, above all Cyprianus Leovitius (1524–1574), to whose famous *De coniunctionibus magnis* (Leovitius, 1564) he refers with admiration, extolling his prognostic of the future of the world in connection with the next great conjunction of Jupiter and Saturn in 1584 and the return of the fiery trigon.[4] Finally, Ramus recorded, full of admiration, the promotion of mathematical disciplines by German princes, with special mention of the Landgrave of Hesse-Kassel Wilhelm IV:

> Germany is especially fortunate and especially blessed for its learned and erudite princes. I was pleased to learn how much importance the princes of Hesse, Saxony, Austria attach to the study of mathematics. Landgrave Wilhelm of Hesse seems to have moved Alexandria to Kassel: Thus, he provided craftsmen to build the necessary instruments for the observation of the stars; thus, he delights in daily observations with the instruments provided, so that Ptolemy seems to have come from Egypt to Germany with his armillary spheres and his instruments.[5]

This testimony illustrates the primacy of Germany in the field of astronomy and cosmology in the sixteenth century, and specifically in its second half and the beginning of the seventeenth century. We verify it in the literature on celestial novelties from 1570 onwards: both novae (often taken for comets at the time; today considered as the galactic supernovae of 1572 in Cassiopeia and 1604 in Serpentarius; also the novae of 1596 in the Whale and 1600 in the Swan, today taken for variable stars) and comets, especially the one of 1577 and following, up to those of 1607 (a transit of Halley's comet) and those that interest us here: the three comets of 1618. This is also true in astrology and in the eschatological interpretation of celestial

[2] Ramus, 1569: 65–66: "Sed illud de civitate singulare [i.e., Nuremberg] est, atque apud omnes civitates praedicandum: Stipendium dare de publico mathematum professori non ei solum qui doctis et eruditis praelegat, sed ei quoque, qui vernacula lingua latinè graeceque ignaros opifices erudiat: hinc etiam nobiles sine literis artifices: imo mathematicae disciplinae etiam apud posteros doctores".

[3] Ramus, 1569: 67: "Siderocrates egregiis ingenii monimentis mihi percharus est". For Eisenmenger and his brief manuscript treatise on the nova of 1572 in Cassiopeia, see Granada, 2014.

[4] Ramus, 1569: 67: "Leovitius, cuius vaticinium illud tanquam de tripode mihi speciatim redditum, equidem sic accepi. Scribis enim, doctissime Leoviti, e magna illa syderum coniunctione, unde et purioris religionis aureum seculum brevi rediturum vaticinaris, artes quoque plurimas, quae adhuc abditae latuerant, emersuras, resque maximas modico impendio perfectum absolutumque iri".

[5] Ramus, 1569: 67: "Verumenimvero Germania [...] principibus doctis et eruditis praecipue fortunata, praecipue beata est. Iucundum mihi fuit didicisse quanto mathematum studio teneantur principes Hessiae, Saxoniae, Austriae. Guilielmus Landgravius Hessiae videtur Casellas Alexandriam transtulisse: Sic Cassellis artifices organorum observandis syderibus necessariorum instruxit, sic quotidianis per instructa organa observationibus oblectatur, ut Ptolemaeus ex Aegypto in Germaniam cum armillis et regulis venisse videatur". On Wilhelm IV and the astronomical program developed in Kassel, see Gaulke, 2007.

phenomena, an interpretation that enjoyed a very wide predicament in Germany because of the apocalyptic and eschatological expectations associated with the Protestant Reformation, of which Melanchthon and his disciples had elaborated a precise theory (Jerratsch, 2020: 103–116; Brosseder, 2004). We see an example in the years immediately preceding the appearance of the comets of 1618, in the Rosicrucian manifestos: the *Fama fraternitatis*, printed in Kassel in 1614, and the *Confessio fraternitatis*, printed in Kassel and Frankfurt the following year. The *Fama* referred to the new period of fiery trigon that had actually begun in 1603, with the great conjunction of Jupiter and Saturn in the fiery sign of Sagittarius, whose "perfection" at the end of September 1604 with the addition of Mars had coincided with the appearance of the nova at the foot of the Serpentarius, just above Sagittarius and the three planets in conjunction. The eschatological significance was echoed in the *Fama* in the following terms:

> Verily we must confess that the world in those days was already big with those great Commotions, laboring to be delivered of them, and did bring forth painful, worthy men, who brake with all force through Darkness and Barbarism [...]: and assuredly they have been the uppermost point in the fiery Trigon whose flame now should be more and more brighter, and shall undoubtedly give to the World the last Light.[6]

The *Confessio*, for its part, pointed out the message of renewal conveyed by the two novas of 1600 and 1604:

> yea, the Lord God hath already sent before certain Messengers, which should testifie his Will, to wit, some new Stars, which do appear and are seen in the Firmament in *Serpentario* and *Cygno*, which signifie and give themselves known to every one, that they are powerful *Signacula* of great weighty matters.[7]

And it was precisely in 1618 when an enthusiastic follower of the message of the Rosicrucians, Daniel Mögling (1596–1635; a native of Böblingen, in the vicinity of Tübingen, the birthplace of the Rosicrucian manifestos) wrote, under the pseudonym of Theophilus Schweighardt, the *Speculum Sophicum Rhodo-Stauroticum*, in which the famous image representing the "Collegium Fraternitatis" appears with the two novas of 1600 and 1604 illuminating the College, whose work is sustained and impelled by Jehovah and His wings (Psalms 17: 8; 91: 4) (Fig. 6.1).[8]

---

[6] Fama Fraternitatis, 1652: 9–10. For the German original text, see Fama fraternitatis: Das Urmanifest, 1998, 78: "Gewißlich wir müssen bekennen, daß die Welt schon damahls mit so grosser Commotion schwanger gangen und in der Geburt gearbeitet, auch sie so unverdrossene rühmliche Helden herfürgebracht, die mit aller Gewalt durch die Finsternuß und Barbareien hindurchgebrochen [...] und freylich der Spitz an dem *Trigono igneo* gewesen, dessen Flammen nun mehr je heller leuchtet und gewißlich der Welt den letzten Brand entzünden wird".

[7] Confessio Fraternitatis, 1652: 47. For the German original text, see Andreae, 1973: 38: "Es hat zwar Gott der Herr schon etliche Botschafften vorhergesandt, die von seinem Willen bezeugeten, nemblich etliche newe Sterne, so am Himmel in Serpentario und Cygno entstanden, welche denn als hoher wichtiger Sachen kräfftige Signaculae menniglich bezeugen und zu erkennen geben".

[8] Cf. Figure 6.1. The *Speculum* was printed in 1618 with no indication of place of publication or printer. See also Gilly, 2012; Bouman and van Heertum, 2014: 99–111. On Daniel Mögling, who had already published the previous year, under the *Pandora sextae aetatis sive speculum gratiae* and the same pseudonym, a first defense of the Society, see also note 38.

Fig. 6.1 Collegium Fraternitatis, from Daniel Mögling, MS. *Speculum Sophicum Rhodo-Stauroticum*, 1618. Courtesy of the Zentralbibliothek Zürich

6.2 The Three Comets of 1618

1618 saw the appearance of three comets: the first, visible in August and September, bears the astronomical identification C/1618 Q1; the second, visible in November and the beginning of December, is known as C/1618 V1 and designated as 1618 III. Both were barely visible for a few weeks and attracted the attention of only a small number of observers. The third and final comet, by far the superior and the main subject of attention throughout Europe, was a very great comet, visible from November to January, and is known as 1618 II and C/1618 W1 (Kronk, 1999, I: 333–341).

The comets appeared when the expectations of a reformation of the human world associated with the Rosicrucian manifestos were reaching a climax and when, in Bohemia, the Thirty Years War had just broken out, a war that would immediately spread throughout Germany, involving the two great European powers (Spain and France). It is not surprising, therefore, that, regardless of the importance that the comets (especially the third) could have in the field of astronomical–cosmological studies 10 years after the last comet was seen in 1607 (a transit of the Halley comet) and now that Galileo had revolutionized astronomy with the application of the telescope and the publication of the *Sidereus nuncius* in 1610, the appearance of the comets aroused in Germany a plethora of interpretations, both in Latin and common language, as well as in printed and manuscript treatises.

Logically, only a minority of the treatises among them were of a rigorously technical–mathematical nature, such as Kepler's *De cometis libelli tres*, in which Kepler (1571–1630) collected and extended the results achieved in his German tract on the comet of 1607; or that of the Dutch Willebrord Snel (1580–1626), the *Descriptio Cometae qui anno 1618 mense Novembri primum effulsit* (Leiden 1619, which included Christoph Rothmann's significant treatise on the comet of 1585) or, to mention a Catholic author and testimony of the science of the Society of Jesus, the treatise by Johann Baptist Cysat (1588–1657), who was one of the few to use the telescope and describes the results of the telescopic observations of the comet's head.[9] A large part of the literature, however, was of an astrological and eschatological character, in this case usually in the German language and addressed to a wide audience. This is the case of the *Kurtzer Discurs Vom Cometen und dreyen Sonnen, so am Ende des 1618 Jahrs erschienen sind, wie auch von der künfftigen Conjunction oder Zusammenkunfft aller Planeten im Krebs Anno 1622* (Alt Stettin 1619) by David Herlitz (1557–1636) or the extensive production of the chiliastic and enthu-

[9] Cysat, 1619: 72–75. On Cysat's treatise, see Schiebert, 2006: 316–325; Ribordy, 2023: 238–253.

siast Paul Nagel (ca. 1580–1624?) among which we note *Stellae prodigiosae seu cometae per oculum triplicem observatio et explicatio. Das ist: des newen Cometen und Wundersterns im October, November und December 1618 erschienen, wahrhafftige Deutung* (Nagel 1619).[10]

An interesting aspect of the German literature on the comet of 1618, specifically that produced by cosmologically innovative mathematicians and astronomers, is its critical reaction to the rearmament of traditional views on comets that had taken place in the first decade of the seventeenth century with the Aristotelian philosopher and Calvinist theologian Bartholomaeus Keckermann (1572–1609) as its fundamental advocate. In a series of works that enjoyed wide circulation and fortune, Keckermann (who nevertheless knew and admired the work of Tycho Brahe) defended the Aristotelian cosmological dualism, the celestial immutability, the doctrine of comets as inflammations in the sublunary world, and harshly criticized the interpretation of celestial novelties (novae and comets) as natural events taking place in the heavens.[11] Keckermann accepted the possibility that the novae were really new stars, but his interpretation of them as 'miracles' of divine omnipotence that transgressed (*de potentia absoluta Dei*) the very natural order freely created by God, allowed to continue preserving the Aristotelian natural philosophy and thus avoided the doctrinal chaos inevitably ensuing, as he feared, the collapse of Aristotelianism.

As far as comets were concerned, Keckermann criticized the mathematical technique (the measurement of the parallax of comets) with which astronomers, after Regiomontanus, intended to demonstrate—with the rank of necessity derived from incontrovertible mathematical proofs—the celestial location of comets. According to Keckermann, the astronomical observations on which the determination of parallax was made lacked credibility for three reasons: imprecision of the instruments of observation and contradictions in the measurements by the astronomers; refraction of the visual rays coming from the cometary phenomena when they pass through the regions of fire and air, which causes that "those celestial bodies cannot be observed in their true places"; the eccentricity of our place of observation on the surface of the Earth. Consequently, Keckermann was persuaded that the geometrical determination by the measurement of parallax could be neither precise nor certain. In order not to deprive astronomical science of all credit, Keckermann declared himself ready to accept that his criticism did not extend to the observation of true celestial bodies (stars and planets, including the miraculous novae), but concerned only the lights seen since 1577, which, as merely comets, were not celestial bodies, perfectly

[10] On the printed literature in Germany about the comet of 1618, see Gindhart, 2006. More recently and focusing on the printed literature in German language, see Jerratsch, 2020: 323–383. On Paul Nagel, see Penman, 2010. On other chiliastic prophets, see Penman, 2019.

[11] On Keckermann, see Gindhart, 2006: 249–257; Granada, 2016: 260–266; Granada, 2020.

spherical and moving uniformly, but sublunary and ephemeral meteorological phenomena.

Keckermann's criticism aimed to discredit and dismantle the pretensions of mathematicians to reach, by means of geometrical demonstrations, cosmological conclusions about the nature of the universe that questioned Aristotelian dualism and celestial immutability. In this way, he was trying to perpetuate the monopoly of celestial physics by natural philosophers of Aristotelian obedience. For their part, Keckermann's detractors defended the involvement of mathematics in the discovery of the essence of celestial substance, in short: the transformation of the so-called middle sciences or "mixed mathematics" into the language of natural science (to use Galileo's statement in the *Saggiatore* of 1623). Keckermann intended to limit mathematics to the rank of an auxiliary discipline, competent in purely accidental realms of reality, and to secure for peripatetic physics the authority in the knowledge of the substance of nature. On the contrary, common to the writings of innovative mathematicians and astronomers was to reduce the Aristotelian doctrine of comets (as expounded by Aristotle in *Meteorologica*, I, 7) to the rank of a mere conjecture and weak hypothesis, due to lack of experience and above all to the absence in Antiquity of the geometrical technique of determining parallax as the only sure means of establishing the distance of celestial phenomena. This is the case, leaving aside Kepler's great work, of treatises such as the following:

Isaac Habrecht, *Kurtze und gründliche Beschreibung Eines Newen ungewohnlichen Sterns oder Cometen [...] im November und December diß 1618. Jahr erschienen*, Strasbourg 1618;

Philip Müller, *De Cometa Anni M.DC.XVIII. Commentatio PhysicoMathematica specialis & generalis*, Leipzig 1619;

Johann Döling, *Discursus Mathematico-Physicus in quo de Cometis contra Aristotelicos, imprimis vero Bartholomaeum Keckermannum Dantiscanum pro Tychone Braheo [...] eorundem ortu, situ, loco, causis [...] disseritur. Ex occasione novi nuper in mense Novembri anni praeterlapsi conspecti Sideris caudati [...] subiectus*, Greifswald 1619;

Peter Crüger, *Uranodromus Cometicus: Ein außführlicher Tractat Vom grossen Cometen deß 1618 Jahrs*, Danzig 1619.

But it is not my aim to speak here about these works, which have been studied to a greater or lesser degree by Marion Gindhart and Anna Jerratsch in their books, but about one manuscript treatise, written in German and coming from the reformed (Lutheran) region of Württemberg and the University of Tübingen. Since I have recently presented the manuscript German treatise by Michael Maestlin (1550–1631), professor of mathematics and astronomy at this university (Granada, 2016) and a critical edition accompanied by an English translation has recently appeared (Maestlin, 2022), I intend to concentrate on another manuscript treatise, also written in German and unpublished, although it is accessible now online in the copy digitized by the Württembergische Landesbibliothek in Stuttgart.

6.3 Wilhelm Schickard and His *Cometen Beschreibung*

The author of this manuscript treatise is Wilhelm Schickard (1592–1635) and its title *Cometen Beschreibung In zwen underschidliche Partes abgetheilt, deren Erster Von denselbigen ins gemein: der Ander Von Allen Insonderheit, sonderlich aber denen dreÿ Jüngsten, Im Abgeloffenen 1618. Jahr erschienen, außführlich handelt.*[12] Schickard, a generation younger than Kepler and two generations younger than Maestlin, is a typical product of the system of education and study support for young people from poor families introduced in the Duchy of Württemberg after the adherence to the Reformation and the dissolution of the monasteries. The young scholarship holders were obliged to serve the Duchy within the ecclesiastical establishment and so Schickard, after studying arts at the University of Tübingen (1607–1613), where he had (like Kepler a generation earlier) Maestlin as his teacher of mathematics and astronomy, served as a deacon in the town of Nürtingen, just as Maestlin had done in the town of Backnang when the nova of 1572 appeared. Only Kepler escaped this fate, against his will and wishes, when he was appointed mathematician of the Protestant community in Graz (Styria) because of the suspicions about his orthodoxy harbored by his superiors at the university.

Indeed, Wilhelm Schickard's career looks like a replica of Maestlin's[13]: Just as Maestlin was appointed—given the mathematical competence he had shown with his treatises on the nova in Cassiopeia and the comet of 1577—professor of mathematics in Heidelberg (1580) and then in Tübingen (from 1583 until his death in 1631), Schickard was deacon in Nürtingen from 1614 and assistant for Hebrew at the *Stift* (Protestant seminary) in Tübingen. In 1617, he entered into a personal relationship with Kepler. In addition, the castle of Nürtingen was the place of residence of the widows of the ducal house, and there resided for many years Duchess Ursula of Württemberg (1572–1635; born Ursula of Pfalz-Veldenz-Lützelstein), widow of Duke Ludwig since 1593. Schickard was in contact with her, who agreed to be the godmother of his first two children.

When the great comet appeared in November 1618, Schickard printed in December or January 1619 for the Duchess an *Einblattdruck* (a broadsheet print)

[12] WLB, Cod. math. qt. 43 (https://digital.wlb-stuttgart.de/purl/bsz307044173). I intend to publish the transcription of Schickard's treatise in a future volume with an introduction and a running commentary. When this essay was already completed, we have had access, thanks to the courtesy of Dr. Friedrich Seck, to Frankenberger, 2012. This valuable work offers, in addition to a preliminary exposition of the genesis and content of Schickard's treatise, a complete transcription of it accompanying, on facing pages, the reproduction of the manuscript.

[13] On Schickard's biography, see Seck, 1978. About the educational program in the reformed Württemberg, see Methuen, 1998: 29-59.

entitled *Presentis cometae motus in coelo*, whose only preserved copy at the Württembergische Landesbibliothek was destroyed during an air raid on Stuttgart in World War II.[14]

When this comet appeared, Schickard was in negotiations with the University of Tübingen to be appointed professor. His good relations in Nürtingen with the dowager duchess Ursula made it possible for her to show the above-mentioned print to the duke Johann Friedrich (1582–1628) on one of her visits to Stuttgart. The result was the transmission (on February 18, 1619) of the wish for Schickard to write a report on the comets and present it to the duke. The manuscript of *Cometen Beschreibung* is the result of this and of Schickard's efforts to turn the ducal request into a springboard for his promotion: a few months later, he was appointed professor of Hebrew at the University of Tübingen. His desire to become a professor of mathematics was blocked by the longevity of his teacher, Maestlin. Only on Maestlin's death in 1631 did he achieve this appointment, which he could only enjoy for a few years, since the plague, the usual companion of the Thirty Years War, took his life and that of his entire family (wife and children) in 1635.

If Maestlin's treatise on the 1618 comet—regardless of the theoretical value of its content, certainly very remarkable—is unfinished and in some respects in a first stage of writing, which is consistent with Maestlin's advanced age and his psychology at that time of his life,[15] Schickard's treatise is at the antipodes: it is a work in which Schickard put all his efforts, brought to completion in the short space of 2 months, written on quality paper in quarto format, with a beautiful calligraphy, and of considerable length (174 pages according to Schickard's running pagination). As an ornament, the paper was provided with a gold edge. The treatise offers to its addressee (the Duke of Württemberg; the dedication is dated April 15, 1619) a general theory of comets, including a thorough explanation of the present comet.[16] This theory is the subject of the first part of the treatise (pp. 8r-69r according to the pencil pagination), consisting of the following 10 chapters, each of which is divided in turn into a greater or lesser number of sections, which unfortunately do not appear in the general index (p. 7v):

1. On the matter of the comet (*Von des Cometen Materj worauß er gemacht werd*), p. 8r.

[14] Seck, 1978: 22, 395 (Anhang 1, 1618.12); Schickard, 2002: vol II, 456; see the bibliographical reference ibid., p. 493, n° 93.

[15] Maestlin, 2022: 4–12. Arguably, it was Maestlin's failure to finish and present his treatise to the Duke that facilitated the request to Schickard for his own treatise.

[16] For a first description of the contents of Schickard's treatise, see Mauder, 1995 and Frankenberger, 2012.

2. On its form and on the generation of its tail (*Von seiner Form, wie sein Schwaiff generirt werd*), p. 15r.
3. On its creation, who illuminates it (*Von seiner Erschaffung, wer in anzünde*), p. 22v.
4. For what purpose it is placed by God in the heavens (*Zu was End er von Gott an himmel gesezt sey*), p. 25v.
5. What are its effects (*Ob, was, und welcher gestalt er ettwas würkhe?*), p. 26v.
6. What is its height in the heavens and what is its disposition with respect to the Sun (*Wie hoch er im himmel droben stehe, von der Sonn abgewendt*), p. 33r.
7. What is their size and whether they are also invisible (*Wie groß sie seyen: und ob auch ohn sichtbare seyen*), p. 46r.
8. About its color, light, shape, and other qualities (*Von seiner farb, liecht, figur und andern qualiteten*), p. 51v.
9. On its course and its own movement (*Von seinem Gang, und Eigner Bewegung*), p. 59v.
10. On its end and destruction (*Von seinem End und der Zerstörung*), p. 66r.

There follows a second part (pp. 70r-94v) which gives a catalog of the comets that appeared from 1475 with indication of their celestial course and observational history. This second part culminates with the observational history of the comets of 1618 distributed in the following terms: first comet (pp. 79r–80v); second comet (pp. 80v–81v); third comet, which logically receives the greatest attention (pp. 81v–94v) and successively presents Schickard's personal observations from November 25 to January 12, 1619 (pp. 82r–86v), with indication of the most important positions of the comet (pp. 86v–87v) and a demonstration of the total absence of parallax (pp. 87v–88v). The treatise concludes with a lengthy discussion of its 'significance' (*Bedeuttung*, pp. 88v–94v), i.e., of its effects on the human world in the present dramatic circumstances.

The theoretical and technical quality of Schickard's manuscript treatise, which was never printed, is enhanced by the fact that its writing is accompanied by a multitude of color illustrations (watercolors) by Schickard himself, who in the immediately preceding years had made engravings for Kepler illustrating his *Harmonice mundi* and his *Epitome Astronomiae Copernicanae*. In this essay, we will examine some of them, but now I will show those illustrating the novae in Cassiopeia and Swan, in the latter case including the Christianization of the Swan as a figure of the crucified Christ (Figs. 6.2, 6.3 and 6.4)[17]:

[17] Schickard contributed to the contemporary attempts on Christianization of the constellations with Schickard, 1623. On this, see Hübner, 1978; Granada, 2021: 156–157, 169–174.

Fig. 6.2 The constellation of Cassiopeia with the nova, p. 39v. Courtesy of the Württembergische Landesbibliothek, Stuttgart

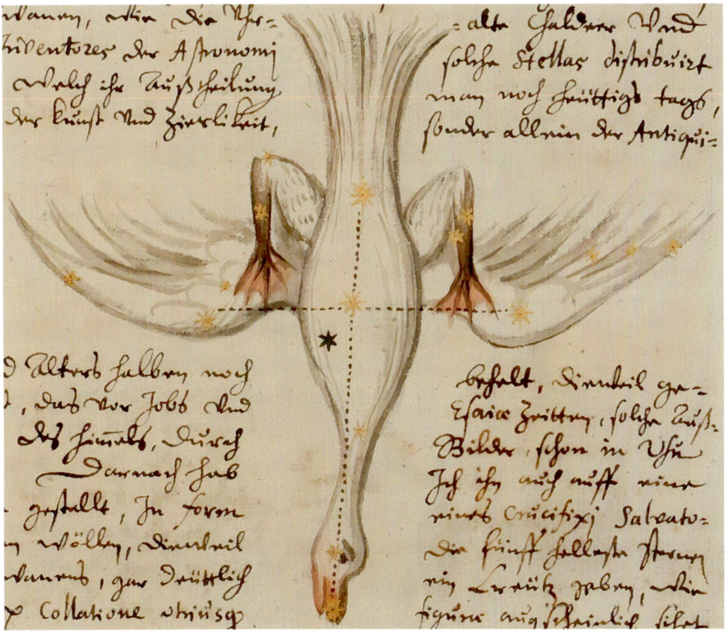

Fig. 6.3 The constellation of Cygnus with the nova, p. 40r. Courtesy of the Württembergische Landesbibliothek, Stuttgart

Fig. 6.4 Transformation of Cygnus into the Crucified, p. 40v. Courtesy of the Württembergische Landesbibliothek, Stuttgart

6.4 Schickard's Cometary Theory

The theory of comets that, in the light of the observation of the comets of 1618, Schickard presents in his *Cometen Beschreibung* is an innovative theory, associated with the new cosmological and astronomical currents that had been making their way for several decades, including heliocentrism and the motion of the Earth. We can get an idea of all this by following the table of contents of the first part and completing it with the different sections within each chapter.

In the first four chapters, Schickard examines the four causes that explain the cometary phenomenon: material, formal, efficient, and final (a usual way of presentation in cometary treatises). Regarding matter, Schickard makes clear, in the first section, that a comet is a physical reality and not a mere optical phenomenon, an appearance or visual illusion, as various authors had maintained since Antiquity, and as Galileo should still maintain for the present comet in his discussion with the Jesuit Orazio Grassi (Granada, 2024). Schickard also denies that the matter of comets is sublunary, that is, he rejects the Aristotelian theory that comets are formed from warm and dry exhalations rising from the Earth (I, 2, p. 9r). On the contrary, the matter of comets is ethereal or celestial, the same matter that constitutes the heavens, but denser and yet transparent (I, 4, p. 14r). This implies, as he will argue later, that comets are not (contrary to the Aristotelian vulgate) sublunary phenomena, but supralunar or celestial. Moreover, Schickard excludes in I, 3 Tycho Brahe's conception (initially raised in connection with the nova of 1572, but later extended by Brahe to comets from 1577 onwards) that comets are formed in the Milky Way and their matter is the condensed ether of the Milky Way.[18] As the title of I, 3 says, the Milky Way is at a much higher altitude than the comets reach[19] (a statement that makes much more sense if we take into account the Copernicanism that Schickard will explicitly assume in later chapters). Comets appear and are formed in every region of the heavens.

Concerning the formal cause, Schickard rejects the second major component of the Aristotelian theory: comets are not inflammations or fires (II, 5, p. 15r), but the result of solar illumination and its tail is the refraction of solar rays as they pass through the transparent head of the comet. In short, Schickard assumes the so-called optical theory of comets' (which goes back to Petrus Apianus' and Girolamo Fracastoro's interpretation of the comets of 1532),[20] adding the rectification introduced by Kepler in his German treatise on the comet of 1607, which Kepler was to repeat and expand in his *De cometis libelli tres* (1619),[21] with the new information obtained from the observation of the comets of 1618: the tail is not only rays of refracted sunlight (a pure optical phenomenon), but matter from the nucleus or head of the comet dragged by the solar rays and illuminated by them. This is clearly stated in the title of section II, 7: "sein Schwaiff, von ihren Stralen, so sich brechen, durchdringen, und ettwas Materialischs mitt führen, geformirt".[22]

---

[18] Brahe, 1916: 305. Schickard introduces here, as a proof of Brahe's wrong concept, a mention of the true nature of the Milky Way as a cluster of a multitude of stars whose light is confused and congregated, departing from the Galilean *Sidereus nuncius*. See Schickard, *Cometen Beschreibung*: 13v: "die himmlische *Galaxia* ist nichts Anders, als ein Versamlung viler gar kleinen Sternlin, welche hart neben einander stehn, und von unserm blöden gsicht nit mögen von einander underscheiden werden, sonder uns als ein *Continuè* Aneinander hangende weisse nüblige Straß fürkommen. Aber mit Hilff der newen Spiegelrohr könden sie absonderlich gesehen und underschiden werden; wie das der berühmte Künstler *Galilaeus in Italia* erstlich erguckt hatt. Das wuste man zu *Tychonis* Zeitten noch nit, Drumb ist es Ihm wol zu verzeihen, das er solches ohnwissend statuirt hatt".

[19] Schickard, *Cometen Beschreibung*: 13r: "So werden die Cometen auch nit auß der *Galaxiâ* oder Milchstraß des himmels geborn; dan dieselbige vil mahls höher ist, als die Cometen steigen".

[20] See Barker, 1993. For Apianus, see Jerratsch, 2020: 122–127.

[21] See Kepler, 1608.

[22] Schickard, *Cometen Beschreibung*: 17r.

As to their efficient cause, Schickard still excludes that their generation is a natural process, much less a casual one. Comets are not generated by "celestial fire" (the presumed sublunary element of fire) nor by any natural heat (II, 12, p. 24r) nor by planetary influences or aspects (II, 10, p. 21r), but are miracles of omnipotent God ("ein wunder Geschöpff des Allmächtigen Gottes", II, 8, p. 18v). Therefore, they are not permanent celestial bodies (as Seneca had argued in his *Naturales quaestiones*), but new creations (II, 9, p. 20r), whose appearance cannot be predicted beforehand (II, 11, p. 23v).[23]

That the comet has the character of a miracle of divine omnipotence, alien to the natural order, allows Schickard to expound, regarding its purpose or final cause, that the comet, besides being an expression of divine omnipotence, goodness and majesty, constitutes a sign of divine anger at the sins of mankind and therefore a warning of future punishments that will befall mankind unless it performs sincere penance (II, 14 and 15, p. 26r). As far as the efficient and final causes are concerned, Schickard still remains within the framework of a Christian concept of comets, without moving on to a purely natural, physical interpretation of them and without conceiving them as permanent celestial bodies or existing beyond their limited visibility over several weeks or months. On the other hand, as it regards their hylemorphic composition, Schickard has already decidedly abandoned the traditional peripatetic doctrine and adopted the thesis of the innovative currents: celestial location and origin as well as brightness due to solar illumination.

Chapter 5, dealing with the effects of the comet (*Würkhung*), rejects (contrary to traditional opinion) that the comet itself causes "war, pestilence, drought, misfortune" (V, 16 and 17, pp. 26v, 28v), although such phenomena may accidentally occur after the comet's appearance (V, 18, p. 29r). In any case, historical experience shows that the appearance of a comet is rarely followed by good things, but rather by great misfortunes (V, 19, pp. 30r ff.), which do not happen immediately, but after a lapse of a few years (V, 20, p. 31v).

More important cosmologically is Chap. 6, devoted to the location of comets in the cosmos. Here Schickard takes up more systematically points that he has already presented in previous chapters: comets are not phenomena in the sublunary region of the upper air (VI, 21, p. 33r). Although some comets may occur below the Moon (VI, 22, p. 35r), most of them occur in the heavens, in the region of the planets (VI, 23, pp. 38v ff.). At this point, Schickard explains to his princely dedicatee, with simple pedagogy and images (see Figs. 6.5, 6.6 and 6.7), that this location and distance of the comet is demonstrated with the necessity of geometric demonstrations by means of the doctrine of parallax (pp. 35v–36r), just as the height of a tower is measured by forming the triangle whose vertex is the peak of the tower and the base two points on the surface of the Earth whose distance from each other is known to us.

Analogically, the distance of the comet is determined by the triangle formed from two points on the surface of the Earth (as its base) and the comet at the vertex,

[23] Schickard confronts here (24r), as Maestlin also does in his manuscript treatise, the claims of the Ulm mathematician Johann Faulhaber (1580–1635) to have predicted the appearance of the first of the comets in 1618. See also Maestlin, 2022: 45–49, 188–193.

Fig. 6.5 Image showing how to determine the height of a tower by establishing a triangle whose base and two angles are known, p. 35v. Courtesy of the Württembergische Landesbibliothek, Stuttgart

or with the triangle's base formed by the radius of the Earth, whose measure is known. Figure 6.7, p. 36r, shows the formation of the parallactic angle, with the distinction between the true place of the comet (at b, extension of the line that starts from the center of the Earth) and its apparent place as seen, at c, from our position on the surface of the Earth.

Comets appear, Schickard says, in all regions of the heavens (VI, 24, p. 42r) and their tail, as he had anticipated, is projected in the opposite direction to the Sun, not in a perfectly straight line, but very close to it (VI, 25, p. 43r). Of course, the tail is formed only by the Sun, not by Venus, as Brahe thought about the comet of 1577, for the light of the planets is not their own, but borrowed from the Sun:

> And even if it gets closer to Venus (the brightest planet after the Sun), its light is nevertheless only a borrowed light, a light received from the Sun, just like the light of the Moon, so it waxes and wanes, as in the case of the Moon, as can be clearly perceived by means of a telescope.[24]

[24] Schickard, *Cometen Beschreibung*, VI, 26, p. 45v: "Und solt er schon Venerj (dem hellesten Planetsternen nach der Sonnen) näher sein, so ist doch sein Liecht nur *mutuatitia lux*, ein entlehnetes, von der Sonnen empfangenes liecht, so wol als des Monds, daher es auch zu: und abnimt, wie der Mond selbsten, welches durch ein *Telescopium* oder guttes Spiegelrohr deuttlich kan wargenommen werden".

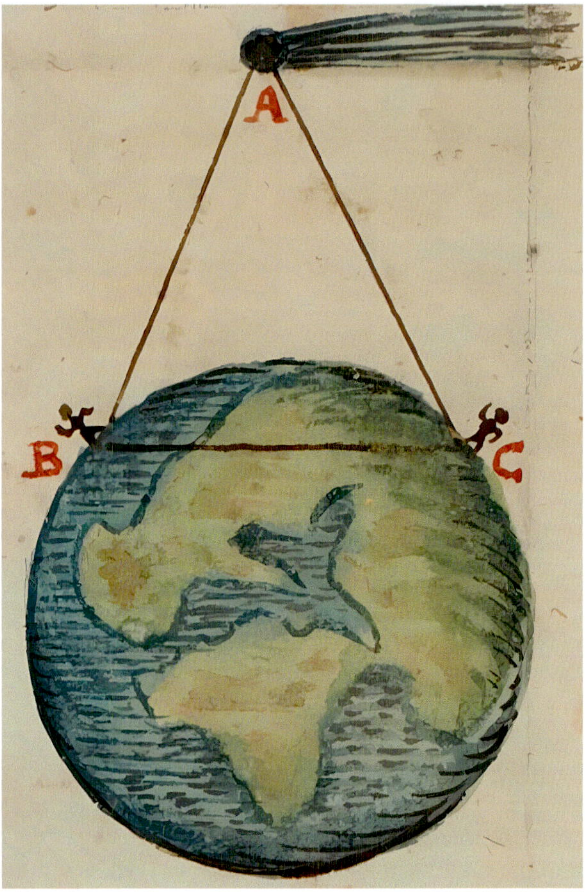

Fig. 6.6 Image showing the application to comets of the parallactic doctrine, p. 36r. Courtesy of the Württembergische Landesbibliothek, Stuttgart

Chapter 7 raises the question of the size and number of comets. Here Schickard points out that, given their great distance, their actual size is enormous (many hundreds of miles; VII, 27, p. 46r). More important is the assertion that two comets have never appeared simultaneously (VII, 28, p. 48r), nor are there any invisible comets (VII, 29, p. 49v). This last opinion is especially interesting, because it is directed (as the critical reference to Seneca shows) against those authors who have thought that comets are permanent and recurrent stars that are only visible to us for a short period of their circular trajectory. It is also directed, although Schickard does not name him explicitly, against Kepler ("ein vornemmer Astronomus heuttigs tags", p. 49v), who in his treatise on the comet of 1607 (and Kepler will repeat it in his 1619 treatise on the comets of 1618) had stated that the celestial ether is

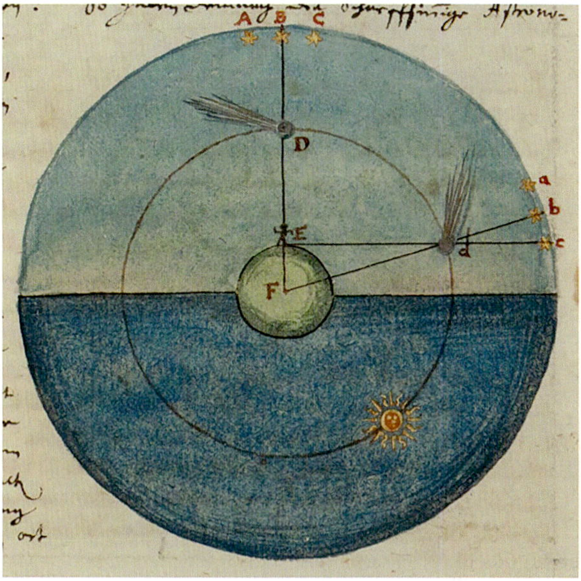

Fig. 6.7 Imagen showing the true and the apparent place of the comet with the formation of the parallectic angle. Courtesy of the Württembergische Landesbibliothek, Stuttgart

full of comets, just as the sea is full of whales that only occasionally come close to the shore.[25]

After an eighth chapter devoted to the comet's light (borrowed from the Sun), its color, and figure, Chap. 9 deals with the question of its motion. It is a very significant chapter, because the examination of this important question gives Schickard occasion to openly expose his adherence to Copernicanism and thus to the daily and annual motion of the Earth. The comet, Schickard says, has a double *apparent* motion: the first is the daily one which it shares with the world and receives from the sphere of the fixed stars; the second is its own motion, by which it moves, longitudinally and latitudinally, from one place to another. The young theologian and mathematician speaks with prudence and restraint in this report addressed to the princely authority, but he does not hide a clear position of global cosmological implications. On the daily motion, which seems to affect the whole world and all its bodies with a rotation around the central and immobile Earth, Schickard points out (IX, 37, pp. 59v ff.) that there is a division within astronomers, old and modern: some ancients, like the Pythagoreans in their esoteric discourse and the moderns since Copernicus, hold that the daily motion is proper to the Earth, which rotates on its axis in the span of 24 hours, so that the comets (like the planets) possess only their proper motion (p. 60r–v). Although Schickard states that it is not his intention to resolve the problem in the present instance, since it would require "a large book devoted to the subject" (p. 60v), he nevertheless asserts that the examination of

[25] Kepler, 1608: 59-60. See also Boner, 2006.

comets speaks in favor of the daily motion of the Earth.[26] In addition to the fact that it is simpler to attribute the motion to a single body (the Earth) than to so many different synchronized bodies and the difficulty of explaining how the comet can make a revolution of increasing diameter in the same time as it moves away from the Earth, Schickard points out another reason: the retrogradation, which comets often exhibit (he mentions the case of the planets of 1580 and 1585), is best explained by attributing it, as in the case of planetary retrogradations, to an appearance produced by the Earth's annual motion.[27] In this way, Schickard shows his full adherence to the Copernican cosmology of a moving Earth.

Finally, he points out that, even if the proper motion of the comet can go in all directions (IX, 38, p. 62r), nevertheless, it is always a regular, proportionate, and orderly motion in all its duration (IX, 39, p. 62v), as is proper of celestial bodies. Against Kepler, however, who had attributed to comets a rectilinear trajectory, like that of a projectile, Schickard goes on to attribute to them a circular trajectory (though incomplete, according to their ephemeral existence) in conformity with the circular figure of the heavens (IX, 40, p. 63v).[28] Schickard concludes this important chapter devoted to the comet's motion with a section (IX, 41, p. 65r–66r) in which he argues that the comet moves of its own accord in the fluid heavens by virtue of an internal impulse granted by God, its efficient cause, and not by a natural agent or external force.[29] However, Schickard gives here a place to Kepler's "celestial physics" in the *Astronomia nova* that provides a reason for the planetary motion: the Sun, not only source of light, but also of motion, may contribute to the motion of the comet in that its motion is slower the farther it is from the Sun.[30]

---

[26] Schickard, *Cometen Beschreibung*: 60v: "Drumb sage ich iez allein so vil, das der Erden Beweglikeit, auch von den Cometen her ettlicher massen zu beweisen, und vil gläublicher sey, das die Erden, dan der Comet, alle 24. Stund herumb lauffe".

[27] "die Cometen, welche in ☍☉ *opposito Solis* der Sonnen entgegen stehn, daselbsten ein Retrogradation haben, und dem Ansehen nach hindersich zu weichen pflegen, als wie zum Exempel *Annis* 1580. und 1585. ist wargenommen worden, und sonsten allen Planeten daselbsten in ☍☉ widerfehret: welches Alles nit anderst dann per *Annuum Terrae motum*, durch die Erdbewegung am füglichsten kan solvirt und entschuldigt werden", 61v–62r.

[28] For a comparison of Schickard's and Kepler's views on the comets, see Mauder, 1995.

[29] "die Cometen auß anerschaffnem von Gott eingebnem Jnnerlichem Trib bewegt werden, wie Gott der Schöpffer jedm Vogel, so bald er auß der Schalen schlupfft, dise Art einpflanzt, das er lernt fliegen, und bedarff nit, das man ihn an einer Schnuor herumb ziehe, oder der Wind ihn fortblase. Also hatt er auch seinem gschöpff dem Cometen seinen *Motum* eingegeben", 65v. This is not very far from Brahe's *scientia infusa* granted by God to the planets in the fluid ether; see Brahe, 1922: 159.

[30] "Doch soll Jch da Nit läugnen, das auch die Sonn hierzu Natürlicher weiß ettwas thuon köndt, und wie sie *Fons lucis* ist, also auch *Fons motus* sein möchte. Wie dan *Keplerus* gar scharpfsinnig erweiset, das sie Magneten Art an sich habe, und die planeten umb sich herumb ziehe. Darinn mich die Erfahrung sterkhet, weil die Cometen, je weitter sie von der Sonnen stehn, auch Je gmächer zu gehn pflegen. Wie Sonderlich *Anno* 1596. geschehen. Darauß zu erachten, das die Bewegung mit der Sonnen ein heimlichen Verstand oder Auffmerkhen, und Gemeinschafft müsse haben", 66r. Schickard, however, does not go so far as to propose, like Kepler, the integration of divine action with the solar *causa motrix* by establishing the God/Sun relationship that Kepler states. See Granada, 2022.

Chapter 10 concludes the first part by reiterating what has already been said in previous chapters: comets are ephemeral bodies that have a birth, by divine will and design, and an end (X, 42, p. 66r). The duration of comets is unequal and proportionate to their size (X, 43, p. 67r), progressively decreasing in magnitude and brightness as time goes by (X, 44, p. 67v). This decrement is caused not only by their incomplete matter,[31] but also by the Sun, which, as it illuminates them and its rays pass through them projecting their tail, drags along part of their matter and progressively annihilates them (X, 45, p. 68v). From this Schickard draws a moral conclusion (inspired by Wisdom 2: 3: "When it is extinguished, the body will turn to ashes, and the spirit will dissolve like empty air") with which he closes the first part: just as the rays and heat of the Sun dissipate and consume a fog, "in the face of the disappearance of a comet, we must think in a Christian way about the fugacity of our life and not miss our end".[32]

6.5 The Three Comets of 1618

As already mentioned, in the second part of his treatise, Schickard studies the comets that have appeared since 1475 (a total of 21 comets), indicating their dates of appearance, their course, and their meaning or effects.[33] Logically, the greatest attention is devoted to the comets of 1618, with whose exposition the treatise concludes (pp. 79r–94v).

Schickard begins by pointing out the exceptionality of three comets appearing in the same year. The first two, being of short duration and difficult to observe, receive a brief exposition. The first (pp. 79r–80v) was visible, before dawn, between mid-August and mid-September. Schickard learned of its existence from peasants, with a certain delay, and was only able to observe it with difficulty, for meteorological reasons, on two occasions: September 1 and 8. The watercolor he drew (see Fig. 6.8) shows its course *in antecedentia signorum* between Virgo and Cancer at the foot of Ursa Major. On its meaning, he does not dare to make any considerations, due to his insufficient observations, and limits himself to indicate that it is certainly no coincidence that its path coincided, except for the opposite direction, with that of the comets of 1531 and 1585, which coincided with important political and religious events in Switzerland and France respectively, but with a global European outreach.

[31] "die Materj ist theils selbs schuldig dran, nit so hoch gradirt und exaltirt, wie die Alchymisten darvon Reden, das sie die Prob der Hitz bstendig erdulden und ausstehn köndte", 69r.

[32] "Drumb wir bey Undergang eines Cometen, auch billig die Flüchtikeit Unsers lebens, und nit außbleiben des End, Christlich zu bedenkhen haben", 69r.

[33] With this Schickard followed the wide diffused contemporary genre of the Cometography, usually beginning with the comets observed since classical Antiquity. On this literature, see Mosley, 2014.

Fig. 6.8 The path of the first comet of 1618, p. 80r. Courtesy of the Württembergische Landesbibliothek, Stuttgart

The second comet (pp. 80v–81v) had a brief appearance, about 2 weeks from early November. It was perceived by forest guards shortly before dawn. Again, Schickard was only able to see it on a few occasions (November 9, 11, and 12). It was a very southern comet that at the northern latitude of Württemberg, was only visible in its long 40-degree tail, stretched along the constellations of the Hydra, the Raven, and Crater. The head always remained below the horizon. The weather prevented Schickard from seeing it again until November 25, when he thought he saw it again, now with its head, in the plausible place of its trajectory according to its previous progress. However, Schickard recognizes that it was in fact the third comet, as pointed out to him by Kepler (p. 81v), who (at the time in Linz, Austria) was able to observe the diminution and final disappearance of this second comet, before the third became visible, at about the same place and dates. From its southern location, Schickard considers its significance to affect the sub-equatorial regions and omits to address the question to focus on the third and great comet of the year (Fig. 6.9).

To this third comet (the one of 1618 par excellence) (Fig. 6.10), Schickard grants dates of appearance between November 18/19 and ca. January 9, 1619. In fact, his last observation was on January 8. In Tübingen, Maestlin said it was still visible in

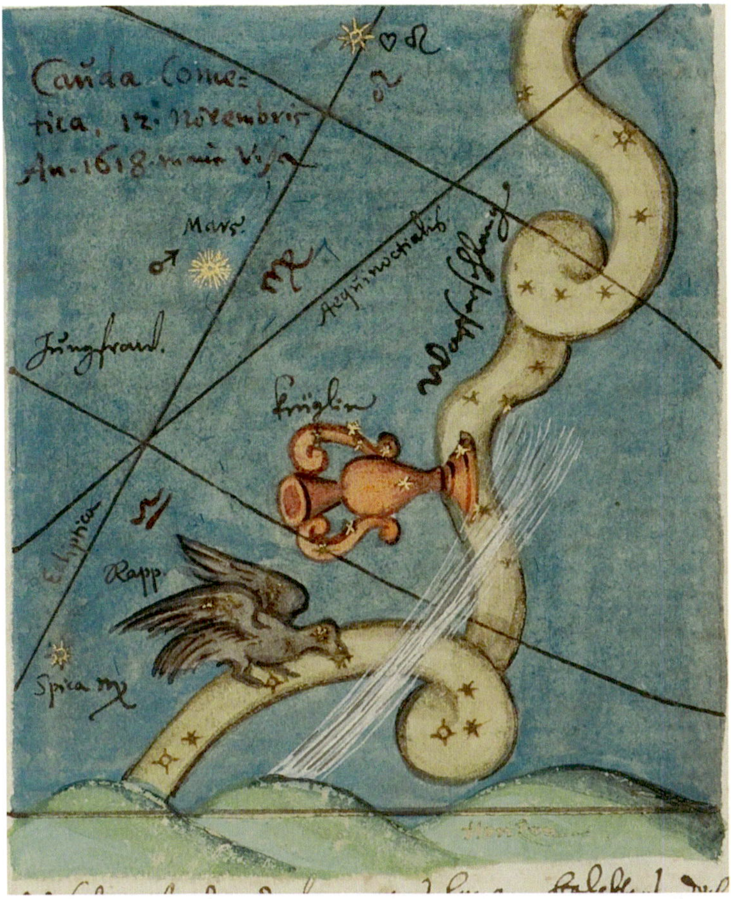

Fig. 6.9 The path of the second comet of 1618, p. 81r. Courtesy of the Württembergische Landesbibliothek, Stuttgart

February.[34] Schickard applied himself to a precise and constant observation, the history of which he describes in detail on pp. 82r–85v. For a better and more accurate observation of the comet's location, Schickard set aside the instrument previoulsy used (an astronomical radius or Jacob's staff with which he observed until November 27) and built himself, taking advantage of several days rendered useless by bad weather, a new radius, larger and with better visors, capable also of discerning up to the minute. Schickard describes in detail its construction, provided with a firm stability by means of a tripod, and adds the beautiful watercolor that paints himself observing with it (pp. 83r–v; see Fig. 6.11).

Schickard delights describing the observation, with the first radius, of November 27, watercolor included, in which he saw "the most beautiful spectacle of the comet

[34] See Maestlin, 2022: Introduction, Chap. 3.

Fig. 6.10 The path of the third comet of 1618, p. 86r. Courtesy of the Württembergische Landesbibliothek, Stuttgart

in the whole time of its duration" (*das aller schönste Spectackhel des Cometen, Jn der gantzen Zeit seiner Wehrung*, p. 82v): the comet's head projecting its long tail (see Fig. 6.12), to one side of the northern Corona and along Bootes, with the star on his right leg (*crus dextrum*) and *Arcturus* visible inside its tail projected toward the north and reaching the Ursa Major (p. 82v):

The entire path of the comet, with its progress along the different days on which it was observed by Schickard or whose position he was able to infer proportionally, is reflected in the beautiful watercolor that occupies the entire page 83r (see Fig. 6.10).

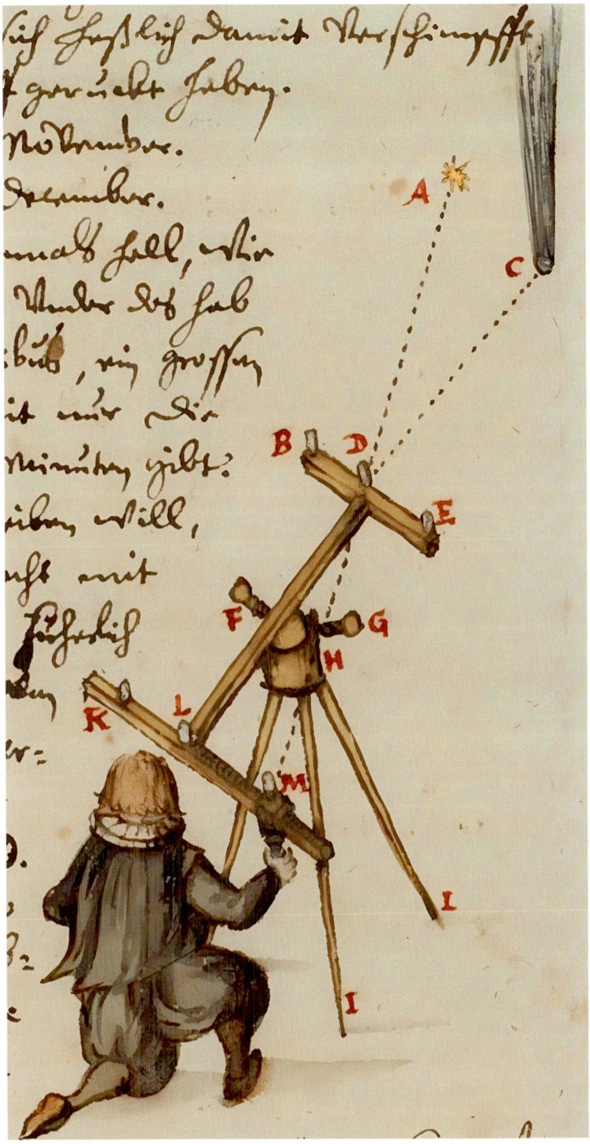

Fig. 6.11 Schickard observing the third comet with his second radius, p. 83r. Courtesy of the Württembergische Landesbibliothek, Stuttgart

Schickard then devotes a few pages (pp. 86v–87v) to a selection of the most important positions of the comet, with geometrical calculation of its longitudes and latitudes, with a view to calculating its parallax and thus to solving the important cosmological question of its sublunary or celestial location. He specifically selects the observations of December 3 and 4, when he made three observations each day

Fig. 6.12 The long tail of the third Comet covering Arcturus in Bootes, p. 82v. Courtesy of the Württembergische Landesbibliothek, Stuttgart

at different times. This allows Schickard to conclude in the section "Parallaxis Cometae" (pp. 87v–88v) that "the comet must have had almost no parallax or at most a small and meager one" (*der Comet schier keine, oder doch gar ein kleine und geringe Parallaxin müsse ghabt haben*, p. 87v), of "a few minutes," so that it has been in its entire path above the Moon and little below the Sun (surely because Schickard still grants the Sun, with tradition, a parallax of 3 min). Schickard shows this by an ingenious procedure (pp. 87v–88r), consisting of showing the portion of the celestial sphere through which the comet moved in those days, between the 16th and 18th degrees of Libra (Fig. 6.13):

If the comet had moved under the Moon, Schickard explains, the longitudinal displacement between the three observations on each day would have been much greater, so that the comet would not have been seen moving in a straight line along

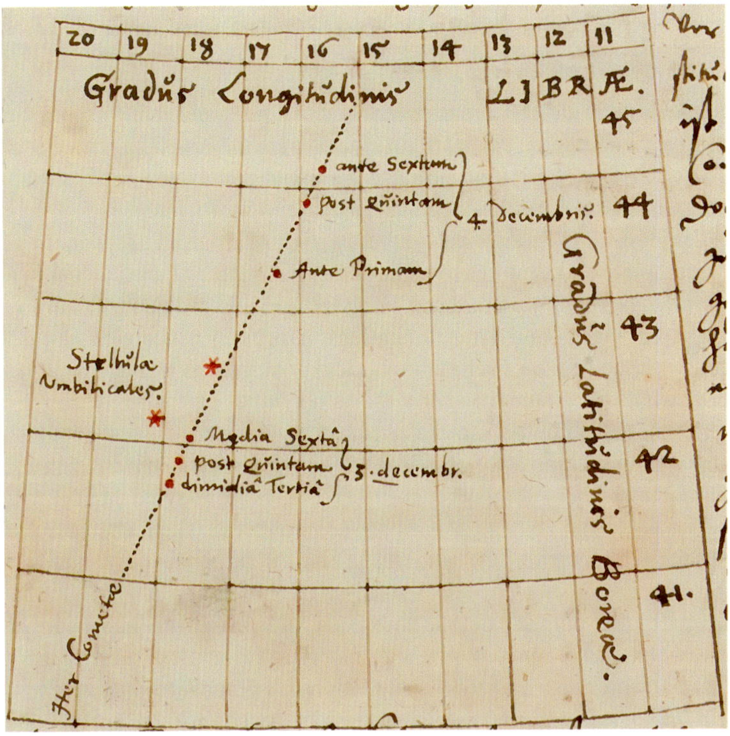

Fig. 6.13 Rectilinear course of the third comet showing the absence of parallax, p. 87v. Courtesy of the Württembergische Landesbibliothek, Stuttgart

those positions, but oscillating by about one degree, which is slightly more than the lunar parallax.[35]

6.6 The Meaning of the Comet

All that remains for Schickard is the question of the meaning of the comet, a fundamental motif for his addressee (the same goes for all the political powers that requested reports on the successive comets and novae of the mathematicians at their universities), and one that Schickard deals with in the last twelve pages (pp. 88v–94v).

[35] "Weil die *observationes*, beedes *prope horizontem*, und *longe supra illum* geschehen, (da es Ja solt Jn *sublunaribus* ein *maximam evariationem visus* gegeben haben) und aber alle solche sechs Ort, beynahend in Eine gerade lini fallen (welches Jm Himmel droben ein *circulus Maximus* ist) so versteth sichs Ja hierauß, es hab die Erddikhin gegen des Cometen Mächtiger höhin ein gar schlechtes, oder schier ein unmerkhlichs außgetragen, und sey er gar weit ob dem Mond, in *ipso aethere* gestanden. Dan were er *sublunaris* gwest, so hetten solche *loca* freylich nimmermehr Jn Eine lini zusamen tretten könden, sonder weren schier ein Grad hin und wider *exorbitirt*", 87v–88r.

He begins by confessing his ignorance on the subject and his conviction that it is impossible for men to predict with certainty the future effects that may follow the appearance of a comet, just as it is impossible to predict its appearance (pp. 88v–89r). But to Schickard, a sincere Christian, there is also no doubt that God, the author of the cometary phenomenon, can instill in chosen men the prophetic knowledge of the events that will follow. Scripture shows such a thing, granted to simple men,[36] as well as the vanity and impiety of heathen astrological prognostication.

However, despite his confessed ignorance, Schickard does not want to leave his treatise incomplete and ventures to make some *conjectures*, based on the hypothesis that the comet, as a creature produced by divine will and providence, can signal, but not cause, great events and future mutations, especially since God is particularly concerned with the health of religion, political government and even the welfare of men, all of which the comet can refer to. As a symbol or "hieroglyphic" of divine wrath (p. 91r), the comet is more associated with negative events (punishments, misfortunes) than with positive ones, for which—Schickard advances; see p. 91r—the Rosicrucian brothers[37] and other chiliasts (Schickard may be thinking of Johann Faulhaber or his friend Daniel Mögling)[38] seem to be mistaken when they associate the comet with the return of the golden age and the fulfillment of Isaiah's eschatological prophecies.[39]

For his part, after pointing out the ambiguity and uncertainty of astrological forecasts, always based on conventional and historical principles (such as the signs of the zodiac and the images of the constellations),[40] Schickard indicates his preference for *a homogeneous interpretation* (p. 92r-v), parallel to the correct interpretation of Scripture by Scripture itself, without letting in heterogeneous factors such as the philosophy of Aristotle.

---

[36] On this, see Bubenheimer, 1995: 69.

[37] Schickard had close acquaintance and friendship with Johann Valentin Andreae, the author of the Rosicrucian manifestos, and with Christoph Besold, a member of Andreae's circle and perfectly informed of the gestation of the manifestos. Schickard had been, at the Stift in Tübingen, a fellow of Andreae, who in his autobiography called Schickard "the epitome of all scholarship" and "the pearl of the university of Tübingen". In addition, Schickard gave a Hebrew course at Besold's home in 1614, while he was repetent for Hebrew at the Stift and participated in Andreae's efforts to form a secret "Societas Christiana". After his appointment as professor of Hebrew at the university, Schickard had to write a report on Rosicrucian writings. His judgment was quite positive, concerning primarily the quest for a reformation of the sciences, and he only reserved strong criticism for the mistaken astronomical calculations in their writings and the chiliast fantasies. On all this, see Van Dülmen, 1971; Brecht, 1978; Kühlmann, 1995: 61; Bubenheimer, 1995.

[38] Daniel Mögling entered the Stift in Tübingen in 1611, one year after Schickard. He maintained a very active correspondence with Schickard. Fifty-two letters from him to Schickard have been preserved, while those of the latter have been lost (the first of April 1625); see Schickard, 2002: 40 f. On Mögling, see Neumann, 1995; Mehl, 2019: *ad indicem*.

[39] Prophecies, according to Schickard (91r), already fulfilled, but "longe alio sensu" than the expected. It cannot be excluded, however, an allusion to Brahe's optimistic prognosis in the conclusion of Brahe, 1916: 312. Cf. note 41 below.

[40] See Granada, 2021: 171.

The first possibility is to posit the possible effects associated with the comet by comparison with the effects that followed another earlier comet whose trajectory had a 'community' (*Gemeinschaft*) with the present one. According to Schickard, the closest case is the comet of 1607, whose course cut the path of the present comet and whose tail at one time also covered the star Arcturus in Bootes. It can be thought that the effects on Earth will be similar and Schickard recalls that, as is well known, the comet of 1607 was followed by the conflict in the Empire between Rudolf II and his brother Matthias, the claim for religious freedom in Bohemia and the 12-year truce between Spain and the United Provinces.

However, it is another possibility that most appeals to Schickard: the prophetic one. Here it is the reference—which Schickard considers practically certain—to the contemporary time and the present comet by the prophecy of the Tiburtine Sibyl that was discovered in 1520 in Switzerland on a marble plaque (p. 93r-v).[41] Schickard finds that the dramatic prophecy ("Haec omnia denique nox erount, Interitous, Rouina, Damnatio ac aeternai Tenebrai"), associated with the submission of France ("prious houic soumittet Cervices Gallia") and England ("ad eious genoua souplex adnatabit classibous Britannia") to Spanish hegemony, announces a dark future ("kein Glückh oder Stern mehr im Reich sein, sonder eitel Nacht, Undergang, ruina und dickhe Finsternuß", p. 94v) in the face of which Schickard can only think of asking divine mercy to avert it: "der barmhertzig Gott vätterlich verhütten wölle" (94v).

Acknowledgments Thanks are given to the Württembegische Landesbibliothek Stuttgart for its kind permission to reproduce the images from the manuscript *Cometen Beschreibung* by Wilhelm Schickard, and to the Zentralbibliotkek Zürich for allowing the reproduction of the image of the *Collegium Fraternitatis* from the manuscript by Theophilus Schweighardt (Daniel Mögling) *Speculum Sophicum Rhodo-Stauroticum*, 1618.

References

Andreae, J. V. (1973). *Fama Fraternitatis (1614), Confessio Fraternitatis (1615), Chymische Hochzeit: Christiani Rosencretuz Anno 1459 (1616)*, eingeleitet und herausgegeben von Richard van Dülmen. Calwer.

Barker, P. (1993). The optical theory of comets from Apian to Kepler. *Physis, 30*, 1–25.

Boner, P. J. (2006). Kepler on the origin of comets: Applying earthly knowledge to celestial events. *Nuncius, 21*, 31–47.

[41] Schickard refers to "Cornelius Gemma and other authors", p. 93r. See Gemma, 1975, vol. ii, pp. 149 f., who quotes the passage. Brahe had also referred to the prophecy of the Tiburtine Sibyl in the conclusion of his *Progymnasmata*, pp. 315 f., with the same quotation as Gemma and Schickard. On this prophecy, see Secret, 1969. As Secret pointed out, the prophecy had been referred to the universal monarchy of Charles V of Spain by some Franciscus Monachus (a franciscan friar) in an epistle *De orbis situ ac descriptione* addressed to Jean Carondelet (archbishop of Palermo) printed in Anvers in 1524 (we thank Saverio Ricci for the information). On this issue see also Ernst, 2008.

Bouman, J., & van Heertum, C. (Eds.). (2014). *Divine wisdom, divine nature: The message of the Rosicrucian Manifestoes in the visual language of the seventeenth century*. In de Pelikaan.
Brahe, T. (1916). Astronomiae instauratae progymnasmata. In J. L. E. Dreyer (Ed.), *Opera omnia* (Vol. III). Libraria Gyldendaliana.
Brahe, T. (1922). De mundi aetherei recentioribus phaenomenis. In J. L. E. Dreyer (Ed.), *Opera omnia* (Vol. IV). Libraria Gyldendaliana.
Brecht, M. (1978). Beobachtungen zum geistigen Hintergrund. In F. Seck (Ed.), *Wilhelm Schickard 1592-1635: Astronom, Geograph, Orientalist, Erfinder der Rechenmaschine* (pp. 41–48). J. C. B. Mohr (Paul Siebeck).
Brosseder, C. (2004). *Im Bann der Sterne: Caspar Peucer, Philipp Melanchthon und andere Wittenberger Astrologen*. Akademie Verlag.
Bubenheimer, U. (1995). Wilhelm Schickard in Kontext einer religiösen Subkultur. In F. Seck (Ed.), *Zum 400. Geburtstag von Wilhelm Schickard: Zweites Tübinger Schickard-Symposion* (pp. 67–92). Thorbecke.
Confessio Fraternitatis. (1652). Confessio Fraternitatis, or The Confession of the laudable Fraternity of the most honorable Order of the Rosie Cross, written to the Learned of Europe. In *The fame and confesion of the fraternity of R: C: Commonly, of the Rosie Cross*, translated by Eugenius Philalethes [= Thomas Vaughan] Printed by J.M. for Giles Calvert.
Cysat, J. B. (1619). *Mathemata astronomica de loco, magnitudine, et causis cometae, qui sub finem Anni 1618 et initium Anni 1619 in coelo fulsit*. Ex typographeo Ederiano.
Ernst, G. (2008). Il linguaggio universale dei cieli. Cornelio Gemma, Tycho Brahe, Tommaso Campanella. In H. Hirai (Ed.), *Cornelius Gemma: Cosmology, medicine and natural philosophy in Renaissance Louvain* (pp. 33–49). Fabrizio Serra.
Fama Fraternitatis. (1652). *Fama Fraternitatis, or, A discovery of the fraternity of the most laudable order of the Rosy Cross*. In *The fame and confesion of the fraternity of R: C: Commonly, of the Rosie Cross*, translated by Eugenius Philalethes [= Thomas Vaughan], Printed by J.M. for Giles Calvert.
Fama fraternitatis: Das Urmanifest. (1998). *Fama Fraternitatis: Das Urmanifest der Rosenkreuzer Bruderschaft zum ersten Mal nach den zeitgenössischen Manuskripten bearbeitet durch Pleun van der Kooij*, mit einer Einführung über die Entstehung und Überlieferung der Manifeste der Rosenkreuzer von Carlos Gilly.
Frankenberger, M. (2012). *Wilhelm Schickards Cometenbeschreibung von 1619*, Magister Artium Thesis in History of Science and Technology. University of Stuttgart (unpublished).
Gaulke, K. (Ed.). (2007). *Der Ptolemäus von Kassel: Landgraf Wilhelm IV. von Hessen-Kassel und die Astronomie*. MHK.
Gemma, C. (1975). *De naturae divinis characterismis*. Christophorus Plantinus.
Gilly, C. (2012). Las novas de 1572 y 1604 en los manifiestos rosacruces y en la literatura teosófica y escatológica alemana anterior a la Guerra de los Treinta Años. In M. Á. Granada (Ed.), *Novas y Cometas entre 1572 y 1618: Revolución cosmológica y renovación política y religiosa* (pp. 275–332). Edicions de la Universitat de Barcelona.
Gindhart, M. (2006). *Das Kometenjahr 1618: Antikes und zeitgenössisches Wissen in der frühneuzeitlichen Kometenliteratur des deutschsprachigen Raumes*. Dr. Ludwig Rauchert Verlag.
Granada, M. Á. (2014). Samuel Eisenmenger on the nova of 1572: A critical edition of his tract *De stella nova apparente in stellato coelo consideratio et obseruatio* (Jan. 1573). In W. R. Dick & D. Fürst (Eds.), *Lebensläufe und Himmelsbahnen, Acta Historica Astronomiae* (pp. 53–70).
Granada, M. Á. (2016). Michael Maestlin and the comet of 1618. In M. Á. Granada, P. J. Boner, & D. Tessicini (Eds.), *Unifying Heaven and Earth: Studies in the history of early modern cosmology* (pp. 239–290). Edicions de la Universitat de Barcelona.
Granada, M. Á. (2020). Bartholomaeus Keckermann and Christoph Hunichius on novas and comets at the beginning of the 17th century: Two opposing views on the relation between natural philosophy and mathematics. In P. D. Omodeo & V. Wels (Eds.), *Natural knowledge and Aristotelianism at early modern Protestant Universities* (pp. 209–234). Harrassowitz.

Granada, M. Á. (2021). The nova of 1600 in Cygnus and the Christianization of the constellations. In P. J. Boner (Ed.), *Kepler's new star (1604): Context and conytroversy* (pp. 145–180). Brill.
Granada, M. Á. (2022). Johannes Kepler. The Sun as the heart of the world. *Journal for the History of Astronomy, 53*, 133–140.
Granada, M. Á. (2024). Nove e comete nel periodo 1572–1623 e il dibattito tra Galileo e Orazio Grassi. In *Il Saggiatore di Galileo a 400 anni dalla sua pubblicazione, (Roma, 23–25 ottobre 2023)* (pp. 11–51). Bardi Edizioni, Accademia dei Lincei.
Hübner, W. (1978). Die Christianisierung der Sternbilder in Schickards Astroscopium. In F. Seck (Ed.), *Zum 400. Geburtstag von Wilhelm Schickard: Zweites Tübinger Schickard-Symposion* (pp. 131–150). Thorbecke.
Jerratsch, A. (2020). *Der frühneuzeitliche Kometendiskurs im Spiegel deutschsprachiger Flugschriften*. Steiner.
Kepler, J. (1608). Ausführlicher Bericht vom dem newlich im [...] diß 1607. Jahrs erschienenen Haarstern oder Cometen. In M. Caspar et al. (Eds.), *Gesammelte Werke*, (Vol. IV, pp. 59–76, 1941). Beck.
Kronk, G. W. (1999). *Cometography: A catalogue of comets. Vol. 1: Ancient—1799*. Cambridge University Press.
Kühlmann, W. (1995). Wilhelm Schickard—Wissenschaft und Reformbegehren in der Zeit des Konfessionalismus. In F. Seck (Ed.), *Zum 400. Geburtstag von Wilhelm Schickard: Zweites Tübinger Schickard-Symposion* (pp. 41–66). Thorbecke.
Leovitius, C. (1564). *De coniunctionibus magnis insignioribus superiorum planetarum, Solis defectionibus et cometis in quarta monarchia cum eorundem effectuum historica expositione. His ad calcem accessit prognosticon ab anno Domini 1564 in viginti sequentes annos.* Emanuel Salczer.
Maestlin, M. (2022). *Michael Maestlin's manuscript treatise on the comet of 1618: An edition and translation of manuscript WLB Stuttgart, cod. Math. 4° 15b, Nr. 8* (M. Á. Granada & P. J. Boner, Eds.). Brill.
Mauder, H. (1995). Die Kometenschriften von Schickard und Kepler. In F. Seck (Ed.), *Zum 400. Geburtstag von Wilhelm Schickard: Zweites Tübinger Schickard-Symposion* (pp. 151–166). Thorbecke.
Mehl, É. (2019). *Descartes en Allemagne, 1619–1620: Le contexte allemand de l'elaboration de la science cartésienne*. Presses Universitaires de Strasbourg.
Methuen, C. (1998). *Kepler's Tübingen: Stimulus to a theological mathematics*. Ashgate.
Mosley, A. (2014). The history and historiography of early modern comets. In M. Á. Granada, A. Mosley, & N. Jardine (Eds.), *Christoph Rothmann's discourse on the comet of 1585: An edition and translation with accompanying essays* (pp. 282–325). Brill.
Nagel, P. (1619). *Stellae prodigiosae seu cometae per oculum triplicem observatio et explicatio.* Das ist: des newen Cometen und Wundersterns im October, November und December 1618 erschienen, wahrhafftige Deutung und Außlegung per Magiam insignem, dergleichen zuvor nicht gesehen: Allen Menschen auff Erden zur guten Nachrichtung und Warnung fürgestellet. Birckner.
Neumann, U. (1995). "Olim, da die Rosen Creutzerey noch florirt, Theophilus Schweighardt genant": Wilhelm Schickards Freund und Briefpartner Daniel Mögling (1596–1635). In F. Seck (Ed.), *Zum 400. Geburtstag von Wilhelm Schickard: Zweites Tübinger Schickard-Symposion* (pp. 93–115). Thorbecke.
Penman, L. T. I. (2010). Climbing Jacob's ladder: Crisis, chiliasm, and transcendence in the thought of Paul Nagel (†1624), a Lutheran dissident during the time of the Thirty Years War. *Intellectual History Review, 20*, 201–226.
Penman, L. T. I. (2019). *Hope and heresy: The problem of chiliasm in Lutheran confessional culture, 1570-1630*. Springer Nature.
Ramus, P. (1569). *Scholarum mathematicarum libri unus et triginta*. Episcopius.

Ribordy, O. (2023). Neue Phänomene am Himmel: Astronomische Beobachtungen des Johannes Baptist Cysat zu den Kometen um 1618-19. In E. Mehl & I. Pantin (Eds.), *De mundi recentioribus phaenomenis. Cosmologie et science dans l'Europe des Temps modernes, XVe-XVIIe siècles. Essais en l'honneur de Miguel Ángel Granada* (pp. 229–253). Brepols.

Schickard, W. (1619). *Cometen Beschreibung In zwen underschidliche Partes abgetheilt, deren Erster Von denselbigen ins gemein: der Ander Von Allen Insonderheit, sonderlich aber denen dreÿ Jüngsten, Im Abgeloffenen 1618. Jahr erschienen, außführlich handelt.* Württembergische Landesbibliothek.

Schickard, W. (1623). *Astroscopium pro facillima stellarum cognitione noviter excogitatum.* Werlinus.

Schickard, W. (2002). *Briefwechsel* (F. Seck, Ed.) (Vol. II). Frommann.

Schiebert, H. (2006). *Die große kosmologische Kontroverse: Rekonstruktionsversuche anhand des Itinerarium exstaticum von Athanasius Kircher SJ (1602–1680).* Steiner.

Seck, F. (1978). Leben und Werk im Überblick. In F. Seck (Ed.), *Wilhelm Schickard 1592-1635: Astronom, Geograph, Orientalist, Erfinder der Rechenmaschine* (pp. 13–40). J. C. B. Mohr (Paul Siebeck).

Secret, F. (1969). Cornelius Gemma et la prophétie de la "Sybille tiburtine". *Revue d'histoire ecclésiastique, 64*, 423–431.

Van Dülmen, R. (1971). Ein Gutachten W. Schickards über Rosenkreutzerschriften. *Zeitschrift für Religions- und Geistesgeschichte, 23*, 254–259.

Miguel Á. Granada is Emeritus Professor of History of Renaissance Philosophy at the University of Barcelona. He has translated into Spanish works of Machiavelli, Bernardino Telesio, Francis Bacon, Tommaso Campanella, Erasmus of Rotterdam and Giordano Bruno. His publications include *El debate cosmológico en 1588: Bruno, Brahe, Rothmann, Ursus, Röslin* (Naples, 1996), *Sfere solide e cielo fluido. Momenti del dibattito cosmologico nella seconda metà del Cinquecento* (Milan, 2002), *Giordano Bruno: Universo infinito, unión con Dios, perfección del hombre* (Barcelona, 2002). He has coedited several astronomical and cometary sources from the period of the cosmological revolution: *Christoph Rothmanns Handbuch der Astronomie von 1589. Kommentierte Edition der Handschrift Christoph Rothmanns "Observationum stellarum fixarum liber primus", Kassel* 1589 (Frankfurt, 2003), with J. Hamel and L. von Mackensen; *Christoph Rothmann's Discourse on the Comet of 1585* (Leiden, 2014), with Adam Mosley and Nicholas Jardine; *Michael Maestlin's Manuscript Treatise on the Comet of 1618: An Edition and Translation of Manuscript WLB Stuttgart, Cod. Math. 4° 15b, Nr. 8* (Leiden, 2022), with Patrick J. Boner.

Chapter 7
The Comets of 1618: Tales from India

R. C. Kapoor

7.1 Introduction

The year 1618 in astronomy was a unique one in that it presented three bright cometary apparitions in quick succession. The comets created sensation all through Europe and belong to the era when Galileo's telescopic observations had created a paradigm shift in our perception of the heavens, and Johannes Kepler was introducing a fundamental change in mathematical astronomy. This chapter is an account of the observations of two of the three great comets of 1618, made from India. This turned into a unique occasion because the same targets of opportunity were followed independently by astronomers from two very different schools, and their observations were recorded quantitatively. Nūr ud-Dīn Jahāngīr (1569–1627), the fourth Mughal Emperor of India who ruled from 1605 to 1627, recorded in the *Tūzūk-i Jahāngīrī* (The Memoirs of Jahāngīr), the appearance of two comets during a Royal journey from the town of Dohad (also Dahod) in Gujarat to Agra, the capital city of the Mughal Empire, via Ujjain in 1027 A.H. (1618 CE; Note 1), the 13th year of his accession. The comets in question are those designated 1618 III and 1618 II that appeared one after the other in November 1618, in that order. From the recorded dates, Jahāngīr turns out to be an independent discoverer of the two comets of November 1618. If told so, Jahāngīr would have been only amused, but his records indicate that he had the requisite scientific equipment and the ability to carry out accurate astronomical observations.

Strikingly, India's tryst with the telescope also began at this time, when a number of Jesuit missionaries observed these same comets. Fr. Venceslaus Pantaleon

R. C. Kapoor died before publication of this work was completed.

R. C. Kapoor (✉)
Indian Institute of Astrophysics, Bengaluru, India

C. J. Cunningham (ed.), *Cosmic Events*, Historical & Cultural Astronomy,
https://doi.org/10.1007/978-3-031-86950-1_7

Kirwitzer (1588–1626) observed Comet 1618 III from Goa, and he was soon joined by other Jesuits in Goa and in Cochin. And then in quick succession, they also observed the second bright comet of November 1618. Their initial observations also coincide with the discovery dates of these comets. While giving phenomenological details of the observations, Fr. Kirwitzer also mentioned that he used an optical device, namely, a *tubo optico* (telescope), to view the comets. We cannot say for certain if it was a Galilean telescope with a mounting that the group had brought from Europe, together with other astronomical instruments and books to be carried further east to Macao. Fr. Kirwitzer and others deserve credit on several counts, as independent discoverers of the two Great Comets in succession; as the first independent users of an optical device for observing comets outside Europe; and for the first modern astronomical observations from India with a telescope within a decade of its invention in Europe. These same comets were also followed by Father Antonius Rubinus from Cochin. Fr. Kirwitzer collated and published these observations in a short treatise in 1620. Our account of the observations by the Jesuits is based on this book that, as per the reference to it in Nature (1878), had possibly been lost soon after publication. However, it surfaced recently when the Austrian National Library and Google Books brought it out in digitized form in 2014.

Figure 7.1 is a map of mainland India with dots corresponding to the various locations mentioned in this article. For reasons of history, old names and spellings have been used.

7.2 The Game Changers

In Aristotle's (384–322 BCE) theory, the world was divided into concentric crystal spheres, namely, the Heavenly Sphere which is in eternal rotation, is immutable, is in perfect order and in which are embedded the stars and the planets, and, beneath it, the terrestrial or Sublunary Sphere where changes could take place. Its border with the Heavenly Sphere was represented by the Moon. Everything in the Elementary (terrestrial) sphere is comprised of four elements—earth, air, fire, and water. The comets, the fuzzy-tailed forms, were viewed as exhalations from the Earth that get ignited in the upper layers of the atmosphere and lie in the Sublunary Sphere (Lee, 1952: *Meteorologica* I: Ch. 3 & 7). In *Almagest*, Ptolemy (ca.100–ca.175 CE) did not discuss comets that he possibly did not regard as celestial. He would have seen Halley's Comet when it appeared in late March in 141 CE and, curiously, the latest observation that he used in *Almagest* (ca. 150 CE) was of 2 February 141 (Toomer, 1984: 1). Aristotle's view prevailed for about two millennia; see more on this in Jervis (1985).

Still, the comets were not disregarded, and the apparitions were recorded and noted with concern. The comets that proved the game changers in our perception rose in the sixteenth and early seventeenth centuries, and these are:

The Comet of 1531
The Great Comet of 1577
The three bright comets of 1618

Fig. 7.1 Map showing localities in Mainland India mentioned in the text (Baba Varghese)

During the European Renaissance, when hardly any theory of comets existed, two important comets appeared that posed the first meaningful questions about their place in the heavens, namely, Halley's Comet seen in the year 1531 and the rise of the Great Comet of 1577.

The comet of 1531 was seen first by the Chinese on the morning of 4.8 August (UT) with a tail over 1° long, while the Japanese saw it in the morning of 8 August in the northeast direction as a bluish white object with a tail 5° long (Kronk, 1999: 298).

The German geographer and astronomer Petrus Apianus (1495–1552), who saw it first on 13 August, estimated its tail to be 15° long (Vsekhsvyatskii, 1964). The apparition became historic when he demonstrated from five consecutive observations during 13–17 August that the comet's tail was always directed away from the Sun. Apian's famous drawings showing the antisolar nature of the tail appeared in a book by him in German, *Ein kurtzer bericht d'Observation unnd urtels des Jüngst erschinnen Cometen im weinmon und wintermon dises XXXII. Jars*, Ingolstadt,

(Apianus, 1532). Figure 7.2 which depicts the comet's path is adopted from Apianus's 1540 book *Astronomicum Caesareum* .... This observed fact did not conform with the Aristotelian picture.

The Great Comet that rose in 1577 stirred viewers like none before. One can make this out from the numerous writings on its sightings in Europe (Hellman, 1944). It was seen first in Peru on 2 November at dusk as a very bright object shining through clouds like the Moon (−7 mag). On 8 November, the Japanese recorded its curved white tail stretching 50° (Kronk, 1999: 317). The glorious form of the comet showed up most impressively in the famous engraving by Jiri Daschitzsky that depicted its passage over Prague on 12 November 1577 (Fig. 7.3).

The Comet of 1577 is a historical one because of its observations made by the Danish astronomer Tycho Brahe (1546–1601). He saw it first on 13 November with a tail 2.5° wide and 22° long. Brahe determined its celestial position by observing from his observatory at Uraniborg on the island of Hven near Copenhagen with large-sized quadrants and compared it with that taken around the same time by another observer Taddæus Hagecius, 600 km away in Prague. Brahe showed that the comet had an inappreciable horizontal parallax so that the comet's distance must be far above the Moon, by over six times; he had taken the Moon's distance 52 Earth Radii (Brahe, 1603: 121–123; Dreyer, 1890: 165–167; know that Aristarchus for the first time had in 270 BCE estimated it to be some 60 times the Earth's radius from the duration of an eclipse of the Moon). For the first time in history, the important question about the place of the comets in the firmament was resolved by positional

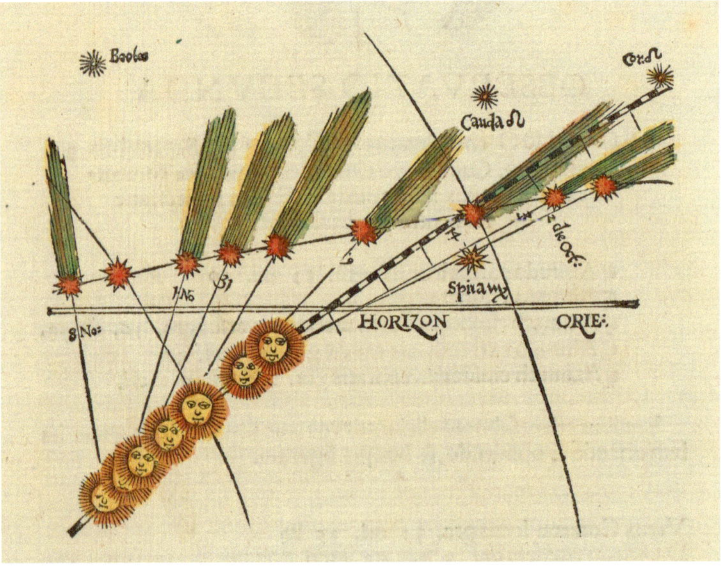

Fig. 7.2 Comet of 1531 on various dates; Peter Apian showed that the tail was directed away from the Sun; image from Apian us (1540): *Astronomicum Caesareum Factum et actum Ingolstadii in ædibus nostris* (Internet Archive)

Fig. 7.3 The woodcut by Jiri Daschitzsky showing the Great Comet of 1577 over Prague on 12 November. (Image courtesy of Wikimedia Commons and judged to be in the public domain)

measurements. Brahe concluded that the comet was superlunar, and it sprang up in the heavenly world in the Æther. It would not have existed in the Elementary World (Brahe, 1603: 457–458).

On the basis of his observations of the comet of 1585 and sensing that, distinct from the planets, the comet had motion of its own, the astronomer Christoph Rothmann (ca. 1560–1601; Granada et al., 2014: 1) argued against the solid and impenetrable celestial spheres carrying the planets (Rothmann, 1619, in Granada et al., 2014). He propounded that between the Earth and the fixed stars there exists nothing but air apart from the seven planets, that the world is heliocentric, and that the matter of the comet was an exhalation from the Earth. Rothmann's treatise got published only in 1619.

Tycho Brahe had also observed the comets of 1580, 1582, 1585, 1590, and 1596. On the comet of 1585, he and Rothmann had also corresponded (Dreyer, 1890: 160–162). Brahe believed that the movement of comets did not conform to the Copernican worldview. Rather, the comets moved in a geoheliocentric world, where the planets move around the Sun in a universe the centre of which is the Earth. Brahe's work *Tychonis Brahe De mundi aetherei recentioribus phaenomenis liber secundus* which dealt with the Comet of 1577 and his system of the construction of the universe, appeared in 1588. It was reissued in 1603 with some added matter.

Figure 7.4, adopted from this book, depicts the Tychonic System, as well as the comet of 1577 (Brahe, 1603: 191).

The year 1618 was a unique one in that three bright comets became visible in the sky within a short span of just 3 months. After the Halley's Comet of 1607, these were the first significant comets to appear in the skies. The comets created a sensation all through Europe and belong to the era when Galileo's telescopic observations had created a paradigm shift in our perception of the heavens, and Johannes Kepler (1571–1630; Fig. 7.5) was bringing in a fundamental change in mathematical

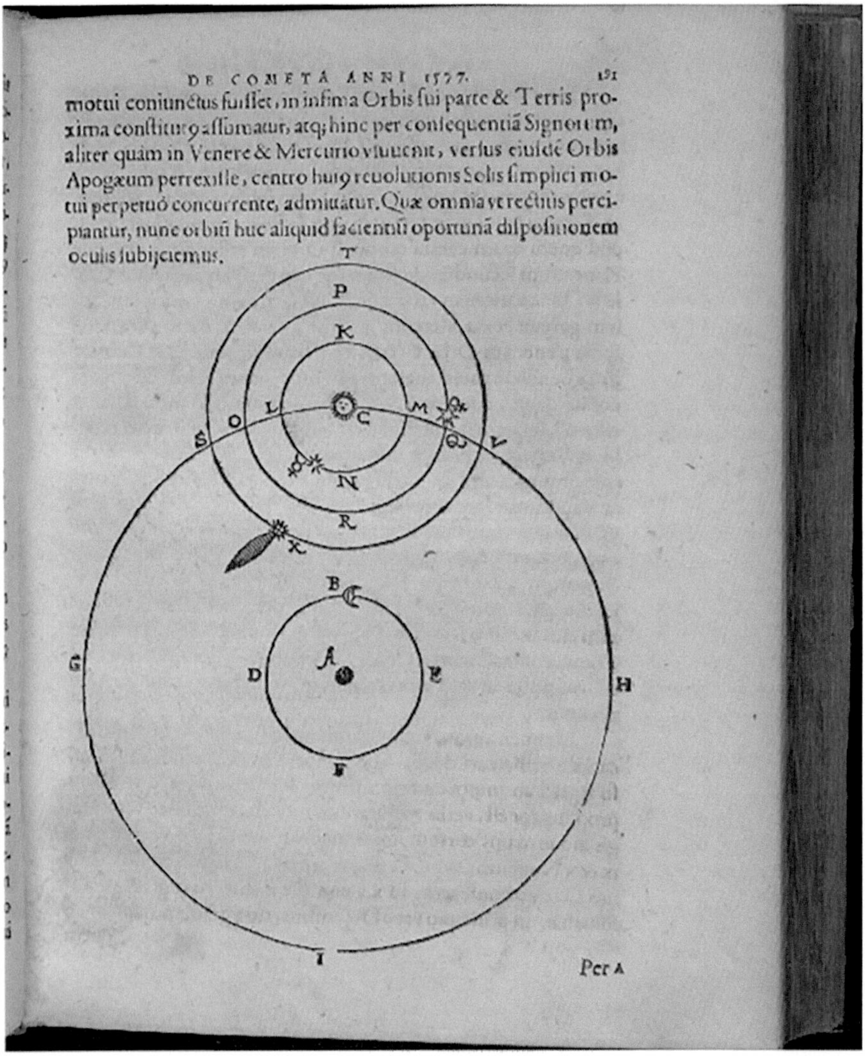
DE COMETA ANNI 1577. 191

motui coniunctus fuisset, in infima Orbis sui parte & Terris proxima constitutus assumatur, atq; hinc per consequentiā Signorum, aliter quàm in Venere & Mercurio viuuenit, versus eiusdē Orbis Apogæum perrexisse, centro huius reuolutionis Solis simplici motui perpetuò concurrente, admittatur. Quæ omnia vt rectiùs percipiantur, nunc orbiū huc aliquid facientiū opportunā dispositionem oculis subijciemus.

Fig. 7.4 The world according to Tycho Brahe (1588/1603. *Tychonis Brahe De mundi aetherei recentioribus phaenomenis liber secundus*: 191; Library of Congress: https://hdl.loc.gov/loc.rbc/General.94796.1, accessed 08 Dec 2021)

astronomy. Moving away from the concept of orbs, or rotating solid celestial spheres carrying the planets along, he introduced the concept of orbits—the paths that the planets take as they move through space. He said that this should result from the action of physical causes (Goldstein & Hon, 2005: 75). He conjectured that the Sun is rotating (a fact discovered later, telescopically, by Fabricius in 1611, from the apparent motion of the sunspots), and that the cause is in its magnetic action (Kepler, 1609: 3: XXXIV: 176):

> When William Gilbert the Englishman points out that the earth itself is a great magnet; and it is asserted by the same author Copernicus, it rotates day by day, as I conjecture that the Sun is turning. And because it has magnetic fibræ cutting the line of motion at right angles therefore those fibræ surround the Earth's poles by parallel motions in various circles so that now, rightfully, the Moon moves from the convolution of the same Earth carried off by the translation of the power of the remains, but thirty times slower (software assisted tr. by RCK).

Kepler, in his astronomical writings, had supported the heliocentric hypothesis of Copernicus (1473–1543) but departed from him by professing a radical change

Fig. 7.5 Johannes Kepler (1571–1630); Portrait by August Köhler, ca. 1910, after 1627 original. (Image: Wikimedia Commons)

that the planets moved around the Sun in elliptical orbits rather than in perfect circles.

In order of occurrence and with their respective modern designations and dates of occurrence, the comets of 1618 are, respectively:

1618 I (C/1618 Q1), discovered on 25 August from Cascheu, Hungary;
Perihelion August 17.627 UT).
1618 III (C/1618 V1), discovered on 11 November by Garcia de Silva y Figueroa from Isfahan;
Perihelion October 27.9.
1618 II (C/1618 W1), discovered on 26 November by Garcia de Silva y Figueroa from Isfahan;
Perihelion November 8.851.

The last two were sighted within a short span of time in the same region of the sky and were visible together for several successive nights. All of these comets were naked eye objects, with tail, motion direct and were noticed after their perihelion passages. Importantly, the year also saw the novel use of telescope for the observations of these three comets.

7.3 The Comets of 1618: *Disputatio Astronomica Publice Habita*

The 'Society of Jesus', founded in 1540 by St. Ignatius of Loyola, a Spanish soldier, had its roots in Rome. Its purpose was to propagate Roman Catholicism. The Collegio Romano (*Collegium Romanum*), which would become the Society's main scientific centre, was founded by him in 1551. The present-day Pontifical Gregorian University is the heir to the College. The Society has male members only, called Jesuits. Soon after its formation, the Society got into educational work.

In those times, the Society had members who were also excellent mathematicians and astronomers. The Jesuit Document, *Ratio atque Institutio Studiorum Societatis Iesu*, 1598–1599 (Method and Arrangement of Studies of the Society of Jesus, 1598–1599; Fig. 7.6) laid down in detail the regulations to be followed by the professors, teachers, and officials in the Jesuit education system. For instance, one of the instructions for the professors of philosophy read 'He shall not depart from Aristotle in matters of importance, unless he find some doctrine contrary to the common teaching of the schools or, more serious still, contrary to the true faith …' (Farrell [tr.], 1970: 40). What is remarkable is that, not many years later, on 13 May 1611 when Galileo Galilei (1564–1642) was in Rome, it was the Jesuits of the Collegio Romano who confirmed his astronomical observations, all contrary to the established knowledge and all that was being imparted in their institutions; see Sect. 7.9.5 also.

When many Jesuits were getting inclined towards the Copernican system giving up upon the old ideas, there came the Anti-Copernican Decree issued by the Sacred Congregation of the Index of the Roman Church on 5 March 1616 condemning *De Revolutionibus Orbium Cœlestium* by Copernicus along with two other works and decreeing that the theories about the mobility of the Earth and the stability of the

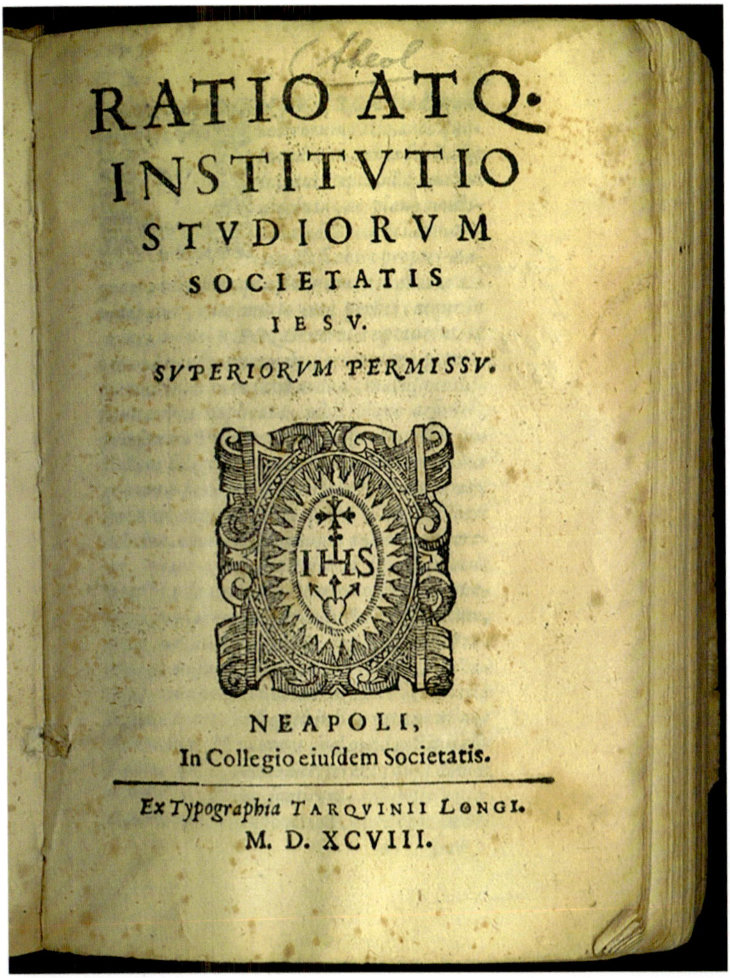
RATIO ATQ.
INSTITVTIO
STVDIORVM
SOCIETATIS
IESV.
SVPERIORVM PERMISSV.

NEAPOLI,
In Collegio eiuſdem Societatis.

Ex Typographia TARQVINII LONGI.
M. D. XCVIII.

Fig. 7.6 The Jesuit Document, *Ratio atque Institutio Studiorum Societatis Iesu*, 1598–1599. (Wikimedia Commons)

Sun were wrong and contrary to Scripture. The Decree unfolded a profound crisis between faith and science. The Jesuits distanced themselves from Copernicanism and turned to the Tychonic system (see Figs. 7.4 and 7.14).

Soon, in 1618, there appeared three comets in quick succession. The bright-tailed forms created a sensation in Europe, stirring men of faith and science alike since they raised more questions than answers. The apparitions became the testing ground for diverse worldviews through a spate of books and pamphlets that began to appear one after the other.

There ensued *Disputatio Astronomica Publice Habita* (publicly held astronomical discussion), and it went on for years. It also drew Galileo into a controversy with the Jesuit mathematician and astronomer Fr. Horatio Grassi (1583–1654) over the nature of comets. Grassi was a professor at Collegio Romano. He had observed the comets of 1618 and, in the book *De tribus cometis Anni MDCXVIII, Disputatio Astronomica Publice Habita* which carried the authorship as by one of the Fathers of the Society of Jesus at the Collegio Romano but is attributed to him (Grassi, 1619; Fig. 7.7), he presented charts of their paths against the backdrop of various constellations, vide Fig. 7.8. The chart at its top left shows the position of the first comet (i.e., 1618 I) as on 29 August and 2 September against Ursa Major. The first of the November-comets (1618 III) was shown as on 18 and 30 November against

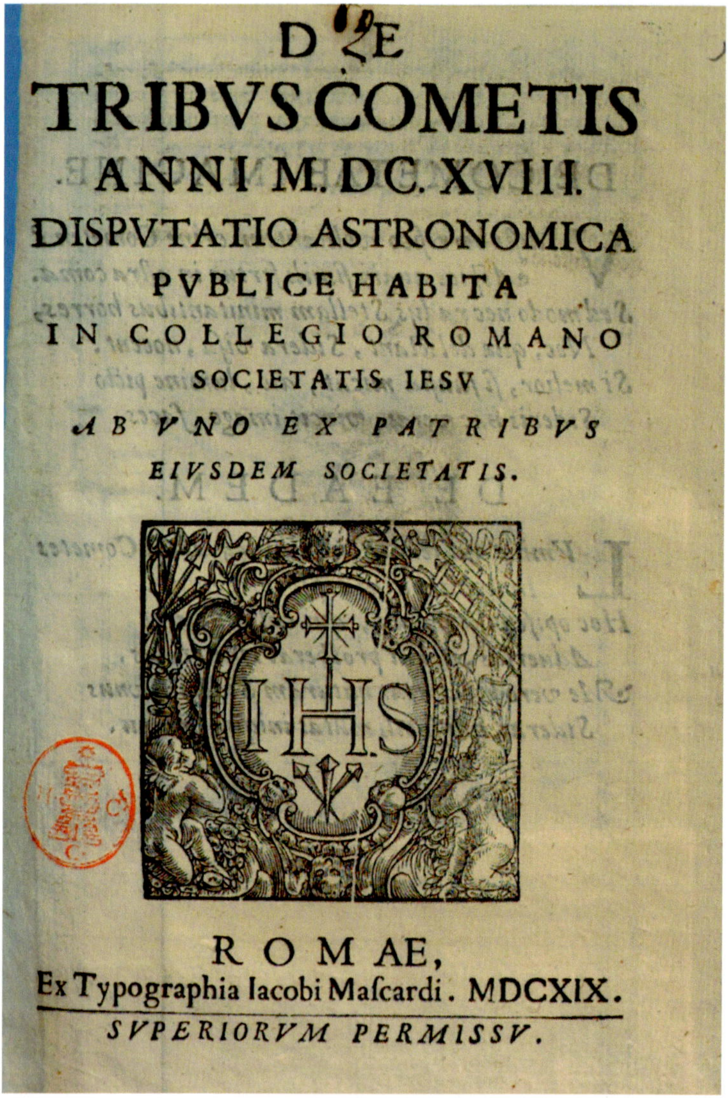

DE
TRIBVS COMETIS
ANNI M.DC.XVIII.
DISPVTATIO ASTRONOMICA
PVBLICE HABITA
IN COLLEGIO ROMANO
SOCIETATIS IESV
AB VNO EX PATRIBVS
EIVSDEM SOCIETATIS.

ROMAE,
Ex Typographia Iacobi Maſcardi. MDCXIX.
SVPERIORVM PERMISSV.

Fig. 7.7 *De tribus cometis Anni MDCXVIII, Disputatio Astronomica Publice Habita* by Oratio Grassi (1619). The book carried the authorship as 'by one of the Fathers of the Society of Jesus at the Collegio Romano'; cover page image from Google Books; accessed 29.11.2021

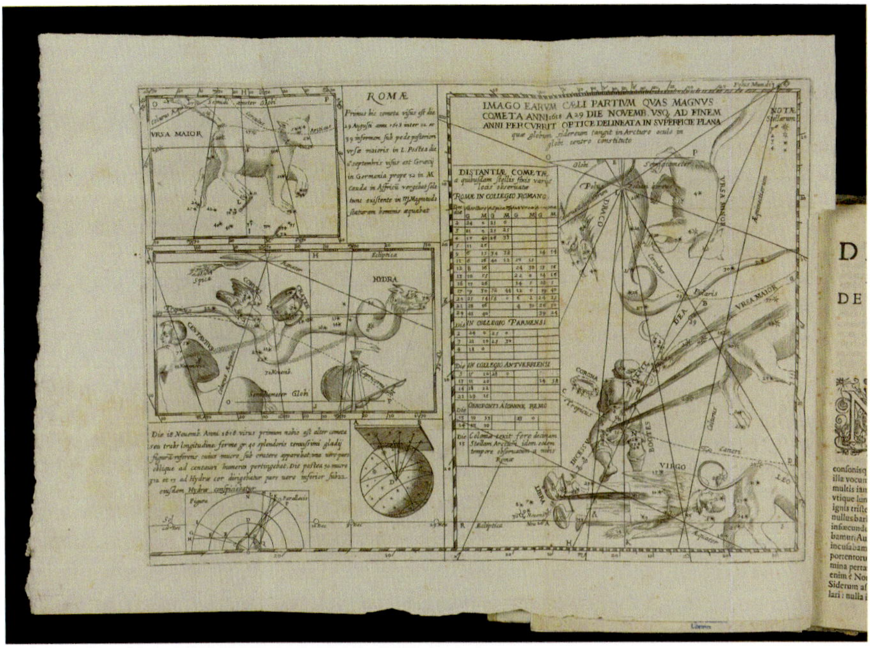

Fig. 7.8 The charts showing the positions of the three comets of 1618 against the various constellations in *De tribus cometis Anni MDCXVIII*; adopted from Lynx open Ed (2021)

the constellations of Corvus-Crater-Hydra. The five positions of the second November-comet (1618 II) from 29 November and 29 December were depicted against the sky between Libra-Virgo and Draco-Ursa Minor. Grassi showed that with no parallax having been detected, the comets lay far away from the Earth and beyond the Moon. He stressed that the apparitions were against the Copernican worldview and espousing the Tychonic view said that the comets moved in circular paths (Grassi, 1619: 13).

Galileo was indisposed at the time, so he responded through the work *Discorso delle comete di Mario Guiducci* (1619), which was published under the signature of his disciple Mario Guiducci (1585–1646) and questioned Grassi's arguments. He argued that if there was no parallax, it was because the comets were not physical bodies but refraction of light of the Sun on the thin vapours rising from the Earth, and therefore parallax had no relevance. Grassi then responded with *Libra astronomica ac philosophica* (1619) under the pseudonym Lothario Sarsi and contended that the comets were not optical phenomena. The book had directly mentioned Galileo and Guiducci by name. In response, Galileo came up with '*Il Saggiatore*' (The Assayer, 1623) in which he weighed the arguments and the criticism of him in *Libra astronomica.*

Kepler was abreast of the disputation and, since Tycho Brahe and his name also had been drawn into it, defending Tycho's name, he discussed the aforementioned works and put forth his own opinions on the subject in an appendix to his book *Tychonis Brahei dani Hyperaspistes* (1625; *The Shield Bearer to Tycho Brahe the*

Dane). Importantly, he professed, the comets differ from the pure substance of the æther:

> ut hæc mea Opinio, esset non mea, sed exponeretur in ipso ortu & abdicaretur, adoptata contraria, quam hactenus & pro mea agnosco & educo: quæ Cometas tam ipsos, quam eorum barbas, crines vel radios corpora statuit, densitatis & raritatis gradibus & inter se & ab ætheris purissima substantia differentia: Caput quidem, veluti nebulam conglobatam, quadamtenus pellucidam: Syrma verò illud seu Comam, effluuium ex capite, per radios Solis expulsum in plagam oppositam, in cuius effusionem continuam caput ipsum denique insumatur & consumatur; ut sit cauda veluti mors capitis (Kepler, 1625: 193).
>
> As this is my opinion, it was not mine, but was set forth in the very beginning and disowned, contrary to what had been adopted, which I hitherto acknowledge and bring up as my own: which determines the comets, as well as their beards, their hair, and their bodies; the degree of density and rarity and the difference between themselves and from the purest substance of the æther: The head, indeed, like a dense cloud, is to a certain degree transparent and that train, or long hair being the outflow from the head, expelled by the rays of the sun, in the opposite direction, into whose continuous outpouring the head itself may eventually be shed and consumed; that the tail may be like the death of the head (Kepler, 1625: 193; tr. by RCK).

Kepler's words were futuristic as they pointed towards the nature of the comets. The dispute was thought-provoking as it involved Galileo Galilei, Horatio Grassi, Mario Guiducci and Johannes Kepler bringing to the fore the highest in intellectual perception. Their works were translated and discussed in detail by Drake and O'Malley (1960); also, see Carolino (2023) and this volume.

7.4 The Comets of 1618: The Public Reception

The three bright comets of 1618 left an indelible imprint on people's minds. They also generated a grave concern among the general population. There were coins and medals cast to commemorate the three comets (see Walter, 1890: 84).

Figure 7.9 shows a German ducat (0.986 gold weighing 0.110902 oz) featuring one of the comets of 1618. The comet is depicted passing from right to left and so is a morning object. Which one of the three comets of the year does the ducat depict? According to Faintich (2007), in Frankfurt, gold and silver pattern ducats with a comet motif were issued on 19 November 1618. This comet was visible through 6–25 September; the comet should therefore be 1618 I. The celebrated comets of 1577 and of 1680—the latter being the first to be discovered with a telescope, also had been commemorated with medals.

The comet 1618 III was even observed by a young John Milton (1608–1674) when he was 10. How it made an indelible impression in Milton's young mind is reflected much later, in his poem *Paradise Lost* (1667, 1674: II: 706–711):

> On the other side,
> Incensed with indignation, Satan stood
> Unterrified, and like a comet burned,
> That fires the length of Ophiuchus, huge
> In the arctic sky, and from his horrid hair
> Shakes pestilence and war.

Fig. 7.9 German ducat of 1618. (After Faintich http://www.symbolicmessengers.com/newfindi.htm accessed in 2009 and on 30.11.2015)

The apparition of Comet 1618 II led King James I (1566–1625) to register royal reaction to the *Angry starre* of late autumn in what is now a famous poem (see Maclean, 1987: 74 and Cunningham in this book) –

Yee men of Brittayne wherefore gaze yee so,
Vpon an angry starre? When as yee knowe
The Sun must turne to darke, the Moone to
bloodde,
And then t'will bee to late to turne to good. …

The poem aimed to alleviate the fear that people then had that it was a sign of God's wrath. However, what is more interesting is what the King actually exclaimed before the Reverend Thomas Lorkin, as recorded in his communication of 1 December 1618 (11 December Greg) to Sir Thomas Puckering (Birch, 1849: II: 110):

> Concerning the blazing star, his majesty, they say, swears it is nothing else but Venus with a firebrand in her___.1 Mr. Briggs conceives it to be a perfect comet and therefore above the moon (so mathematicians have demonstrated Aristotle's tenet in this point to be false), and thinks the comet to be a hundred thousand miles in length – I say, thinks, because, as yet, his observation hath found no ground for geometrical demonstration.

In a footnote on the same page, Thomas Birch (1705–1766) added:

> The word omitted, if proper for a king to utter and a clergyman to repeat - of which we cannot entertain doubts - is certainly too objectionable to be printed.

The 'Mr. Briggs' referred to above was the English mathematician Henry Briggs (1561–1630), well-known for his pioneering work on common (base 10) logarithms. We may appreciate his perception of the comet's tail length and its place in the heaven, quite profound for the times.

The comet of late November 1618 was, in prognostications and in retrospection, also linked to the Great Thirty Years War of 1618–1648, a religious war that led to death and destruction and suffering, greatly affecting life in the Holy Roman Empire. As the comet blazed through the sky, it caused dismay and terror all through Europe. Figure 7.10 shows the cover of Paul Nagel's 1619 book, *Prognosticon astrologo-harmonicum*, which gave in detail the prognostications

Fig. 7.10 Comet of November–December 1618 in prognostications: featured on the cover of Paul Nagel's 1619 book—*Prognosticon astrologo-harmonicum, super tres vel plures etiam annos conscriptum: ausführliches Prognosticon uber drey oder mehr Jahr beschrieben von 1620 an zu rechen, dorinn erwiesen wird, in was für harmonia die Revolutiones erfunden warden* .. (Detailed Prognostications described over 3 or more years from 1620 onwards, it proves for what harmony the revolutions were invented ...). (Source: The Bavarian State Library/Google Books)

for 3 or more years from 1620 onwards. Nagel's was not the only work, and there appeared a flood of pamphlets, over a hundred in Germany itself, warning people of God's anger and asking them to repent and mend their sinful ways (see Theibault, 1994; Meinel, 2009). In the light of the above, we may call it a War Comet.

7.5 The Comets of 1618: In Brief

A brief description of the three comets is in order. For a detailed account of the observations and the contemporary discussions, one should refer to Nature (1878), Vsekhsvyatskii (1964), and Kronk (1999). The orbital elements of the Comets are listed in Table 7.1 (JPL, 2021).

7.5.1 Comet 1618 I

Comet 1618 I (C/1618 Q1) was discovered at Caschau in Hungary on 25 August in the morning skies at magnitude 2–3, and by Johannes Kepler (1571–1630) at Linz on 27 August. On 1 September, its tail stretched up to 5° long. It passed closest to the Earth at 0.5162*au* on 19 August and was last seen on 25 September. This comet holds the distinction of being the first to be observed with a telescope, on 6 September. This observation was made by Kepler, who described the view as a large object that resembled a cloud (Kronk, 1999: 333–335).

Table 7.1 Orbital elements of the 1618 comets (heliocentric ecliptic J2000)

	C/1618 Q1	C/1618 V1	C/1618 W1	
Element	1618 I	1618 III	1618 II	Units
e	1.0	1.0	1.0	
a	n/a	n/a	n/a	au
q	0.51298	0.744	0.38954	au
i	21.494	8.4	37.196	deg
node	298.844	229.5	81.001	deg
peri	24.802	155.9	287.436	deg
t_p	2312251.127	2312322.4	2312334.351	
	(1618-Aug-17.627)	(1618-Oct-27.900)	(1618-Nov-08.851)	
period	n/a	n/a	n/a	d/yr
Q	n/a	n/a	n/a	au

Notation: e = eccentricity, a = semi-major axis, q = perihelion distance, i = orbital inclination, node = longitude of the ascending node taken eastward from the vernal equinox, peri = angle of the perihelion from the ascending node measured in the ecliptic plane, t_p = perihelion passage, Q = aphelion distance

7.5.2 *Comet 1618 III*

The Great Comet designated 1618 III (C/1618 V1) was spotted earlier than 1618 II, on 11.04 November (UT) by Garcia de Silva y Figueroa (1550–1624) from Isfahan in Iran. It was visible in the south-eastern sky, and had a tail that was about 60° long (Kronk, 1999: 335–338). Figueroa (1967), who had travelled extensively through the country, including to the city of Shiraz and the ruins of Persepolis, was the ambassador of Philip III, the King of Spain and Portugal, to the court of the renowned Persian ruler Shāh Abbās I (1571–1629; r. 1588–1629). His observations are also mentioned in the *Cométographie* of Pingré (1784: II: 6–8).

Figueroa (1967: 272–273) put down his impressions of the two comets of November 1618 in the account (in French) of his travels through Persia,

> Vers le 8. ou 10 jour de Nouembre de cette année, quelques Domestiques de l'Ambassadeur, qui s'estoient levez de plus grand matin que le autres, vinrent rapporter, que depius deux nuicts ils avoient veu trois heures deuant le jour un grand Meteore au Ciel, & parce que le lendermain les autres Valets & plusieurs Armeniens assevroient avoir veu la mesme chose, sans s'accorder neantmoins en la figure, ny en la grandeur de cette nouvelle lumiere, l'Ambassadeur voulut se leuer le deux nuicts suivantes, quoy qu'il fist grand froid, & s'estant enveloppé de sa robbe, il se mit à considerer de dessus une grand terrasse, & à admirer ce nouveau Comete, qui estoit si grand, qu'il occupoit presque la quatriesme partie du Ciel. Il n'auoit pas le mesme esclat, que l'on voit aux autres Cometes, mais sa couleur estoit sombre & cendrèe, se rapportant à peu prés à celle, que l'on voit en la fumèe de la bonne poudre à Canon, ou en de la cendre fort claire. Il representoit en sa figure un coutelas à la Turque, fort courbé vers la pointe, & plus large, mais beaucoup plus estroite vers la garde. A ce bout l'on voyoit la matiere, de laquelle le Comete prenoit sa nourriture, & du costé oѝ il se penchoit, l'on vid la premiere nuict sortir de temps en temps de petites flammesches de feu. Depuis l'un des bouts, qui estoit vers l'Est, il alloit tousiours en s'clargissant de plus en plus, insqu'à ce qu'il demeurast fort large & courbé vers la pointe, paroissant ainsi comme le rejetton de la Palme, qui sette ses sueilles en arriere, Il occupoit la plus grande partie des signes de la Vierge, de la Balance & du Scorpion, & à sa main gauche, vers le Levant, il yauoit un autre Comete plus petit, & des communs, mais bien plus vis & plus enflammé que l'autre, avec lequel il se perdit quelques jours apres, à cause du voisinage du Soleil.
>
> Towards the 8th or 10th of November of this year, a few of the Ambassador's servants, who got up earlier in the morning before the others, came to report, that for two nights, they had seen three hours before the daylight a great Meteor in Heaven, and because the next day the other valets and several Armenians asserted seeing the same thing, without, nevertheless, agreeing in the figure, nor in the grandeur of this new light, the Ambassador wished to rise the following two nights, though it was very cold, and having wrapped himself in his robe, he began to view from above a large terrace, and to admire the new Comet, which was so large that it occupied almost the fourth part of the Sky, that one sees in the other Comets, but its color was dark and ash-like, roughly related to that which is seen in the smoke of the good gunpowder, or in very clear ashes, it represented in its face a Turkish cutlass, strongly curved towards the point, and wider, but much narrower towards the hilt. At this end we saw the material from which the Comet took its nourishment, and from the side where it leaned, we saw small dry flames of fire emerging from time to time during the first night. From one end, which was towards the east, it went all the way, widening more and more, until it remained very wide and bent towards the point, thus appearing as the offshoot of the Palm, who pulls his sweat backward, it occupied the greater part of the signs of the Virgo, Libra, and Scorpion, and to its left hand, toward the East, there was another smaller Comet, but

more viscous and more inflamed than the other, which he lost a few days afterwards, because of the proximity of the Sun (tr.: RCK).

Comet 1618 III is not listed in Vsekhsvyatskii (1964). It also does not figure in the lists of comets given by Hind (1852: 128), although it is mentioned on p. 21. The Roman College Jesuits followed the traverse of this comet across the sky from 28 November until 9 December, during which time it may have been visible for a large part of the night (Kronk, 1999: 337). This comet was also observed by Kepler from Linz, from 20 to 29 November. On the latter date, he saw it share the morning skies with the Comet 1618 II, the third bright comet of the year.

7.5.3 *Comet 1618 II*

As we gather it from Figueroa (1967: 272–273), the Comet 1618 II (C/1618 W1; Great Comet) probably was first seen on 23 or 24 November by him from Isfahan toward the east as a diffuse form and having the colour of Venus.

There is some confusion about the discovery date of this comet. In 1619, Fr. Horatio Grassi argued that among the most popular dates, namely 14, 26, and 29 November, the earliest observations that fitted well could be of 26 November only, although there were independent reports of its discovery on 24 November. The Danish astronomer Longomontanus also had some reservations about the correct dates and the chronology presented by Garcia de Silva y Figueroa (Kronk, 1999: 338–341).

As deduced by Kronk (1999: 338), the first to record the comet were the Chinese who found it in the constellation of Libra on 25.9 November with a tail more than 10° long. Elsewhere, the comet was observed, amongst others, by Fr. Johannes Cysat S.J. (1587–1657; Fig. 7.11), John Bainbridge, Johannes Kepler, the Koreans, and the Japanese. It passed closest to the Earth (0.358 *au*) on 6 December. Four days later, it exhibited an unusually long tail displaying coruscations which Longomontanus measured to be 104° (Hind, 1852: 15, 106; Nature, 1878: 247). During the month of December, its brightness reached first magnitude, and subsequently it dropped to the third magnitude (Vsekhsvyatskii, 1964: 113). François Arago (1786–1853), the French mathematician and astronomer, mentioned that the head of the comet split into several parts during December, a phenomenon observed by Cysat (Hind, 1852: 10; Lynn, 1889: 408) and also by Wendelin and Christoph Scheiner. It then appeared as a cluster of bright stars, each with its own tail and travelling together. Comet 1618 II was last seen on 22 January 1619, by which time it had faded to between magnitude 5 and 6 (Kronk, 1999: 339).

Fr. Johann Cysat S.J. (also Cysatus) at Ingolstadt was the first to use a telescope to observe this comet and detect structure in a comet's head (Cysat, 1619). He used 6 ft and 9 or 10 ft *tubum opticum* and could resolve the head into a nucleus surrounded by nebulous envelope (coma), with yet another luminous, though relatively fainter, appearance around it. Cysat (1619: 74) in his *Mathemata astronomica …*

Fig. 7.11 Fr. Johann Baptist Cysat (1587–1657); image from: *Wissensspeicher zur Geschichte von Ingolstadt* https://www.ingolstadt.de/stadtmuseum/scheuerer/personen/cysat-01.htm accessed 3.9.2021

drew the fine structure of the comet's head. Figure 7.12, adopted from his book, shows the nucleus and coma as on 1 December (top left), on 8 December (top right), on 17, 18, 20 December (lower left), and on 24 December (lower right). Cysat found the nucleus consisting of many bright spots. In his observation of 8 December, he depicts three bright stars one to the north, the other to the south, of the fifth and sixth magnitude, and each of the above stars, he said was bright like Jupiter's brightest companions.

Fr. Cysat also noticed unusual movements in the tail, and he noted that the path of the comet began to deviate from a stipulated straight line, commenting that

> This curvature (of the orbit) would be a phenomenon of great importance, if it could be confirmed by more observations' (Schreiber, 1904: 100).

One may gauge the significance of this observation from the fact that cometary orbits are so acutely elliptical or parabolic that Kepler and Galileo believed that comets travelled in straight lines.

From his observations of Halley's Comet in 1607, Kepler drew in his 1619 work *De cometis libelli tres*, the inference that this comet travelled in a straight line (see Yeomans et al., 1986: 82). In a beautiful diagram, reproduced in Fig. 7.13, he traced the path taken by the comet with respect to the Sun, giving its position on various dates of observations through 23 September–26 October, 1607 (Kepler, 1619: 8–9).

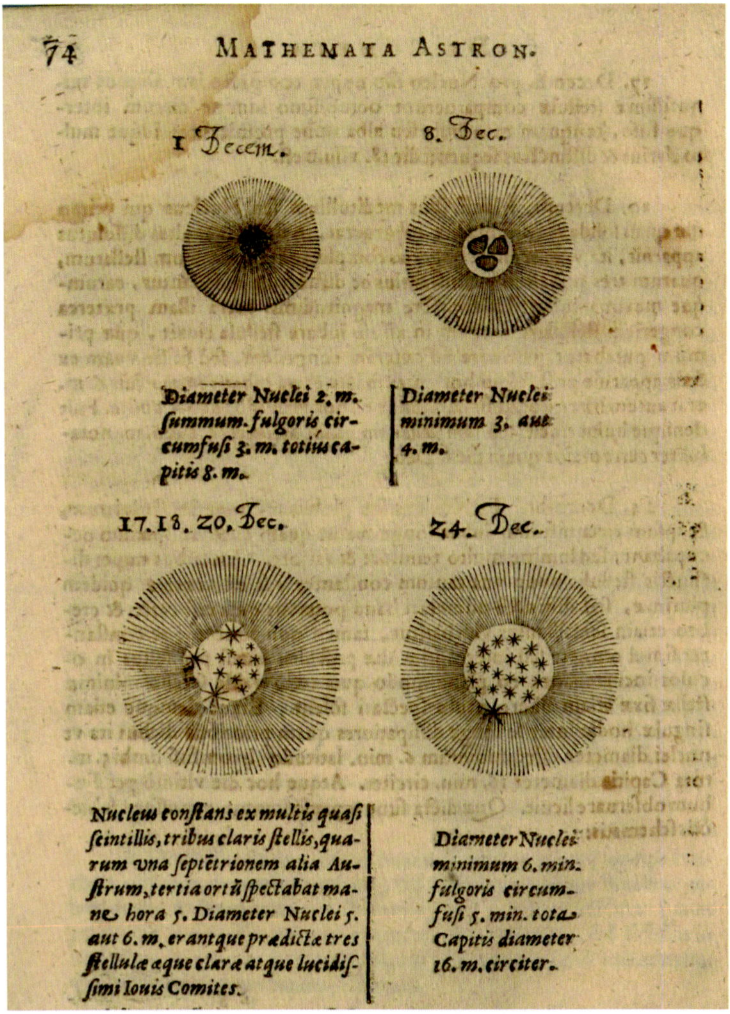

Fig. 7.12 The fine structure of the comet's head as drawn by Cysat (1619: 74) in his *Mathemata astronomica*. It shows its nucleus and coma as on 1 December (top left), on 8 December (top right), on 17, 18, 20 December (lower left), and on 24 December (lower right). (After; Hathi Trust, Google Books); accessed 13.11.2021

It shows the comet passing through the orbits of the Earth, Venus, and Mercury in a heliocentric arrangement. One may notice how the tail's position angle changed through the period. In his book *Astronomia Nova* (Kepler, 1609), Kepler had shown how the planets really moved in their orbits. He professed that the planetary orbits around the Sun are not perfect circles but ellipses with the Sun at one of the foci, and that the area swept out by a line connecting the planet to the Sun is always the same in a given time interval, irrespective of its position in the orbit. These statements now constitute the first two of Kepler's laws of planetary motion. A decade later,

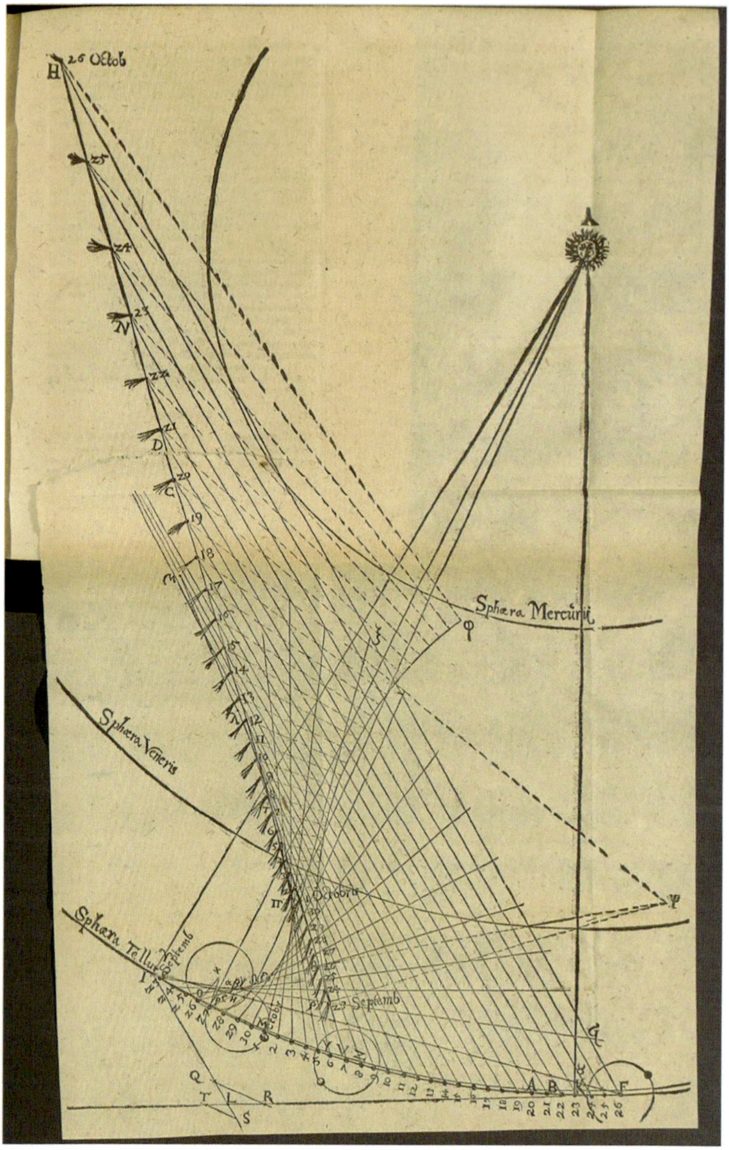

Fig. 7.13 Kepler drew in his 1619 work *De cometis libelli tres* (pp. 8–9) this impressive diagram to show the path taken by the comet of 1607 through the orbits of the Earth, Venus and Mercury. (Adopted from the copy of Kepler's book with ETH-Bibliothek Zürich, www.e-rara.ch)

Kepler proposed the third law that related the average orbital distance of a planet from the Sun to its orbital period, published in his book *Harmonices Mundi* in 1619.

Fr. Cysat (1619: 57) also depicted the Comet of 1618 in a diagram in his book *Mathemata astronomica* as according to the Tychonic system (Fig. 7.14). Here, the Moon and the Sun are shown moving about a stationary Earth and around

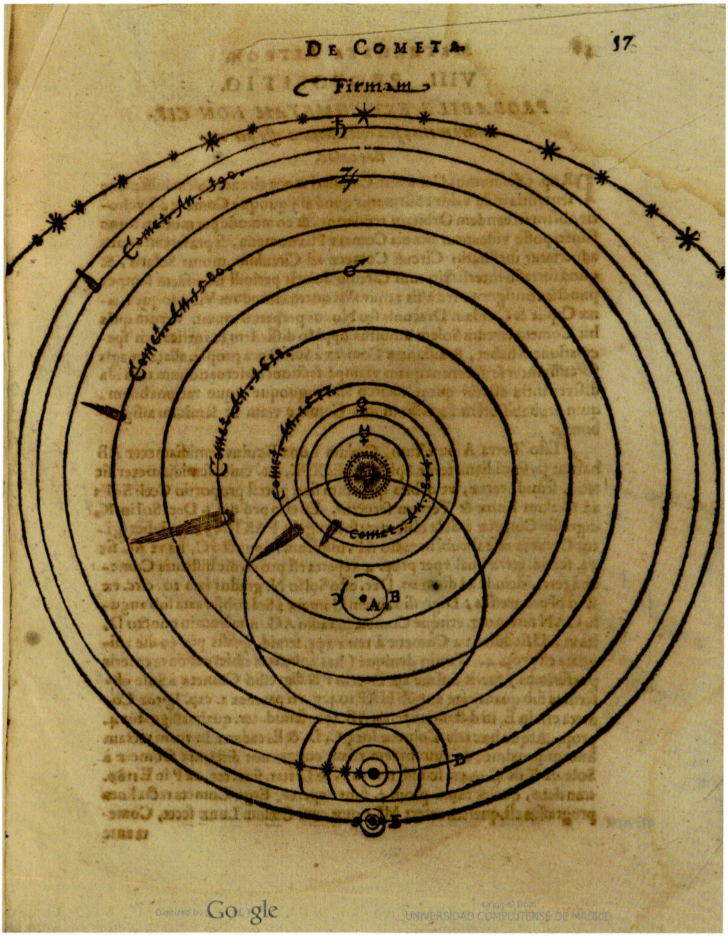

Fig. 7.14 Cysat's representation of the Tychonic system in *Mathemata astronomica de loco*: Earth A, Moon B, and the Sun C (Cysat, 1619: 57). (After: Hathi Trust, Google Books; accessed 13.11.2021)

these the planets and the comets, those that appeared in 390, 1580, 1618, 1577, and 844. Around the Sun are the sunspots, shown as a belt and considered by Cysat to be small planets. Jupiter is shown with the four Galilean satellites. The planet Saturn is shown with two satellites—that should be the rings, to be resolved much later in 1659 by the Dutch astronomer Christiaan Huygens (1629–1695).

It was only in 1687 that Isaac Newton (1642–1727) showed that cometary orbits took the form of a conic section, rather than a rectilinear one. Jeremiah Horrox (1619–1641) was, however, the first to conclude that the comet of 1577 followed a curvilineal path—'in an elliptical figure or near it' (Whatton, 1859: 15–16).

7.6 The Comets of 1618: A Few Asian Records

The bright comets of 1618 were duly recorded by the Chinese, Japanese, and the Koreans. These observations have been covered in Kronk (1999). Below are some from the Asian region.

Kennedy (1980) mentions a book titled *Tanbīhāt al-munajjimīn* (*Admonitions to Astrologers*) by Muzaffar b. Muḥammad Qāsim Junābādī (Gunābādī) which refers to several comets seen in history and also provides a classification. The author of the book cites Arabic, Egyptian, Greek, Indian, and even Iranian sources. The Indian works mentioned are –

> *Barahi (?), an Indian book, No. 17, f. 192r; Bistiham (?) the Indian, No. 25,ff. 197v, 200r.* is a name otherwise unknown to us. The author reports his views on the motion of tailed stars; Brahma and the Indian astrologers, ff. 184r, 192v

Barahi, mentioned above, is probably the Indian astronomer Varāhamihira (485–587 CE) and *Brahma* is Bṛhmagupta (b. 598 CE).

Junābādī points out that he witnessed the great comet of 1577 that appeared in the west towards 'the latter part of Sha'ban (ca. 2–12 November, 1577)'. Lastly, he refers to two consecutive comet sightings in the year 1027 A.H. The reference is made at two places. The first one reads as follows:

> in the beginning of Dhu al-Hijja 1027 H. (ca. 21 November 1618), while the royal court was at Qazwin (northwest of Tehran), a *harbah* appeared in the east, in the sign of Libra.

Then in a later chapter, the author describes how

> On the morning of Monday, 8 Dhu al-Hijja (26 November 1618) of the (above-) mentioned year, a comet (*Au dhawaba*) appeared in the east in the middle of the sign of Scorpio and lasted for about forty days.

The Royal Court referred to above is that of Shāh Abbās I of Iran to whom *Tanbīhāt al-munajjimīn* was dedicated. As for the 1027 A.H. dates, Kennedy (1980) observed that

> The two dates are practically the same, and the zodiacal signs are adjoining, but different names are used for the category of tailed star observed. We have no explanation.

We find that the references in the book *Tanbīhāt al-munajjimīn* are correct and made in respect of two different comets, which are now designated 1618 III and 1618 II (Kapoor, 2016).

One also comes across records of a number of comets in some Balinese historical sources. These chronicles are called *babad*, meaning opening up or clearing (Hägerdal, 2006). In the *babads*, many records use dates according to the Śaka calendar. The chronicle by the name *Pangrincik Babad* refers to a comet in the year 1618. About this, the description in entry 27 in Hägerdal (2006: 151) reads:

> 27. Babad lintang kukusan sumembur mamenjor, nora banyu bhudo sasih [1540].
> A comet gushes forth, making like penjor, 1618

Hägerdal explains the word *penjor* in the quote above thus:

a penjor is an offering that consists of a bamboo pole with decorations; its shape might have had a superficial likeness with the comet in question.

A comet from the same year is also mentioned in another chronicle named *Korn 4* (after the Dutch colonial administrator and scholar, Victor E. Korn [1892–1969]). According to Hägerdal (2006: 110), the comet alluded to in its entry 31 is possibly the last of the three comets of that year, which means 1618 II.

There are numerous references to be found about the natural phenomena in an early Annamese Historical source, '*Đai-Viêt Su'-ky Toan-Thu*', vide Ho Peng-Yoke (1964). These include the events such as the eclipses, planetary movements, meteors, comets and sunspots, of which many are attributed astrological connection. The Annamese had adopted the Chinese sexagenary cycle in their calendar. About the bright apparitions of 1618, there are two entries, reading as follows (Ho Peng-Yoke, 1964: 133–134):

During the last decade of the ninth month (of the nineteenth year of Hoăng-đinh reign-period) (7 November to 16 November, 1618) a 'black star' was seen during dan double-hour (03:00 to 05:00 hours) at the south, measuring over ten feet in length.

On the eleventh day (of the tenth month of the nineteenth year of the Hoăng-đinh reign-period) (27 November, 1618) at the fifth watch (03:00 to 05:00 hours) a *hui* comet appeared in the east. After more than a month it went out of sight.

As can be made out in the above, the comets being alluded to are 1618 III and 1618 II, respectively. The length 'ten feet' is supposedly 10° in angular measures.

Trinh Lords were the feudal chiefs who ruled Đại Việt, the Annam country which is the region around the present-day Hanoi, from 1545 to 1789. From 1545, they held all the power of the Lê Court. Trinh Tung (1550–1623) was the first Trinh Lord who ruled from 1572 to 1623. The comets of 1618 appeared in his time and were seen by the Annamese as signs from Heaven.

Dos Santos (2005) has brought out the observations of the November 1618 comets reported in some Jesuit communications and the stir the apparitions caused in Japan. There are details of the comets and their movement in a letter by Fr. Christovão Ferreira, from Japan. He was a Jesuit of Portuguese descent who had been taught astronomy at the Jesuit college in Coimbra by Christoph Grienberger (1561–1636). In 1600, he boarded in Lisbon to come over to Goa, but in 1601, he headed for Macao. Post his ordination, Ferreira headed in 1609 to Japan. In one of his annual letters to the M.R.P. General of the Society of Jesus, Ferreira writes in detail about

two Eastern stars, which throughout November appeared one after the other in the same month, as the newest and lightest that have been seen in these times. . .. on the 11th of November of the year 1618, at 10 am in this city of Manila there was seen a Comet towards the East tailed with a silvery color and a light gray of extraordinary shape……..

The comet had an extraordinary form, first like a trumpet, then a Japanese catan (resembling a cutlass) with its edge towards the southwest, and then assumed palm-shape.

On Saturday the 24th of the same month, another tailed comet appeared, more beautiful and resplendent than the first with the slope of a flaming star; It was seen from the East creeping towards the middle … (Annual Letters, 1621, pp. 332–333).

The comets lasted about 3 months and were also seen in Japan.

Fr. Christoforo Borri S.J. (1583–1632), an Italian Jesuit missionary, also had observed the two bright comets while in Vietnam. To this we shall return later.

7.7 The Comets of 1618: The Mughal References

The thirteenth century was a turning point in the history of astronomy in India with the entry of Islamic astronomy and its adoption, which flourished along with Hindu astronomy until around the nineteenth century. It led to amalgamation of the observational techniques and instruments of the former and the computational techniques of the latter, which paved the way for accurate astronomical observations. This overlapped with several stray instances of telescopic use in the seventeenth and eighteenth centuries for celestial events and geographical surveys, and the establishment of a modern astronomical observatory at Madras (now Chennai) in 1786 CE.

In the Middle Ages, astronomy in north India was mostly under the patronage of the Delhi Sultanate (1206–1526) and the Mughal emperors (1526–1858). Some of these rulers were interested in astronomy. In the Mughal writings of the sixteenth and seventeenth centuries, in the royal journals and the chronicles of the same period, there are accounts of serendipitous celestial and terrestrial phenomena such as comets, eclipses, meteors, and earthquakes.

Of the numerous comet apparitions during the Mughal period, we find brief accounts of only the Great Comets of 1577 and 1618, respectively, in *Akbarnāmā* and in *Tūzūk-i Jahāngīri* as also in a few chronicles. Abū'l Faḍl (1551–1602), the Prime Minister of the Mughal Emperor Jalāl ud-Din Muḥammad Akbar (1542–1605; r. 1556–1605), has recorded in *Akbarnāmā*, the highly acclaimed biographical account of Akbar, the appearance of a comet during the 22nd year of his reign, in 985 A.H. From the recorded date, Abū'l Faḍl turns out to be an independent discoverer of one of the most famous comets in history—the Great Comet of 1577 (C/1577 V1; Kapoor, 2015). These references assume significance in view of the fact that observations from Europe of the very same comets made a decisive impact on the course of astronomy.

7.7.1 *Mughal Emperor's Interest in Astronomy*

In the Muslim world, comets were regarded as ill omens for the rulers and so were routinely monitored by historians and chroniclers. Astronomical observations were mainly required for astrological purposes, to precisely determine the auspiciousness of the events and their timings, like Royal births, or when to embark on missions or campaigns, etc. Some of these records have little astronomical content, but they underline how these types of events got the rulers seriously concerned about their auspiciousness and possible consequences, so they even sought counsel for remedial measures. There was generally no clear division between astrology and astronomy, and superstition remained intertwined with both. Many medieval scholars considered astrology as part of astronomy, while others were opposed, saying that this did not conform to the principles of Islam.

So, what does one say when a ruler records, virtually in his own hand, accounts of some 'evil' celestial phenomena that occurred during his reign but with hardly a comment on their ominous side? Nūr ud-Dīn Jahāngīr (1569–1627), the fourth Mughal Emperor of India (r. 1605–1627) was such a person. He was just 8 when the Great Comet of 1577 appeared, and he probably witnessed it. Jahāngīr was a great naturalist, a gifted author, and in his writings, he displayed an excellent command of language. We note how his father, Emperor Akbar, '… paid very great attention to the education of his sons and grandsons, and appointed learned men of very high reputation to superintend their studies' (Law, 1916: 160). Apart from ornithology, biology, and lexicography, Jahāngīr had an interest in astronomy, and he maintained records of his observations in his journal *Tūzūk-i Jahāngīrī* (Memoirs of Jahāngīr; Rogers & Beveridge, 1909, 1914). In the other memoir that he wrote, *Wāki'āt -i Jahāngīrī*, we come across very similar descriptions (Elliot, 1873, 1875, 1975). In these *Memoirs*, we find descriptions of a few solar and lunar eclipses, a meteorite, and two Great Comets that he observed.

Jahāngīr's *Memoirs* clearly demonstrate his interest in astronomy and the level of precision reached in the observations. We find the recorded information in excellent agreement with modern computation. Jahāngīr's astronomers used instruments such as *ghati-yantra* (water-driven clocks; clepsydras), astrolabes, sundials, and hourglasses. One can find a depiction of many of the astronomical instruments then in use in a number of Mughal miniature paintings of the sixteenth and seventeenth centuries. One such example is presented here in Fig. 7.15, a margin drawing from the folio of Jahāngīr's Album depicting an astrologer surrounded by his equipment.

Jahāngīr has written in *Tūzūk-i Jahangīrī* about the circumstances of fall of an iron meteorite in a village in Jalandhar district in Punjab on 30 *Farwardin*, in the sixteenth regnal year, that is 19 April 1621 Greg (but Howe, 1896: 296 cites the year as 1620; also see Khan, 1935). The iron that was extracted from the site weighed 1.93 kg (Blochmann, 1869: 167–168), and the meteorite broke under the hammer. The sword-maker got Emperor two sword blades, a dagger, and a knife using three parts of the meteorite for metal to which the sword-maker added one part of terrestrial iron. One of the coveted artifacts, the dagger, is with the Smithsonian Institution's Freer and Sackler Galleries (Sabri, 2012).

There is mention in *Tūzūk-i Jahāngīrī* of the total lunar eclipse of 1018 A.H. (Rogers & Beveridge, 1909–14: I: Chap. 6). The solar eclipses of 1019 A.H. and 1024 A.H. (15 December 1610 and 29 March 1615 Greg, respectively; both annular) also were recorded. In the matter of the solar eclipse of 1024 A.H., he states also that on the occasion,

> Alms of all kinds, and things in the shape of metals, animals, and vegetables, were given to fakirs and the poor and people in need. On this day the offering of Rāja Sūraj Singh was laid before me; what was taken was of the value of 43,000 rupees. The offering of Bahādur Khān, the governor of Qandahar, was also laid before me on this day; its total value came to 14,000 rupees. (Rogers & Beveridge, 1909–1914 (I): Chap. 12).

Fig. 7.15 An early seventeenth century margin drawing from the folio in Jahangir's Album showing an astrologer surrounded by his equipment—an astrolabe, zodiac tables and an hour glass. (Courtesy: Werner Forman Archive/Naprestek Museum, Prague)

In connection with the marches, too, an auspicious hour mattered.

Zodiacal gold coins and silver rupees struck in the names of Jahāngīr and his queen Nūr Jahān respectively, were issued from the 13th year of his reign (1027 A.H.). These coins have a zodiacal sign on the obverse and a Persian verse on the reverse. Jahāngīr ordained that rather than carry his name on the obverse and the name of place and the month and year of his reign on the reverse, each coin would feature a zodiacal symbol for the name of the solar month it belonged to, with the Sun emerging from this symbol (Brown, 1922: 95–96). On the obverse, these symbols are depicted with the (rising) Sun in the background. These coins, which clearly reflect Jahāngīr's inclinations, are fine examples of numismatic art and quickly became collectors' items.

Some very fine Mughal miniatures were made by Jahāngīr's court artists, like the one by Bichitr reproduced here in Fig. 7.16 where the Emperor is shown proferring a Sufi Shaikh to kings. Jahāngīr is seated on an hour-glass throne, and the halo behind his head has a golden Sun and a crescent Moon.

Fig. 7.16 Jahāngīr Proferring a Sufi Shaikh to Kings. Jahāngīr is seated on an hour-glass throne and presents a book to the Sufi. This miniature by Bichitr is part of an album made for Emperor Jahāngīr circa 1615–1618; from the St. Petersburg album—Google Art Project.jpg. (Wikimedia Commons)

7.7.2 Emperor's New Find

In 1910, when Halley's Comet was yet to reach naked eye visibility, it had already stirred up the cognoscente of India. Jivanji Jamshedji Modi (1854–1933), a renowned scholar of Sanskrit, Persian, and the Avesta, carried out a comprehensive examination of the accounts of Muslim scholars and those in the books of ancient

Persians of the records and sightings of comets over the ages. On 9 February 1910, he presented his findings at a meeting of the Bombay Branch of the Royal Asiatic Society. Modi (1917) gave a remarkable exposition on the 'Mahomedan view of comets' and he listed comets that the Muslim scholars of the Middle Ages came to know about through historical documents, literary works, and records, or they had actually observed during their own lifetimes. Modi (1917: 84–86) discussed at length Abū'l Faḍl's discourse on comets, his record in *Akbarnāmā* of the appearance of the Great Comet of 1577, and, Jahāngīr having recorded two comets that occurred in the year 1618. Modi consulted Russel Hind's (1852) book on comets and Ferguson's Astronomy (Brewster, 1811 (II): 360) to identify the comets described in Jahāngīr's Memoirs. Modi identified the first of the comets as the one seen in November 1618 with perihelion on 8 November. For the other, the only candidate he could find was the comet of August 1618. These comets respectively are 1618 II (C/1618 W1) and 1618 I (C/1618 Q1). However, the comet 1618 I had appeared months before Jahāngīr's dates of observations; it was discovered on 25 August and last seen on 25 September by Kepler. The period of visibility does not match Jahāngīr's whereabouts at the time and his sighting of the two comets within a span of 16 days.

More recently, Mousavi (2000: 114–115) and Ansari (2002: 257) mentioned that Jahāngīr recorded a comet that was visible in early November 1618. Their references correspond to the comet 1618 III. Ansari (ibid.) cites the period of Jahāngīr's observations as 11 November–9 December, 1618, whereas Mousavi (2000: 115) cites the date of first observation as 'a few days before 18 Ābān 13 R.Y. [= 3 November 1618, Greg.]'. The converted date of 3 November, although a Saturday—just as in Jahāngīr's account—is not consistent with the course of the comet 1618 III. On 3 November, 1618, the comet 1618 III was still an evening object (no records of its having been spotted), whereas Jahāngīr's are morning observations. On the other hand, Ansari (2002: 257) first record date as 11 November, could be acceptable except that it was a Sunday.

So, which of the three comets of 1618 did Jahāngīr observe and on which dates? As the present study concludes, the comets in question are the ones designated 1618 III and 1618 II, appearing in that order and the right dates of Jahāngīr's first record are, respectively, 10 November for the comet 1618 III and 26 November for comet 1618 II. Muslim historians are acknowledged for their chronological precision, and Jahāngīr's dates can be relied upon (**Note 1**). In what follows, we shall deliberate upon Jahāngīr's description. We then move on to the observations made by other observers from India (at times with a telescope), and then to India-related stories of the two great comets.

About the march of an advance of the *Lashkar* (troops) from Gujarat to Agra in the year 1027 A.H., Jahāngīr writes:

> On Thursday, the 7th, with great joy and congratulation, the advance camp was started towards Agra. The astrologers and astronomers had already fixed the auspicious hour for the march. As excessive rain fell, the main camp could not cross the river of Maḥmūdābād (the Vātrak) and the Mahī at this hour. Out of necessity, the advanced camp was started at

the appointed hour, and the 21st Shahrīwar was fixed for the march of the main camp (Rogers & Beveridge, 1914 (II): 25).

The Mahi flows north from Vadodara and eventually enters the Arabian Sea, whereas the Vartak flows through Kheda, a district bordering Ahmedabad and Vadodara. The first date above corresponds to 29 August, 1618 (Greg.). That is just days after the Comet 1618 I was discovered on 25 August. It is surprising that no chronicler mentions this object in his writings. No one recalled it either when, later during the month of November, two more bright comets appeared.

7.7.3 The Many Versions of Jahāngīr's Observations

The following passage from Jahāngīr's Memoirs *Wāki'āt -i Jahāngīrī* (Elliot, 1867–1877, 1875: VI: 363; Elliot, 1975: 88–89), pertains to the account of the 13th year of his reign, i.e., 1027 A.H.:

> Saturday, 17th Zi-l Ka'da: Several nights before this, a little before dawn, a luminous vapour, in the form of a column, had made its appearance, and every succeeding night it arose half an hour earlier than on the preceding night. When it had attained its full development, it looked like a spear with the two ends thin, but thick about the middle. It was a little curved like a reaping-sickle, with its back towards the south, and its edge towards the north. On the date above mentioned, it rose three hours before the sunrise. The astronomers measured its size with their astrolabes, and, on an average of different observations, it was found to extend 24 degrees. Its course was in the empyrean heaven, but it had a proper motion of its own, independent of that firmament, as it was retrograde- first appearing in the sign of the Scorpion, then in that of the Scales. Its declination was southerly. Astrologers call such a phenomenon a spear, and have written that it portends evil to the chiefs of Arabia, and the establishment of an enemy's power over them. God only knows if this be true!
>
> Sixteen nights after its appearance, a comet appeared in the same quarter, having a shining nucleus, with a tail in appearance about two or three yards long, but in the tail there was no light or splendour. Up to the present time, nearly eight years have elapsed since its first appearance, and when it disappears, I shall take care to record it, as well as the effects which have resulted from it.

Modi (1917: 86) interpreted the above translation in Elliot to mean that it is not that the comet continued to be seen for 8 years, but rather the reference alludes to the supposed disastrous and unlucky influences the comet had lasting as long. However, the mysterious '8 years' appear elsewhere too, in an account of the apparitions by another historian. Modi (1917: 78) cites an account of the sighting of the two comets of 1618 from Mū'tamad Khān's *Iqbālnāmā-i Jahāngīrī* that is also a part of Elliot's History of India (Elliot, 1875: VI: 406–407). Mū'tamad Khān's description matches Jahāngīr's in the *Wāki'āt-i Jahāngīrī*. Only the date, 'On the 16th of De', differs. According to Modi (1917: 86), this may have arisen from Mū'tamad Khān mistaking 'Mahomedan month Zi-l Kada for the "Ilahi De"'. The mismatch also could be a result of paraphrasing or translation. However, Mū'tamad Khān adds the evil aspects of the comet's appearance:

> It was in consequence of its appearance that a pestilential disorder (*waba-o-ta aun*) spread throughout this extensive country of Hindustan, which exceeded everything known and recorded in former ages, nor is there any mention made of such in the authentic works of the Hindus. The pestilence arose in the count cne year before the appearance of the phenomenon and continued to rage for eight years, It was also through the effects of this phenomenon that a misunderstanding arose between His Majesty and the fortunate Prince Shah Jahan. The disturbances which thus originated lasted seven or eight years. What blood was shed in the country! and what families were ruined!

Mū'tamad Khān (d. 1049/1639) was a courtier and favourite of Jahāngīr and had taken up the task of continuation of the Royal *Memoirs* after the seventeenth regnal year, under the Emperor's supervision. *Iqbālnāmā-i Jahāngīrī* is his independent work finished in 1029/1620 (Hadi, 1995: 443).

From Rogers and Beveridge (1914 (II): 48; see also Baber, 1996: 83), a translation of the part on the sighting of the comets from the *Tūzūk-i Jahangīrī* is reproduced below that does not contain the confusing 'eight years' in the crucial part of the narration:

> On Saturday the 18th (Aban), the camp was at Ramgarh. For some nights before this there appeared, at three *gharis* before sunrise, in the atmosphere, a luminous vapour in the shape of a pillar. At each succeeding night it rose a *ghari* earlier. When it assumed its full form, it took the shape of a spear, thin at two ends, and thick in the middle. It was curved like a sickle, and had its back to the south, and its face to the north. It now showed itself a watch *(pahar)* before sunrise. Astronomers took its shape and size by the astrolabe, and ascertained that with differences of appearance it extended over twenty-four degrees. It moved in high heaven, for it was first in Scorpio and afterwards in Libra. Its declination (*harakat-i-arz*) was mainly southerly. Astrologers call such a phenomenon a spear (*ḥarba*) in their books, and have written that its appearance portends weakness to the kings of Arabia, and points to their enemies prevailing over them. God knows! Sixteen nights after this phenomenon, a star showed itself in the same quarter. Its head was luminous and its tail was two or three yards long, but the tail was not luminous. It has now appeared for eight nights; when it disappears, the fact will be noticed, as well as the results of it.

A *ghari* (*ghati*) as a unit of time is equivalent to 24 min (1 *ghati* = 1/60 of a day). It is obvious from the above quotation that the observations refer to the sighting of two comets in quick succession.

7.7.4 The Royal Traverse and the First Observations of the Comets

Let us ascertain the Gregorian dates of the observations from the chronology of the traverse of the Royal *Lashkar* as entered in the *Memoirs*, keeping in mind the paths of the comets, and see if these reconcile with the observations made elsewhere. The recorded dates differ but 'Saturday' is common in the English translations of both the *Memoir*-extracts. A clue to the chronology could come from another part of the *Tūzūk-i Jahangīrī* wherein, while referring to a particular day, Jahāngīr enters in the same passage a date according to two calendars together. He notes that:

On Saturday, the 11th, the auspicious equipage alighted in the pargana of Dohad…On the eve of Sunday, the 12th of the Ilāhī month of Ābān, in the thirteenth year from my accession, corresponding with the fifteenth Z̲ī-l-Qa'da of the Hijrī year 1027, in the nineteenth degree of Libra, the Giver of blessings gave my prosperous son S̲h̲āh-Jahān a precious son by the daughter of Āṣaf K. I hope that his advent may be auspicious and blessed to this everlasting State. (Rogers & Beveridge, 1914 (II): 47).

That is about the birth of Aurangzeb, who would be the sixth Mughal Emperor in the times to come (ruled 1658–1707) and where Āṣaf K is for Abdul Hasan Āṣaf Khān, the father of Arjumand Bānu Begum (Mumtāz Mahal). Corresponding to the date 15, *Dhu-al-Qa'dah*, 1027 A.H., not only the day is *yawm as-sabt* (Saturday), the equivalent Persian date 12 *Ābān*, 997 also is a *Shanbeh* (Saturday). The corresponding Gregorian date is 3 November, 1618 CE, Saturday (October 24, 1618 Julian). It was a Full Moon on the night of 2/3 November, 1618. Making 12 *Ābān* a Sunday would be in conflict with the timetable. As for the phrase '… in the nineteenth degree of Libra …', that might be about the Ascendant. Now, note that while a '… formidable sign …' appeared in the heavens in a week after the Royal birth, followed 16 days later by another one and as the two went on to dominate the morning skies, neither Emperor nor any chronicler connected these to the Royal birth.

We see that the double entry of the crucial dates in the *Memoirs* still leaves some ambiguity. Considering that the date 15 *Dhu-al-Qa'dah*, 1027 A.H. can still be our reference, the dates in the *Memoirs* help us follow the course of the Royal traverse from Ahmedabad to Agra, through Ujjain in Madhya Pradesh. Beginning with Dohad (Saturday, the 11th day of *Ābān*), the Royal party halted on the way at the village of Samarna (or Samarni/Tamarna; on the 15th day), Ramgarh (on the 18th), Sitalkhera (or Sambhalkhera; on the 20th), the parganas of Madanpur (Badhnur or Badnawar; on the 22nd) and Nawari (on the 25th), the banks of the Chambal River (on the 26th) and the banks of Kahnar River (27th) (Rogers & Beveridge, 1914: II: 48–49). Jahāngīr further notes:

On Tuesday, the 28th, the royal standards were raised in the neighbourhood of the city of Ujain. From Ahmadabad to Ujain is a distance of ninety-eight *kos*. It was traversed in twenty-eight marches and forty-one halts—that is, in two months and nine days.

The first observations of the comet would have been made in this period. Some of the halts mentioned in the foregoing account are not readily identifiable. Dohad, about 200 km east of Ahmedabad and about the same distance from Ujjain by road (see Fig. 7.17), is where Jahāngīr's grandson, Aurangzeb, was born. It is near the border of the state of Gujarat with Madhya Pradesh.

In the Survey of India's State Map of Madhya Pradesh (scale 1:1,000,000; 1978), Ramgarh, Nawari, and the river Kahnar do not figure, but we do find the Chambal and Gambhir Rivers near Ujjain, on the stipulated route. Notably, the Memoirs do not refer to the Kshipra (Sipra) River that flowed to the immediate west of the city of Ujjain and fell on the Royal route. We looked through the District Census Handbooks (DCH) of the Census of India 2011, released in 2015 by the Registrar General & Census Commissioner, India, for the districts of Dohad in the state of Gujarat and for Jhabua, Dhar, and Ujjain (DCH Ujjain, 2015) in western Madhya

Pradesh. It is these districts, spread along a west-east corridor, through which the stipulated route lay. These DCHs list every village and town in the districts, along with the census data.

We did not find any place of interest in the DCH of Dohad (DCH Dohad, 2015). There is a state highway, SH 18, running through the last three districts, and this connects to its counterpart in the State of Gujarat. We cannot say if this highway or a part or parts of it actually represents the medieval route between Dohad and Ujjain, but we can identify two major places where halts may have occurred. SH 18 passes by the Thandla Tehsil (a block of a district) of Jhabua and the Badnawar Tehsil of Dhar, eventually touching Ujjain. Figure 7.17 shows certain modern locations that lie on or near the route that the Royalty would have followed in 1618.

In the census listings of Jhabua, there is a 'Bherugarh alias Ramgarh' in Thandla block (DCH Jhabua, 2015: 60), and, a Ramgarh (23.0588°N, 74.8256°E) in the adjoining Petlawad block (DCH Jhabua, 2015: 66) near the SH 18. In Fig. 7.18, Ramgarh is in the middle near the top. It is situated near the Mahi River, a natural place for the troops to halt and set up camps at. Also, about 30 km east of Ramgarh, there is a place by the name Semal Khera on the SH 18 towards Badnawar (DCH, Dhar, 2015: 66). That is reminiscent of the place Sitalkhera/Sambhalkhera mentioned in *Tūzuk-i-Jahangīri.*

On the west side of Jhabua, the SH 18 stretches from the town of Thandla to join a road at the state border with Dohad. The names Madanpur and Nawari, as such do not figure in any of the DCHs. However, Rogers and Beveridge (1914: 49) suggested that Madanpur is the present-day Badhnawar, which does fit into the sequence.

The Chambal River splits into the Chamla and Chambal Rivers in the Khachrod Block in the northern part of the Ujjain district. The western component, the Chamla River, passes close to Badnagar in the district of Ujjain, and the eastern component,

Fig. 7.17 A map showing certain modern locations in the states of Gujarat (on the left) and Madhya Pradesh that might have been on or near the route Jahāngīr took from Dohad to Agra through Ujjain in 1618 (Location map by the late Dr Baba Varghese, prepared in 2015)

Fig. 7.18 Ramgarh (23.0588°N, 74.8256°E; 500 m) near the top in the middle in the Petlawad Block; Petlawad is at lower left. This place is near the SH 18, and also close to the branch of Mahi River, a natural place to set up a camp at (Google Maps)

the Chambal River, flows between Badnagar and Kharent, about 10 km west of Ujjain on the SH 18. Considering the geography of the region, we believe Badnagar can be identified with the Nawari of the *Memoirs*. The Gambhir River passes by Kharent. Finally, Samarna could be identified. Only, it should be west of Ramgarh. The districts of Dohad and Jhabua both have hilly, undulating terrain. Only modern on-site explorations, coupled with a re-look at the place names in the *Memoirs* and the route, will allow us to confirm the identifications suggested above.

We now proceed to place the Sun and the Comet 1618 III in the sky of Ramgarh by computation. Any discrepancy with respect to the location of the actual Ramgarh of the *Memoirs* would be too small to influence the scenario we now present.

In *Tūzuk-i-Jahangīri*, the celestial positions of the comets are given zodiac-wise only, but the computed ephemerides can help us fix the dates of the first sightings. We take the first date of observation at Ramgarh to be Saturday, 21 *Dhu-al-Qa'dah* 1027 A.H. (corresponding to the 18th of *Ābān*, Saturday). The date converter www.islamicfinder.org gives this as Friday, 9 November, 1618 (Greg). The converter indicates that there is a small probability of a day's error in conversion but this date is the same as those derived using other converters (such as www.imcce.fr, www.iranchamber.com and the CalendarHome.com). The last converter also gives the corresponding Persian date as *Jomeh* 18, *Ābān*, 997. As the date 21 *Dhu-al-Qa'dah* begins at sundown (Friday, 9th November until midnight) and the observation is made about 12 h later, the morning of 21 *Dhu-al-Qa'dah* 1027 A.H, Saturday corresponds to the morning of Saturday 10 November 1618 (Greg). The observation was made before the sunrise at Ramgarh.

Comet 1618 III had passed perihelion on 27.9 October ($q = 0.744$ au), about 2 weeks before its discovery. Jahāngīr records that the comet moved westwards from Scorpio into Libra. The computations place the Sun on 10 November a few degrees east of α Librae and in the middle of the zodiac. In terms of the signs, that is being ~17° in Scorpio.

The sky view in Fig. 7.19 corresponds to 10 November 1618 at sunrise (01:13 UT) at Ramgarh; north is at the top. The field of view is 45° and the bull's eye at RA 14 h 59 h 27 s, Dec. –19°14′03″, the apparent position of the comet with respect to the true equator and equinox of date as computed with the Horizons system. The blue line is the equator and the red line the ecliptic; north is at the top. A line through the stars ζ Herculis, the Sun, and ε Centauri (not within the field) roughly defined the eastern horizon at the time of sunrise. The ecliptic (the red line) is roughly vertical to the line defining the eastern horizon.

As per the computation, the comet began to lead the Sun on 9 November at 23:38 UT. At this time, the Sun was below the horizon, at an altitude −22°. As the nautical twilight (Note 2) passed, in a dark sky, the comet's tail would begin to be noticeable.

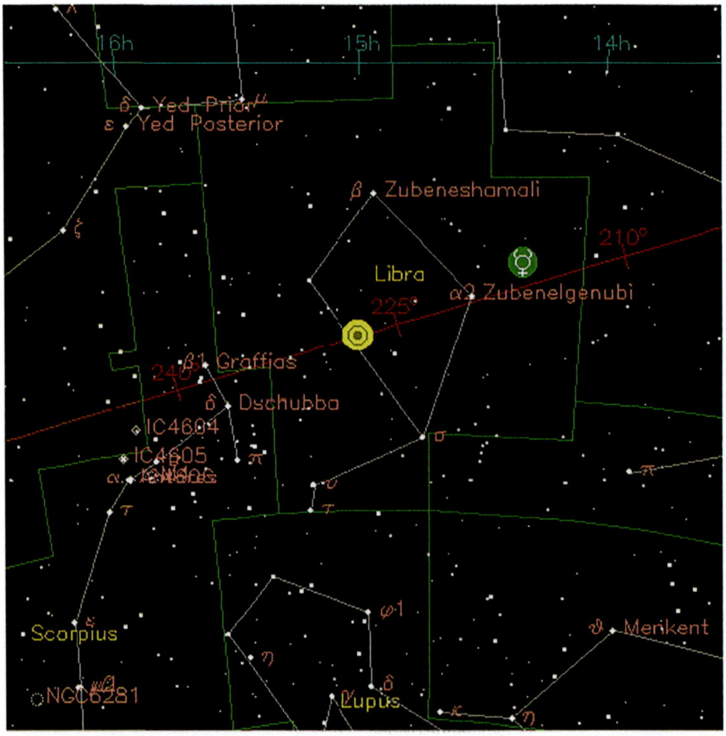

Fig. 7.19 The 45° field of the apparent sky on Saturday, 10 November 1618, 01:13 UTC, the sunrise time at Ramgarh, centred at RA $14^h\ 59^m\ 27^s$, December −19°14′03″—the computed position of the comet at that moment; the blue line is the equator and the red line the ecliptic; north is at the top. (Generated from Fourmilab's Your Sky)

At sunrise (01:13 UT) by which time the UT had changed to 10 November, the comet head, azimuthally 2° south of the Sun, was ~1° below the horizon; as per the *Memoirs*, its tail would have been curved and placed in the north-west quadrant of the sky.

This scenario can be better visualized by referring to an observation of this comet by an observer in Rome on 11 November, where the head was not noticed but the tail extended to the constellation Corvus and remained visible until twilight. However, see Kronk (1999: 335) on ascertaining a date according to the convention of the beginning of an official day in those times, and also **Note 5** below. The elongation of the star, say, β Corvi from the Sun that morning was ~40°, which points to the length and position angle of the tail. In fact, between 10 November and 11 November, the PA of the tail, and its length, would change noticeably. However, the perceived length depended in part on haze and twilight. At sunrise on 10 November, the star β Corvi had reached altitude 30° and azimuth 133°. With the head of the comet still very near the Sun, the tail had to be dense and bright in order to fit the description that it was long and magnificent.

In 1985, Landgraf computed the orbital elements of the comet C/1618 V1 on the basis of observations made from 11 November to 9 December 1618. He assigned 225°, −3° to the ecliptic position of the comet's head for the observation of 11.15 November. Landgraf also claimed that the perihelion date could be wrong by ±2.5 days (Kronk, 1999: 335–337). The traverse of the comet computed with the Horizons system that uses Landgraf's orbital elements is as follows: in Libra (10 November), Libra (14 November), Libra-Virgo-Hydra (15 November), Virgo-Hydra-Corvus (19 November), Corvus (23 November), Crater-Hydra (29 November), indicating that it moved mainly westwards. In the orbit determination, Jahāngīr's observations (as also those of the Jesuit missionaries in Goa; see later) are not part of the initial data on the comet and could affect its elements, even though by a trifling amount. Note how Ravene (1897: 205) remarked, while presenting elements of the comet of 141 CE on the basis of the Chinese records, that diminishing dates of observation by about one third of a day to about 1 day would bring the computed orbit closer to Halley's.

The epoch of the observation, as read off Jahāngīr's *Memoirs* is 72 min before sunrise. Since the Sun rose at 01:13 UT at Ramgarh that day, this would make the observation at 00:01 UT, 10 November 1618. The timing is possibly linked with the *Al Fajr* prayer that is offered when the sky begins to lighten up (the Sun 19°0.5 below the horizon; see Bobrovnikoff, 1984: 18). The prayer is offered facing the direction of the *Ka'aba* and lasts about 5 min. The refraction effect ~35 arcmin, so that the Sun would have risen about 2 min earlier, has hardly any consequence here.

Sixteen nights after the 18th of *Ābān*, and therefore on the morning of 26 November, Jahāngīr observed the second comet, now designated 1618 II (C/1618 W1). It then lay in Libra. This was also a 'Great Comet' and was the third of the year. According to Yeomans (2007), it became brightest on 29 November at magnitude 0^{m} -1^{m}. Kronk (1999: 338) gives its path as follows: Libra when discovered, Virgo (2 December), Boötes (5 December), Ursa Major (22 December), and Draco (31 December).

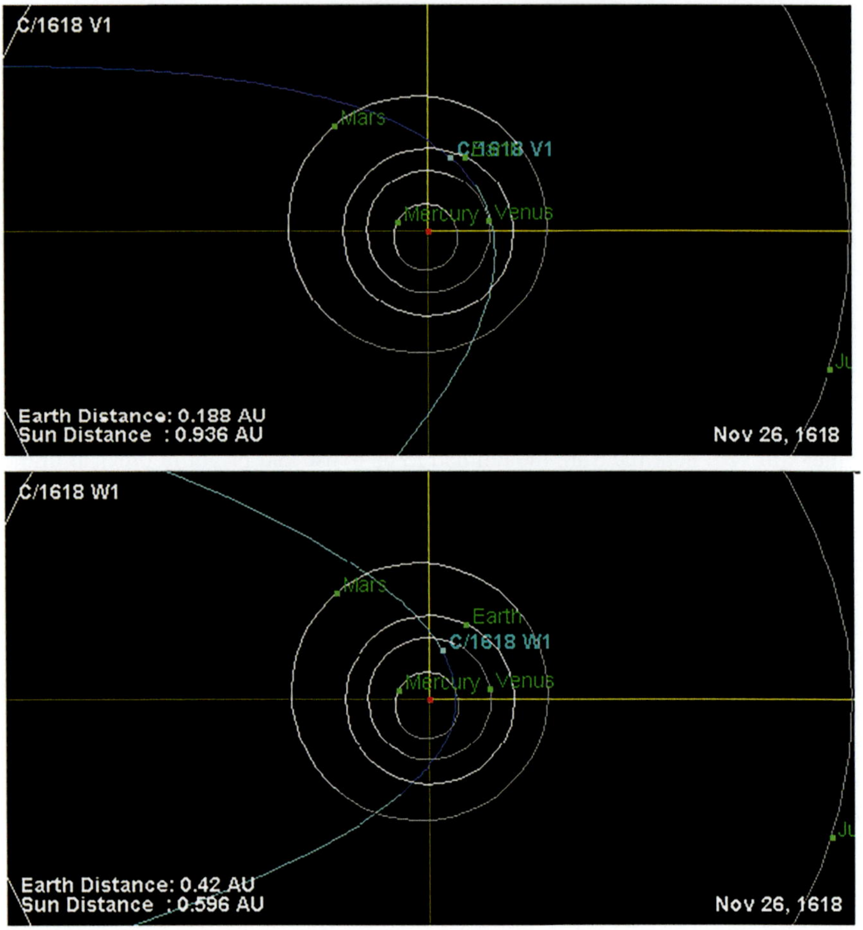

Fig. 7.20 Comets 1618 III and 1618 II as on 26 November 1618; a bird's-eye view. (Generated from Horizons System)

On 28 *Ābān* (19 November), Jahāngīr encamped near Ujjain. Five days later, on 3 Azar, he marched from Kāliyādaha in Ujjain. The month following *Ābān* is *Azar* and the date Jahāngīr observed the second comet should be 5 *Azar*, 997 (≡ 8 *Dhu-al-Hjjah*, 1027 AH) which is 26 November. We can only guess the spectacle it may have been when two bright comets adorned the sky.

There are two images in Fig. 7.20 generated with the Horizons System. These are views of the inner Solar System from the top and allow us to pinpoint the two comets when they were in the morning sky of 26 November 1618. At that time, the comet 1618 II moved faster in its orbit (direct motion) than Comet 1618 III, and began to lead the Sun on 16 November at 15:00 UT, the moment of its inferior conjunction, while 0.587 au from the Earth. It passed its closest to the Earth from 0.358 au on 6 December at 15:00 UT.

7.7.5 Observing Comets at Dawn

If a comet is visible in the sky when the nucleus is below the horizon, this is because it has: (1) an active nucleus; (2) the right combination of the r and Δ values; and (3) an absolute magnitude H_{10} on the brighter side. To that we can add the enhancement in brightness as a result of forward scattering of sunlight off the comet's dust grains when the comet passes between the Earth and the Sun (Marcus, 2007: 119). Post-discovery, Comet 1618 III was nearing the Earth and growing into a spectacular object. It passed closest to the Earth from 0.1706 au on 18 November at 21:00 UT.

Figure 7.21 shows an early morning view of Comet Lovejoy (C/2011 W3) post-perihelion in December 2011. There are many examples of comets seen just around the time of their solar conjunction. For example, Marsden (2005) wrote about the comet of 1668 (C/1668 E1):

Fig. 7.21 Comet Lovejoy (C/2011 W3) after its perihelion passage in December 2011 (http://astronomy-vlm.blogspot.in/2011_12_01_archive.html)

> Cassini (1668) was the first in the western world to publish observations—made from Bologna soon after sunset in March 1668—of what appeared to be the bright tail of a comet extending to more than 30° above the southwestern horizon. Neither he nor other observers located farther to the south (including some in Brazil and South Africa) were able to locate the comet's head, which evidently passed very close to the sun.

Some other comets seen in bright twilight are the sungrazer X/1702 D1, the Great Comet C/1927 XI, Seki-Lines C/1962 C1, and Ikeya Seki C/1965 S1 (see Bortle, 1998). According to R. Vasundhara (2012, personal communication; 29 December 2012):

> At twilight, there is surface brightness gradation from horizon to zenith. Fainter regions of the distant parts of the 60 deg tail should show up against the relatively darker (compared to horizon) sky. At horizon the sky is brighter but the tail closer to the head is more dense (larger surface brightness) and should show up despite bright sky. In case the comet head is close to the Sun, in my opinion, the inner region will be washed out in the glare, however rest of the tail should be visible out to great distances. Comet 1618 III (C/1618 V1) appears to be similar to last years comet "comet Lovejoy" which had grazed the sun and then was visible in the morning …….. Elevation of 60 Deg can be closely approximated to the zenith under conditions of negligible air pollution as in 1618.

An idea of the precision that Jahāngīr's astronomers reached in their observations can be formed from the following few examples. In *Akbarnāmā*, Abū'l Faḍl underlined Emperor Akbar's interest in astronomy saying, a 'part of the Astronomical Tables of Ulugh Beg which we have noticed in Bābar's reign was translated under the supervision of Amīr Fatḥullāh Shīrāzī…' (Law, 1916: 150), and it was

> Maulāna Chānd, the astrologer, who was possessed of great acuteness and thorough dexterity in the science of the astrolabe, in the scrutinizing of astronomical tables, the construction of almanacs and the interpretation of the stars. (Beveridge, 1897–1939(1): 69).

Maulāna Chānd's astronomical tables, *Tahsilāt-i-Akbar Shāīi*, were later used even by Mahārājā Sawāi Jai Singh II (Beveridge, 1897–1939: 1: 69). Sheikh Alāhādād's family in Lahore flourished during the period 1570–1660 CE and was highly acclaimed for the production of high-precision astrolabes and other equipment; Fig. 7.22 shows his creation, an astrolabe ca. 1601. An astrolabe has plates to address different latitudes; each plate is engraved with a grid marked with degrees from 0° to 90°.

Muslim astronomers measured time by measuring altitudes of the Sun or Moon or a bright clock-star that they reduced to local time using astrolabes and the *zījes*. About an individual equipment's precision, Stephenson and Said (1991: 196–197) cite an account of the solar eclipse of 17 August, 928 CE where the observers were able to measure the altitude of the eclipsed Sun, observing its reflection in water, to a third of a division of the measuring ring that itself was graduated in thirds of a degree. We may also note that the medieval Muslim astronomers were able to make eclipse predictions from *zījes* accurate to a fraction of an hour (Yazdi, 2008: 79). The obscuration would be expressed as the fraction of the area of the disc of the Sun occulted by the Moon, and astronomers could calculate this with great accuracy, with errors averaging to 0.05 of the Sun's disk (Stephenson & Said, 1991: 206).

Fig. 7.22 An astrolabe by Alāhādād, ca. 1601 (Accession No. N-69, Adler Planetarium, Chicago. (Image: Wikimedia Commons)

In the matter of the solar eclipse of 1024 AH (29 March 1615 Greg) that he observed from Agra, Jahāngīr wrote down that the maximum magnitude attained during the eclipse was four out of five parts of the Sun (0.8) and it lasted 8 *gharis* (3 h 12 min; Rogers & Beveridge, 1909–14: I: Chap. 12). These figures are very close to those calculated using Fred Espenak's (2015) eclipse predictions, namely, 0.794 for the magnitude and nearly 3 h duration as at Agra. One cannot help but admire Jahāngīr's astronomers for their observational abilities, and for their attempt also to 'measure' astronomical objects other than the Sun and the planets.

To conclude, we deduce from Jahāngīr's records that the comet 1618 III was first sighted on 10 November before sunrise at Ramgarh, around 00 UT, and the comet 1618 II on the morning of 26 November, both post perihelion passage. These dates make Jahāngīr an independent discoverer of the two comets in succession. In case the Islamic date converters err by a day, for the date pair 11/27 November 1618, Jahāngīr is still an independent discoverer of the two comets, notwithstanding the fact that the weekdays do not suggest this. We believe that there is lesser room for ambiguity about the pair of the dates 10 and 26 November. The sunrise times on 10 November and 11 November 1618 at Isfahan (32°0.658 N, 51°0.669 E) are 3:01 UT and 3:02 UT respectively, so, Jahāngīr would have sighted the comet 1618 III hours before Garcia de Silva y Figueroa.

7.7.6 Comets in Jahāngīr's Perception

The heliocentric worldview of Copernicus (*De Revolutionibus*, published in 1543) arrived in India much later. The Jesuits missionaries possessed this knowledge at the time, but, involved as they were in their missions at several places in India, their primary goal was to spread the Christian faith. The miniature in Fig. 7.23, by Nar Singh ca. 1605, shows Mughal Emperor Akbar and the Jesuit missionaries from Goa, Rodolfo Acquaviva (1550–1583) and Francisco Henriques (in black), in his *Ibādat Khānā* (House of Worship), Fatehpur Sikri, where he used to preside over cross-religious discussions.

When Jahāngīr ascended the throne in 1605, Father Jerónimo Xavier, a grand-nephew of St. Francis Xavier, had already spent 10 years in the Royal Court as head of their mission (Guerreiro, 1930: xvi). In fact, Abd us-Sattār (d. unknown; Hadi, 1995: 28, 181–182), son of Qāsim Firishtā, the historian (d. 1623) and courtier since Akbar's times, had been asked by Akbar to learn the language of the Franks, particularly Portuguese. Abd us-Sattār closely interacted with Fr. Xavier with a view to translating into Persian certain Latin and Greek works about science, including astronomy. Jahāngīr also interacted with the Jesuits on matters of faith and even joined in the sessions between his Moors and the priests. In 1606, Jahāngīr had shown interest in establishing a printing press to print books in Persian, and he was assured by Fr. Xavier on this matter (Alam & Subrahmanyam, 2009: 471, 477).

The cometary sightings of 1618 provided a rare opportunity for widespread discussions among astronomers and other scholars. While many past apparitions were recorded in various Muslim texts, and now the comets of 1618, there is nothing in Jahāngīr's *Memoirs* to suggest any interaction between Jahāngīr's astronomers and the Jesuits.

To Islamic astronomers, comets and meteors were atmospheric rather than heavenly phenomena, and so they were usually ignored. However, Ja'far b. Muḥammad Abū Mash'ar (787–886 CE), the famous Persian astronomer, astrologer, and philosopher, considered comets to be celestial. Whether Jahāngīr knew about this or not, his writings do take a departure from the conventional viewpoint. That is exceptional as there is no concern expressed about any untoward consequences due to the apparitions, and the royal traverse continued.

For the first comet, Jahāngīr wrote that

> It moved in high heaven, for it was first in Scorpio and afterwards in Libra. Its declination (*harakat-i-arz*) was mainly southerly.... and it had a proper motion of its own, independent of that of firmament, as it was retrograde - first appearing in the sign of the Scorpion, then in that of the Scales.

Fig. 7.23 Mughal Emperor Akbar and the Jesuit missionaries Rodolfo Acquaviva and Francisco Henriques (in black) in the *Ibādat Khānā* (House of Worship), Fatehpur Sikri; illustration to the *Akbarnāmā* by Nar Singh, ca. 1605. (Wikimedia Commons)

An orbital calculation indicates that the comet entered the sign Libra on 16 November. Jahāngīr's reference to the positions is according to the tropical system, and he also took due notice of the retrograde nature of the movement. It is clear that he and his astronomers had conducted observations on several dates in addition to on 10 and 26 November.

While the comet's tail was rising above the horizon, bright objects like *Simāk* (Spica), *Al Ḥāris al Simāk* (Arcturus), *Al Jabhah* (Regulus), Mars, and the Moon were up in the sky with Mercury hovering near the horizon. There are no *manāzil*, no bright stars near the comet's position, or any planets in the description by Jahāngīr. There is nothing more on the tail direction whereby one could ascertain the position angle of the tail. Jahāngīr also wrote that 'At each succeeding night it rose a *ghari* earlier' and that the comet had appeared a few nights prior to that date (18 *Ābān*/10 November). Those 'few nights', the comet was actually trailing the Sun. For example, on 8 and 9 November, the calculations give its altitudes at sunset (12:18 UT, at Ramgarh) as ~3.5° and 0.1° respectively. On 10 November, around 00 UT, the comet transited the Sun, passing ~2° below it, but gave a spectacular appearance. Therefore, the word translated to mean 'night' may also mean the dark phase soon to be twilight, and what Jahāngīr referred to as the appearances a few nights earlier could also mean the evening sightings of the same object. The key question is why there are no reports of sightings in the evening from anywhere, even though the comet would have been nearly as bright, the change in the r and Δ values over these few days being small only. Georg Dorffel, a student of Hevelius, first suggested that the two bright comets, seen in quick succession in late 1680 and early 1681, were the same comet (now designated C/1680 V1). By saying that the comet made its appearance before *and* after its perihelion passage and that it followed a parabolic path with the Sun at its focus, Dorffel, for the first time, provided an explanation for the pair of comets (Festou et al., 1993: 366). Though earlier, Petrus Apianus (1495–1552) had noted the disappearance of comets as they orbited close to the Sun and their subsequent reappearance (Hellman, 1944: 90). Jahāngīr had noticed that same thing.

Jahāngīr had no knowledge of the developments that were taking place in astronomy in Europe at the time. He used astrolabes for his observations, but he did not use the prospective glass [telescope] gifted to him by Sir Thomas Roe (1581–1644) in early 1616—see below. Jahāngīr had ascended the throne a week after Akbar's death on 17 October 1605 (Julian; Majumdar et al., 1967: 450, 456). He had been taught by Abū'l Faḍl's brother, Abū'l Faiẓ Faiẓī (d. 1595), a scholar and poet-laureate in Akbar's court. Jahāngīr had access to astronomical literature in the *kitābkhānā* (the Royal Library), including *Akbarnāmā* and *Ā'in-i Akbari*—wherein Abū'l Faḍl had provided a scholarly exposé on astronomical phenomena, his own sighting of the Great Comet of 1577, and the current knowledge and his views on the nature of comets. Ironically, Abū'l Faḍl was assassinated in 1602 at Jahāngīr's instance (Elliot, 1975: 13–14), much to his father Akbar's anguish. As a keen student of astronomy, we may wonder why he missed noting Kepler's Nova in 1604, the appearance of Halley's Comet in 1607, and the first bright comet of 1027 A.H. (1618 I) that appeared only a few days before Jahāngīr's entourage left Dohad for Agra.

Apart from Jahāngīr, there were others too who observed the November 1618 comets from India.

7.7.7 The Reverend Edward Terry on Two Bright Comets

In his travelogue *A Voyage to East India,* originally published in 1655, the Reverend Edward Terry (1590–1660; Fig. 7.24) records the occurrence of the two comets of 1618. Terry came to India during the reign of Jahāngīr, and was the 'Then Chaplain to the Right Hon. Sir Thomas Row, Knt, Lord Ambassador to the Great Mogul' (Terry, 1655: 393–394). He reported:

> In the year 1618, when we lived at that court, there appeared at once, in the month of November, in their hemisphere, two great blazing stars, the one of them north, the other south; which unusual sight appeared there for the space of one month. One of those strange comets, in the north, appeared like a long blazing torch, or launce fired at the upper end; the other, in the south, was round, like a pot boiling out fire. The Mogul consulted with his

Fig. 7.24 Edward Terry. (After Foster, 1921: 288)

> flattering astrologers, who spake of these comets unto the King, as Daniel sometimes did of Nebuchadnezzar's dream, Dan. iv. 19. "My lord, the dream is to them that hate thee, and the interpretation thereof unto thine enemies"; for his astrologers told him, that he needed not trouble himself with the thought thereof, for it concerned other places and people, not him nor his. But not long after this, their season of rain (before spoken of) which was never known to fail 'till then, failed them, and this caused such a famine and mortality in the south parts of his empire, that it did very much unpeople it.... (ibid.).

7.8 India's Tryst with the Telescope

The invention of a device to see an enlarged version of objects at a distance is attributed to the German–Dutch spectacle maker Hans Lippershey in 1608, who lived in the Netherlands. It consisted of a concave eyepiece and a convex objective lens. By the end of 1608, the so-called Dutch 'perspective glass' had entered the public domain, and in April 1609, perspective glasses magnifying 3× could be purchased in spectacle makers' shops in Paris, and soon thereafter in Italy.

Here was one of the most important inventions in history, which in the hands of Galileo Galilei, the Professor of Mathematics at the University of Padova, would soon evolve into a high-powered instrument (*occhiale* as he called it), and usher in a revolution in science, transforming our worldview forever. News of the device and Galileo's exciting findings—described in his book *Sidereus Nuncius* (*The Starry Messenger*), published on 12 March 1610 spread more quickly than anyone could have imagined at that time. The term 'telescope' was coined in 1611 by the Greek poet and theologian Giovanni Demisiani from two Greek words *tele* (far away) and *skopeo* (to look).

In August 1610, Johannes Kepler acquired a telescope, and he confirmed Galileo's observations of the Jovian satellites (see van Helpern et al., 2010). In 1611, he went on to propose in his book *Catoptrics* a variation to the optics by using only convex lenses, which would produce an inverted image but give a wider field of view. Based on this innovation, the first astronomical telescope in the true sense was developed by the German mathematician Jesuit Christoph Scheiner (1573–1650) between 1613 and 1617 (Mitchell, 1916: 345). The telescope found favour with the astronomers by 1630, but it was only when the English astronomer William Gascoigne (1612–1644) inserted a crosswire in the eyepiece and installed a micrometer (by 1640) that it was transformed into an astronomical measuring instrument.

Early in 1616, Emperor Jahāngīr also came to possess a telescope. One of the most important visitors to the Mughal Empire at that time was Sir Thomas Roe (1581–1644), who came to Jahāngīr's Court as Ambassador of King James I for the years 1615–1619. Sir Thomas landed at Surat in September 1615 and proceeded to have an audience with the Emperor, who was then at Ajmer in Rajasthan. Sir Thomas arrived in Ajmer at Christmas and was first presented to Jahāngīr in his court on 10 January 1616 (Wheeler, 1881: 68). Of the many valuable gifts that he had brought along to India was a 'spyglass', which he presented to Jahāngīr (Huff, 2010: 12). A spyglass is actually a small hand-held telescope that extends when pulled out and can be adjusted to the viewer's eye (see Warner, 1998). Sir Thomas' exotic gift apparently did not impress Jahāngīr very much, so he passed it to Āṣaf Khān (the brother of Nūr Jahān and father of Mumtāz Mahal) in his court. This device then joined Khān's collection of other fascinating optical objects like spectacles and prospective glasses [sic.] that had been purchased from a Venetian merchant in 1616 (Foster, 1928: 83). Āṣaf Khān was a scholar of astronomy (Khan Ghori, 2000: 33) but he did not realize the astronomical potential of this gift.

However, a telescope was eagerly sought after by a Jesuit missionary in India soon after the news of the invention reached Indian shores. We will meet him below, but first let us learn some history of the Jesuits in the Indian Subcontinent and in the Far East.

7.8.1 *Jesuits in India*

In the sixteenth century, Goa initially was an important centre of long-distance trade, but it soon became the political, cultural, and religious 'central engine' of Portuguese India (Fig. 7.25). In due course, the city of Goa became headquarters to many religious orders, the Franciscans, Jesuits and Dominicans being the major

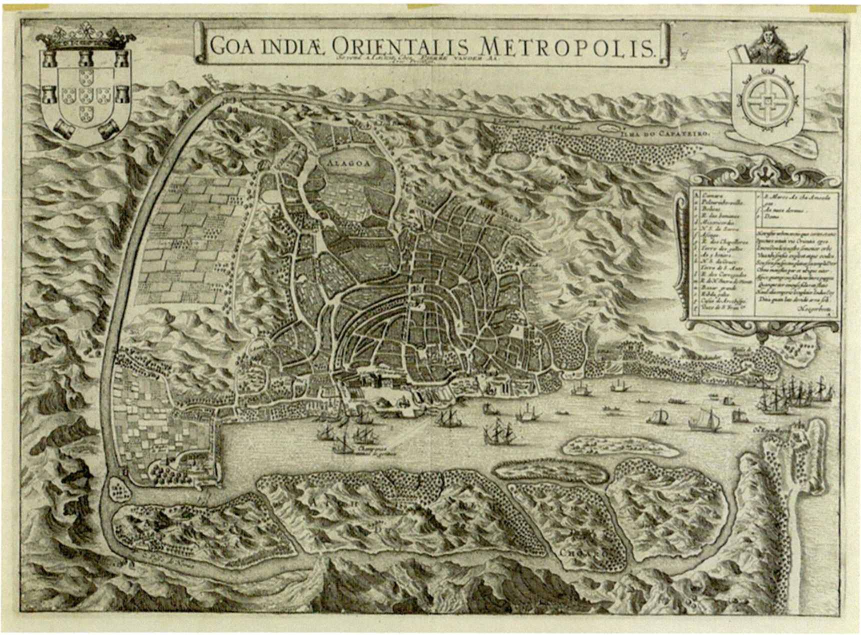

Fig. 7.25 A 1719 pictorial map of Goa by Pieter Boudewyn van der Aa (fl. 1700–50). Divar Island is in the middle of the lower half of the image and Chorão (pr. Shorão) Island to its right; the Mandovi River flows from left to right into the Arabian Sea. (Adapted from British Library Online Gallery http://www.bl.uk/onlinegallery/onlineex/apac/other/019pzz000002417u00000000.html)

ones, so much so that in 1554–1555, the Portuguese King assigned different parts of Goa to them, namely Baldez (Bardez) in the north to the Franciscans, Salcete in the south to the Jesuits, with the contiguous islands of Divar and Chorão to be shared by the Jesuits and the Dominicans. The Dominicans had come to Goa in 1548, and they soon established houses at various places; the Augustinians arrived much later, in 1575 (de Mendonça, 1958: 80–81). The activities of all the religious orders that arrived in India in quick succession to spread the Christian faith received the backing of the Portuguese government. Missionaries heading for the Far East needed to set sail from Lisbon, with a stopover at Goa, the capital of Portuguese India, and Macao, the Portuguese commercial base further east in China. Voyages from Lisbon to Goa took 6 months, and missionaries headed for Macao had to stop over in Goa for a further 6 months until winds became favourable again.

The first Jesuit mission in India was established in Goa in 1542 under St. Francis Xavier (1506–1552), one of the co-founders of the Society of Jesus. The Jesuits developed an economic framework and acquired land and houses so that they could continue their mission to spread the faith and pursue scientific interests. In due course, this was followed by more missions which were established at various places in India, but initially in the coastal regions in the south, such as Malabar, Cochin, and the fishery coast of Tamil Nadu. However, the missions soon spread inland. Figure 7.26 is a seventeenth-century Mughal miniature where we can see how a Jesuit missionary dressed up.

Goa is landscaped with innumerable churches, which are affiliated with various congregations and reflect an exquisite architectural splendour. An interesting feature of the Portuguese-style Goan churches built in the sixteenth and seventeenth centuries is that they incorporated certain astronomical aspects in their design. There were sundials for timekeeping (Fig. 7.27), while an east-west orientation allowed sunlight to enter the eastern or the western entrance at sunrise or sunset on the day of the equinox or the solstice, or on some other important day (Borkar, 2016).

The present Holy Spirit Church in Margao, built by the Jesuits in 1675 and the most beautiful one in Salcete, is a fine example of this aspect. One can witness in here the illumination of the emblem of the Holy Spirit Dove on the main altar for brief moments by the rays of the setting Sun that enter the Church through an oculus on the west-oriented façade on the equinox days. At the same time, the statues of Sacred Heart of Jesus and the Immaculate Heart of Mary on the side altars are also illuminated by the rays of the Sun let in through the smaller side windows on the façade.

During the sixteenth century, many of the Jesuits were trained in mathematics, geography, and astronomy, and they carried the latest developments in European science to their missions worldwide. A member of the mission in Goa from 1606 was the cartographer Giovanni Antonio Rubino (Antonius Rubinus, 1578–1643), who perhaps was the first Jesuit in India to show an interest in astronomy. After learning about the invention of the telescope, Rubino wrote on 2 November 1612 to Christof Grienberger (1561–1636), a Jesuit astronomer at the Collegio Romano and a Galileo sympathizer, asking for astronomical literature and other equipment (Sharma, 1982: 346). The letter makes for wonderful reading as it shows the

Fig. 7.26 A Jesuit Priest in a seventeenth-century Mughal miniature. (Acquired by Henry Walters; The Walters Art Museum, Baltimore https://art.thewalters.org/object/W.695/)

excitement of the Jesuits to scientific developments and the pace with which the news of these spread. This was at a time when communication with Europe took anywhere from 6 months to 2–3 years. An excerpt from Rubino's letter, originally written in Italian, follows:

> Somebody wrote me from Italy that certain occhiali (eye-pieces) have been invented by means of which objects 15 or 20 miles away are seen clearly and many discoveries have been made in the heavens, particularly in the planets. Your Reverence will do me a great favor by sending me these, together with a little treatise on such occhiali, if there is demonstration of the things one sees by them. But if Your Reverence does not have the occasion or the money to send me these, please send me in writing and in figures, as clearly as possible, the manner of their construction, so that I may have them made in this land of many officials and abundance of crystals. (Leitao, 2008: 118).

On the basis of this letter, Father Rubino has been credited with introducing the telescope to India (Udias, 2003; Baldini, 2015). According to Baldini (pers. commn., 2 Aug 2022),

Fig. 7.27 A sundial, in the campus of the *Se Cathedral* (sixteenth–seventeenth Centuries), Velha Goa (RC Kapoor, 11 Nov 2018)

> Borri was in Goa from late 1615 to early 1616, together with Jan Wremann, another jesuit coming from Rome, and both had telescopes, which then they brought to China and Vietnam. Thus, it's practically certain that they made observations in Goa, and showed the instrument to Rubino.

However, no published account on this part is available.

7.9 The Comets of 1618: Observations by a Few Jesuits in India

In its astronomical column in the issue of 24 January 1878, *Nature* (1878) carried a contribution titled 'The comet of 1618' by an unnamed author stating that the Jesuit astronomer Giovanni Riccioli (1598–1671) had mentioned observations made from Goa, India, by another Jesuit astronomer Kirwitzer of the comet of 1618. Riccioli taught in Bologna, and is acknowledged for his star catalogue, the book *Almagestum Novum* and a detailed map of the Moon. Wilhelm Olbers (1758–1840) also had received from H.G. Brandes a work by Father Kirwitzer, which by then had become scarce, about the observations of the second comet of 1618 (Nature, 1878: 247–248). Olbers was a German physician and astronomer who discovered Comet 13P/Olbers and the minor planets Pallas and Vesta. He was also the first to devise a suitable method for calculating cometary orbits. However, he is best known in astrophysics for posing the fundamental question—"Why is the sky dark at night?", the famous Olbers Paradox, that led to great debates in astronomy. In 1821, Olbers wrote to Franz Xaver von Zach (1754–1832) about Kirwitzer's account of the comet, adding that the work he had received from Brandes unfortunately was in too bad a shape due to copying or printing for him to deduce a realistic orbit. One may consult Zach (1822: 369–371) for more information on this.

7.9.1 A Mission to the Orient with a Passage Through India

The exchanges above were about the missionary and astronomer Father Venceslaus Pantaleon Kirwitzer (1588–1626), originally from Kadaň in Bohemia (present-day Czech Republic). He first saw the second comet of the year on 14 November and noted down his observations, subsequently joined in the observations on 26 November by the German Jesuit missionary Johann Adam Schall von Bell, S.J. (Nature, 1878). Fr. Adam Schall (1591–1666; Fig. 7.28) was a mathematician

Fig. 7.28 A portrait of Johann Adam Schall von Bell wearing mandarin attire while missionary to China during 1622–1666, accompanied by an astrolabe, a celestial globe, and an armillary sphere. (https://commons.wikimedia.org/wiki/File:Adam_Schall.jpg)

and astronomer, and an expert on calendrical science. Olbers knew that in his work, Zach had stated that 14 volumes of Fr. Schall's works were available in the library of the Vatican. Believing that these might contain information on the second comet of 1618, Olbers urged Zach to examine that aspect. However, subsequently, a search by Conti revealed nothing of the sort. Further, Nature (1878) noted that

> It does not appear that a more accurate copy of Goa observations has been found since Olbers wrote on the subject. There are two works by Kirwitzer in the British Museum, but they afford no assistance. It thus happens that there is as yet no orbit of the comet in question. (Nature, 1878).

Father Kirwitzer was a member of the *Collegium Romanum* at the time when, in May 1611, Galileo Galilei was in Rome to state his case for the Heliocentric vs. Ptolemaic systems. The Collegium was then open to the Copernican views and was warm to Galileo.

Fr. Kirwitzer was among a group of missionaries led by Nicolas Trigault (1577–1628) destined for China that included Giacomo Rho (1592–1638), Johannes Schreck-Terrentius (also Terrenz; 1576–1630), and Adam Schall. In April 1618, they set sail from Lisbon aboard the San Carlos (Leitao, 2008: 107) and braving the rigours of the voyage, sickness, and the death of 5 of the 22 China missionaries, they sailed into Goa on 4 October 1618. The group was carrying a few telescopes and some measuring instruments, along with a large number of books. Incidentally, when on tour to Milan in 1616, Terrenz had received a Galilean telescope from the Cardinal Federico Borromeo, and eventually he would take this precious gift to China. Along with Terrenz and Adam Schall, Kirwitzer subsequently proceeded to China, setting sail on 15 May 1619 and reaching Macao on 22 July 1619 (Zettl, 2008: 29–36). In 1621, Terrenz presented the Emperor with a telescope as a gift (Udias, 1994: 467). Fr. Kirwitzer's life and activities, and his ideas in astronomy, are covered in good detail in Baldini (2015) and this volume.

What kind of instruments did the Jesuits bring with them? Baichun (2003) has provided illustrations of several astronomical instruments taken to China by Trigault et al., as well as those that subsequently were made there. One may look up Bolt and Korey (2010) for the examples of early seventeenth-century telescopes starting with the earliest-known surviving one, which dates to the year 1617 (Fig. 7.29). This consists of a main tube and a number of draw tubes.

In Fig. 7.30, we reproduce the sketch of a Galilean telescope that Fr. Adam Schall drew in his book *Yi hai zhu cen* (Pearl dust of artistic sea). The inscription says 'Compiled by Tang Ruowang'—which was his Chinese name, and his preface is dated 1626. The book is about astronomy and discusses Galileo's findings on the Solar System and the Milky Way made with such a device. It also delves into general optics, optics of concave and convex lenses and their combinations, and the making and use of telescopes. The telescope that is illustrated has an alt-azimuth mount; recall that telescopes with equatorial mountings lay in the future.

Christoph Scheiner who had been using an alt-azimuth mounted telescope to follow sunspots, found it difficult to follow them properly, so on a suggestion from his brother Christof Grienberger, he constructed what is regarded as the forerunner of the equatorial mounting. With it, Christoph Scheiner was able to observe sunspots

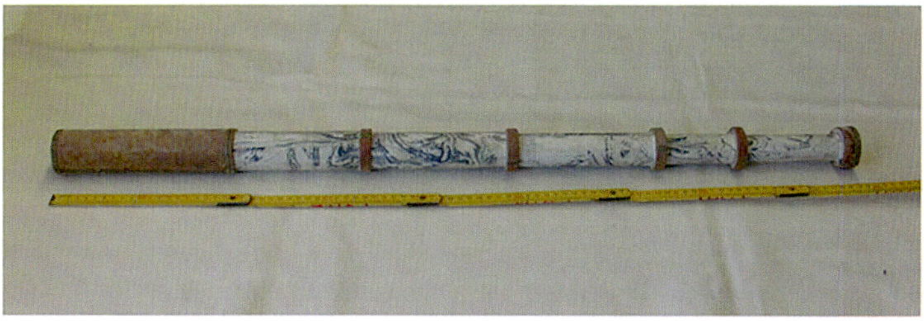

Fig. 7.29 'The world's oldest surviving, securely-datable telescope', 1617, Kunstgewerbemuseum/ Staatliche Museen zu Berlin; 'cloth and paper'; maker unknown. 'Main tube and five draws of pasteboard. Each draw covered in marbled paper; end tube and ring stops in silk velvet embroidered by gold thread. Collapsed, rings and end tube form cylinder with common outer diameter (48mm), common feature of other early scopes'; picture and description after http://dioptrice.com/ telescopes/782, accessed 2016

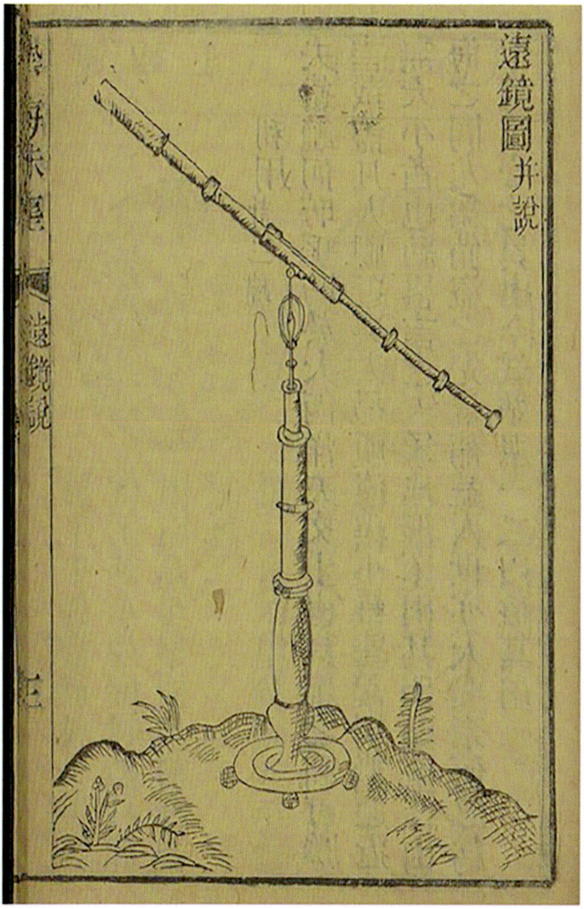

Fig. 7.30 A drawing of a Galilean telescope by Adam Schall von Bell. (After World Digital Library http://www.wdl.org/en/item/11434/)

and determine their position and motion conveniently and with precision. As outlined in the *Rosa Ursina* (Scheiner, 1630), his great work on sunspots, Scheiner put the device to use from 4 March 1627 (Mitchell, 1916: 347–348; Woods, 2005: Chap. 5).

7.9.2 Fr. Kirwitzer's Treatise on the Comets of 1618

Father Kirwitzer presented a detailed description of his observations of the spectacular comets that appeared in the morning skies in November 1618 in a monograph titled *Observationes Cometarvm Anni 1616 [*sic.*]. In India Orientali Factæ A Societatis Iesv Mathematicis in Sinense Regnum Nauigantibus ex itinere eo delatis.* It was published in 1620.

As per the reference in Nature (1878), the book had possibly been lost. A search through the internet in 2010 revealed that there was only one copy of the book with a WorldCat member library. Beginning in December 2015, while working to conclude my paper on the comets of November 1618 where I had expressed the possibility that a telescope would have been used for observations, I hit upon the digitized version of this book in January 2016 (Kapoor, 2016). It had been digitized in October 2014 by The Austrian National Library & Google Books. The cover page, bearing the seal KAISERLICHE KOENIGLICHE HOFBIBLIOTHEK, WIEN, is reproduced in Fig. 7.31. In a recent search, the relevant page of OCLC WorldCat (www.worldcat.org accessed 01.12.2021), showed

> 1 edition published in 1620 in Latin and held by 4 WorldCat member libraries worldwide
> 2 editions published in 1621 in Latin and held by 2 WorldCat member libraries worldwide

The treatise is short, consisting of 24 pages only and signed ex 'Goæ in India Orientali 11. Febr. 1619'. In the *Præfatio*, Fr. Kirwitzer refers to being assigned to India by Muzio Vitelleschi (1563–1645), the Sixth General of the Society from 1615–1645, to bring the light of the Gospel to the great empire of the Chinese. Fr. Kirwitzer (1620) also mentions how fellow Jesuit astronomers met with very unkind circumstances (also see Golvers, 1992: 391).

The significance of Fr. Kirwitzer's treatise lies in the fact that it reports the first ever modern astronomical observations carried out in India and the first-ever use in India of the telescope for astronomical observations soon after the introduction in Europe (see Fig. 7.32). Hereinafter, our description closely follows the relevant parts of the treatise, and we cite the dates just as given in the treatise. The dates match those given in Nature (1878) and also agree with the chronology of the apparitions given in the *Tūzūk-i Jahangīrī*. In the account that follows, the phrases 'first comet' and the 'second comet' refer specifically to the November 1618 comets only, just the way Fr. Kirwitzer described them (i.e., to Comet 1618 III and 1618 II, respectively).

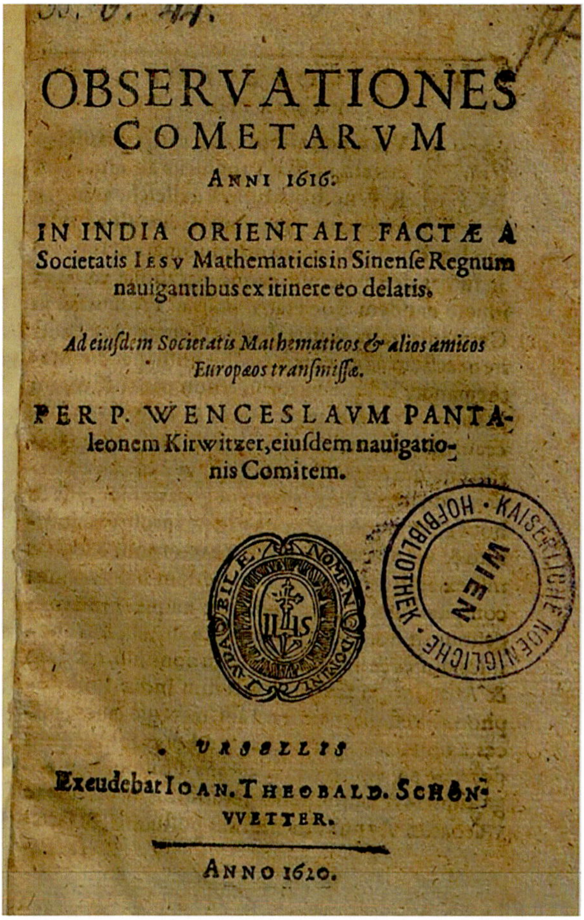
OBSERVATIONES
COMETARVM
ANNI 1618.
IN INDIA ORIENTALI FACTÆ A
Societatis IESV Mathematicis in Sinense Regnum
nauigantibus ex itinere eo delatis,
*Ad eiusdem Societatis Mathematicos & alios amicos
Europæos transmissæ.*
PER P. WENCESLAVM PANTA-
leonem Kirwitzer, eiusdem nauigatio-
nis Comitem.
VRSELLIS
Excudebat IOAN. THEOBALD. SCHÖN-
WETTER.
ANNO 1620.

Fig. 7.31 The cover page of Fr. Kirwitzer's treatise. (After Google Books)

In the *Præfatio*, Fr. Kirwitzer (1620) states that he was not prepared to make any observations when a bright comet suddenly appeared in the sky. His first record is dated 10 November when

> various people reported to us that the sky vehemently flared up in the morning hours, and some of the more prudent, having suspected a comet, asked us to observe the sky the following day to explain what it was. On that day we were all gathered in Goa because in that year we had all come from Portugal to India, and by the orders of the superiors, on the same evening, we spread out to various places so that we could all see this phenomenon in the best way, too. As the new instruments were not available, still we proceeded so that in one joint effort everyone would bring the observations together. Fr. Giacomo Rho then wrote to Fr Antonio Rubinus at Cochin on the same day to view this new star from Cochin (Kirwitzer, 1620: 5; tr. RCK).

Unfortunately, their baggage at that time was still on the ship, and no instruments or books were to hand but later this would change.

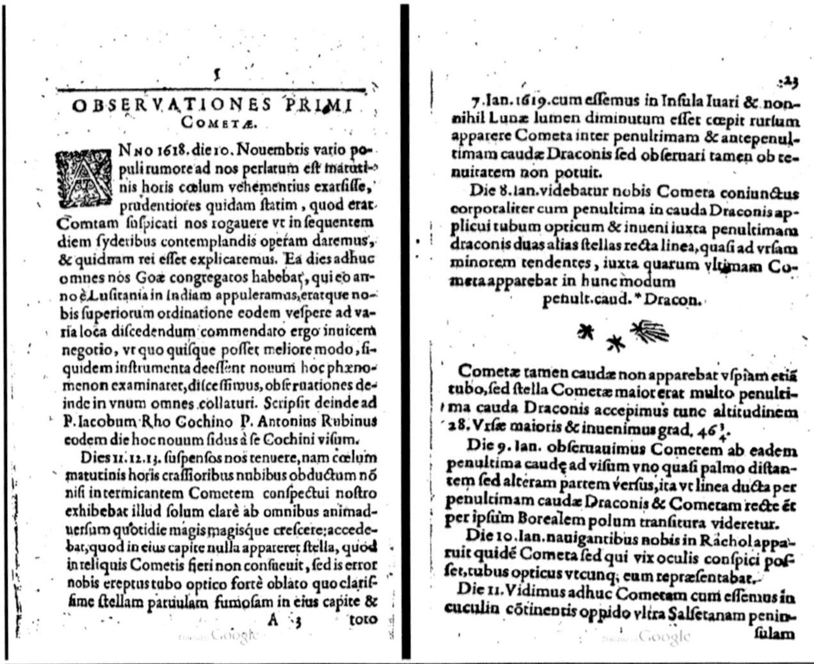
5

OBSERVATIONES PRIMI
COMETÆ.

ANNO 1618. die 10. Nouembris vario populi rumore ad nos perlatum eſt matutinis horis cœlum vehementius exarſiſſe, prudentiores quidam ſtatim, quod erat Comtam ſuſpicati nos rogauere vt in ſequentem diem ſyderibus contemplandis operam daremus, & quidnam rei eſſet explicaremus. Ea dies adhuc omnes nos Goæ congregatos habebat, qui eo anno è Luſitania in Indiam appuleramus, eratque nobis ſuperiorum ordinatione eodem veſpere ad varia loca diſcedendum commendato ergo inuicem negotio, vt quo quiſque poſſet meliore modo, ſiquidem inſtrumenta deeſſent nouum hoc phænomenon examinaret, diſceſſimus, obſeruationes deinde in vnum omnes collaturi. Scripſit deinde ad P. Iacobum Rho Cochino P. Antonius Rubinus eodem die hoc nouum ſidus à ſe Cochini viſum.

Dies 11. 12. 13. ſuſpenſos nos tenuere, nam cœlum matutinis horis craſſioribus nubibus obductum nō niſi intermicantem Cometem conſpectui noſtro exhibebat illud ſolum clarè ab omnibus animaduerſum quotidie magis magiſque creſcere; accedebat, quod in eius capite nulla appareret ſtella, quod in reliquis Cometis fieri non conſueuit, ſed is error nobis ereptus tubo optico fortè oblato quo clariſſime ſtellam paruulam fumoſam in eius capite &

A 3 toto

23

7. Ian. 1619. cum eſſemus in Inſula Iuari & nonnihil Lunæ lumen diminutum eſſet cœpit rurſum apparere Cometa inter penultimam & antepenultimam caudæ Draconis ſed obſeruari tamen ob tenuitatem non potuit.

Die 8. Ian. videbatur nobis Cometa coniunctus corporaliter cum penultima in cauda Draconis applicui tubum opticum & inueni iuxta penultimam draconis duas alias ſtellas recta linea, quaſi ad vrſam minorem tendentes, iuxta quarum vltimam Cometa apparebat in hunc modum

penult. caud. * Dracon.

Cometæ tamen caudæ non apparebat vſpiam etiā tubo, ſed ſtella Cometæ maior erat multo penultima cauda Draconis accepimus tunc altitudinem 28. Vrſæ maioris & inuenimus grad. 46¾.

Die 9. Ian. obſeruauimus Cometem ab eadem penultima caudę ad viſum vno quaſi palmo diſtantem ſed alteram partem verſus, ita vt linea ducta per penultimam caudæ Draconis & Cometam rectè ēt per ipſum Borealem polum tranſitura videretur.

Die 10. Ian. nauigantibus nobis in Rachol apparuit quidē Cometa ſed qui vix oculis conſpici poſſet, tubus opticus vtcunq; eum repræſentabat.

Die 11. Vidimus adhuc Cometam cum eſſemus in cuculin cōtinentis oppido vltra Salſetanam peninſulam

Fig. 7.32 Pages 5 (left) and 23 (right) in Fr. Kirwitzer's book, which discuss his observations of the two bright comets that appeared in November 1618. Both pages mention that a telescope was used for viewing the comets. (After Google Books)

7.9.3 The Jesuits' Observation Posts

Where in Goa and Cochin were the Jesuits stationed? In his treatise, Fr. Kirwitzer refers to a few places from which the Jesuits made astronomical observations. One named Rachol (pronounced Rashol) is mentioned as being 5 leagues from Goa (i.e., Old or Velha Goa). It is a town in the Salcete Peninsula south of Panjim (now Panaji) and 7 km north-east of Margao. The Portuguese had occupied Salcete in 1543, and they fortified Rachol and placed Salcete in the care of the Jesuits (Fig. 7.33). The Jesuits took up residence in a small house in Rachol where the present-day parish church is situated (15°18′29″N, 74°00′19″E). Rachol has been home to the Patriarchal Seminary of Rachol since 1610, which was built by the Jesuits atop a small hillock. With the passage of time the Seminary evolved into a multipurpose institution (Patriarchal Seminary of Rachol, 2016). Figure 7.34 shows the main building at the Seminary.

As for his own location, Fr. Kirwitzer writes of being at Insula Ivari and sometimes at S. Paul. The Insula Ivari must be Divar Island, which is in North Goa 10 km north of Panjim, and was among the first places in Goa that the Portuguese strove to spread the Catholic faith. A major portion of the island is rather flat, thickly forested, with a 49-m high hillock. At the base of the hillock is a Goan village named

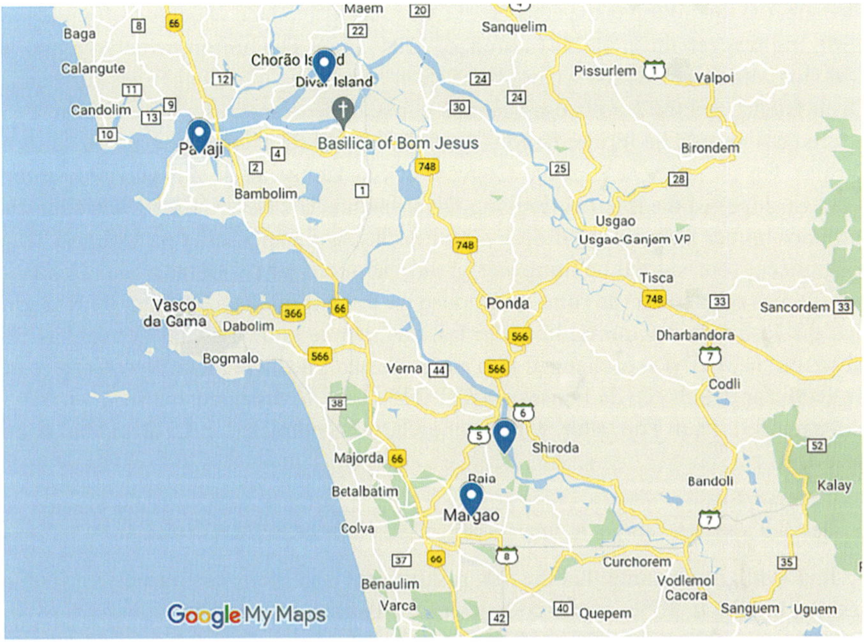

Fig. 7.33 The various locations in Goa; the lower right-side marker is over Rachol. (Copy of Untitled Map—Google My Maps; map modification by R C Kapoor)

Fig. 7.34 Patriarchal Seminary of Rachol, Goa (RC Kapoor, 11 November 2018)

Piedade. On the hilltop is the Church of Our Lady of Piedade which was founded in 1541 and was rebuilt in 1625 (Lourenço, 2005: 158). The hilltop, which overlooks Old Goa, may have served as an observation point. The Mandovi River surrounds Divar Island, and the Jesuits used canoes to reach the island.

'S. Paul' would be for the famous New College of St. Paul in Velha Goa, east of Panjim. On the other side of the Convent of St. Augustine, it was situated on the western slopes of the hill, overlooking Divar Island. The New College was an institution of higher learning in theology, philosophy, and many other disciplines. This magnificent four-storied building was greatly admired for its vastness and architecture. It was erected on the ruins of a house on the hill of Nosa Senhora de Rozario that the Jesuits had acquired in 1578. Initially, it was known as the Convent of St. Roch, but in 1610 it was changed into a college and given the name New College of St. Paul (for details see da Fonseca, 1878: 315–320), to distinguish it from an older College of St. Paul. The latter, with its church, was completed in 1542 and has been described as

> the chief institution of the disciples of Loyola in India, to which more than three hundred churches with their colleges in different parts of Asia were subject (da Fonseca, 1878: 262).

It is worth mentioning that the first printing press in India was introduced by the Jesuits, in Goa in 1556, in the colleges of St. Paul and Rachol (da Fonseca, 1878: 58). Because of an epidemic in 1570 that afflicted this locality including 58 priests, in 1578 the Jesuits decided to move to a new house in a healthier location so that they could more effectively care for their sick. Activity at the Old College then reduced, although it remained the prime institution of the Jesuits for some time (da Fonseca, 1878: 264). All that remains today of the New College of St. Paul is a magnificent gateway on a small road south of St. Cajejan's Church in Velha Goa (Fig. 7.35), A signboard describes it as the *Arch of Conception*. Fr. Kirwitzer mentioned S. Rochi a few times in his treatise.

After Goa, Cochin became a stronghold of the Catholic faith. The residence of the Society of Jesus came up in Cochin in 1550, and the College of Mother of God (*Madre de Deus*) and the Seminary were established in 1560. Later, both grew to become major cultural centres. Father Antonius Rubinus, who carried out the observations of the November 1618 comets in Cochin, was probably based here, although there is no corroborating evidence for this.

7.9.4 The Instruments Used by the Jesuits

Fr. Kirwitzer records in detail what the observers saw and measured, namely, the altitude and azimuth of the comet, its angular distance from stars like Spica, etc. in *grad* (degrees). He also reports the individual observers' visual impressions, sometimes gained with difficulty because of illumination by the Moon or sunlight. The observations are presented systematically and divided into two sections for each comet, incorporating those by brother-priests at the other Jesuit establishments.

Fig. 7.35 The *Arch of Conception*, Velha Goa (R.C. Kapoor, 24 Feb. 2016)

In those times, astronomers observed the position and direction of the tail, and they determined the position of the comet with respect to many nearby fixed stars whose position was already known. The main objective was to determine the position of the comet on the ecliptic and its motion. Whilst indicating the positions, Fr. Kirwitzer also refers to the *informi.* These are field stars not grouped into any certain figure or into a constellation and are called unformed stars (*stellæ informes*; Long, 1742: 176).

Astronomical instruments that these Jesuits were able to access were an astrolabe (*astrolabium*) and an astronomical radius or cross-staff (*radium astronomicum*) belonging to the Goa College. A cross-staff consists of a staff with a smaller, sliding transversal arm, generally made of wood but sometimes of brass, and bearing a scale that could be read directly in degrees. One measured altitudes and angular separations between two objects by pointing it at the object and moving the transversal arm until the angle was covered.

The introduction of the device is attributed to a Jewish mathematician, Levi ben Gerson (1288–1344; see Goldstein, 2011 for details), but according to Joseph Needham, the Chinese had developed one much earlier. The cross staff was also called Jacob's Staff, and when used for astronomical observations, the term astronomical radius was used. The cross-staff was introduced by the Portuguese as aid to

Fig. 7.36 The cross-staff, as depicted in Pedro de Medina's 1552 treatise *Regimento de Navegacion*. (Image from Regimento de Navegacion edicion 1563; Wikimedia Commons)

navigation in the mid-sixteenth century. Figure 7.36 from Pedro de Medina's (1552) work *Regimento de Navegacion* shows one such. The earliest depiction of the device in an astronomical observation is found in Petrus Apianus's book (Apian, 1532), showing the anti-solar nature of a comet's tail.

Even Tycho Brahe had used one. The instrument was handy for seafarers and astronomers of the time, even though viewing added to errors that limited the precision. Notably, the measurements given by Fr. Kirwitzer are to a fraction of a degree or minutes of arc. At this time, astronomers determined time by measuring the altitude of the Sun or Moon or a bright clock-star that they then reduced to local time by using astrolabes and astronomical tables. Finally, these were not the only instruments that Fr. Kirwitzer and his colleagues used. Most importantly, a telescope (*tubo optico*) was also used by him to view the comets.

7.9.5 Observations of the Second Comet of 1618 (1618 III)

As Fr. Kirwitzer writes, dark clouds that were present before sunrise on 11, 12, and 13 November prevented them from making any observations, but providence intermittently showed the comet grow day-by-day. Fr. Kirwitzer looked for a star in the head as was typical of comets, but found none. However, when he fetched a *tubo optico* (telescope), it clearly revealed a little star, with 'smoke' in the head that appeared pale in colour. The comet's form was best described as like a palm leaf, and it stretched as a straight, smoky column from the east to the midst of heaven, with the tip a little turned to the north.

For this comet, the observations extend from 10 to 30 November, with no observations possible on 19, 20, 22, and 25 November due to clouds. The date 10 November is included here since the mysterious phenomenon was first noticed that morning and suspected to be a comet, which turned out to be right. Observations were made by Fr. Kirwitzer at Divar Island and at *S. Rochi*, and by Fr. Jacobus Rho at Rachol and at *S. Rochi*. Fr. Rho used a cross-staff for his measurements. The altitude and azimuth of the comet and the first magnitude star Spica, along with the length of the comet's tail, were given to the fraction of a *grad*. At times, there is reference to the constellations of Corvus, Crater, Hydra, Crux, and Centaurus, and how the angles to these varied. On page 7 in the Treatise, Fr. Kirwitzer depicts the position of the comet (1618 III) on 17 November with respect to two nearby stars (which unfortunately are not readily identifiable). Nowhere does Fr. Kirwitzer mention the time when the observations were made, even though he writes in a phrase '... the clock of our College ...'.

Detailed observations began on 14 November. At Rachol that morning, the comet altitude was recorded as 11°, its horizontal distance *ab ortu* as 24°, while at the same time *canicula* stood 44° above the horizon. The phrase *ab ortu* in the text is a reference point for the azimuth. Literally, it means 'from rising' and 'from east'. As confirmed by Dr. Paul Gabor (pers. commn., 30 Nov 2023), it refers to azimuth measured from the East. The word *canicula* mentioned in the text refers to the Dog-star (Sirius); see, e.g., Allen (1899: 121).

For the observations on 15 November, Fr. Kirwitzer refers to using star charts:

> Day 15. There with Spica of Virgo raised on the horizon 18.30 degrees, the comet was 5 degrees distant [alt.] from the same horizon, in fact 27 degrees from the East. The tail went up to the second star [star 2] of Corvus in the right wing, if the stars are numbered according to the images [diagrams] of the celestial constellations by P. Christophorus Gruenberger, a distinguished mathematician of our Society. In fact, by chance they were on hand outside the baggage and which we made use of throughout the duration of our observations.

Christopher Grienberger (1561–1636; MacDonnell, 2014) was an Austrian Jesuit with an interest in astronomy and mathematics, and a supporter of the Copernican worldview and a Galileo sympathizer. He joined the Society of Jesus in 1580 and later came over to the Collegio Romano to assist Christopher Clavius (1538–1612) and succeeded him in 1612 as mathematics professor. The College attracted the best of the scientists at the times. Apart from Clavius and Grienberger, the other notable mathematicians at the College were Paolo Lembo and Odo van Maelcote. It is these Jesuit mathematicians who signed a declaration with favourable verdict on the new celestial phenomena that Galileo had claimed to have seen with his telescope. Soon the Society of Jesus developed a rigid stance against Galileo's views that even Grienberger could not go against and had to side with the Society (Museo Galileo, 2021).

Grienberger's acclaimed work on the constellations, *Catalogus veteres affixarum Longitudines, ac Latitudines conferens* cum *novis. Imaginum Caelestium Prospectiva duplex ... Romae, Apud Bartholomaeum Zanettum, MDCXII*, was published in 1612 (Fig. 7.37). It was duly illustrated, and contained astronomical tables of the positions of fixed stars and maps relating to 21 numbered northern constellations, 12 zodiacal constellations and 15 numbered southern constellations

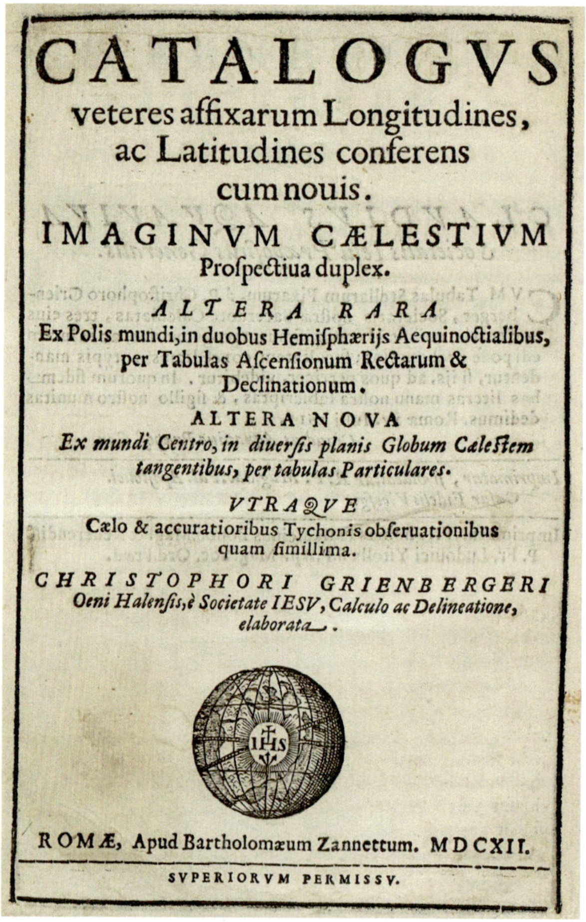
CATALOGVS
veteres affixarum Longitudines,
ac Latitudines conferens
cum nouis.
IMAGINVM CÆLESTIVM
Proſpectiua duplex.
ALTERA RARA
Ex Polis mundi, in duobus Hemiſphærijs Aequinoctialibus,
per Tabulas Aſcenſionum Rectarum &
Declinationum.
ALTERA NOVA
Ex mundi Centro, in diuerſis planis Globum Cæleſtem tangentibus, per tabulas Particulares.
VTRAQVE
Cælo & accuratioribus Tychonis obſeruationibus
quam ſimillima.
CHRISTOPHORI GRIENBERGERI
Oeni Halenſis, è Societate IESV, Calculo ac Delineatione, elaborata.

ROMÆ, Apud Bartholomæum Zannettum. MDCXII.
SVPERIORVM PERMISSV.

Fig. 7.37 Catalogus veteres affixarum longitudines, ac latitudines conferens cum novis, 1612 (Biblioteca Europea di Informazione e Cultura https://www.beic.it; image: Wikimedia Commons; accessed 17.03.2025)

(Grienbergeri, 1612: 56; Fig. 7.38). It also included the new northern constellations of 'Antinous' and 'Berenices Crinis'; these, however, were not numbered (see also Stoppa, 2016). These new constellations were first added to the existing 48 Ptolemaic constellations by the cartographer Casper Vopel (1511–1561) in 1536, but Antinous

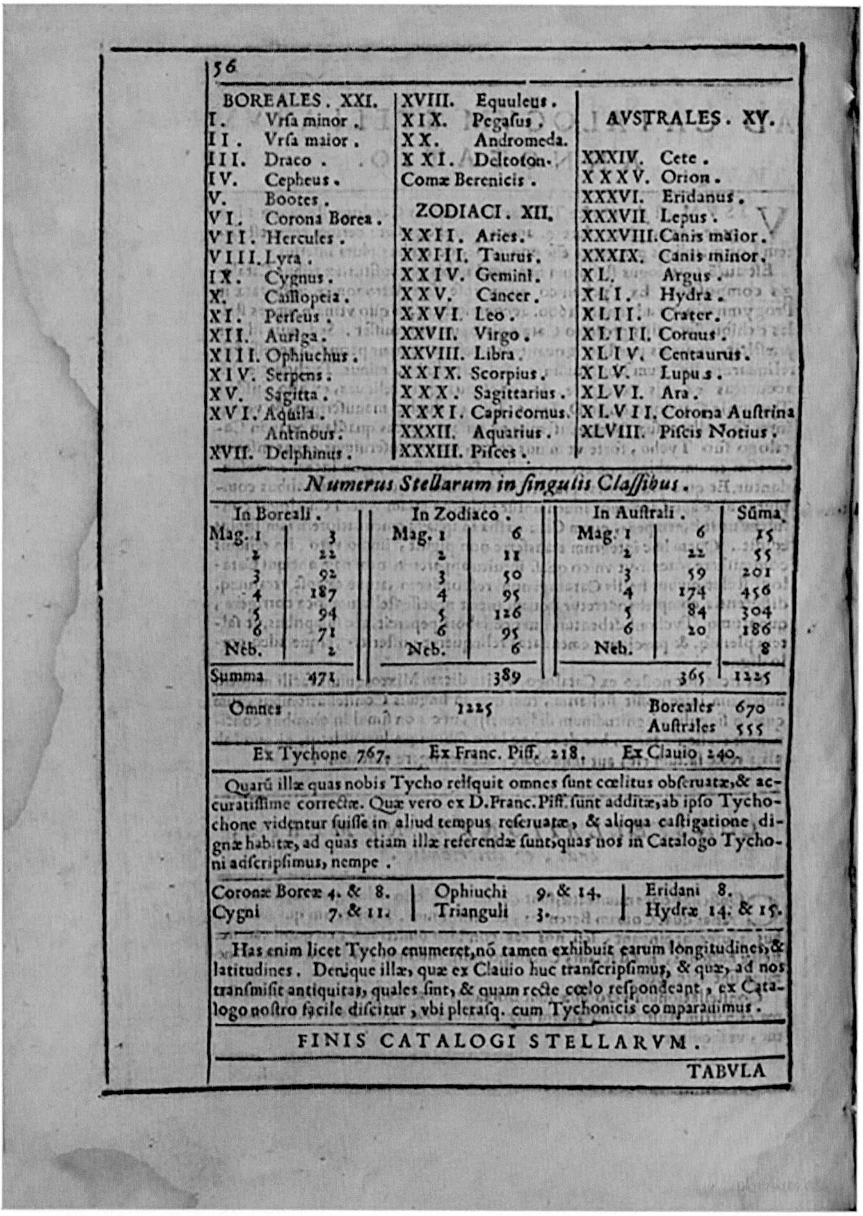

56

BOREALES . XXI.
I. Vrſa minor .
II. Vrſa maior .
III. Draco .
IV. Cepheus .
V. Bootes .
VI. Corona Borea .
VII. Hercules .
VIII. Lyra .
IX. Cygnus .
X. Caſſiopeia .
XI. Perſeus .
XII. Auriga .
XIII. Ophiuchus .
XIV. Serpens .
XV. Sagitta .
XVI. Aquila .
Antinous .
XVII. Delphinus .

XVIII. Equuleus .
XIX. Pegaſus .
XX. Andromeda.
XXI. Deltoton.
Comæ Berenicis .

ZODIACI . XII.
XXII. Aries .
XXIII. Taurus .
XXIV. Gemini .
XXV. Cancer .
XXVI. Leo .
XXVII. Virgo .
XXVIII. Libra .
XXIX. Scorpius .
XXX. Sagittarius .
XXXI. Capricornus.
XXXII. Aquarius .
XXXIII. Piſces .

AVSTRALES . XV.
XXXIV. Cete .
XXXV. Orion .
XXXVI. Eridanus .
XXXVII. Lepus .
XXXVIII. Canis maior .
XXXIX. Canis minor .
XL. Argus .
XLI. Hydra .
XLII. Crater .
XLIII. Coruus .
XLIV. Centaurus .
XLV. Lupus .
XLVI. Ara .
XLVII. Corona Auſtrina
XLVIII. Piſcis Notius .

Numerus Stellarum in ſingulis Claſſibus .

In Boreali .		In Zodiaco .		In Auſtrali .		Sūma
Mag. 1	3	Mag. 1	6	Mag. 1	6	15
2	22	2	11	2	22	55
3	92	3	50	3	59	201
4	187	4	95	4	174	456
5	94	5	126	5	84	304
6	71	6	95	6	20	186
Neb.	2	Neb.	6	Neb.		8
Summa	471		389		365	1225

Omnes 1225 Boreales 670
Auſtrales 555

Ex Tychone 767. Ex Franc. Piſſ. 218. Ex Clauio 240.

Quarũ illæ quas nobis Tycho reliquit omnes ſunt cœlitus obſeruatæ, & accuratiſſime correctæ. Quæ vero ex D. Franc. Piſſ. ſunt additæ, ab ipſo Tychone videntur fuiſſe in aliud tempus reſeruatæ, & aliqua caſtigatione dignæ habitæ, ad quas etiam illæ referendæ ſunt, quas nos in Catalogo Tychoni adſcripſimus, nempe .

Coronæ Boreæ	4. & 8.	Ophiuchi	9. & 14.	Eridani	8.
Cygni	7. & 11.	Trianguli	3.	Hydræ	14. & 15.

Has enim licet Tycho enumeret, nõ tamen exhibuit earum longitudines, & latitudines . Denique illæ, quæ ex Clauio huc tranſcripſimus, & quæ, ad nos tranſmiſit antiquitas, quales ſint, & quam recte cœlo reſpondeant , ex Catalogo noſtro facile diſcitur , vbi pleraſq. cum Tychonicis comparauimus .

FINIS CATALOGI STELLARVM .

TABVLA

Fig. 7.38 Index of the Constellations in Grienberger's (1612: 56) *Catalogus*, with specifications of bright stars therein, etc. (Google Books)

was later abandoned (ibid.). Fr. Kirwitzer uses the term 'crines Berenices' several times in his treatise while describing the location of the second comet. See Note 3 on *Elementa Trigonometrica* of Grienberger (1630).

On 26 November, Fr. Adam Schall von Bell joined Fr. Kirwitzer to observe the comet from Divar Island while Fr. Rho observed it on the 27th from Goa. The observing campaign continued until 30 November. From that day, 'the light of the Moon diminished the Comet and we could no longer observe it'. Fr. Antonius Rubinus' first measurement of this comet from Cochin was made on 28 November, noting a tail 40° long and a maximum width ~3°. His last measurements were on 18 December when he noticed that the tail had increased in length to 44° from an initial 25°.

7.9.6 Observations of the Third Comet of 1618 (1618 II)

The second part of the treatise contains observations of the second comet, i.e., 1618 II. Fr. Kirwitzer (1620: 11–12) begins thus:

On 24 November before the rising of the sun, in the dawn, we saw this brilliant Comet on the Island of Ivar. Its star was brighter even compared to Venus, with a small tail, and a straight line drawn through Arcturus and Mars passed through the Comet star, so much that the distance between Arcturus and Mars was three times greater than that between the star of the Comet and Mars.

On the same day, Fr. Joannes Terrentius saw the comet from the agro-fields of Rachol College in Salcete. The Jesuits' discovery date of the Comet 1618 II, therefore, is 24 November.

The next day, before the sunrise, Fr. Kirwitzer and many other Jesuits saw the same Comet brighter and with a longer tail. They also marvelled at the sky glowing with so many new bright stars. From Goa, the first to observe the Comet on 25 November was Fr. Jacobus Rho at S. Rochi, whereas Fr. Antonius Rubinus in Cochin saw it first on the 26th November.

Fr. Kirwitzer writes (ibid. p. 12),

While we observe the new Comet, another very bright remote star very similar to Venus appeared in the East, not far from it in color and the magnitude: this produced no small confusion many bore…

Actually, Venus was then the 'Evening Star' but Mercury was visible in the morning hours. The confusion among the Jesuits arose when they saw it close to the Comet. Fr. Kirwitzer sought to explain this by quoting from the *Prutenic Tables* (see below), which on that day placed Mercury at 14° in the sign of Scorpio. On 25 November, the (Comet-) star's magnitude as observed by the Jesuits appeared to be on the much brighter side than Mercury, so that there was no doubt about its identity.

Observing the Comet on the 26th from Goa Fr. Giacomo Rho found it as a bright object with a tail ~2° long, and it resembled a beard. From 28 November, two sets of observations for each comet were taken from a given location, and the comets were observed together until 30 November.

On 8 December, in Goa, Fr. Rho noted the Comet 1618 II 'turned away' from the bright Arcturus. Its column extended through almost the middle of the *Catalogus* stars 12, 13, 14 in Boötes (Fig. 7.39; respectively, η, τ, and υ Bootis). That actually means the Comet's tail has passed over Arcturus. That this happened may be guessed from the Korean observation of 7 December that spoke about the tail passing through the southern Coma Berenices (Kronk, 1999: 339); the particular star field was then vertically up. As for the tail length, Cysat, who was observing from Ingolstadt had found the comet tail to be 55° on 8 December and 70° on 9 December (Kronk, 1999: 339).

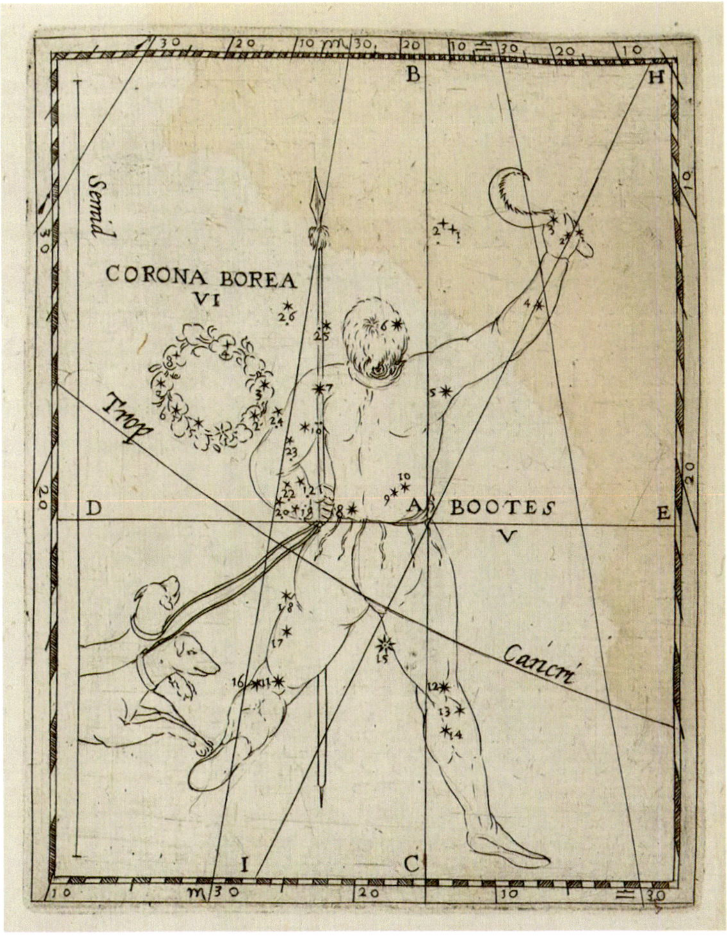

Fig. 7.39 The constellation of Corona Borealis and Boötes as depicted in Grienbergeri (1612); Google Books

Now, that is interesting. Fr. Rho's observation means that the Comet transited Arcturus on 8 December morning. However, a computation suggests that this would have happened later, on the morning of 9 December, from about 6° below it.

That the Comet's tail swept past Arcturus can also be inferred from the drawing on the cover of Peter Cruger's (1619) *Uranodromus cometicus, Dantiscum,* depicting the Comet's path on the celestial hemisphere for various dates of December 1618, vide Fig. 7.40. Mathematician and astronomer, Crüger was a student of Brahe and Kepler, teacher of Hevelius (1611–1687), and had books on trigonometry,

Uranodromus
COMETICUS.
Ein außführlicher Tractat
Vom grossen Cometen
deß 1618 Jahrs/
Darinnen seine erscheinung vnd Lauff/seine
Höhe von der erden vnd andere daraus folgende sachen
durch Astronomische rechnung dargethan/ vnd seine bedeu-
tungen durch gebührliche muthmassung gesucht
werden.
Neben eingefügtem Discurs von Cometen in gemein.
Beschrieben von
M. PETRO CRüGERO, Königl. Stadt
Dantzigk Mathematico.

Dantzigk/
Gedruckt vnd Verlegt durch Andream Hünefeldt/
Im Jahr 1619.

Fig. 7.40 Part of the cover page of Peter Cruger's (1619) *Uranodromus cometicus*, Dantiscum; After https://www.digitale-sammlungen.de; accessed 2.8.2021

astronomy and geography to his credit. The instance of the Comet's tail sweeping past Arcturus would have made quite a spectacle, much like the event that had glued innumerable curious eyes in Europe to the western skies on 5 October 1858 when Donati's Comet (C/1858 L1) was to transit Arcturus.

Fr. Cysat (1619: 2 and pp. 12–13) also seems to have observed a transit, that of the third star in the leg of Boötes, possibly ζ Bootis:

> **7. Decembr.**
>
> Hora 6.1/8 mat. altitudo Cometæ 38.g. distabat ab Arcturo 9.g.48.m. à lucida Coronæ 17.g. 46.m.
>
> Diameter Capitis seu stellæ Cometæ libero oculo per radium Astronomicum obseruata fuit 6. aut 7.m.
>
> Hora 5.1/2 cum altus esset Cometa 32.g 50.m. accepta est per Radium distantia Cometæ à stella tertiæ magnit. in tibia Bootis & ferè in via Cometæ, 2.g.1.m. eiusque altitudo fuit 32.g.circ.
>
> Hora 7. mat. cum altitudo Cometæ esset 44.g. peridem Instrumentum reperta est Cometæ à prædicta stella distantia 1.g.58.m. altitudo proximè eadem quæ Cometæ. Porro inter 4. & 5. mat. fuit eadem stella in mediis ferè Comæ radiis, hora 7. ferè in extremis prope Caput Cometæ.
>
> **7.Decembr.**
>
> At Hour 6.1/8 in the morning. the height of the comet 38°. its distance from Arcturus 9°.48′. from bright Coronæ 17°. 46′.
>
> The diameter of the head or star of the comet was observed by the free eye with an astronomical radius 6. or 7′.
>
> At Hour 5.1/2. Comet 32° 50′ altitude. the radius measured the comet distance to the third star. in the leg of Boötes and almost in the way of the comet, 2°.1′. and its altitude was circa 32°.
>
> At Hour 7. In the morning. when altitude of the comet was 44°. the same instrument found the comet at the distance of 1°.58′ from the aforesaid star. the same height as that of the comet. And between 4 and 5. morning. the same star was almost in the middle of the Coma's rays, at hour 7. near the edge of the comet's head (tr. RCK).

The Jesuits in Goa continued to record positions of Comet 1618 II, but from the end of December, they could not observe it because of the bright Moon. On 7 January 1619, Fr. Kirwitzer, observing from Divar Island, noticed that the comet was positioned between the *penultimate* and *ante-penultimate* stars in the tail of Draco but could not be observed because it was not dense and was diminished by the overwhelming light from the Moon.

On 8 January, the comet seemed to be joined to the penultimate star of the tail of Draco (Fig. 7.41: *Catalogus* star 30; κ Draconis). Viewing through the telescope, Fr. Kirwitzer found two new stars near it that were aligned towards Ursa Minor. He depicted these on page 23 in his treatise, reproduced here in Fig. 7.42, where the comet's tail extends to the right from a star-like object. At this time the comet's tail was not visible in the telescope with which the penultimate star of the Dragon's tail was magnified. The apparent position of the comet computed with the Horizons system for 8 January 1619, 00:00 UT suggests that it was about 0°0.5 away from the star κ Draconis (precessed), and, just the way depicted in the treatise (Fig. 7.42), the two new stars fall nearly on a line directed towards Ursa Minor. On 10 January, Fr. Kirwitzer had to use the telescope to have a view of the Comet. On 12 January, it could no longer be seen, so it was pursued no longer (Kirwitzer, 1620: 23–24).

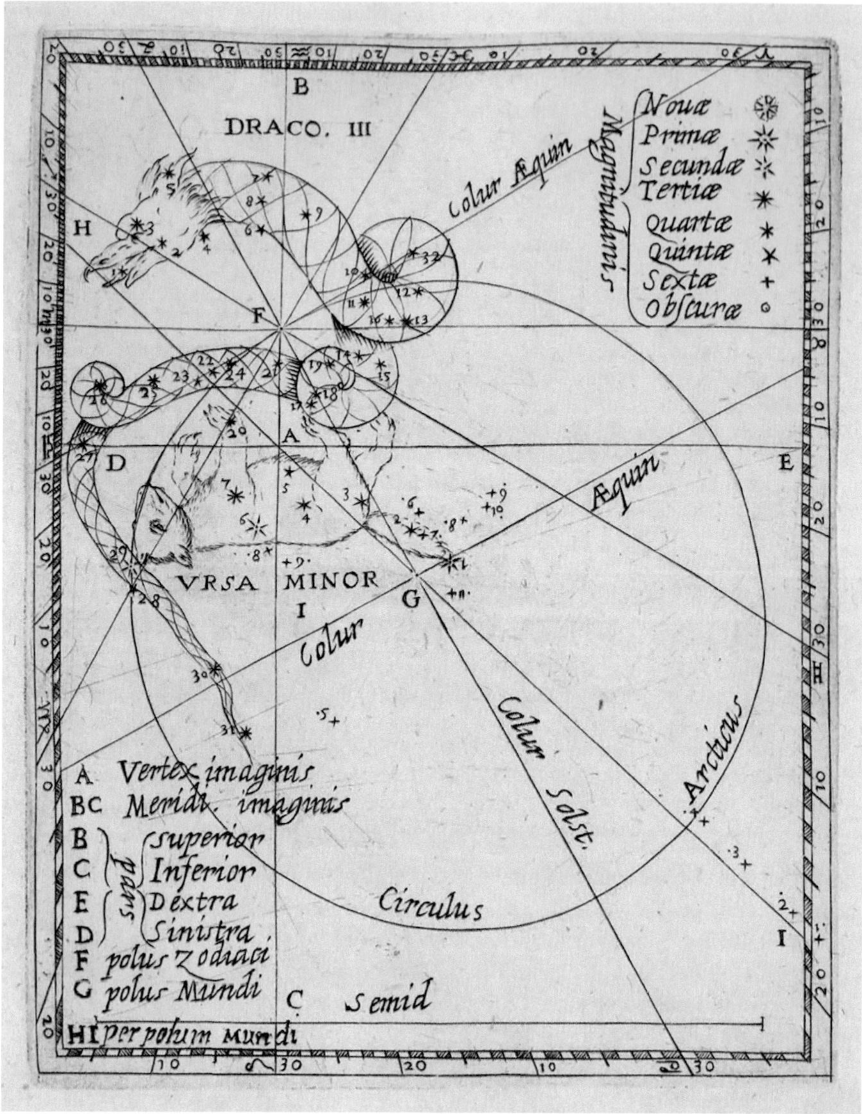

Fig. 7.41 The constellation of Draco as depicted in Christoph Grienberger's *Catalogus…* (After Stoppa, 2016)

The *Prutenic Tables* that Fr. Kirwitzer referred to were an ephemeris that was prepared by the astronomer Erasmus Reinhold (1511–1553) in 1551 to compute planetary positions, eclipses, phases of the Moon, etc., based on Copernicus' Heliocentric Model of the Universe as portrayed in *De revolutionibus Orbium Coelestium*. The tables helped to substantiate the Copernican model, but as circular orbits were used, accuracy was limited. Elliptical planetary orbits were introduced in Kepler's Rudolphine Tables in 1627, which enabled one to compute solar and

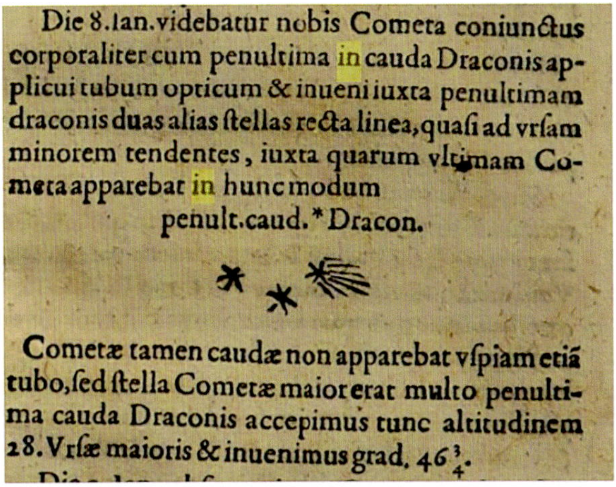
Die 8.Ian.videbatur nobis Cometa coniunctus
corporaliter cum penultima in cauda Draconis ap-
plicui tubum opticum & inueni iuxta penultimam
draconis duas alias ſtellas recta linea, quaſi ad vrſam
minorem tendentes, iuxta quarum vltimam Co-
metaapparebat in huncmodum
penult.caud.* Dracon.

Cometæ tamen caudæ non apparebat vſpiam etiã
tubo,ſed ſtella Cometæ maior erat multo penulti-
ma cauda Draconis accepimus tunc altitudinem
28.Vrſæ maioris & inuenimus grad. 46$\frac{3}{4}$.

Fig. 7.42 A depiction of the comet 1618 II near the penultimate stars in the tail of Draco (Kirwitzer, 1620: 23; Google Books; accessed 14.11.2021)

planetary positions in the sky for any date. It was the Rudolphine Tables that led to the prediction and successful observations of the transits of Mercury and Venus in the years to come.

Inclined towards astronomy as the Jesuits were, as also was Jahāngīr, one wonders if anyone noticed the Leonid meteor shower earlier in November, which in 1618 would have peaked around 8 November. Dick (1998: Table 1) has shown that the Leonids can be traced back to 902 CE. Table 7.1 in Brown (1999: 289) provides details of Leonid showers between 1799 and 1999, where a shift by 3 days in the epoch of the peak every hundred years can be easily noticed.

7.9.7 Fr. Kirwitzer's Concluding Remarks

All through, Fr. Kirwitzer's treatise contains observations and descriptions of the two comets, but there is no theorizing about comets, or, where they belong in the Universe. Fr. Kirwitzer (1620: 24) summed up the Jesuits' campaign

> These are roughly the things that were observed around these comets except that P. Ant. Rubin adds that he observed in Cochin the tail grown to 24 degrees and then gradually decreased. For a fuller understanding of those observations, it remains to make known the true longitude and latitude of the places where the observations have been done. However, we have not seen yet any Moon Eclipse and from others nothing we learned that we can accept with confidence, we will work diligently so that no latitude and longitude of this or other places of Asia remain unknown. In the mean time we will rely on Johannes de Barros, prince of Portuguese Historians.

In fact, a lunar eclipse had taken place on 31 December 1618. However, it was penumbral and with shadow over India. Whether the Jesuits were aware of this upcoming eclipse is not indicated, although among them was Fr. Adam Schall, who, years later in Macao, would determine the precise details of the lunar eclipses of 8 October 1623 and 9 September 1624 (Udias, 1994: 467). As we know, eclipses and certain other records were used by astronomers to refine eclipse computations and determine longitude differences between the places of observations of the predicted eclipses.

In the indented quotation presented above, Fr. Kirwitzer refers to 'Johannes de Barros'. João de Barros (1496–1570) was the Portuguese Royal Historian. He is best known for his multi-volume classic *Décadas da Ásia,* which was published in 1552 and describes the early history of the Portuguese in India and in Asia. Fr. Kirwitzer quotes from this work a solar eclipse that was seen in Cochin,

> in the Year of Christ 1506, 13 January, series 4, the time prior to the second half and it was so clear that daytime saw the stars.

There *was* an eclipse of the Sun on 13 January in 1507. It was annular, not total. The path of annularity passed across Sri Lanka, not India, and at Cochin it was a partial eclipse at magnitude touching 0.83. The partially eclipsed Sun would be too bright to allow stars in the daytime, except possibly for Venus at the maximum eclipse, and only if one knew where to find it. Burke-Gaffney (1944: 127–128) adds an interesting dimension to this eclipse reference:

> In 1618 Remus wrote to Kepler that he had heard from Father Schreck that when Venus was observed in conjunction with the moon on June 6, 1617, it was further from the moon than Kepler predicted. "This observation," Father Schreck wrote from Lisbon, "was made by Father Lembo, who is now in Naples, and by Father Pantaleon, who is sailing with me to China." Father Pantaleon's full name was Wencelas Pantaleon Kirwitzer; in his signature, he omitted his surname. He, too, was a bit of a thorn in Kepler's side. He wrote to Father Ziegler (who was European Procurator for the Chinese Mission), from Goa, in February, 1619, saying that he had seen a comet in India, and, going back a hundred years, telling of an eclipse seen in Cochin on January 13, 1507. It was this eclipse which bothered Kepler. He was not sure of the fact. Father Kirwitzer's authority was Joao de Barros's *History of the Portuguese in India*. Kepler had not heard of de Barros or his history. In his *Rudolphine Tables*, he harped back on the possibility of the eclipse reported by Father Kirwitzer.

As for the Venus–Moon conjunction, it surely was on 6 June 1617 and the duo came within 16′ of each other at around 19:30 UT. It would be interesting to find out what difference in the time and the minimum separation was noted by Fr. Kirwitzer from the prediction in Kepler's *Ephemerides Joan. Kepleri annorum 1617, 1618, 1619 et 1620*. The ephemerides appeared in 1617 and were more accurate than any other astronomical tables at that time.

Fr. Kirwitzer's treatise got published in 1620 but it did not get a wide circulation. There was no commentary from anyone who had some say on the nature of the comets. It now appears that the treatise made no impact on the theories then being advanced on the nature of the apparitions of 1618. Two hundred years later, only when Franz Xaver von Zach and Olbers corresponded in 1821 about Fr. Kirwitzer's work on the comet of 1618 that some information came out that later found mention in Nature (1878). One may find some details on this in Zach (1822) and Baldini (2015).

The Jesuits passing through or on sojourn in India were sufficiently equipped and knowledgeable but seem to have made no attempt to disseminate the new knowledge—say, the Copernican ideas or the works of Galileo and Kepler. It was all fresh and compelling but there are no indications of any dialogue with the natives.

7.10 The Comets of 1618: Fr. Borri's Observations

To the Indian part of the story of the comets of 1618, we may add some from Father Christophoro Borri S.J. (1583–1632). He was an Italian Jesuit missionary known for his magnetic observations in the Oceans. These provided an ingenious way to determine longitudes as acknowledged by his peer Athanasius Kircher (1602–1680), himself an acclaimed mathematician and astronomer (Dror & Taylor, 2006: 40). Borri entered the Society of Jesus in 1601, and could speak Latin and Portuguese with ease. He left Lisbon in April 1615 with Jan Wremann (1583–1621) and Giovanni Battista Bonelli for Macao, stopped over at Goa for 6 months, and left in 1616 for Macao (Carolino, 2007: 189; Baldini, 2017: 92). He had brought along a telescope also. The mission soon faced unfavourable circumstances, and he moved over to Cochinchina (the central region of Vietnam). It is from here that he observed the two bright comets of 1618. Unfortunately for him, Fr. Borri could not adapt to the circumstances, so he decided to return to Europe. He went to Macao in 1622, and, later that year left for Goa. He stayed in Goa until February 1624, where he eventually set sail for Lisbon, together with Garcia de Silva y Figueroa (Dror & Taylor, 2006: 42).

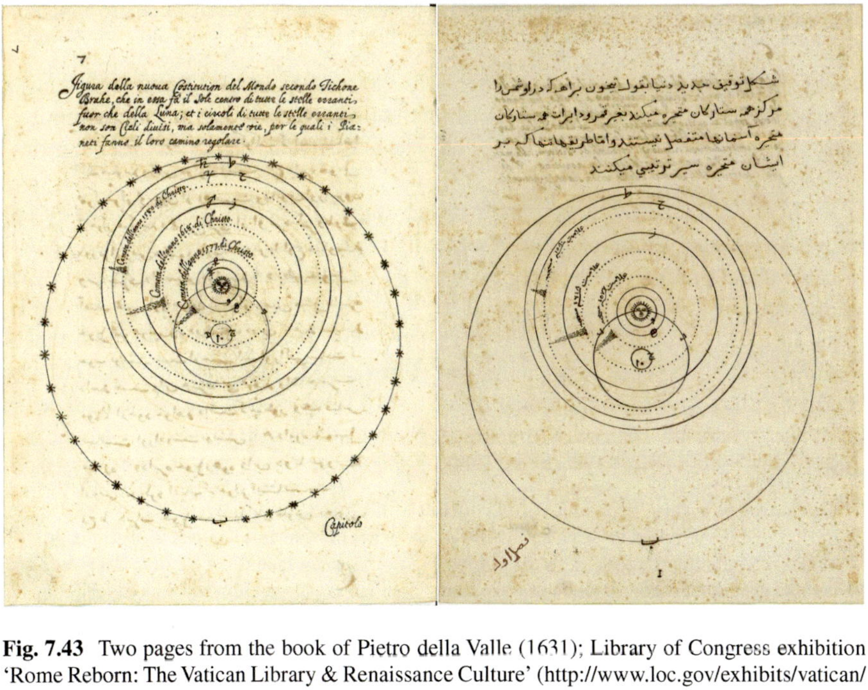

Fig. 7.43 Two pages from the book of Pietro della Valle (1631); Library of Congress exhibition 'Rome Reborn: The Vatican Library & Renaissance Culture' (http://www.loc.gov/exhibits/vatican/images/orient18.jpg; accessed 30.11.2015)

While he was in Goa in 1623, Fr. Borri befriended a well-to-do and knowledgeable Roman nobleman Pietro della Valle (1586–1652), who had travelled far and wide and knew many languages. The celebrated traveller described Borri as a great mathematician who shared with him the Tychonic worldview and also his own perception of the three heavens. Fr. Borri had been developing his theory of tenuous heavens and recorded his views and observations in a book. An impressed Pietro translated this work in 1624 in Persian, with the title *Risalah-i- i Padri Khristafarus Burris Isavi dar tufiq-i jadid dunya* (Compendium of a tractate of Father Christoforo Borri Giesuita on the new model of the universe according to Tycho Brahe and the other modern astronomers); see Fig. 7.43. Note its similarity with Fig. 7.14 as drawn by Fr. Cysat.

Later, in 1631, della Valle was in Rome and he proceeded to translate Borri's work into Italian: *Compendio di un trattato del Padre Christoforo Borro Giesuita della nuova costitution del mondo secondo Tichone Brahe e gli altri astologi modern.* This book is preserved in the Vatican Apostolic Library (Vat. pers. 10 fols. 7 recto—6 verso; www.ibiblio.org/expo/vatican.exhibit/overview.html) and carries a depiction of Tycho's hybrid Universe with the planets revolving around the Sun and the Sun revolving round the Earth. A few comets in orbits are depicted too, with tails pointing away from the Sun to demonstrate that comets could not be carried by the crystalline spheres. According to Fr. Borri, the book was well received by scholars in Persia, Armenia, and Arabia. He refers to his observations of the two comets of 1618 also in a letter to Mutius Vitelleschi, General of the Society of Jesus, stressing that his precise observations substantiate

> the tenuousness and corruptibility of the Heaven, which I already in Europe demonstrated for the sake of modern observations. This phenomenon was observed not only by myself but also by Father Giovani Vremano in China, and Father Manuel Dias in India,

and by mathematicians in Europe (Dror & Taylor, 2006: 40). Fr. Borri (op. cit) was confident of his precision, which according to him provided ' ... the unique proof of truthfulness of observations, when it is found in different territories and countries, far from each other'.

Fr. Borri had received accounts of observations of the first comet made by the Jesuit astronomer Vremano (Jan Wremann) from Macao, and by Fr. Emmanuel Diaz (Manuel Dias) of the Society of Jesus and a Portuguese philosophy professor who according to Borri had observed the first comet from the city of Cochin in India. In these accounts, the Jesuits concurred with Borri about the comet: that it was undoubtedly a celestial phenomenon (Borri, 1631: 116; Carolino, 2007: 190). Fr. Borri refers to this communication in his 1631 work *Collecta Astronomica ex Doctrina*, where he speaks of the first sighting of the comet as on 9 November (Borri, 1631: III: 115–117):

> **De Cometis qui post Tychonem apparuerunt anno. 1618.**
>
> Duo Cometæ dederunt sevidendos illo anno: prior illorum fuit ille maior australis qui exhibebat figuram Harpes, cuius magnitudo tanta erat, ut perueniret secundum longitudinem ad gradus 45. ocuparetque totum spatium signi Libræ, & medium Virginis (quanquam isti gradus non erant ex maioribus, quales fiunt in circulis maximis, erant tamen ex medijs, quandoquidem versabatur in media distanta à Zodiaco ad polos) cæpit autem videri ad 9.

Novembris, & durauit usque ad 22. Dezébris: statim autem atque apparuit extendebat se ex Canobo per Hydra, Crateré, & Corvum usque ad Centaurum. Mouebatur secundum signa Zodiaci, quasi uno gradu singulis diebus, inclinando semper ad Austrum.

About the Comets that appeared post Tycho in the year 1618

Two comets followed that year, the first inclined southerly, showed a figure of sword/scimitar, whose size was so great that it reached 45 degrees in length occupying the entire space of the sign of Libra, and the middle of Virgo ... but it began to be seen from 9 of November, and it lasted until 22 December: it extended from Canobo through Hydra, Crater and Corvus to as far as Centaur. It was moving according to the signs of the Zodiac, as it were, by one degree each day, always inclining to the south (tr. RCK).

The date for the first sighting as in Fr. Borri's account needs caution—see Note 5 below. Fr. Borri (ibid.) does not refer to an evening sighting, and he states that the comet was in Libra and Virgo *totum spatium signi Libræ, & medium Virginis* (Borri, 1631: 115), and with the Sun in Scorpio, this would have to be a morning observation on 10 November. There are no further details provided, and Fr. Borri quickly focuses his discussion on the comets' *supra Lunam* position in the heavens. The object *Canobo* is not identifiable. It could not be *Canobos*, which was for Canopus (Crossen & Procházka, 2007: 92).

Further, Borri (1631: 380) also observed the split nucleus that Cysat (1619: 74) had:

Non tantum totum Cometae caput (nempe Nucleus una cum circumfuso iubare) sed solitarius quoque Nucleus, & iam duplo maior Arcturom diametro 3. aut 4. min videbatur (cum primo die longe esset minor) neque amplius rotundus, sed difissus internos, aut quaternos irregularis figurae globus inter se cohaerentes, quales solent apparere Saturni comites.

Not only did the whole head of the comet (that is, a nucleus together with a surrounding beam) but also the solitary nucleus, and now twice as large as the diameter of the Arcturus, 3 or 4. min (when on the first day it was far smaller) and not rounder, but split inward, or a group of four irregular figures attached to one another, such as the companions of Saturn usually appear (tr. RCK)

From his parallax estimations derived from triangulation calculations of the observations of the comets of 1618, Fr. Borri argued that comets were located far beyond the Moon. This cast doubt on the impervious nature of the celestial spheres. Borri thus tried to combine mathematical astronomy and canonical cosmology, fruitfully using it in carrying forward the Tychonic worldview that he later put forth in his work, written while in India, *De Nova Mundi Constitutione juxta Systema Tychonis Brahe aliorumque recentiorum mathematicorum* (Carolino, 2007: 190).

We come across reference to the comets observed in Asia by Jesuits in a manuscript by the German Jesuit Johann Chrysostomus Gall (1586–1643) from early 1620s, who taught mathematics at the Colégio de Santo Antão, Lisbon. He specifically points to the motions of the comets that are unprecedented and defy any comprehension:

Auermos de dar à estes cometas distintos orbes dos orbes dos planetas nos constrangem a isso seos mouimentos desimilhantes a todos os mouimentos dos planetas como uimos neste ultimo cometa [de 1618] e se pode uer assi nos mathematicos alegados, como tambem nas obras de Tycho o que confirmam cartas de nossos padres escriptas da Etiopia, China e India porque de Etiopa se escreue que hum daquelles dous cometas que a menos de dous annos aparecerão se mouia para o Sul, o outro para o Norte: porem da China só se fas menção do mouimento de hum delles, a saber daquelle que se mouia para o Sul, o outro nos o vimos mouerse para o norte. Os quais mouimentos nunqua forão obseruados, nem nos planetas, nem nas estrellas fixas, como bem se nota em huma carta, que este anno nos escreuerão de Cochim, porque os mouimentos ou de tripidação ou de libração, não sam tam aprecados, nem tam grandes, nem se fazem iuntamente para o sul e para o norte alem doutras muitas couzas em que differem. (Gall, C., 1621. *In Sphaeram Ioanis de Sacro Bosco Commentarius*, B.G.U.C., MS 192, fols 18r.-v).

Our desire to give these comets orbs different from the orbs of the planets constrains us to do so if the movements are dissimilar to all the movements of the planets as seen in this last comet [of 1618]. It is said this way in the works of mathematicians mentioned above, as well as in the works of Tycho. This is confirmed by letters from our priests written from Ethiopia, China and India. As per letters from Ethiopia, it is believed that one of those two comets that has appeared within the last two years is moving towards the South, the other towards the North. However, in letters from China, only one mentions the movement of one of them, namely the one which moved to the South, the other we saw moved to the North. Such movements have never been observed, neither in the planets nor in the fixed stars, as is well noted in a letter which they wrote to us this year from Cochin, because the movements of either trepidation or libration, are not so appreciated, nor so great, nor are they made equally for the south and the north, besides many other things in which they differ. (RCK: software assisted translation).

A doubt may arise here because we come across references to Father Manuel Dias Sr. (1561–1639) as well as to Father Manuel Dias Jr. (1574–1659). Father Dias Sr. was a Portuguese philosopher who was in Macao at the time we are discussing (see Pina, 2007: 90). Father Dias Jr., also a Portuguese philosopher and well known for his 1615 work *Tianwenlue* (in Chinese; Epitome of Questions on the Heavens), was in Goa during 1601–1604, in Macao during 1604–1610 before entering China in 1610. He came back to Macao in 1616. His book contained description of Galileo's findings, though without mentioning his name, and that he possibly learnt about these being in communication with Fr. Rubinus in Cochin. He had observed the comet of November 1618, from Macao (Leitao, 2008: 101, 115–116). So, Fr. Manuel Dias Fr. Borri was referring to could be the Father Sr. If Fr. Dias Sr., also a Jesuit, was at Cochin, he and Fr. Antonius Rubinus would have known about each other and their respective cometary observations. But Fr. Kirwitzer mentions Fr. Rubinus only, who he knew was in Cochin and had carried out cometary observations. In the communications from Cochin, only observations by Fr. Rubinus find place in Fr. Kirwitzer's treatise.

Incidentally, in the extensive account of his travels, Pietro della Valle too refers to sighting the comet 1618 III on 21 November while he was journeying through Persia (della Valle, 1843: I: 838–841).

7.11 Concluding Remarks

This paper has presented India-related stories of two bright comets that became visible in November 1618 in quick succession. Their observations in Europe, including those of the bright comet of August 1618, had a major impact on our thinking about the nature of comets. In the Indian context, their significance relates to the first-ever use of an optical device, a *tubo optico* (telescope) for astronomical observations, within a decade of its invention in Europe. This was a time when the telescope was a fascinating gadget that made for a wonderful gift—only later would it become an indispensable instrument of the astronomer.

The Latin phrase (*tubo optico*) in Fr. Kirwitzer's (1620) treatise translates as *telescope* but, was it a Galilean telescope that the Jesuits had brought along from Europe like the one Fr. Adam Schall von Bell depicted in his 1626 book (see Fig. 7.30)? In the *Preface* to his treatise, Fr. Kirwitzer mentions that their baggage with instruments and books still lay in the ship and in unsafe condition, but they could use an *astrolabium* (astrolabe) and a *radium astronomicum* (Astronomical Radius; cross-staff) from the local Jesuit establishments. That explains how the angles, as measured, came to the fraction of a degree. Fr. Kirwitzer does not mention the instruments Fr. Antonius Rubinus used at Cochin. Whether he also used an optical device to view the comets is not stated, but we do know that he was eager to obtain one. Fr. Kirwitzer also does not mention the bright comet of August 1618 (1618 I) that the missionaries would have noticed while still at sea. Once he was in Goa and armed with a telescope, Fr. Kirwitzer may have used it to demonstrate Galileo's observations to fellow Jesuits and others.

Until now, the credit for the first use of a telescope in India for astronomical observations has rested with Jeremiah Shakerley (1626–1655), who specially came to Surat in Gujarat to observe the transit of Mercury of 3 November 1651. The following year, he also observed a comet, most probably C/1652 Y1, of which he was an independent discoverer too (Kapoor, 2018a). However, we know nothing about his telescope, timing device, or observing methods (Kochhar, 1989: 188). While these seventeenth-century examples did not prove to be trend setters for modern astronomy in India, they were the first ones nevertheless (Note 4).

The orbits of the Great Comets of November 1618 were determined on the basis of observations that were made elsewhere, between 11 November and 9 December for Comet 1618 III and between 30 November 1618 and 22 January 1619 for Comet 1618 II (JPL, 2021). As we have seen, Jahāngīr's and Fr. Kirwitzer's observations were unknown at the time, and thus could not form part of the initial datasets assembled for these two comets. We noticed some differences in Fr. Kirwitzer's observations and the numbers obtained from the Horizons system, particularly in the matter of the transit of Comet 1618 II against Arcturus observed on 8 December. The question of whether these observations can be suitably used to revise the orbital elements even minutely remains.

Recent investigations of comet C/1618 V1 (1618 III) by Maik Meyer and Gary W. Kronk during their revision of Cometography, Vol. 1, have shown that the orbit by Landgraf for this comet is not correct. Meyer and Kronk (2026, in press) were able to extract positions for this comet mainly from the tract by Kirwitzer, but also from other observers. The resulting orbit shows that the comet had a rather small perihelion distance and an inclination around 150° and is in agreement with most of the observation reports. The observations can likewise be represented by an orbit of the Kreutz group on sungrazing comets, which would explain the rather long tail, so this comet might well be a member of this famous comet group. Between November 9 and 20, the difference between Landgraf and reality would be about 10° on the 9th and 12° on the 20th. Around the 15th, Landgraf's orbit would be within 5° of the new orbit. But after November 20, large differences arise: 33° on Nov 30 and 43° on Dec 20. Dr. Kapoor did not compare observations from such late dates, which were only given by Kirwitzer.

7.12 Notes

7.12.1 The Hijri Calendar

Most dates listed in this paper are Gregorian, and the years are CE or BCE. However, some dates, like 1027 A.H., are given using the *Hijri* or Islamic Calendar. This is a lunar calendar that consists of 12 months in a year of 354 days. The first year (1 A.H.) of the *Hijri* Calendar began in CE 622, when Muhammad moved from Mecca to Medina. For example, the Islamic year 1445 A.H. is from 19 July 2023 to 7 July 2024.

7.12.2 Defining Twilight

To refresh, the morning civil twilight commences when θ_s, the elevation of the centre of the Sun's disc, is 6° below the horizon, and lasts until its top shows. During the civil twilight, the Sun is down but the sky is lit up by the sunlight scattered from the upper layers of the Earth's atmosphere, and the brightest stars and planets can be seen. It is nautical twilight when θ_s lies between −6° and −12°. The sky is deep blue, the horizon is still visible and the navigator's guide stars can be sighted. It is astronomical twilight when θ_s lies between −12° and −18°. When θ_s is $< -18°$, it is night and the sky is dark for regular astronomical observations to be made (U.S. Naval Observatory, 2012). These definitions are for a geometrical horizon that is 90° from the zenith. The length of the twilight depends on latitude and the time of the year.

7.12.3 Trigonometrical Tables of Grienberger

Grienberger took up the task to produce the most accurate trigonometric tables, giving values in increments of 1 arcminute to 22 places after the decimal. That would mean 5400 entries for just the sine values up to 90°. The value of sin 1′ he computed was 0.0002908882045634245911 that is correct to 19 places. From the value of sine of 1′ he computed the value of cosine 1′ from the identity $\cos x = \sqrt{1 - \sin^2 x}$. That led him to sin 2′ and cos 2′ using sin 2′ = 2 sin 1′cos 1′ leading to cos 2′ = 1–2 $\sin^2 1$, and so on. Grienberger completed his sine table on 14 December 1596. The finished manuscript is with the Biblioteca Nazionale Centrale di Roma. Decades later, the tables in their completed form with the values set with respect to a large number, namely 100,000, being for the full sine, were published in *Elementa Trigonometrica* in 1630; an excerpt is in Fig. 7.44. The number stands for a unit circle of radius 100,000.

Elementa Trigonometrica also carried Grienberger's evaluation of π that is accurate to 36 places after the decimal and another notable from among his contributions

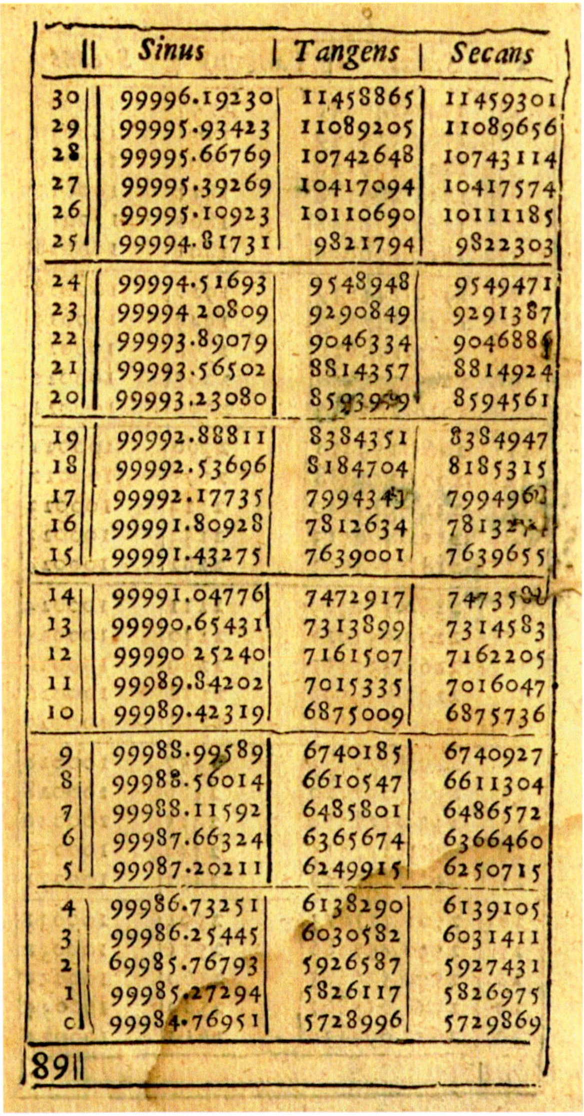

	Sinus	*Tangens*	*Secans*
30	99996.19230	11458865	11459301
29	99995.93423	11089205	11089656
28	99995.66769	10742648	10743114
27	99995.39269	10417094	10417574
26	99995.10923	10110690	10111185
25	99994.81731	9821794	9822303
24	99994.51693	9548948	9549471
23	99994 20809	9290849	9291387
22	99993.89079	9046334	904688[illegible]
21	99993.56502	8814357	8814924
20	99993.23080	8593[illegible]9	8594561
19	99992.88811	8384351	8384947
18	99992.53696	8184704	8185315
17	99992.17735	799434[illegible]	79949[illegible]
16	99991.80928	7812634	78132[illegible]
15	99991.43275	7639001	7639655
14	99991.04776	7472917	74735[illegible]
13	99990.65431	7313899	7314583
12	99990 25240	7161507	7162205
11	99989.84202	7015335	7016047
10	99989.42319	6875009	6875736
9	99988.99589	6740185	6740927
8	99988.56014	6610547	6611304
7	99988.11592	6485801	6486572
6	99987.66324	6365674	6366460
5	99987.20211	6249915	6250715
4	99986.73251	6138290	6139105
3	99986.25445	6030582	6031411
2	69985.76793	5926587	5927431
1	99985.27294	5826117	5826975
c	99984.76951	5728996	5729869
89			

Fig. 7.44 Sine values, in Grienberger's (1630) *Elementa Trigonometrica*

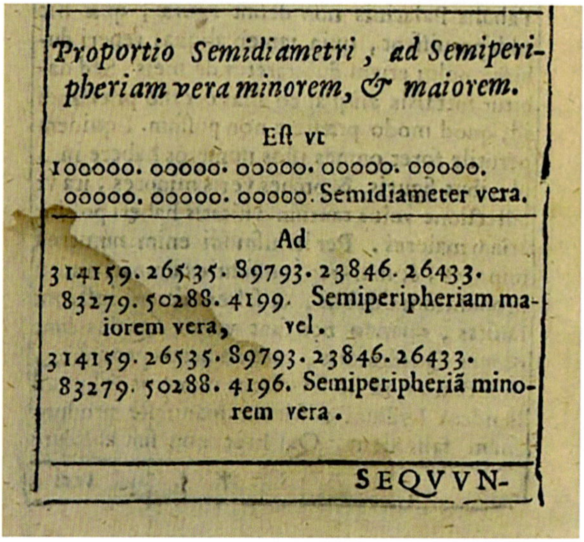

Proportio Semidiametri, ad Semiperipheriam vera minorem, & maiorem.

Eſt vt

100000. 00000. 00000. 00000. 00000.
00000. 00000. 00000. Semidiameter vera.

Ad

314159. 26535. 89793. 23846. 26433.
83279. 50288. 4199. Semiperipheriam maiorem vera, vel.
314159. 26535. 89793. 23846. 26433.
83279. 50288. 4196. Semiperipheriã minorem vera.

SEQVVN-

Fig. 7.45 Screenshot from Grienberger (1630: 5) showing the value of Pi

(Fig. 7.45). A conventional method evaluated π from the Polygon Approximation attributed to Archimedes (287–212 BCE) where one found the perimeters of inscribed and circumscribed polygons of *n*-sides. Grienberger computed π to 39 places of decimal using a polygon of 1040—sides.

7.12.4 Fr. Jean Richaud

An astronomically significant incident in the history of Indian and international astronomy was the discovery that the brightest star in the constellation of Centaurus, α Centauri was a double star. The discovery was made by a French Jesuit priest Jean Richaud (1633–1693), on 19 December, 1689 from Pondicherry, with a 12 ft telescope. Although Fr. Kirwitzer's and Shakerley's were innovative uses of the telescope whilst in India, Fr. Richaud made a systematic effort to introduce telescopic astronomy. He practiced and taught astronomy, at the Jesuit school in São Tomé in Madras until his last (Kameswara Rao et al., 1984). Fr. Richaud also was one of the discoverers of the comet of 1689, a sungrazer now designated C/1689 X1. He made his discovery on 8 December from Pondicherry (Vsekhsvyatskii, 1964: 121).

7.12.5 The Time Convention

Many dates of the cometary sightings given in here are in UT. Hence, these appear in fractions. Any correction to the dates needs care. In this paper, we have cited the dates just as they were reported in the original treatises.

Note that the dates prior to 15 October 1582 are Julian. The Gregorian calendar was introduced in 1582 to reform the Julian calendar by neutralizing the accumulated error of 13 centuries of time-keeping amounting to 10 days. The reformed calendar began when the last day of the Julian calendar Thursday, 4 October 1582 was followed by the first Gregorian day Friday, 15 October 1582, without affecting the cycle of weekdays. The Catholic countries in Europe were the first to adopt the Gregorian calendar.

As is clear from his description, Fr. Borri (1631: 115) used Gregorian dates. The date 9 November as given by Fr. Borri for the earliest sighting of the first comet does not align with the dates of first sighting the contemporary European astronomers came to accept. Therefore, the convention of division of time that Fr. Borri used needs to be ascertained. In an early nineteenth century guide for navigators (Moore, 1807: 218), we do find reference to a particular convention where the civil day begins at midnight and the astronomical day begins at the noon of the civil day. Kronk (1999: 335) also has pointed to this convention when giving dates for the Great Comets of 1618. Fr. Borri's observation of 9 November would have been in the morning and therefore the correct date is 10 November.

7.12.6 Kirwitzer Astronomical Day

The city of Kadaň observes a Kirwitzer Astronomical Day in the Jesuit astronomer's honour, in the month of October every year since 2004. Fr. Kirwitzer hailed from Kadaň. So, that took me to Kadaň, after the IAU in Vienna in August 2018. The trip was facilitated by Prof. Vladimir Karas, Astronomical Institute, Czech Academy of Sciences, Prague, and Dr. Petr Hadrava from the same Institute, and Mr. Jan Losenický, Vice Mayor, Kadaň, who played a wonderful host. In a lecture on 30 August 2018, I presented Fr. Kirwitzer's astronomical observations in India at the Franciscan Monastery of the 14 Holy Helpers, which dates back to the fifteenth century. Figure 7.46 is an artistic impression of a Jesuit priest observing a comet in Goa with a telescope. The situation is of 11 November 1618, as it shows Mars and the Moon. The painting by Dr. Richa Kapoor, a neurobiologist and hobbyist, was given in the hands of Jan Losenický as a gift to the city of Kadaň.

The year 2018 marked 400 years since the first astronomical use of the telescope in India. On 4 November 2018, Kadaň celebrated 400 years of the independent discovery of the Great Comets by Fr. Kirwitzer and brother Jesuits as a part of the annual Kirwitzer Astronomical Day, vide the Poster in Fig. 7.47. In India, the Astronomical Society of India celebrated Mahatma Gandhi's 150th Birth Anniversary year starting from 2 October 2018 as 'Bapu Khagol Mela' (Mahatma Gandhi's Astronomy Fair) to mark his interest in astronomy, and 400 Years of the astronomical use of the telescope in India. In Goa, this was celebrated during 11–18 November 2018 through an astronomy festival at Birla Institute of Technology and Science (BITS), Goa, a lecture on the first astronomical use of the telescope in India by R C Kapoor at the Goa Science Centre and Planetarium, Panaji, and sky watch through telescopes at the Patriarchal Seminary, Rachol and at the Miramar beach in

Fig. 7.46 An artistic re-creation of a Jesuit in Goa observing the Comet 1618 III with a telescope; the red object in the middle represents Mars, with the Moon to its right. (Art by Richa Kapoor, July 2018)

Panaji, organized by Nehru Planetarium, Nehru Memorial Museum and Library, in collaboration with Goa Science Centre and Planetarium, Association of Friends of Astronomy, Goa and Public Outreach Committee, Astronomical Society of India. See also Kapoor (2018b, 2018c).

7.12.7 An Interesting Painting

In 2009, while beginning a search for old records of comet observations from India, I came across an interesting painting on the website of Amateur Astronomers Association, Delhi. It depicts the Emperor Jahāngīr observing a comet in the skies of October 1618 (Fig. 7.48). According to Dr. C.B. Devgun (personal communication, 19.03.2010), the painting featured in a book by Dr. Nirupama Raghavan was published around when the Comet Hale-Bopp made its appearance. As the present research evinces, the depiction does not agree with the observations in *Tūzūk-i Jahangīrī* but is beautiful.

Fig. 7.47 Kirwitzer Astronomical Day 2018. (Source: Jan Losenický)

For a feel of the style of Jahāngīr's court artists, one may look up Bichitr's Jahangir proferring a Sufi Shaikh to Kings (Fig. 7.14), or, Abū'l Hasan's in—'Jahangir Embraces Shāh Abbās While Standing on a Globe'. The former shows Jahāngīr seated on an hour-glass throne and presenting a book to the Sufi. Abū'l Hasan had painted some of the best-known illustrations of the Emperor in the year 1618. Several of these paintings are presented and discussed in Bailey's (2001) paper.

Fig. 7.48 'Jehangir observing a comet in the skies of October 1618. It sported a tail of 24° and was observed for 16 days'. (Adapted from: 'Are there any ancient references to comets in India?', http://delhiamateur.tripod.com/anc_refs.htm)

Acknowledgements The present work is an extension of my paper 'Kapoor, R.C., 2016. Nur Ud-Din Jahangir and Father Kirwitzer: the independent discovery of the Great Comets of November 1618 and the first astronomical use of the telescope in India, *Journal of Astronomical History and Heritage*, 19(3), 264-297'. I am grateful to the Editors, *Journal of Astronomical History and Heritage* for their kind permission to use it.

I am grateful to Dr. Sabino Maffeo, S.J., Professors Anna Reale and Armando Reale, who translated on request about half of Fr. Kirwitzer's treatise from Latin to Italian and to Dr. Richard Boyle, S.J., Vatican Observatory, for all the help in this matter. Thanks are due to Dr. Paul Gabor, S.J., Vice-Director, Vatican Observatory, Tucson, and Dr. Ilaria Cristofaro, Università della Campania Luigi Vanvitelli, for clarification on certain Latin phrases. Several passages that are translations from Latin and French into English are software-assisted, and after several iterations; for any error or imperfection, the responsibility is solely mine. My thanks are due to Dr. Ian Favell for introducing me to a few useful papers and the Jesuit Document, *Ratio atque Institutio Studiorum Societatis Iesu, 1599*, and very insightful correspondence. I thank Dr. R Vasundhara for correspondence on critical visibility of comets. Thanks are due to the late Dr. Baba Varghese, Indian Institute of Astrophysics (IIA), for preparing the outline maps of India marked with specific locations. At Goa, I had helpful discussions in February 2016 with Dr. Savio Abreu, S.J. and Dr. Anthony da Silva, S.J. at the Xavier Centre of Historical Research, and with the late Mr. Percival Noronha (1923–2019), Founder of The Indian Heritage Society (Goa chapter) and Association of Friends of Astronomy, Goa, then 94 and very alert, on the history of Goa. Conversations with the late Prof.

R. K. Kochhar on the yesteryears of Indian astronomy and his suggestions have been valuable. Prof. T. E. Huff, Harvard University, kindly made his paper 'The Telescope and Scientific Curiosity in Non-Western Societies' available prior to its publication, for which I am thankful to him. I thank the Werner Forman Archive/Naprestek Museum, Prague, for providing the image of a margin drawing from the folio of Jahangir's Album (IDS Ref: 55403395-67).

I am grateful to the Director, Indian Institute of Astrophysics (IIA), for the facilities. I thank Drs Christina Birdie, A. Vagiswari, Mr. P. Prabahar, and B.S. Mohan at the Library, IIA for help with accessing various references. This work has gratefully made use of 'The SAO/NASA Astrophysics Data System', the 'On-Line Solar System Data Service' of the Jet Propulsion Laboratory, 'Eclipse Predictions by Fred Espenak (NASA's GSFC)', 'Eclipse Predictions by Fred Espenak, EclipseWise.com', the websites of Chandra X-ray Center, Juhani Kaukoranta and Raahen lukio, the Google Maps, the Fourmilab of John Walker, Walker, J. (2013), and CalendarHome.com. The classical reference material was gratefully used from many websites, particularly the Internet Archive, Google Books, Hathi Trust, al Hakawati, the Packard Humanities Institute, Project Gutenberg, Scribd, the British Library, the Library of Congress, Wikimedia Commons, and various institutions. I am happy to thank Dr. Richa Kapoor for the painting depicting a Jesuit observing a comet with a telescope in Goa, and Mr. Manu Kapoor for the crucial help at the computer.

References

Alam, M., & Subrahmanyam, S. (2009). Frank disputations: Catholics and Muslims in the court of Jahangir (1608-11). *The Indian Economic and Social History Review, 44*, 457–511.

Allen, R. H. (1899). *Star-names and their meanings*. G.E. Stechert.

Annual Letters. (1621). *Annual Letters from Japan, China, Goa, and Ethiopia: Written to the M.R.P. General of the Society of Giesu. By fathers of the same society in the years 1615, 1616, 1617, 1618, 1619. Popularized by Father Lorenzo delle Pozze of the same Society.* After her: By Paeifiaco Pontio, & Gio. Battista Piccaglia Stampatori Archiepiseopali.

Ansari, S. M. R. (2002). Practical astronomy in Indo-Persian sources. *Indian Journal of History of Science, 37*(3), 255–265.

Apian, P. (1532). *Ein kurtzer bericht der Observation und urtels des jüngst erschienenen Cometen jm Weinmon vnd Wintermon dises 32.* Jars. no publisher given.

Apianus. (1540). *Astronomicum Caesareum.*

Baber, Z. (1996). *The science of empire: Scientific knowledge, civilization and colonial rule in India.* State University of New York Press.

Baichun, Z. (2003). The introduction of European astronomical instruments and the related technology into China during the seventeenth century. *EASTM, 20*, 99–131.

Bailey, G. A. (2001). The end of the 'Catholic era' in Mughal painting. *Marg, 53*, 47–59.

Baldini, U. (2015). Scienza, missioni, destino: Wenceslas Pantaleon Kirwitzer. In L. F. Barreto & W. Zhiliang (Eds.), *Macau past and present* (pp. 283–361). Centro Científico e Cultural de Macau, I.P.

Baldini, U. (2017). Le scienze matematiche a Milano nell'età dei Borromeo: il noto e l'ignoto, in Nell'età di Galileo. Milano, l'Ambrosiana e la nuova scienza, edd. E. Bellini - A. Rocca, Milano-Roma 2017 (Studia Borromaica, 30) (pp. 3–106).

Beveridge, H. (Trans.). (1897–1939). *The Akbarnama of Abu-L-Fazl*, three volumes. Asiatic Society of Bengal (translated from the 1596 Persian original).

Birch, T. (1849). *The court and times of James the first*, 2 volumes. Henry Colburn.

Blochmann, H. (1869). Note on the fall of a Meteorite at Jullunder in April A.D. 1621, according to the *Iqbālnāmah-i Jahāngīrī. Proceedings of the Asiatic Society of Bengal, 38*, 167–168.

Bobrovnikoff, N. T. (1984). *Astronomy before the telescope* (Vol. I). Pachart Publishin House.

Bolt, M., & Korey, M. (2010). The world's oldest surviving telescopes. In A. van Helden, S. Dupr, R. van Gent, & H. Zuidervaart (Eds.), *The origins of the telescope* (pp. 231–255). KNAW Press.
Borkar, S. S. (2016). *The astronomical mysteries of Goan churches*. Retrieved December 3, 2016, from http://planetgoaonline.com/places-to-visit/Culture-n-Heritage/Goan-churches
Borri, C. (1631). *Collecta Astronomica ex Doctrina*, Ulysipone, Apud Matthiam Rodrigues. Anno M.DC.XXXI.
Bortle, J. E. (1998). The bright comet chronicles. *International Comet Quarterly*. Retrieved July 15, 2012, from http://www.icq.eps.harvard.edu/bortle.html
Brahe, T. (1603). *Tychonis Brahe De mundi aetherei recentioribus phaenomenis liber secundus*. Typis inchoatus uraniburgi Daniae, absolutus Pragae Bohemiae (Library of Congress: http://hdl.loc.gov/loc.rbc/General.94796.1)
Brewster, D. (1811). *Ferguson's astronomy*, 2 volumes. John Ballantyne and Co.
Brown, C. J. (1922). *The coins of India* (pp. 95–96). Association Press.
Brown, P. (1999). The Leonid Meteor Shower: Historical visual observations. *Icarus, 138*, 287–308.
Burke-Gaffney, M. W., S. J. (1944). *Kepler and the Jesuits*. The Bruce Publishing Company.
Carolino, L. M. (2007). Cristoforo Borri and the epistemological status of mathematics in seventeenth-century Portugal. *Historia Mathematica, 34*, 187–205.
Carolino, L. M. (2023). The burden of Galileo's controversy: The Jesuit revisiting of the Aristotelian cosmos in Collegio Romano (1618-1677). *Galilæana, XX*(2), 33–60.
Crossen, C., & Procházka, S. (2007). The seven sleepers and ancient constellation traditions—A crossover of Arabic dialectology with the history of astronomy. *Wiener Zeitschrift Für Die Kunde Des Morgenlandes, 97*, 79–105. http://www.jstor.org/stable/23861409
Cysat, J. (1619). *Mathemata astronomica de loco, motu, magnitudine et causis cometae qui sub finem anni 1618 et initium anni 1619 in coelo fulsit*. Volpert.
Da Fonseca, J. N. (1878). *Historical and archaeological sketch of the city of Goa*. Thacker & Co. Limited.
De Mendonça, D. J. (1958). *Conversions and citizenry: Goa under Portugal 1510–1610*. Concept Publishing Company.
della Valle, P. (1631). *Risalah-i Padri Khristafarus Burris Isavi dar tufiq-i jadid dunya. Compendio di un trattato del Padre Christoforo Borro Giesuita della nuova costitution del mondo secondo Tichone Brahe e gli altri astologi moderni* (Compendium of a tractate of Father Christoforo Borri, S.J. on the new model of the universe according to Tycho Brahe and the other modern astronomers), in Persian and Italian. Naqsh script in the hand of della Valle, 1624, 1631.
Della Valle, P. (1843). *Viagi di Pietro della Valle, il pelegrino. Descritti da lui medesimo in lettere familiari all'erudoito suo amico Mario Schiapano, divisi in tre parti cioè: La Truchia, La Persia, e l'India, colla vita dell'autre* (Vols. I & II). G. Gancia.
Dick, S. J. (1998). Observation and interpretation of the Leonid meteors over the last millennium. *Journal of Astronomical History and Heritage, 1*, 1–20.
District Census Handbook, Dhar. (2015). Registrar General & Census Commissioner, India. *Census of India 2011 Madhya Pradesh*, Series—24, Part XII—B, Directorate of Census Operations, Madhya Pradesh, Government of India.
District Census Handbook, Dohad. (2015). Registrar General & Census Commissioner, India, 2015. *Census of India 2011 Gujarat*, Series—25, Part XII—B, Directorate of Census Operations, Gujarat, Government of India.
District Census Handbook, Jhabua. (2015). Registrar General & Census Commissioner, India. *Census of India 2011 Madhya Pradesh*, Series—24, Part XII—B, Directorate of Census Operations, Madhya Pradesh, Government of India.
District Census Handbook, Ujjain. (2015). Registrar General & Census Commissioner, India. *Census of India 2011 Madhya Pradesh*, Series—24, Part XII—A, Directorate of Census Operations, Madhya Pradesh, Government of India.

dos Santos, J. M. P. (2005). *The "Kuroda Plot" and the legacy of Jesuit scientific influence in seventeenth century Japan, Bulletin of Portuguese/Japanese Studies, June-December, numero 10-11* (pp. 97–191). Universidade Nova de Lisboa.

Drake, S., & O'Malley, C. D. (Trans.). (1960). *Galileo Galilei, Horatio Grassi, Mario Guiducci and Johan Kepler: The controversy on the comets of 1618*. University of Pennsylvania Press.

Dreyer, J. L. E. (1890). *Tycho Brahe: A picture of scientific life and work in the sixteenth century*. Adam and Charles Black.

Dror, O., & Taylor, K. W. (Eds. and annotators). (2006). *Views of seventeenth century Vietnam: Christoforo Borri on Cochinchina and Samuel Baron on Tonkin*. Cornell Southeast Asia Program Publications.

Elliot, H. M. (1873). *The history of India, as told by its own historians, The Muhammadan period* (Vol. V). Trubner and Company.

Elliot, H. M. (1875). *The history of India, as told by its own historians. The Muhammadan Period* (Vol. VI). *Akbar and Jahangir*. Trubner and Company.

Elliot, H. M. (1975). *Wāki'āt-i Jahāngīrī*. Hafiaz Press.

Espenak, F. (2015). *Eclipse predictions by Fred Espenak, NASA/GSFC*. http://eclipse.gsfc.nasa.gov

Faintich, M. (2007). *Astronomical symbols on ancient and medieval coins*. McFarland & Company, Inc. Symbolic Messengers. http://www.symbolicmessengers.com/newfindi.htm

Farrell, A. P. (1970). *Ratio atque Institutio Studiorum Societatis Iesu, 1599*. Conference of Major Superiors of Jesuits.

Festou, M. C., Rickman, H., & West, R. M. (1993). Comets I. Concepts and observations. *The Astronomy and Astrophysics Review, 4*, 363–447.

Figueroa, G. d. S. (1967). *L'ambassade de D. Garcias de Silva Figueroa en Perse*. Chez Louis Billaine.

Foster, W. (1921). *Early travels in India, 1583–1619*. Humphrey Milford, Oxford University Press.

Foster, W. (1928). *A supplementary calendar of documents in the India office relating to India or to the home affairs of the East India Company, 1600–1640*. Printed by His Majesty's Stationary Office, for the India Office.

Goldstein, B. R. (2011). Levi ben Gerson and the cross staff revisited. *Aleph, 11*(2), 365–383.

Goldstein, B. R., & Hon, G. (2005). Kepler's move from orbs to orbits: Documenting a revolutionary scientific concept. *Perspectives on Science, 13*(1), 74–111.

Golvers, N. (1992). Scientific literature in Latin by the Jesuits in XVIIth-century China. *Revista de la Academia Colombiana de Ciencias, XVIII*(70), 389–402.

Granada, M. A., Mosley, A., & Jardine, N. (2014). *Christoph Rothmann's discourse on the comet of 1585*. Brill.

Grassi, O. (1619). *On the three comets of 1618. Rome (in Latin)*. Google Books. Retrieved December 24, 2015, from https://galileo.ou.edu/exhibits/

Grienberger, C. (1630). *Elementa Trigonometrica, id est Sinus Tangentes Secantes In Partibus Sinus totius 100000*. Per Haered. Barthol. Zan.

Grienbergeri, C. (1612). *Catalogus veteres affixarum Longitudines, ac Latitudines conferens cum nouis. Imaginum Caelestium Prospectiva duplex...* Romae, Apud Bartholomaeum Zanettum, MDCXII (Biblioteca Nazionale, Napoli; Google Books).

Guerreiro, F. F. (1930). *Jahangir and the Jesuits* (E. D. Ross & E. Power, Eds.). George Routledge & Sons.

Hägerdal, H. (2006). *Candrasangkala: The Balinese art of dating events*. University of Växjö, Department of Humanities.

Hadi. (1995). *Dictionary of Indo-Persian literature*.

Hellman, C. D. (1944). *The comet of 1577 its place in the history of astronomy*. A.M.S. Press.

Hind, J. R. (1852). *The comets: A descriptive treatise upon those bodies. With a condensed account of the numerous modern discoveries respecting them; and a table of all the calculated comets, from the earliest ages to the present time*. John W. Parker and Son.

Howe, H. A. (1896). *Study of the sky*. Mead Ville Penna Flood and Vincent.

Huff, T. E. (2010). The telescope and scientific curiosity in non-Western societies. In L. Stenberg (Ed.), *Science, technology and entrepreneurship in the Muslim world*. University of Utah Press.
Jervis, J. L. (1985). Cometary theory in fifteenth-century Europe. In *Studia Copernicana* (Vol. 26). Springer.
Jet Propulsion Laboratory (JPL). (2021). *Small-body database browser*. https://ssd.jpl.nasa.gov/sbdb.cgi
Kameswara Rao, N., Vagiswari, A., & Louis, C. (1984). Father J. Richaud and early telescope observations from India. *Bulletin of the Astronomical Society of India, 12*, 81–85.
Kapoor, R. C. (2015). Abū'l Faẓl, independent discoverer of the great comet of 1577. *Journal of Astronomical History and Heritage, 18*(3), 249–260.
Kapoor, R. C. (2016). Nur Ud-Din Jahangir and Father Kirwitzer: The independent discovery of the Great Comets of November 1618 and the first astronomical use of the telescope in India. *Journal of Astronomical History and Heritage, 19*(3), 264–297.
Kapoor, R. C. (2018a). Comet tales from India. 1: Ancient to medieval. In M. Sôma & K. Tanikawa (Eds.), *Proceedings of the Fifth Symposium on 'Historical Records of Astronomy and Modern Science'* (pp. 14–61). National Astronomical Observatory of Japan.
Kapoor, R. C. (2018b). The first astronomical use of the telescope in India. In W. Orchiston, A. Sule, & M. N.Vahia (Eds.), *Growth and development of astronomy and astrophysics in India and the Asia-Pacific region. Proceedings of the 9th International Conference on Oriental Astronomy* (pp. 34–39). Tata Institute of Fundamental Research.
Kapoor, R. C. (2018c). First astronomical use of the telescope in India. *Science Reporter, 55*(11), 41–43.
Kennedy, E. S. (1980). Astronomical events from a Persian astrological manuscript. *Centaurus, 24*, 162–177.
Kepler, J. (1609). *Astronomia Nova*. G. Voegelinus.
Kepler, J. (1619). *De cometis libelli tres*. ETH-Bibliothek Zürich. www.e-rara.ch
Kepler, J. (1625). *Tychonis Brahei dani Hyperaspistes*. Apud Godfridum Tampachium.
Khan, M. A. R. (1935). A forgotten Indian meteorite. *Nature, 135*, 39.
Khan Ghori, S. A. (2000). Development of Zīj literature in India. In S. N. Sen & K. S. Shukla (Eds.), *History of astronomy in India* (pp. 24–54). Indian National Science Academy.
Kirwitzer, V. (1620). Observationes Cometarvm Anni 1618. In *India Orientali Factæ a Societatis Iesv Mathematicis in Sinense regnum Nauigantibus ex itinere eo delatis*. Schönwetter.
Kochhar, R. K. (1989). Transit of mercury 1651: The earliest telescopic observation in India. *Indian Journal of History of Science, 24*, 186–192.
Kronk, G. W. (1999). *Cometography, a catalogue of comets, Vol. 1: Ancient—1799*. Cambridge University Press.
Law, N. N. (1916). *Promotion of learning in India during Muhammadan rule*. Longmans, Green & Co.
Lee, H. D. P. (1952). *Aristotle Meteorologica*. Harvard University Press.
Leitao, H. (2008). The content and context of Manuel Dias' *Tiawenlue*. In L. Saraiva & C. Jami (Eds.), *The Jesuits, the Padroado and East Asian sciences* (pp. 1552–1773). World Scientific.
Long, R. (1742). *Astronomy in five books* (Vol. I). Printed for the Author.
Lourenço, J. (2005). *The Parish churches of Goa*. Amazing Goa Publications.
Lynn, W. T. (1889). The great comet of 1618. *The Observatory, 12*, 407–408.
Lynx open Ed. (2021). Exhibit guide navigation > Oratio Grassi, De tribus cometis anni MDCXVIII (Rome, 1619), *On the three comets of 1618*. Retrieved November 29, 2021, from https://lynx-open-ed.org/node/328
MacDonnell, J. F. (2014). Grienberger, Christopher. In T. Hockey et al. (Eds.), *Biographical encyclopedia of astronomers* (2nd ed., pp. 853–854). Springer.
Maclean, A. (1987). King James' poem on the great comet of 1618. *Journal of the British Astronomical Association, 97*, 74.
Majumdar, R. C., Raychaudhuri, H. C., & Datta, K. K. (1967). *An advanced history of India*. The MacMillan Company of India Limited.

Marcus, J. N. (2007). Forward-scattering enhancement of comet brightness. II. The light curve of C/2006 P1 (McNaught). *International Comet Quarterly, 2007*, 119–130.
Marsden, B. G. (2005). Sungrazing comets. *Annual Reviews of Astronomy and Astrophysics, 43*, 75–102.
Meinel, C. (2009). *Grenzgänger zwischen Himmel und Erde: Kometen in der Frühen Neuzeit*. Universitaetsverlag Regensburg.
Mitchell, W. M. (1916). The history of the discovery of the solar spots. *Popular Astronomy, XXIV*, 341–354.
Modi, J. J. (1917). *Asiatic papers, part II*. Papers read before the Bombay Branch of the Royal Asiatic Society, The Times Press.
Moore, J. H. (1807). *The new practical navigator*. Printed for J. Johnson et al.
Mousavi, S. (2000). Science and superstition under Akbar and Jahangir: The observation of astronomical phenomena. In I. Habib (Ed.), *Akbar and his India* (pp. 109–120). Oxford University Press.
Museo Galileo. (2021). *Christoph Grienberger.* Retrieved November 20, 2021, from https://portalegalileo.museogalileo.it/igjr.asp?c=36931
Nature. (1878). The comets of 1618. *Our Astronomical Column, 17*, 247–248.
Patriarchal Seminary of Rachol. (2016). *The heritage of Rachol*. Retrieved March 1, 2016, from http://racholseminarygoa.org/heritage-of-rachol
Peng-Yoke, H. (1964). Natural phenomena recorded in the Đai-Viêt Su'-ky Toan-Thu', an early Annamese historical source. *Journal of the American Oriental Society, 84*(2), 127–149.
Peter Cruger's, (1619). Uranodromus cometicus: Ein ausführlicher Tractat vom grossen Cometen des 1618. Jahrs, Danzig.
Pina, I. (2007). Manuel Dias Sénior//Li Manuo. *Bulletin of Portuguese—Japanese Studies, 15*, 79–94.
Pingré, A. G. (1784). *Cométographie: ou Traité historique et théorique des comètes*, Tome Second, A Paris de Limprimerie Royale.
Ravene, G. (1897). The appearance of Halley's comet in A.D. 141. *The Observatory, 20*, 203–205.
Rogers, A., & Beveridge, H. (Trans.) (1909). *The Tuzuk-i-Jahangiri* (Vol. I). Royal Asiatic Society.
Rogers, A., & Beveridge, H. (Trans.) (1914). *The Tuzuk-i-Jahangiri* (Vol. I). Royal Asiatic Society.
Sabri, B.-A. (2012). *Smithsonian displays mysterious dagger from the stars*. https://www.facebook.com/FreerSackler/posts/428758037160077
Schreiber, J. (1904). Jesuit Astronomy. II. *Popular Astronomy, 12*, 94–112.
Sharma, V. N. (1982). The impact of the eighteenth century Jesuit astronomers on the astronomy of India and China. *Indian Journal of History of Science, 17*(2), 342–352.
Stephenson, F. R., & Said, S. S. (1991). Precision of Islamic eclipse measurements. *Journal for the History of Astronomy, XXII*, 195–207.
Stoppa, F. (2016). *Christof Grienberger.* Retrieved March 22, 2016, from www.atlascoelestis.com
Terry, E. (1655). *A voyage to East India; wherein some things are taken notice of, in our passage thither...* (p. 393). J Wilkie.
Theibault, J. (1994). Jeremiah in the village: Prophecy, preaching, pamphlets, and penance in the thirty years' war. *Central European History, 27*(4), 441–460.
Toomer, G. J. (1984). *Ptolemy's ALMAGEST*. Gerald Duckworth & Co. Ltd.
The Assayer. (1623). *Il saggiatore*. [*The Assayer*].
U.S. Naval Observatory. (2012). *Rise, set and twilight definitions*. Retrieved July 17, 2012, from https://aa.usno.navy.mil/faq/RST_defs
Udias, A. (1994). Jesuit astronomers in Beijing 1601-1805. *R.A.S. Quarterly Journal, 35*, 463–478.
Udias, A. (2003). *Searching the heavens and the Earth: The history of Jesuit observatories*. Kluwer Academic Publishers.
Van Helpern, A., Dupré, S., van Gent, R., & Zuidervaart, H. (Eds.). (2010). *The origins of the telescope*. Royal Netherlands Academy of Arts and Sciences.
von Zach, F. X. (1822). *Correspondance Astronomique, Géographique, Hydrographique et Statistique*. A. Ponthenie.

Vsekhsvyatskii, S. K. (1964). *Physical characteristics of comets*. Israel Program for Scientific Translations.
Walker, J. (2013). *Your sky & solar system live* Fourmilab. https://www.fourmilab.ch/
Walter, D. L. (1890). Medallic memorials of great comets, and the popular superstitions connected with their appearance. *American Journal of Numismatics, and Bulletin of the American Numismatic and Archaeological Society, 24*(4), 84–90.
Warner, D. J. (1998). Telescopes for land and sea. *Rittenhouse, 12*(2), 33–54.
Whatton, W. R. (1859/2010). *Transit of venus across the Sun; a translation of the celebrated discourse thereupon by Jeremiah Horrocks*. General Books.
Wheeler, J. T. (1881). *Tales from Indian history: Being the annals of India retold in narratives* (p. 68). W Thacker & Co.
Woods, T. E. (2005). *How the Catholic Church built Western civilization*. Regnery Publishing, Inc.
Yazdi, H.-R. G. (2008). Solar eclipses in medieval Islamic civilization: A note on cultural and social aspects. *Iranian Journal for the History of Science, 6*(2008), 75–82.
Yeomans, D. K. (2007). *Great comets in history*. https://ssd.jpl.nasa.gov/?great_comets
Yeomans, D. K., Rahe, J., & Freitag, R. S. (1986). The history of comet Halley. *Journal of the Royal Astronomical Society of Canada, 80*, 62–86.
Zettl, E. (2008). *Johannes Schreck-Terrentius Wissenschaftler und China-Missionar (1576-1630)*. Hochschule für Technik, Wirtschaft und Gestaltung Konstanz Studiengang Wirtschaftssprachen Asien und Management.

R. C. Kapoor began his career in 1971 at the Uttar Pradesh State Observatory (now Aryabhatta Research Institute of Observational Sciences, ARIES) in Nainital, India, focusing on observational astronomy. His main interest was flare stars. From March 1974 until September 2010, he was with the Indian Institute of Astrophysics (IIA), Bangalore, where he worked on various topics in relativistic astrophysics, including observational aspects of black holes, white holes, quasars, and pulsars. Currently, his interest lies in the history of astronomy in the Indian region, resulting in a succession of papers in various journals. He has participated as an observer and organizer in a few solar eclipse expeditions of IIA and also went to observe a few on his own. He has all along been active in popularizing astronomy. He died on August 21, 2025.

Chapter 8
Celestial Portents in 1615–1619 and the Manchu Overthrow of the Ming Dynasty (1368–1644)

David W. Pankenier

8.1 Introduction

In China, based on the earliest written records from at least the mid-second millennium BCE on deviations from the normal order of things, that is, anomalies and prodigies in nature, were viewed with great concern. Among the phenomena that appear in the oracle-bone divinations from the late Shang dynasty (ca. 1562 BCE to 1046 BCE) are eclipses of the sun and moon. Apart from noteworthy meteorological phenomena, other celestial irregularities have yet to be reliably deciphered. It is likely that comets and meteor showers also caused disquiet, but such records have not been identified. From these beginnings, celestial signs became most prominent among the omens bearing on the rulership, and as such, they were routinely divined about by the scribe-astrologers.

The most comprehensive narrative history of events of the long-lived Zhou dynasty (1046–256 BCE), the laconic *Spring and Autumn Annals,* and especially its extensive *Zuo Commentary*, routinely report eclipses, comets, meteors, planetary phenomena, major floods, landslides, prodigies, and so on, often together with their implications for the rulership and the state. Along with the institutionalization of omen interpretation and divination, by the beginning of the Zhou dynasty, an ideology of kingship—the Mandate of Heaven—had taken shape in which celestial omens played a particularly prominent role in legitimizing the rulership. As Christopher Cullen put it, by the beginning of the empire centuries prior to the Common Era, "a Chinese emperor was theoretically responsible for the orderly functioning of the entire cosmos as well as that limited part of it that constituted human society" (Cullen, 1991: 114). Thereafter, not much changed beyond the

D. W. Pankenier (✉)
Lehigh University, Bethlehem, PA, USA
e-mail: dwp0@lehigh.edu

C. J. Cunningham (ed.), *Cosmic Events*, Historical & Cultural Astronomy,
https://doi.org/10.1007/978-3-031-86950-1_8

rigorous bureaucratization of the office of Grand Scribe-Astrologer until the fall of the last imperial dynasty, the Qing (1644–1911).

Cullen (1991: 117) analyzes a particularly illustrative case of a momentous cometary apparition, that of Comet Halley in 12 BCE (return −26) which was taken at the time to portend Heaven's withdrawal of the Mandate from the Han dynasty (206 BCE–220 CE) at a most vulnerable juncture:

> BCE 12 Aug 26–Oct 20: First year of the *Yuanyan* reign period of Emperor *Cheng* of the Han Dynasty (12 BCE), seventh month, day *xinwei* [Aug 26]; a star bristled in Eastern Well [8 stars, marked by *mu* Gem]; it trod on Five Lords [five stars, *theta* Gem]; it appeared to the north of *River Garrison* [*alpha* Gem] and passed through the *Yellow Emperor* [*alpha* Leo] and amidst the *Grand Tenuity Enclosure* [10 stars, *beta* Vir]; later traveling more than six *du* daily. In the morning, it appeared in the east. On the evening of the thirteenth day [Sep 7], it was seen in the west; it trespassed against the *Secondary Concubine* [*zeta* Leo], the Empress's Palace [*alpha* Leo], the Northern Dipper, and Saturn; its swarm of flames again penetrated the *Vermilion Palace* (circumpolar region), with Great Fire [*alpha* Sco] right behind it. It reached the *Celestial River* [Milky Way)] sweeping the area of the *Imperial Concubine* [*gamma* Leo] and the *Empress* [*alpha* Oph] and then it moved south, crossing and trespassing against *Great Horn* [Arcturus] and the *Assistant Conductors* [*omicron & eta* Boö]. When it reached *Celestial Marketplace* [22 stars forming a ring in Oph], it moved slowly at a regular pace, its flames entering the *Marketplace*. After another ten days, it proceeded westward. On the fifty-sixth day [Oct 20], it went out of sight together with the [setting] *Cerulean Dragon* [*Vir~Sgr*] (Cullen, 1991: 137, tr. modified).

Proffering his opinion on the meaning of the apparition, a prominent official *Gu Yong*, and the highly influential scholarly member of the imperial family *Liu Xiang* said:

> 'This is [an omen of] extreme disorder such as has rarely been seen since high antiquity. If we examine the rapid movement [of this object], the variations in the length of its flaming rays, and the [constellations] on which it has trespassed successively, it [clearly signifies] harm to the women of the rear palace within, and the curse of rebellion amongst the people without.' Likewise, *Liu Xiang* said, 'when the Three Dynasties [Xia, Shang and Zhou] fell [in antiquity], the Assistant Conductors changed direction, and the destruction of the Qin dynasty (221–206 BCE) and [*Xiang*] *Yu* [before the establishment of the Han dynasty] there was a bristling star at Great Horn.' (Pankenier, 2013: 507).

This is an allusion to the comet of Aug–Sep 204 BCE that appeared as the Qin dynasty (221–206 BCE) was being overthrown. Prognostications along such lines are indicative of the concern at the highest level provoked by cometary apparitions such as codified in the precedent-setting *Treatise on the Celestial Offices* by the Grand Scribe Astrologer *Sima Qian* (145/135–86 BCE) a century earlier. According to Sima Qian, the appearance of major comets always portended either serious internal rebellion or major military conflict along the frontiers (Pankenier, 2013: 506–508; Loewe, 1994). They were never auspicious. Subsequent directors of the imperial Bureau of Astrology and the Calendar followed these precedents, as evidenced by the astral prognostication manuals produced through the centuries. Such material was closely held due to its sensitive nature, and as in *Sima Qian's* own case, the position of scribe-astrologer was often hereditary. Officials in the bureau were not allowed to transfer to other positions within the imperial administration.

8.2 The Comet(s) of 1618–1619

Having laid the basic groundwork, we can move on to the events of some 17 centuries later in 1618 when the Great Comet appeared as well as other anomalies and omens, including a few more comets. As Chris Cullen noted regarding the inauspicious omens associated with Comet Halley's appearance in 12 BCE (an eclipse, a meteor shower, landslides), "the last years of the first century BC were marked by an increasing sense of imminent political crisis" (Cullen, 1991: 114). If anything, the last years of the Ming dynasty *Wanli* Emperor's reign (1573–1620) must have occasioned an even greater sense of foreboding. Beset on both foreign and domestic fronts, the huge but decrepit Ming dynasty (1368–1644) was on its last legs. The Ming would soon fall to the Manchus, the non-Chinese tribal people known as the Jurchens who had earlier established a short-lived dynasty incorporating parts of north China (1125–1234). The Jurchens came from the northeast, outside the Great Wall, in present-day Liaoning Province (Manchuria). The Manchus established the last imperial dynasty, the Qing, in 1644. The Qing ("Pure") dynasty was to last for 267 years until 1911.

8.3 Discussion

The first thing to note is that between the late fall of 1616 and the spring of 1619 five or six impressive comets were sighted and recorded in China, Japan, and Korea, so there was an unusual spate of omens that portended multiple internal and external threats to the Ming dynasty (see Appendix; Swope, 2013, 11). For a number of years, the Ming had been beset by military challenges along the northern frontier posed by Mongol and Jurchen raiders from outside the Great Wall, which extended across the Bohai Sea coast in the northeast to the Tarim Basin in the far west. The threats were even more serious because both Mongols and Jurchens had previously invaded and toppled (or conquered) vast areas of China: these were the Jurchen *Jin* dynasty (1125–1234) and the Mongol *Yuan* dynasty from 1261 to 1368. In the second decade of the seventeenth century, the serious challenges to the Ming in the north coincided with a series of mutinies among Ming garrisons there, defections of local administrators, and sporadic rebellions both great and small in the interior provinces (Dardess, 2012: 123ff.).

Managing such challenges proved nearly impossible for the imperial administration, which was rife with nepotism, corruption, and incompetence. Elite troops were in short supply, funds were lacking to equip and train recruits, the burden of conscription on the populace had been increased several times, and levies on the agricultural economy were already high. The imperial treasury was not able to adequately provide the millions of ounces of silver needed to enable the armies to resist the increasingly aggressive military challenges in Manchuria. Nurhaci (1559–1626), posthumously Qing Emperor *Taizu*, who had risen to overall leadership of the Jurchen tribal forces ("banners") chose 1618 to mount large-scale cavalry attacks against the Ming garrisons in the northeast with the objective of overthrowing the Ming and establishing a new *Jin* Dynasty. In 1618, the campaign he initiated scored several major victories over the beleaguered and often outnumbered and outclassed Ming forces guarding the frontier towns and cities (Swope, 2013: 12ff.).

At the same time, with the aid of capable Chinese collaborators who could read the handwriting on the wall, Nurhaci waged a propaganda campaign to assure the Chinese populations that they would not be mistreated or their towns pillaged as had been the raiders' practice in the past. A major challenge was the conviction of ethnic superiority on both sides, which Nurhaci made concerted efforts to address (Roth, 1979: 9–11). The propaganda campaign also included efforts to exploit the omens to persuade the Han-Chinese that the celestial portents confirmed that the Ming dynasty had clearly lost the support of Heaven and was nearing collapse—the withdrawal of Heaven's Mandate. A major challenge in this regard was to convince the ethnic Chinese populace that Heaven had chosen a non-Chinese, *Nurhaci*, to undertake the founding of a new dynasty (Roth, 1979: 8). A fascinating precedent was the rebellion of General *An Lushan* in 755 which nearly brought down the *Tang* dynasty (618–907) (Pankenier, 2019: 335–360).

Contemporary pre-imperial Manchu records for the period are scarce and the surviving Ming records do not disclose much in the way of rationalization of the ominous omens. Most of officialdom would not have been privy to such national security deliberations in any case. However, the subsequently compiled *Veritable Records* of the two emperors (the day-to-day diaries of official acts, memorials to the throne, and imperial rescripts), in addition to citing the official reports of the portents, also hint at the importance of the astral signs in the minds of the leading actors. So, for example, Gertraude Roth (1979: 8) wrote:

> Strange lines of light in the sky were said to have appeared in 1612 and 1614. They supposedly occurred twice in 1615 as a proper prelude to the establishment of the Latter Chin [*Jin*] state in 1616. But the notion that the emperor is endowed by Heaven and receives Heaven's approval or disapproval was adopted by the Manchus with the greatest zeal at the beginning of the period of conquest. During the year 1618 an overwhelming number of reports of such heavenly signs aided the Manchus and the Chinese both to gradually become used to the idea that a dynastic change was drawing near. Lines of light in the sky appeared nearly every month, once staying for the length of an entire month. At other times, *Nurhaci's* scholarly aide *Erdeni* interpreted good weather, or a rainstorm as having been sent by Heaven to facilitate Manchu victories. Because the Manchus may not have begun writing the documents now collected in the *Man-chou chiu-tang* until 1618, the records for the earlier years seem to have been written from recollection of the events. References to heavenly omens before 1618 therefore may have been construed later. The phenomena recorded for 1618, however, most likely occurred in some form and were interpreted in the above way at that time.

As early as 1616 Ming Emperor was already greatly distressed by the growing threat from the Jurchens: "There were various portents that year including floods and crimson snow falling in Nanjing. Wanli apparently even had dreams about Nurhaci's state replacing the Ming" (Swope, 2013: 19). This is borne out by the record. In addition to the official report of the Great Comet on November 26th, the *Veritable Records of Ming Shenzong* includes the text of a memorial to the throne dated 1 week later:

> On day *renshen* (Dec 3, 1618) Academician *Fang Congzhe* said: "since the third month, fierce winds, darkness in the daytime, and gleaming flames have lit up the heavens. Subsequently, monkey aberrations and monstrosities among swine appeared in Liaodong and Shanxi Provinces. In the fourth month there was heavy snow, and the frozen corpses of horses and mules were found within the passes. As for the area of the ancestral tumuli, in the same month there were reports of cries emanating from heaven and earthquakes. In Liaodong, in the eighth month, at the same time as the earth quaked, in the imperial palaces of the Three Capitals very close to the emperor there was an earthquake at the beginning of autumn and again at the end. Before long, a white vapor appeared in the southeast, a broom star emerged in lunar lodge Base [Libra], and demonic signs and strange omens constantly appear and unfold. How could these be mere happenstance? It is my hope that officials whose service in office is ineffectual will be removed or rehabilitated. Your majesty should urgently renovate the suburban altar and the temple [of the ancestors] and personally direct administrative deliberations to renew all under Heaven [sc. the imperial administration] once again. Promptly refill vacancies among the officials in the cabinet, bureaus, and imperial secretariat, do away with incompetents by terminating and suspending. Issue commands in a steady stream to delight the hearts of the people, and in accord with the will of Heaven achieve pacific auspiciousness for ten thousand generations, in a trice turning things around. In short order all officials and the imperial secretariat will comply" (*Veritable Records of Shenzong*, ch. 575, §55).

The only element missing in this recommendation is that the emperor should issue the requisite mea culpa, taking upon himself the blame for causing Heaven's displeasure and promising to immediately undertake the reforms recommended by *Fang* in his memorial. However, by this late date in his long reign, Emperor *Shenzong* had become reclusive, detached from the administration of the empire, and possibly senile. This contributed to the demoralization and degeneration of the imperial administration. *Shenzong* died in mid-1620.

8.4 Conclusion

The *Veritable Records* of the first Qing emperor make it abundantly clear that Nurhaci's advisors appreciated the ideological implications of the Great Comet of 1618 and exploited the omens in their propaganda. Traditionally, the interpretation of the meaning of a broom star's appearance could be deduced from the asterisms swept by the tail or "business end" of the comet. The Supernal Lord of the cosmos was thought to dwell at the Pole and the emperor was his counterpart at the temporal end of the *axis mundi*. Nathan Sivin has referred to the emperor's role as that of a "vibrating dipole" responsible both to humanity and the Supernal Lord. Consequently, more than any other factor it was the Great Comet's trespassing on and sweeping the stars of the Dipper and the circumpolar region (whose asterisms all correlated with the imperial court, officialdom, and the person of the emperor himself) that revealed to the world that the end of the Ming dynasty was near. Take note of the language in the official records of celestial anomalies compiled by the Qing officials (Pankenier et al., 2008: 246–248). Normally, such reports were strictly objective, dry accounts of what was observed.

8.4.1 1618 Nov 27 [China]

(a) 3rd year of the *Tian ming* [Heavenly Mandate] reign period of Emperor *Taizu* of the *Qing* Dynasty, 10th month, 11th day [*bingyin*]; during the 5th watch, in the southeast sector a white vapor 5 ft wide emerged from a star and **surged straight at the Ming state**; after the 15th day it no longer appeared. The star that produced it gradually moved toward *Northern Dipper* each night, until the 29th day when **it went right over the handle of *Northern Dipper***, after which it was no longer seen. (*Da Qing Manzhou shilu Veritable Records of the Great Manchu Qing Dynasty*).

(b) 3rd year of the *Tian ming* [Heavenly Mandate] reign period of Emperor *Taizu* of the *Qing* Dynasty, 10th month, 11th day [*bingyin*]; a broom star appeared in the east, to the east of the previously sighted white vapor that was 5 ft wide, **with its tail pointing right at the Ming state**; each night it gradually moved toward *Northern Dipper* for 19 days in all until it passed over the tail of *Northern Dipper*, and only then was it was extinguished. (*Qing Taizu Gao huangdi shilu Veritable Records of Emperor Taizu Gao*, ch. 5).

Compare this with the standard form of such reports contained in a passage from the *Veritable Records of Ming Shenzong*:

8.4.2 1618 Nov 26 [China] (Ho & Ang 77)

(a) 46th year of the *Wanli* reign period of Emperor *Shenzong* of the Ming Dynasty, 10th month, day *yichou* [Nov 26]; a grayish-white broom star as big as a hen's egg and over one rod long emerged in *Base* [*alpha*^{1} Lib], its tail pointing southeast. After 10-odd days, it turned and pointed northwest, sweeping and trespassing against *Guardian of the Great Yang* [*chi* UMa], then entered *Neck* [*kappa* Vir] at seven *du*, 10 marks. After another 3 days, it gradually traveled northwestward, its tail sweeping Northern Dipper, *Celestial Pivot* [*beta* UMa Merak], *Celestial Armil* [*gamma* UMa Phecda], *Imperial Secretariat* [*theta* UMa], *Five Chariots* [*iota* Aur], etc., and bore down on the *Palace of Purple Tenuity* at a location six *du* in *Neck* [*kappa* Vir]. Not until the 19th day of the following month was it extinguished. (Ming *Shenzong shilu*, ch. 575).

Even though these historical records were compiled after the dynastic transition, the usurpation by the Manchus was thought to be condoned by the Supernal Lord in accordance with the ideology of the Mandate of Heaven. It was an article of faith that this transfer of the Mandate was premised on the portents of 1618 as evidenced by the naming of the first dynastic reign period of the Qing as "Heaven's Mandate" *Tian ming*.

Appendix

East Asian Records of Years 1614–1619

The following is a representative sample of reports from Chinese, Japanese, and Korean sources concerning the ominous celestial events of the penultimate years of the *Wanli* Emperor's reign. The reports may seem tediously repetitive, and they are to a large degree, of course, except that each adds one or more additional details that together with the diversity of sources may be of interest in interpreting or dating reports from other parts of the world. The translated records are taken (with minimal revision) from Pankenier et al. (2008).

AD 1614 Mar 14 [Japan]

(a) 19th year of the *Keichō* reign period, 2nd month, 4th day; a broom star was seen in the east; the court and the land verged on confusion, taking it to be inauspicious. [*Isshi*].
(b) 19th year of the *Keichō* reign period, 2nd month, a certain day; someone recorded that there was a broom star several rods in length whose rays pointed to the northwest, which lasted until summer before disappearing. [*Zoku shigushō*].

1616 Winter [China]

44th year of the *Wanli* reign period of Emperor *Shenzong* of the Ming Dynasty, in winter; a broom star appeared. From *Eastern Well* [LM 22] it entered the *Vermilion Palace Enclosure*, pointing south with a gleam. Qing Dynasty *Kangxi* period ed. Shanxi (*Yuci xian xu zhi,* ch. 12).

1617 Jun 3–Jul 2 [China]

45th year of the Wanli reign period of Emperor *Shenzong* of the Ming Dynasty, 5th month; a broom star over three rods long emerged in Southern Dipper [LM 8] for more than a month. Qing Dynasty *Kangxi* period ed. *Zhejiang* (*Changxing xian zhi*, ch. 4).

1618 Jan 26–Feb 24 [China]

46th year of the *Wanli* reign period of Emperor *Shenzong* of the Ming Dynasty, 1st month; a white broom star emerged in *Base* [*alpha*^{1} Lib] and *Neck* [*kappa* Vir]. (*Er shen ye lu*, ch. 6).

1618 Mar 26–Apr 24 [China]

46th year of the *Wanli* reign period of Emperor *Shenzong* of the Ming Dynasty, 3rd month; a broom star emerged in *Barrens* [*beta* Aqr]. Qing Dynasty *Qianlong* ed. (*Shandong tongzhi*, ch. 33).

1618 Spring [China]

46th year of the *Wanli* reign period of Emperor *Shenzong* of the Ming Dynasty, spring; a broom star emerged in the southeast, thrust toward the *Palace of Purple Tenuity* [15 stars, *kappa* Dra]. After 50 days it was extinguished. Qing Dynasty *Qianlong* period ed. Shanxi (*Jiezhou Ruicheng xian zhi*, ch. 11).

1618 Aug 28–Dec 20 [Japan]

(a) 4th year of the *Genna* reign period (1615–1624), 7th month, 9th day [Aug 28]; at dawn a broom star was seen in the northeast, its tail 3 ft long … 11th month, 2nd day [Dec 20]; at dawn the broom star was seen moving from the east to the north. (*Takasue nichi jiki*)
(b) 4th year of the *Genna* reign period, 7th month, 9th day *yiwei* [Aug] clear sky; this dawn a broom star 3 ft long was seen between azimuth *chou* [NNE] and *yin* [ENE]. .. 10th month, 28th day *guiwei* [Dec 14], clear sky; after midnight a broom star was seen at the boundary between azimuths *chou* [NNE] and *yin* [ENE]. (*Takasue sukune hinami ki*)
(c) 4th year of the *Genna* reign period, 8th month, 8th day (Sep 26); a broom star was seen. (*Sōto nikki)*.
(d) 4th year of the *Genna* reign period, 10th month, 11th day *bingyin* [Nov 27]; a broom star was seen in the east (at daybreak); by sunrise it was over one *rod* long. (*Zoku shigushō*)

1618 Sep 19–Oct 17 [China]

(a) 46th year of the *Wanli* reign period of Emperor *Shenzong* of the Ming Dynasty, 8th month; a broom star emerged in the southeast, its tail sweeping the north-west for more than 80 days, then it was extinguished. Qing Dynasty *Shunzhi* period ed. Hebei (*Raoyang xian zhi*, ch. 5)

1618 Nov 14 [China]

(a) 6th year of the *Wanli* reign period of Emperor *Shenzong* of the Ming Dynasty, 9th month, day *guichou*; in *Guangzhou* a broom star appeared, emerging in sector *chen* [ESE] in the space of *Horn* [*alpha* Vir] and *Neck* [*kappa* Vir], its tail thrusting toward the space of *Swine* [*zeta* And], *Hillock* [*beta* Ari], and *Game Net* [*epsilon* Tau].

(b) 46th year of the *Wanli* reign period of Emperor *Shenzong* of the Ming Dynasty, 8th month; a white vapor extended from east to west; after several evenings there was a giant star over one rod long that appeared every night in the east while suspended in the air there was a white vapor shaped like long-handled sword, or an elongated rectangular serpentine spear, or again like white cotton cloth; emerging at sector *chen* [ESE] in the space of *Horn* [*alpha* Vir] and *Neck* [*kappa* Vir], its tail thrusting toward the space of *Swine* [*zeta* And], *Hillock* [*beta* Ari], and *Game Net* [*epsilon* Tau]; it spread across the whole sky from the north to the south, and was not extinguished until the 10th month. Qing Dynasty *Kangxi* period ed. Guangxi (*Pingle xian zhi*, ch. 6)

1618 Nov 15–16 [China]

(a) 3rd year of the *Tianming* reign period of Emperor *Taizu* of the 16 Qing Dynasty, 9th month, day *jiayin* [Nov 15] at night; a white long-star shaped like a bolt of cloth and more than 1 ft wide by more than two rods long appeared in *Wings* [*alpha* Crt] and *Chariot Platform* [*gamma* CrV] for 19 days in all before being extinguished. (*Guo que*, ch. 8)

(b) 46th year of the *Wanli* reign period of Emperor *Shenzong* of the Ming Dynasty, 9th month, day *yimao* [Nov 16]; a long star appeared in the southeast, a swath of white vapor shaped like a bolt of cloth, more than 1 ft wide by more than two rods long; to the east it reached *Chariot Platform* [*gamma* CrV] and on the west entered *Wings* [*alpha* Crt]. It lasted 19 days in all before being extinguished. (Ming *Shenzong shilu*, ch. 574)

1618 Nov 14 [China]

(a) 46th year of the *Wanli* reign period of Emperor *Shenzong* of the Ming Dynasty, 9th month, day *guichou* [50]; in Guangzhou a broom star appeared, emerging in sector *chen* (ESE = LM 1–2) in the space of Horn [*alpha* Vir] and *Neck* [*kappa* Vir], its tail thrusting toward the space of *Swine* [*zeta* And], *Hillock* [*beta* Ari], and *Game Net* [*epsilon* Tau]; several nights previously a white vapor extended from east to west like a sword, then appeared together with the broom star, its pointed rays like a broom; after 2 months, it was extinguished. (*Er shen ye lu*, ch. 6)
(b) 46th year of the *Wanli* reign period of Emperor *Shenzong* of the Ming Dynasty, 9th month, day *guichou* [50]; a broom star several *zhang* long and shaped like an arrow appeared in the east; emerging at azimuth *chen* (ESE = LM 1–2) in the space of *Horn* [*alpha* Vir] and *Neck* [*kappa* Vir], its tail thrusting toward the space of *Swine* [*zeta* And], *Hillock* [*beta* Ari], and *Game Net* [*epsilon* Tau]. Qing Dynasty *Kangxi* period ed. Guangdong [*Sanshuixian zhi*] ch. 1, Qing Dynasty Kangxi period ed. Guangxi (*Zhaoqing fu zhi*, ch. 3)

1618 Nov 15–16 [China]

(a) 3rd year of the *Tianming* reign period of Emperor *Taizu* of the Qing Dynasty, 9th month, day *jiayin* [Nov 15] at night; a white long star shaped like a bolt of cloth and more than 1 ft wide by more than two rods long appeared in *Wings* [*alpha* Crt] and *Chariot Platform* [*gamma* CrV] for 19 days in all before being extinguished. (*Guo que*, ch. 83)
(b) 46th year of the *Wanli* reign period of Emperor *Shenzong* of the Ming Dynasty, 9th month, day *yimao* [Nov 16]; a long star appeared in the southeast, with a swath of white vapor shaped like a bolt of cloth, over 1 ft wide by over two rods long; on the east it reached *Chariot Platform* [*gamma* CrV] and on the west entered *Wings* [*alpha* Crt]; it lasted 19 days in all before being extinguished. (*Ming Shenzong shilu*, ch. 574)

1618 Nov 26 [China]

(a) 46th year of the *Wanli* reign period of Emperor *Shenzong* of the Ming Dynasty, 10th month, day *yichou* [Nov 26]; a grayish-white broom star as big as a hen's egg and over one rod long emerged in *Base* [*alpha*[1] Lib], its tail pointing southeast. After 10-odd days, it turned and pointed northwest, sweeping and trespassing against *Guardian of the Great Yang* [*chi* UMa], then entered *Neck* [*kappa* Vir] at seven *du*, ten marks. After another 3 days, it gradually traveled northwestward, its tail sweeping Northern Dipper, *Celestial Pivot* [*beta* UMa Merak], *Celestial Armil* [*gamma* UMa Phecda], *Imperial Secretariat* [*theta* UMa], *Five Chariots* [*iota* Aur], etc., and bore down on *Palace of Purple Tenuity*, at a location six *du* in *Neck* [*kappa* Vir]. Not until the 19th day of the following month was it extinguished. (Ming *Shenzong shilu*, ch. 575)

(b) 46th year of the *Wanli* reign period of Emperor *Shenzong* of the Ming Dynasty, 10th month, day *yichou*; a broom star over one rod long emerged in Base [*alpha*^{1} Lib], pointing southeast, then gradually pointed northwest. It swept and trespassed against *Guardian of the Great Yang* [*chi* UMa], then entered the space of Neck [*kappa* Vir]; then to the northwest it swept Northern Dipper, *Celestial Armil* [*gamma* UMa Phecda], *Imperial Secretariat* [*theta* UMa], *Five Chariots* [*iota* Aur], and bore down on the right side of the *Palace of Purple Tenuity*. By day *jiachen* [Jan 4] of the 11th month, it was extinguished. (*Ming shi: tianwen zhi*, ch. 27)

(c) 46th year of the *Wanli* reign period of Emperor *Shenzong* of the Ming Dynasty, 10th month, day *yichou* [Nov 26], at night, a greenish-white broom star as big as a hen's egg and over one rod long emerged in lunar lodge *Base* [*alpha*^{1} Lib], its tail pointing southeast. After several evenings it turned and pointed northwest, trespassing against *Guardian of the Great Yang* [*chi* UMa]; after another 3 days it gradually headed northwestward, its tail sweeping the bowl of Northern Dipper, *Celestial Armil* [*gamma* UMa], *Imperial Secretariat* [*theta* UMa], *Five Chariots* [*iota* Aur], and so forth, and bore down on *Palace of Purple Tenuity*; by the 19th day of the following month it was extinguished. (*Guo que*, ch. 83)

(d) 46th year of the *Wanli* reign period of Emperor *Shenzong* of the Ming Dynasty, 10th month, day *yichou* [26 Nov]; a broom star appeared, emerging in the azimuth of *mao* [E], in the space of *Base* [*alpha*^{1} Lib] and *Chamber* [*pi* Sco]; its tail thrusting up and pointing toward the azimuth of *shen* [WSW], [i.e.,] the spaces of *Beak* [*phi* Ori] and *Triaster* [*delta* Ori]. Seen from below, it was more than one rod long and shaped like a sword; the flames of its rays were not extinguished for more than a month. Qing Dynasty Kangxi period ed. Guangdong (*Sanshui xian zhi*, ch. 1), Qing Dynasty *Kangxi* period ed. Guangdong (*Zhaoqing fu zhi*, ch. 3)

1618 Nov 26–1619 Jan 13 [Korea]

10th year of *Gwanghaegun* of the *Joseon* Dynasty, 10th month, day *yichou* [Nov 26]; a broom star more than one rod long was seen in the east within the space of *Base* [*alpha*^{1} Lib], its color yellowish-red; when the sky brightened it was no longer seen ... on day *xinsi* [Dec 12], during the 3rd watch, when the broom star was first seen, it was about 20 rods long and more than 4 or 5 ft wide. It trespassed close to the second and third stars of the *Northern Dipper* [Merak, Phecda] and went straight through the stars of the *Celestial Secretariat* (*theta* UMa), its size somewhat reduced compared to before and its color gradually fading ... in the 11th month, on day *dingyou* [Dec 28], the broom star's vaporous rays dimmed. .. on the night of day *guichou* [Jan 13], the body of the broom star was faintly visible, its rays almost completely extinguished. (*Joseon wangjo sillok: Kwanghaegun*, ch. 133–134).

1618 Nov 27 [China]

(a) 3rd year of the *Tianming* [Heavenly Mandate] reign period of Emperor *Taizu* of the Qing Dynasty, 10th month, 11th day [*bingyin*]; at the time of the 5th watch, in the southeast sector a white vapor 5 ft wide emerged from a star and **surged straight at the Ming state**; after the 15th day it no longer appeared. The star that produced it gradually moved toward the *Northern Dipper* each night, until the 29th day when it went right over the handle of the *Northern Dipper*, after which it was no longer seen. (*Da Qing Manzhou shilu* ***Veritable Records of the Great Manchu Qing Dynasty***)

(b) 3rd year of the *Tianming* [Heavenly Mandate] reign period of Emperor *Taizu* of the Qing Dynasty, 10th month, 11th day [*bingyin*]; a broom star appeared in the east, to the east of the previously sighted white vapor that was 5 ft wide, **with its tail pointing right at the Ming state**; each night it gradually moved toward *Northern Dipper* for 19 days in all until it passed over the tail of *Northern Dipper*, and only then was it was extinguished. (*Qing Taizu Gao huangdi shilu,* ***Veritable Records of Emperor Taizu Gao***, ch. 5)

1618 Dec 17–1619 Jan 15 [China]

47th year of the *Wanli* reign period of Emperor *Shenzong* of the Ming Dynasty, at the end of the 1st month; a broom star several [sic] hundred feet long appeared in the east, shooting by beneath the moon, its tip curved and pointed; it appeared a while in the northeast, and then again in the west. (Ming *Shenzong shilu, Veritable Records of Emperor Shenzong of Ming*, ch. 578).

1619 Spring [China]

47th year of the *Wanli* reign period of Emperor *Shenzong* of the Ming Dynasty, spring; a broom star appeared at azimuth *wu* [S = LM 24–26] and later pointed north, then it entered Northern Dipper; it was not extinguished until after 20 days. Qing Dynasty *Kangxi* period ed., Guangdong (*Wenchang xian zhi*, ch. 9).

References

Cullen, C. (1991). Halley's comet and the 'ghost' event of 10 B.C. *Quarterly Journal of the Royal Astronomical Society, 32*, 113–119.

Dardess, J. W. (2012). Outlaws. In *Ming China 1368–1644: A concise history of a resilient empire*. Rowman & Littlefield.

Loewe, M. (1994). The Han view of comets. In *Divination, mythology and monarchy in Han China* (pp. 61–84). Cambridge University Press.

Pankenier, D. W. (2013). *Astrology and cosmology in early China: Conforming Earth to heaven*. Cambridge University Press.

Pankenier, D. W. (2019). Parallel planetary astrologies in medieval China and inner Asia. *Journal of the International Society for the Critical Study of Divination and Prognostication, 1*(2), 335–360.

Pankenier, D. W., Xu, Z.-t., & Jiang, Y.-t. (2008). *Archaeoastronomy in East Asia: Historical observational records of comets and meteor showers from China, Japan and Korea*. Cambria Press.

Roth, G. (1979). The Manchu-Chinese relationship, 1618-1636. In J. D. Spence & J. E. Wills (Eds.), *From Ming to Ch'ing: Conquest, region, and continuity in seventeenth-century China* (pp. 9–11). Yale University Press.

Swope, K. M. (2013). A gauntlet is cast down: The rise of the Latter Jin, 1618–21. In *The military collapse of China's Ming Dynasty, 1618-44* (pp. 10–33). Routledge.

David W. Pankenier earned his PhD in Chinese from Stanford University in 1983. He is an International Fellow at the Swedish Collegium for Advanced Study, Uppsala, Sweden, and a member of the Institute for Advanced Study in Princeton. He has served on the Executive Committee of the International Conferences on the Inspiration of Astronomical Phenomena (INSAP), the European Society for Astronomy in Culture (SEAC), and the advisory board of the *Journal of the History of Astronomy*. He is the author of numerous articles and monographs on ancient Chinese astronomy, astrology, cosmology, chronology, and history. His book, *Astrology and Cosmology in Early China: Conforming Earth to Heaven*, was published by Cambridge University Press in 2013. A Chinese translation published by Jiangsu renmin chubanshe appeared in 2025.

Chapter 9
Comets over Iberia: Astronomical Observations, Cometary Theory, and Astrological Prognostications in Spain and Portugal (1618)

Luís Miguel Carolino

9.1 Introduction

In a seminal paper addressing the celestial novelties in Spain from 1572 to 1618, Víctor Navarro Brotons assessed the Spanish contributions to the comets of 1618 as follows: "A good number of texts on the comets of 1618 were published in Spain, at least 18, but of very little astronomical-cosmological value" (Navarro Brotons, 2012: 38).[1] This evaluation applies broadly to the Portuguese case, with one noteworthy exception. Several decades earlier, Spanish astronomers distinguished themselves in the cosmological and astronomical discussions following the appearance of a new start in 1572, capturing the attention of audiences across Europe. One prominent figure in this context was Jerónimo Múnoz, a professor of mathematics and Hebrew at the universities of Valencia and Salamanca. Muñoz not only advocated for the celestial location and nature of comets but also introduced a revolutionary cosmology grounded in the Stoic tradition.[2] However, by 1618, the astronomical–cosmological setting had undergone a significant shift in the Iberian Peninsula. The majority of publications on the comets of 1618 were astrological booklets that took the Aristotelian cosmos for granted, reflecting a departure from the earlier focus on astronomical and cosmological inquiry.

[1] Navarro Brotons provided additional details, specifying that his assessment was limited to the copies he was able to examine, as some of them appeared to be missing or lost over time.

[2] On Muñoz's biography and scientific achievements, see Navarro Brotons (2019).

To Víctor Navarro Brotons.

L. M. Carolino (✉)
ISCTE - Instituto Universitário de Lisboa, CIES-IUL, Lisbon, Portugal
e-mail: Luis.Miguel.Carolino@iscte-iul.pt

C. J. Cunningham (ed.), *Cosmic Events*, Historical & Cultural Astronomy,
https://doi.org/10.1007/978-3-031-86950-1_9

Thus far, the Iberian publications addressing the comets of 1618 have not captured the attention of historians, except for scholars concentrating on astrological prognostications.[3] In the rare instances when they were examined, national barriers, however, appeared to constrain historians, leading them to study Spanish and Portuguese materials separately.[4] Nevertheless, in the context of the first half of the seventeenth century, the Iberian Peninsula constitutes an integral analytical unit. Notably, the integration of the Portuguese kingdom into the Spanish Habsburg empire in 1580, following the death of Portuguese King Sebastião in the Battle of El-Ksar el-Kebir, Morocco, in 1578—an event associated by early-modern astrologers with the appearance of the comet of 1577—alongside the military pressure exerted by Philip II, underscores the interconnected historical dynamics of these two countries. Portugal remained under Spanish rule until 1640, contributing to the political and social tensions that strained the Spanish monarchy during the initial decades of the seventeenth century.

The political unification of the various Iberian kingdoms stimulated the flow of political intermediaries, scholars, books, and ideas throughout the Iberian Peninsula. An astrological almanac calculated for the Valencia meridian and written in Castilian by a Catalan-speaking astronomer and astrologer, such as Francisco de Guzman, for instance, could be printed in Lisbon at the beginning of the seventeenth century (Guzman, 1609). This phenomenon extends to the booklets released after the observation of the comets of 1618. Notably, 2 out of the 12 booklets authored by Spanish individuals, as known today, were published in Portugal in the Castilian language.[5] Additionally, Portuguese authors, publishing shortly thereafter, engaged in discussions regarding the interpretations put forth by their Spanish counterparts concerning the comets of 1618.[6]

Furthermore, analyzing the Iberian booklets on the comets of 1618 as a whole allows for a more profound understanding of the cultural and scientific exchanges between the Spanish and Portuguese kingdoms in the early seventeenth century. This approach not only sheds light on those influences but also provides meaning to specific works, such as Manuel Bocarro Francês's *Tratado dos cometas que apareceram em Novembro passado de 1618*, which stands out as undoubtedly the most innovative book authored by an Iberian scholar during this period.

This article focuses on the works published in the Iberian Peninsula subsequent to the observations of the comets of 1618. Additionally, it examines a selection of

---

[3] These are the cases of Lanuza Navarro (2005), Lanuza Navarro and Navarro Brotons (2013): esp. 52–54, and Carolino (2003): 167–198. An exception should be made for Manuel Bocarro Francês, whose work on the 1618 comets was the most innovative among Iberian authors. For insights into Bocarro's cosmological thoughts and the role played by comets in its development, see Carolino (2012).

[4] For example, Carolino (2003): 271–281 and Rosselló Botey (2000): 42–45.

[5] I identified 12 booklets discussing the comets of 1618 authored by Spanish scholars, along with four contributions from Portuguese scholars and one translation derived from a French source. See Bibliography.

[6] This was the case, for example, of Brito (1619): fol. 18r., not numbered, henceforth n.n.

manuscript materials produced during that period. The study specifically scrutinizes the astronomical observations conducted in the Iberian kingdoms and the mechanisms by which they achieved widespread publication. Following the appearance of the three comets in 1618, Iberian astrologer-astronomers, physicians, and *literati* raced to print their observations and astrological predictions, making it an editorial success. While, as discussed in this paper, astrologers held distinct perspectives on comets compared to natural philosophers, authors in the Iberian Peninsula who published on the comets of 1618 tended to adhere to a conventional Aristotelian cosmology, complemented by Arabic-Ptolemaic astrology. One conspicuous exception was Manuel Bocarro Francês, a Portuguese physician, astronomer, and astrologer who, having studied in Spain, proposed a cosmological framework akin to that of Jerónimo Muñoz. After examining this noteworthy case that encapsulates the cultural integration of the Iberian Peninsula, the article concludes by briefly analysing the astrological prognostications made by Iberian observers of the comets of 1618. As emphasized by Tayra Lanuza Navarro and Víctor Navarro Brotons (2013), these prognostications frequently encompassed references to wars, governmental difficulties, and, as I would argue, political expectations as well.

9.2 The Observation of the Comets of 1618 in the Iberian Peninsula

The comets observed in 1618 garnered widespread attention across the Iberian Peninsula, with a particular focus on the second (C/1618 V1) and third (C/1618 W1) comets, visible in Europe in November and early December and from November to late January 1619, respectively.[7] Conversely, the first comet (C/1618 Q1) received scant mention, and that only indirectly. The Portuguese astronomer/astrologer Mendo Pacheco de Brito recorded a comet observed briefly in August but conceded his inability to make detailed observations, stating that it "did not rise sufficiently high above the Earth" (Brito, 1619: fol. A3r.).[8] Another account comes from Onofre Pelechá, a Valencian priest and university professor (Lanuza Navarro, 2005: 99), who reported that "a very curious mathematician" from Valencia, whose identity he did not disclose, purportedly claimed to have observed the comet in late August in Leon constellation (Pelechá, 1619: 27–28).[9]

The second comet was allegedly sighted in the early days of November within the Iberian Peninsula. While various authors acknowledged its observation during

[7] For accurate details concerning the time and location of the 1618 observations, see Kronk (1999): 333–341. See also Seargent (2009): 110.

[8] There is a Portuguese manuscript whose title mentions the "three comets that appeared in November" but focus only on the second and third comets. *Exposisão breue dos tres Cometas que aparesserão em Nouembro*, Biblioteca da Ajuda, COD. 46-VIII-16, fols. 9r.-10r.

[9] Juan de Soto reported the observation of a comet (which he identified as the second comet) from 12 to 15 September 1618. Soto (1619): fol. 3v n.n.

this period,[10] Manuel Bocarro Francês stands out as the first to officially document his own sighting, which occurred on November 9. According to his account,

> On the ninth of November in the year 1618, around four o'clock after midnight, a heavenly signal emerged above our horizon from the eastern vernal region (where the sun rises on the summer equinox). It appeared to our eyes to be more than two spears long and two rods wide, with a transparent, whitish, clay light, the colour of Saturn and Venus. The luminous body was not excessively dense, allowing the observation of stars through its body which was similar to a palm branch. [...] As I began my observation on November 9th, our comet, Palma, traversed stars in four southern constellations and continued until nearly the conclusion of another, namely the Ship, Hydra, Vessel or Crater, Raven, and Centaurus, where it ended as it was moving from east to west. (Bocarro Francês, 1619: fols. 7v-8r and 9v.)

Despite Bocarro Francês consistently claiming to have observed the comet on November 9, it is highly improbable that such an observation took place on that specific day, as the C/1618 V1 comet does not appear to be observable in Europe before November 11 (Kronk, 1999: 335). This time discrepancy could be explained by considering the possibility that Bocarro Francês and probably the Iberian observers were still adhering to the Julian calendar. If this were the case, it would correspond to November 19, a more plausible date.[11] Like many other astronomers, Bocarro continued his observation of the comet until late December, specifically until the 22 (or December 31/January 1), a period during which the comet had already initiated fragmentation.[12]

In early December, the comet manifested considerable luminosity and substantial dimensions, with its tail extending across the constellations of Libra and a segment of Virgo (Bocarro Francês, 1619: fols. 8r.-11v.) (Fig. 9.1).

Lisbon assumed a prominent role as a central hub for the observation of the second comet of 1618, marked by great activity. Following the observations of Bocarro Francês, Portuguese observers Mendo Pacheco de Brito and António Najera documented alleged sightings on November 10 and 13, respectively. Furthermore, the

[10] While the Portuguese Miguel de Silveira, who collaborated with the professor at the Academia de Matemáticas de Madrid and chief-cosmographer Juan Cedillo Diaz, reported that the comet was first seen on November 4, Juan Bautista, who does not appear to have made direct observations of the comet, reports its visibility in Valencia beginning on November 5. Silveira/Cedillo, *Observaciones e informe sobre el cometa de 1618* in Cedillo Diáz (2019: 432; Bautista, 1618: frontispiece, fol. A2r). Other Spanish authors similarly documented the comet's observation in the early days of that month, specifically on November 6 or 7 (Pedro, 1618: fol. 1v. n.n.) and 8 or 9 (Casiano, 1618: fol. A5r.). Likewise, an anonymous Portuguese manuscript notes that, according to certain accounts, this comet had reportedly become visible as early as November 7 or 8, although the observer himself only documented its presence on the 25 of that month. *Observações do Cometa que foi visto em Novembro do anno de 1618*, Biblioteca da Ajuda, COD. 46-VIII-16, fol. 11r., *Obseruaçois do Cometa que foi uisto em Nouembro do anno de 618*, Biblioteca da Ajuda, COD. 51-VI-2, fol. 45r.

[11] If this were the case, it would challenge the perspective asserting that "the adoption of the Gregorian calendar in Portugal proceeded without complications." Leitão and Carvalho (2020): 412.

[12] In his words, "This comet was visible from November 9th, when I first observed and began tracking it, until November 29th, when it started disintegrating. Despite its disintegration, I can still observe remnants of the comet as of today, December 22nd." Bocarro Francês (1619): fol. 14v.

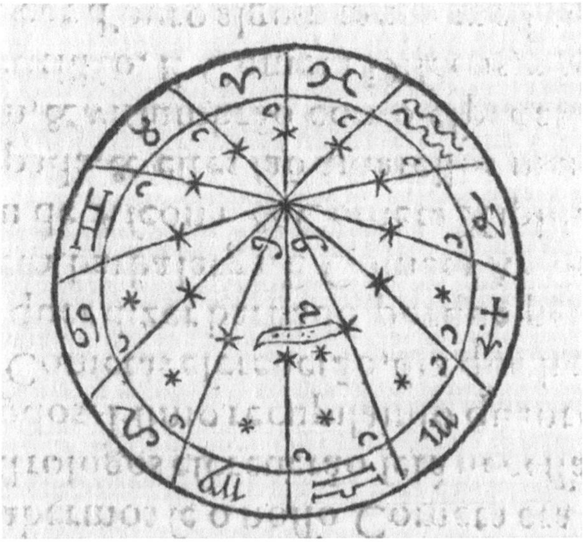

Fig. 9.1 The second comet of 1618 crossing the skies according to Manuel Bocarro Francês (1619: fol. 11r., Biblioteca Nacional de Lisboa)

Spanish astronomer/astrologer Pedro Mexia recorded the comet while in Lisbon on November 14.[13] In the Eastern Iberian region, António Luciano noticed the same comet during a similar timeframe, specifically on November 12, while located in Valencia. Advancing further north, in Zaragoza, Miguel Pedro recorded the comet on November 16. Observations were also chronicled in Andalusia, with astrologer Juan Casiano observing the comet in Sevilla on November 24.[14] In Granada, Vespasiano Geronimo Bargas y Heredia and Cristóbal de Montalvo reported sightings in November, albeit without specified dates (Brito, 1619: fol. A3r.; Najera, 1619: fol. A3r.; Mexia, 1619: fol. 8r. n.n.; Luciano, 1618: frontispiece, fol. 1v. n.n.; Pedro, 1618: fol. 1v. n.n.; Casiano, 1618: fol. A5r.; Bargas y Heredia, 1618: fols. 8v.-9r. n.n.; Montalvo, 1618: frontispiece).

The first documented observations of the third comet of 1618 (C/1618 W1) by Iberian astrologers and astronomers took place between November 24, when Luciano sighted it in Valencia, and November 28, when Bargas y Heredia reported it in Granada, though mentioning that some observers had sighted it as early as November 17. It was observed in Lisbon on November 25 by Najera and then on 26

[13] In Portugal, the comet was also noticed in Coimbra by the mathematics university professor André de Avelar on November 20th. Avelar, *Discurso Astronomico e Astrologico do cometa que apareceo por novembro de 1618 em Coimbra. Pello Me. Andre de avellar lente jubillado em mathematica*, Biblioteca da Ajuda, COD. 46-VIII-16, fol. 1v.

[14] This comet might not be the C/1618 V but rather the C/1618 W1. However, given Casiano's description of the comet having a palm branch shape and his report of some individuals sighting it on November 8 or 9, it seems more likely that he was referring to the C/1618 V. Casiano (1618): fol. A5r.

by Mexia, Pacheco de Brito, and Bocarro Francês (Luciano, 1618: frontispiece, fol. 5r. n.n.; Bargas y Heredia, 1618: fol. 11r. n.n.; Najera, 1619: fol. A3r; Mexia, 1619: fol. 11v.; Brito, 1619: fol. A3r.; Bocarro Francês, 1619: fol. 18v.).

Not all the authors of the booklets on the comets of 1618 directed their attention to the C/1618 W1 comet. This was likely not due to a lack of observation but rather because they had already submitted their works for printing by the time the comet became visible. One such case is that of the Valencian author Onofre Pelechá. In other cases, the appearance of the third comet coincided with the completion of the writers' works. Typically, upon observing this "new" comet, they briefly referenced it in the concluding pages of their cometary treatises. This was exemplified by the Valencian astrologer Antonio Luciano, who, having already prepared his work for printing, appended a brief notice regarding the "second comet" (in this case, the third, as he had missed observing C/1618 Q1) (Luciano, 1618: fol. 5 n.n.).

However, some authors gave preference to this comet over the C/1618 V1, with Najera being a notable example. According to this Lisbon-based astrologer and cosmographer, the comet first became visible on the night of November 25, positioned "between the shells of Libra, at 14° of Scorpio, and a northern latitude of just over 1°, adjacent to a 4th magnitude star." Its initial trajectory went against the order of the signs, moving from southeast to northwest, covering less than 2° a day during the initial phase. In the second week, its speed increased, the comet moving 3° and 8 min per day. By mid-December, the comet's motion gradually decelerated. During the last 6 days, it moved only 50 min each day before finally extinguishing on January 2. The comet's demise occurred against the backdrop of the constellation of Ursula Major, "in the first degree of Virgo with 61° of northern latitude, 64° of declination, and 198° of right ascension. At this point, it reached the middle of the sky, with the *spica virginis* positioned 132° from the Sun." (Najera, 1619: fols. C3r.-C4r.).

Najera provided a relatively detailed description of the cometary motion; however, his approach deviates from Iberian scholars' typical procedure. In general, references to astronomical observations are notably vague, merely specifying the initial day of comet observation and, in some instances, visual sighting, without providing any technical details associated with these observations. The same holds true for the measurements concerning the altitude of the 1618 comets. Apart from Bocarro Francês, whose argument, as we will explore, relied on parallax computations to support the celestial positioning of the comet, Iberian astrologers–astronomers generally accepted, without critical examination, that the comets traversed below the Moon, offering no accompanying calculations.

Regarding instruments, certain authors, such as the Spaniard Vespasiano Bargas y Heredia, allude to the utilization of "mathematical instruments" in their observations. However, they refrain from providing specific details about these instruments (Bargas y Heredia, 1618: fols. 8v. and 11r. n.n.). Bartolomé del Valle, a physician and professor of astrology at the University of Salamanca, who observed the comet in Granada, illustrates in his work *Explicacion y pronostico de los dos cometas* a gathering of scholars employing astrolabes for angular measurements of two comets on the outskirts of Sierra Nevada. Notably, there is no mention of the use of

Fig. 9.2 Astrologer-astronomers observing the C/1618 V1 and C/1618 W1 comets in the outskirts of Granada. (Bartolomé del Valle, 1619: fol. 2r., Biblioteca Nacional de España)

telescopes by the authors who published booklets on the comets of 1618 in the Iberian Peninsula (Fig. 9.2).

9.3 The Cometary Business

Upon the observation of the 1618 comets, authors hastily endeavoured to publish their works. A comparative analysis of the printed output following the sightings of the three bright comets in 1618 and that related to the 1572 new star and the 1577 comet unmistakably demonstrates that, despite the comparatively modest significance from an astronomical and cosmological perspective of those works, the comets of 1618 garnered considerably more attention from authors, printers, and, one would argue, readers than their predecessors.

The primary target audience for the booklets disseminated in the Iberian kingdoms following the apparition of the 1618 comets likely comprised the urban middle classes with a keen interest in astrology. Predominantly, the content of these pamphlets centred around astrological themes. Predictions encompassing weather patterns, ailments, civil unrest, the demise of individuals from the upper echelons of society, and even prospective earthquakes dominated the pages of these publications. Typically, authors asserted that the impetus for their short treatises stemmed from numerous requests by individuals seeking prognostications about the effects of the comets. As expressed by the Valencian astrologer Luciano,

> Compelled by the persistent inquiries of countless individuals, who, cognizant of my proficiency in mathematics and celestial knowledge, approach me incessantly to seek insight into the significance of this comet that has stirred so much tumult [...] it has occurred to me, in order to address them collectively, to employ the written word to respond to their inquiries. (Luciano, 1618: fol. 2r. n.n.)[15]

Devised for an audience lacking proficiency in Latin, the selection of Castilian and Portuguese languages is unsurprising, with only one publication available in Latin. As explained by Bargas y Heredia to his prospective readers, "I have chosen, curious reader, to compose these discourses in our native language so that you may easily comprehend the information about these comets." (Bargas y Heredia, 1618: fol. 3r. n.n.).

The Iberian treatises on the comets of 1618 are generally of small size. In Spain, these treatises typically comprised eight 8° folio-pages, exemplified by the works of Bautista, Ferragut, Pedro, Puget de la Serre, and Salaçar. Conversely, the booklets printed in Portugal varied in length, ranging from sixteen 4° folio-pages (Mexia) to twenty (Avelar, Bocarro, Brito), and in some instances, thirty-two 4° folio-pages (Najera).[16] However, there were exceptions with shorter booklets in Portugal, such as Casiano's, also consisting of eight 8° folio-pages. In the Spanish context, there existed treatises of varying lengths, including Luciano's succinct four folio-page *Discurso* and the more extensive *Juyzio* by Soto. Particularly noteworthy was Pelechá's treatise, comprising 80 pages.

These booklets were generally affordable. The pricing of these materials varied. Shorter booklets like Casiano's *Breve discurso*, printed in Lisbon, were priced at six *réis*—roughly equivalent to the cost of an astrological almanac. In contrast, longer treatises, such as those by Mexia and Najera, were priced at 15 or 16 *réis*, respectively (Casiano, 1618: fol. 2v. n.n.; Mexia, 1619; Najera, 1619: fol. 2v. n.n.).

---

[15] Similar statements can be found in Brito (1619): fol. A2v., Casiano (1618): fol. A3r., Ferragut (1618): fol. A2r., Najera (1619): fol. A3r., Soto (1619): fol. 2r. n.n.

[16] Nevertheless, Bargas y Heredia's booklet (1618), printed in Granada, also consisted of sixteen 4°folio pages, while Valle's booklet (1619) comprised twenty-five 4°folio pages.

These printed materials probably achieved editorial success. If that were indeed the case, it would elucidate why Pedro Craesbeeck, a member of a prominent printing family in Lisbon, succeeded in monopolizing the publication of treatises on the 1618 comets in Portugal. He oversaw the publication of all the treatises, with the exception of Avelar's *Nox Attica*, which was published in Coimbra.[17]

While not a strict rule, a considerable number of authors sought the patronage of individuals within the Spanish and Portuguese nobility and clergy, to whom several booklets on the comets of 1618 are dedicated. The dedication of books has been recognized as a strategic means to enter the "clientage system," serving as both a public declaration of patronage connections and, at times, a means of seeking protection (Feingold, 1984: 209; Pumfrey & Dawbarn, 2004: 152). This is exemplified by Bocarro Francês, a crypto-Jew, who dedicated his unorthodox work *Tratado dos cometas* to the Chief Inquisitor in Portugal, Francisco Martins Mascarenhas, with the aim of securing protection from detractors: "because when they learned that this treatise is accepted by Your Excellency, they will not dare to go against it." (Bocarro Francês, 1619: fol. A2r.).

A cluster of cometary booklets is dedicated to noblemen and clergymen holding prominent positions in the civil and religious administration of both Spain and Portugal. Such were the cases of Fernão Martins Mascarenhas, Agusti Mesia Trece, a member of the king's Council of State and War, to whom Vespasiano Bargas y Heredia dedicated his treatise on the comets of 1618, and Don Francisco Lesdesma, a Church authority in Granada and Palencia, who supported the publication of Cristóbal de Montalvo's *Discurso a los dos cometas.* Other nobles, including André de Almada and Rodrigo Sarmiento de Ulloa Villandrando y Lacerda, prominently served as patrons of cometary treatises, particularly those by André de Avelar and Pedro Mexia. However, the zenith of support came from Onofre Pelechá, who secured the backing of none other than the king himself, Philip III. Typically, when cometary treatises garnered support from a noble or a clergy dignitary, their frontispieces featured the supporters' coat-of-arms.

In contrast to instances in Europe where there is evidence of authors being rewarded with monetary compensation upon presenting a book,[18] the practice in the early modern Iberian Peninsula raises ambiguity regarding whether the dedication of a book signified that the patron had indeed made a financial contribution or simply facilitated its publication. Despite the absence of explicit evidence of monetary transactions, Iberian booklets supported by the patronage of nobles or clergy distinguished themselves through distinctive paratexts. Unlike conventional pamphlets, these patronized works featured not only the coat-of-arms of their sponsors but also incorporated elements such as Latin poems, frequent depictions of comets against a stellar backdrop, engravings of celestial bodies (particularly the Sun and the Moon), and floral motifs. A case in point is Luís de Avelar's *Nox Attica* dedicated to André

[17] On Pedro Craesbeeck and his family, see Dias (1996).

[18] For example, for the English case, Mordechai Feingold informs us that "Prince Henry was in the habit of giving between £3 and £5 for each dedication". Feingold (1984): 209.

de Almada, a noble and professor at the University of Coimbra. As the sole Latin booklet published following the observation of the comets of 1618 in the Iberian Peninsula, *Nox Attica* included Almada's coat-of-arms and a captivating representation of the second comet traversing the celestial scenery. Additionally, the entire book featured an elegant adornment, contributing to its visual appeal.

9.4 Astrologers Versus Natural Philosophers on Comets

Nox Attica stemmed from an informal academy of sciences enlivened in Coimbra by André de Almada. The author, Luís, the son of the university professor and astrologer André de Avelar, was a student of canon law at the local university and an active participant in Almada's academy.[19] In this refined treatise, presented in the form of a dialogue, three characters took centre stage: a student of law (*scholasticus*), an astrologer (*astrologus*), and a natural philosopher (*philosophus*). The *scholasticus*, who, weary of studying law, opted to dedicate a night to discuss the recently appeared comet with an astrologer and a philosopher, is indeed Luís de Avelar. One may speculate whether the character of the *astrologus* was inspired by his father André, who had a lifelong commitment to astronomical and astrological studies, including various observations of comets.

In this fictional debate, the student assumed the role of directing the conversation by posing questions, guiding its progression in line with the typical themes covered in cometary treatises. *Nox Attica* delves into topics such as the nature, composition, location, cause, and influence of comets, representing pivotal questions within the cometary discussion. Additionally, Avelar stage-managed the natural philosopher and the astrologer to respond based on their respective areas of expertise. He proposed that the nature of comets fell within the purview of the philosopher, while the astrologer was tasked with detailing their motion.

In the discourse on the cause of comets, a divergence emerged. The philosopher argued that the instrumental "efficient cause" consisted of the fire excited by planetary movement, whereas the astrologer attributed comet formation to the occurrence of eclipses and planetary conjunctions (Avelar, 1619: fols. 2r., 8v.). This perspective found resonance in Spain, as articulated by Onofre Pelechá, who asserted that "astrologers say that the efficient cause is the constellations that usually engender and elevate these exhalations; but philosophers say that this efficient cause is the element of fire that ignites the exhalation or the movement of the heavens." (Pelechá, 1619: 15).

The distinction also extends to the comet's tail, where the astrologer seeks its shape to identify the leading planet, whereas the philosopher interprets it as a rarefaction of celestial matter, as well as to its colour. While the philosopher attributes the comet's colour to the nature of its exhalations, the astrologer attributes it to the

[19] For more information on this academy, see Carolino (2016).

influence of the planets producing and governing the comet (Avelar, 1619: fols. 4v-5.).[20] The disparity in the roles of natural philosophers and astrologers is further evident in their interpretation of the comet's effects. According to Avelar, the philosopher tends to confine his analysis to the natural consequences of a comet's appearance, such as adverse weather, dry conditions, strong winds, resulting sterility, and subsequent human fatalities. In contrast, the astrologer delves into the specific effects purportedly caused by planets and celestial constellations (Avelar, 1619: fols. 10r.-13v.).

By delineating the respective domains of activity for philosophers and astrologers, Avelar provided insights into the disciplinary framework within which he operated. This classification was rooted in a traditional Aristotelian system of sciences, wherein mathematics held a subordinate position relative to natural philosophy. Not fulfilling all the requirements of scientific syllogism, the Aristotelian tradition considered that astronomy or astrology, as a mixed science, was primarily concerned with describing quantitative aspects—such as the trajectory, dimensions, and distances of comets—without offering an explanation for these quantitative properties, a task assigned to the natural philosopher.

Avelar's booklet illustrates that this epistemological taxonomy had direct implications for the study of comets.

Avelar likely became acquainted with this scientific classification while pursuing philosophy studies at the College of Arts, University of Coimbra. Founded in 1548 under the patronage of King John III, the college transitioned to the Society of Jesus in 1555 before being integrated into the university in 1561. Operating within the traditional educational model, the institution offered philosophical training to students aspiring to enter advanced faculties such as theology, law, canon law, and medicine. Luís de Avelar exemplified one such student. The predominant philosophical orientation during this period was Aristotelian.[21]

It is highly probable that most scholars who authored treatises on the comets of 1618 in the Iberian Peninsula shared Avelar's perspective on the classification of sciences. The majority of these scholars possessed a university education. Aside from those identifying themselves as astrologers and/or mathematicians, such as Juan Bautista, Juan Casiano, Antonio Luciano, and Pedro Mexia, three distinct professional categories contributed to the writing of cometary treatises in late 1618 and early 1619. These categories comprised physicians, including Bargas y Heredia, Juan de Soto, Bartolomé del Valle, and Bocarro Francês, priests (Onofre Pelechá, Juan Salaçar, and Miguel Pedro), and even law students. This circumstance applied not only to Luís de Avelar but also to Bernardo Ferragut, who was pursuing studies in law and canon law at the University of Huesca when the comets of 1618 made their appearance.[22] In each of the three instances, philosophical training was

---

[20] Similar views are found in Pelechá's (1619): 14–15.

[21] For an introduction on the teaching of natural philosophy and mathematics at the Portuguese universities, with a close look at the University of Coimbra, see Carolino and Leitão (2006).

[22] In addition, Najera was trained as cosmographer.

instrumental in advancing university studies or obtaining ordination. Therefore, it is highly probable that they became acquainted with this Aristotelian-based epistemological taxonomy at some point during their educational pursuits. The cometary theories articulated in their treatises further substantiated their reliance on the Aristotelian philosophical framework.[23]

9.5 Under the Sign of Aristotle's Cometary Theory

The authors of most cometary booklets not only embraced the division between philosophers and astrologers' areas of expertise but also cast their lot with the astrologers. Indeed, their works primarily delved into topics traditionally associated with astrologers, namely the prognostications derived from the astrological influences of planets, celestial constellations, and, of course, comets.

However, they unanimously embraced the Aristotelian theory of comets, with a singular exception.[24] According to the Iberian astrologers and astronomers who authored treatises on the comets of 1618, these phenomena were incandescent vapours composed of diverse exhalations originating from the Earth's surface and moving within the boundaries of the terrestrial region. In the words of Cristóbal de Montalvo,

> Comets are generated and originate from the exaltations of the Earth. Upon their formation, they ascend to the third region of air, where they ignite and burn. The nature and quantity of their materials influence the intensity and duration of this phenomenon. With a diurnal motion, they move from East to West, occasionally exhibiting variations in movement, as recounted in various narratives [*historias*]. (Montalvo, 1618: fol. A3r.).

Celestial bodies play a pivotal role in the genesis of comets. The motion of celestial bodies enables the concentration of terrestrial exhalations at specific spots, where, owing to their proximity to the sphere of fire, they undergo deflagration, thereby giving rise to a comet. Planetary influence not only accounts for the diurnal motion observed in comets but also contributes to the diverse movements alluded to by Montalvo. António Najera proposed a more comprehensive theory regarding the planets' role in comet production. According to the Portuguese cosmographer, in addition to motion, two other instrumental factors play a critical role in comet formation: light and occult influence. In his words,

> The celestial bodies, although they may appear of small size to our sight due to the vast distance from us, are, in fact, much larger than the entire Earth. Because they are so immense and numerous, they exert considerable effect on things below through motion, light, and influence. They pull and lift moist vapours and dry exhalations from the Earth, directing them into the region of the air at various altitudes, depending on their qualities and consistencies. These substances transform into rain, snow, and hail, and the more oily and warm ones result in various fiery phenomena such as lightning, rays, and flashes. The

[23] Exception should be made to Bocarro Francês (1619).

[24] See the previous note.

stickier substances ascend to the highest region, from which comets are formed. (Najera, 1619: fol. B1v.).

By presenting this perspective, Najera heavily relied on a theory cherished by Aristotelian natural philosophers, which posited that celestial bodies wielded a comprehensive influence over the terrestrial realm. The prevailing view maintained that the three factors identified by Najera—motion, light, and influence (*influentia* or *influxus*)—enabled celestial bodies to impact the terrestrial region by generating the four primary qualities: hotness, coldness, wetness, and dryness. The interaction of these qualities, particularly hotness and coldness, was thought to give rise to both generation and corruption in the sublunar region, encompassing terrestrial life.[25] Regarding comets, Najera argued that while motion, considered the "source of life for every terrestrial," gathered terrestrial exhalations at a specific location, light influenced the primary qualities, contributing to the ascent of exhalations that would ultimately result in the formation of a comet. However, for the concentration of such a substantial volume of vapours, the intervention of celestial occult influence was deemed necessary (Najera, 1619: fols. B1v.-B2r.).

However, astrologers delved deeper into exploring the influence of celestial bodies on the formation of comets. Following Ptolemy, Abu Mas'shar, and the rich Arabic astrological tradition, Iberian astrologers attributed a crucial role to solar and lunar eclipses, along with planetary conjunctions—especially those involving superior planets—in the genesis of comets. For the prediction of their effects, they advocated for a meticulous examination of atmospheric colour and meteorological conditions during eclipses. Furthermore, planetary conjunctions were considered determinants of the comet's ruling celestial body.[26]

In this way, celestial bodies were intimately linked to the generation of comets, with the influences of the planets being, strictly speaking, their efficient causes. This paved the way for astrological speculation based on comets. As Pedro Mexia recognized, "the most essential and necessary task is to know which planet is the true ruler of the comet." (Mexia, 1619: fol. 6v. n.n.). As these phenomena resulted from planetary action, the qualities of their "causal agents," along with the astrological features of the celestial constellations in which they had appeared and later crossed, would also be present in comets. In other words, a comet generated by the action of Saturn, for example, would assume Saturnine qualities. Hence, referencing the second comet of 1618, Manuel Bocarro Francês confidently asserted, "given that Saturn produced this comet, it must inherently reflect Saturn's nature, as comets are likened to portraits and bodies akin to the planets that give rise to them." (Bocarro Francês, 1619: fol. 16v.).

[25] On the theory of celestial influence, see Grant (1987) and North (1986). For the Portuguese case, see Carolino (2003): 45–77.

[26] See, for example, Avelar, *Discurso Astronomico e Astrologico do cometa*, Biblioteca da Ajuda, COD. 46-VIII-16, fol. 3v., Bargas y Heredia (1618): fols. 4r.-4v., Luciano (1618): fol. 3r n.n.

9.6 The Iberian Critics of Aristotle: From Muñoz to Bocarro Francês

Historians have successfully linked the debate that ensued from the appearance of the new stars and the comets of 1577 and 1618 with the decline of the Aristotelian worldview, to paraphrase the title of Tabitta van Nouhuys's remarkable book (1998). Manuel Bocarro Francês actively engaged in this extensive and intricate process, thereby encapsulating a rich cosmological discussion that originated in the Iberian Peninsula in the early 1570s.

Following the discovery of a new star in the constellation of Cassiopeia in 1572, Jerónimo Muñoz published the seminal work *Libro del nuevo cometa*, which wielded substantial influence in the Iberian Peninsula over the following decades. Born in Valencia, Muñoz likely studied in Paris and Leuven under Oronce Finé and Gemma Frisius, respectively, before pursuing a career as a professor of Hebrew at the University of Ancona in Italy. Later, he held positions at the universities of Valencia and Salamanca, where he also became a professor of mathematics. Muñoz authored several significant works, although the majority remained confined to manuscript form. *Libro del nuevo cometa* stands as one of the few works he published during his lifetime. Nevertheless, Muñoz's contributions to the understanding of the 1572 supernova reached a wider audience through his correspondence with various European scholars. Noteworthy among his correspondents were the Viennese physician and mathematician Bartholomaeus Reisacherus and the Bohemian imperial astronomer and physician Thaddaeus Hagecius. Hagecius, in particular, provided Tycho Brahe with data received from Muñoz regarding the supernova of 1572, later commended by Tycho in his work *Astronomiae instauratae progymnasmata* (Navarro Brotons, 2019).

In his *Libro del nuevo cometa*, Munõz maintained that the comets were produced by the concentration of celestial matter, thus occurring in the celestial region. The celestial location and nature of comets prompted him to challenge the Aristotelian traditional worldview. He therefore not only sustained the idea that stars and planets moved "as fishes in the sea or birds through the air" driven by an intrinsic force but, more significantly, he argued that the celestial realm was susceptible to processes of corruption. Additionally, the Valencian scholar advocated for the epistemological superiority of mathematics over philosophy in unravelling the mysteries of cosmological science.

Although Muñoz's cosmological theories did not hold a dominant position in the Iberian Peninsula, they exerted a considerable influence across various Iberian kingdoms. Notably, Antonio Núñez Zamora, a student of Muñoz at the University of Salamanca and later his successor in the chair of mathematics and astronomy, adopted the perspectives of the Valencian scholar. Núñez Zamora upheld the notions of the fluid and elemental nature of the heavens, as well as the epistemological primacy of mathematics over other scientific disciplines. Evidenced in his work on the *nova* of 1604, he posited that comets originated from terrestrial exhalations reaching the celestial realm under the influence of planetary conjunctions.

Muñoz's cosmological ideas also left an imprint on other scholars, such as Marco Antonio Palau and Diego Pérez de Mesa. A disciple of Antonio Juan Ripollés, who succeeded Muñoz at the University of Valencia, Palau diverged from Aristotle in his analysis of the new star of 1604, asserting that comets move well above the terrestrial region (Navarro Brotons, 2012: 32–38). Pérez de Mesa, a former student of Muñoz, taught mathematics and astronomy at the universities of Alcalá de Henares and Seville. There, he expounded on the theories of celestial fluidity and corruptibility, drawing on observations of the *nova* of 1572 by Muñoz and the cosmological views of the Valencian astronomer (Navarro Brotons, 2014: 59, 117–118).

Jerónimo Muñoz's cosmological and epistemological ideas also found fertile ground in Portugal, as Manuel Bocarro Francês's treatise on the comets of 1618 demonstrates. Bocarro Francês was born into a crypto-Jewish family around 1593 in Lisbon. After studying philosophy and likely mathematics at the Jesuit college of Santo Antão in his hometown, he furthered his education at the University of Alcalá de Henares and later at the University of Sigüenza, where he obtained a degree in medicine. Returning to Lisbon in 1618, he observed the comets that appeared in the skies that year.[27] Despite being back in Portugal, he kept alive the ideas acquired during his time in Spain, particularly at the University of Alcalá de Henares, where Diego Pérez de Mesa, a disciple of Muñoz, had previously taught before Bocarro Francês's own studies there. It is plausible that Pérez de Mesa's influence in astronomy teaching persisted in Alcalá during Bocarro's tenure as a student.

Indeed, Bocarro Francês's *Tratado dos Cometas* resonates with some of the fundamental ideas of Muñoz and his followers. Similar to the Valencian professor, he also recognized the epistemological significance of mathematics in studying the structure of the heavens. It was through the application of mathematical principles that he reached the conclusion that the second comet of 1618 traversed the celestial realm. In his attempt to calculate its parallax, the Lisbon astronomer noticed the absence of any perceptible parallax. As Bocarro Francês pointed out, the apparent equinoctial elevation of the comet corresponded to the true equinoctial elevation of the Sun in Lisbon, indicating its movement in the celestial region. As he stated,

> In accordance with the rules above [exposed] to calculate the parallax of any comet, I observed our comet in this city of Lisbon, where we have the altitude of north [pole] of 39 degrees 38 minutes and the equinoctial line elevates above our horizon 59 [sic., 50] degrees 22 minutes. By means of the abovementioned rules, I found that the apparent equinoctial elevation [of the comet] presented the same 59 degrees, and consequently our comet had no parallax. Thus, I am certain that it was in the heavens, above the Sun, and not in the air (Bocarro Francês, 1619: fol. 13r.).

[27] Later, Bocarro Francês departed from Portugal for good following a denunciation accusing him of being a secret Jew to the Inquisition. Over the subsequent decades, he travelled and resided in various locations, including Rome, Amsterdam, Hamburg, and Leghorn, where he eventually passed away around 1662. During his time in Northern Europe, Bocarro embraced his Jewish religious beliefs, adopting the name Jacob Rosales, particularly within Jewish circles. However, for the focus of this paper on the cosmological theories formulated by Bocarro during his period in Portugal—prior to adopting his Jewish identity—and intended for the wider community of scholars, I employ the name Manuel Bocarro Francês rather than Jacob Rosales.

Indeed, Bocarro Francês advanced that the movement of the comet in 1618 occurred within the region of Mars (Bocarro Francês, 1619: fols. 14r.-14v.). Departing from Aristotelian doctrines, the Portuguese astronomer argued that comets did not conform to the motion of the "first mover" (whose existence he denied), but instead followed the motion patterns of their "generators." In the case of the 1618 second comet, a conjunction between Mars and Saturn played a decisive role in its genesis within the Martian region, causing the comet to exhibit an east-west motion analogous to the planets (Bocarro Francês, 1619: fols. 13v.-14v.).

Conjunctions of superior planets were believed to aggregate celestial matter, which, upon condensation, ignited and gave rise to a comet (Bocarro Francês, 1619: fols. 5r.-5v.). Thus, according to Bocarro Francês, a comet not only traversed the space above the Moon but also constituted a celestial body made of heavenly matter. Regarding the 1618 comet, he affirmed, "The matter of this comet was part of the heavens (which I believe was that of Mars), altered and inflamed, which gathered together in that fire, took form, and produced such a solid, massive, and not very compact body" (Bocarro Francês, 1619: fol. 13v.). Celestial region was therefore subject to processes of coming to be and passing away as postulated by the Stoic philosophers, a perspective embraced by Bocarro Francês who asserted, "From here [the Earth's surface] to the Empyrean heaven everything is air" (Bocarro Francês, 1619: fol. 4v.).[28]

Bocarro Francês's treatise on the comets of 1618 seemed to have circulated in seventeenth-century Portugal, though it also attracted criticism from his contemporaries.[29] One of the mathematicians who reacted to Bocarro Francês's treatise on the comets of 1618 was Mendo Pacheco de Brito. Brito's *Discurso em os dous phaenominos aereos do anno de mil e seiscentos e dezoito* likely neared completion when he came across Bocarro Francês's work. In this singular publication, Brito showcased his competence in mathematics and familiarity with contemporary astronomical literature by citing various authors, including Cardano, Gemma Frisius, Pedro Nunes, Jerónimo Muñoz, and Tycho Brahe.

Brito's objections fell into two categories: critiques of Bocarro Francês's mathematical methodology for determining the celestial location of the 1618 comet and challenges to Bocarro Francês's cosmological theories. In terms of methodology, Brito focused on Bocarro Francês's parallax estimation, arguing that despite their shared belief in the epistemological superiority of geometric demonstrations over Aristotelian reasoning, Bocarro Francês's mathematical procedures were flawed. Brito identified inaccuracies in Bocarro Francês's values for terrestrial and celestial coordinates, specifically highlighting the incorrect latitude of Lisbon. He concluded

[28] Bocarro Francês subsequently elaborated on his cosmological ideas in greater depth in his *Vera mundi compositio* (1622). For additional information, see Carolino (2012).

[29] Bocarro Francês's treatise on the comets of 1618 is available in various Portuguese public libraries, and there is additional evidence suggesting its circulation in manuscript form during the seventeenth century. Manuscript copies of the *Tratado dos Cometas* are preserved, notably at the Biblioteca Pública de Évora (COD. CX/1–5, n° 14) and the Arquivo Nacional da Torre do Tombo (Manuscritos da Livraria, n.° 1736).

that his failure in mathematical calculations led to an erroneous estimation of the comet's true parallax angle. While unable to provide an accurate parallax estimation himself due to a lack of precise instruments, Brito critiqued Bocarro Francês's cosmological conclusions. He censured Bocarro Francês's Stoic-inspired theories, which posited the self-movement of planets and a shared airy matter across terrestrial and celestial realms, leading to celestial corruptibility. In contrast, Brito defended an Aristotelian worldview, emphasizing the perfection of celestial matter and the clear division between celestial and terrestrial regions (Carolino, 2012: 216–220).

Brito's objections, presented as seven "doubts" in the final part of his work, reveal his astute criticism of Bocarro Francês's ideas. Additionally, other astronomers, such as Luís de Avelar, echoed reluctance towards Bocarro Francês's concepts, criticizing his parallax calculation (Avelar, 1619: fols. B1r-B1v. n.n.).

9.7 Astrological Prognostications and Political Expectations

Despite their cosmological differences, Iberian astronomers and astrologers unanimously believed that comets held a significant influence over meteorological phenomena, human lives, and the course of history. In booklets documenting the comets of 1618, some authors exclusively delved into the natural causes of comets, while others underscored what they perceived as divine intervention. As Pedro Mexia emphasized, "Comets derive from the exhalations of the Earth; their effects from the will of God" (Mexia, 1619: fol. A2r.). These phenomena were seen as a means through which God warned people of their misconduct, providing an opportunity for correction (Mexia, 1619: fol. A3r.). Nevertheless, contrary to perceiving the appearance of comets as a manifestation of the *potentia Dei absoluta*, Iberian astrologers generally maintained that comets were produced by natural means but held a metaphysical significance, much like everything in nature. Decades later, the Portuguese astrologer Manuel Gomes Galhano Lourosa synthesized this perspective by stating, "God employs comets as natural instruments to reveal misfortunes to the Magnates" (Lourosa, 1666: 38).

The prediction of their effects necessitated recourse to the astrological theory of comets. As previously indicated, a crucial step in this process involved determining the ruling planet of the comet. To achieve this, astrologers were required to attentively observe preceding eclipses and planetary conjunctions: "It is thus a standard procedure that upon the observation of a comet's appearance, one must record the conjunction of superior planets or the eclipse that preceded such manifestation" (Mexia, 1619: fol. 7r. n.n.). Comets were thought to embody the astrological attributes of their respective "causal agents."

Additional factors should be taken into account to comprehensively understand the primary effects induced by a specific comet. These factors encompassed characteristics such as colour, shape, size, duration, the position of the nucleus and tail, the time and cardinal orientations of the comet's visibility, as well as the constellations it traversed during its trajectory (Pedro, 1618: fol. 5r n.n.; Soto, 1619: fol. 6v n.n.).[30] For instance, Miguel Pedro, in reference to the second comet of 1618, asserted that "the colour showed that it partakes much of Saturn's nature, little of Mars, and something of Venus. From this, as well as from what has been said, it follows that this first comet [i.e., the C/1618 V1, as Pedro did not observe the C/1618 Q1] announces misfortunes for merchants," etc. (Pedro, 1618: fol. 5v n.n.).

Astrologers typically forecasted an array of adverse effects following the appearance of comets, including severe weather conditions, earthquakes, plagues, wars, and the death of kings and people of elevated social status. Bernardo Ferragut extensively elucidated the natural processes believed to underlie the production of such ominous outcomes by comets. According to Ferragut, the ascent of poisonous exhalations could lead to the outbreak of plagues by deteriorating the atmosphere, triggering widespread epidemics. Furthermore, these intensely hot and dry exhalations significantly impacted the natural humours within human and animal bodies, creating an imbalance that often resulted in severe diseases and fatalities (Ferragut, 1618: fols. A3v-A4r.). Nobles were particularly vulnerable, as their bodies were deemed more fragile (Ferragut, 1618: fol. A5r.). Given that these hot and dry exhalations typically incited human anger and rage, they were believed to contribute to constant conflicts, revolts, and wars (Ferragut, 1618: fols. A4v-A5r.).

History, as often recalled by astrologers, seemed to confirm the pernicious effects of comets. Within this framework, particular attention was devoted to the comet of 1577, an event that a majority of Iberian astrologers linked to the demise of the Portuguese King Sebastião in the Battle of El-Ksar el-Kebir, Morocco, the subsequent year. This event had profound consequences for the political history of the Iberian Peninsula, paving the way for the integration of the Portuguese kingdom into the Spanish Habsburg Empire in 1580. Onofre Pelechá, situated in the eastern region of the Iberian Peninsula in Valencia, was among the scholars who associated the appearance of the 1577 comet with the death of King Sebastião and the higher ranks of Portuguese nobility in the Moroccan desert (Pelechá, 1619: 43). Indeed, according to Pelechá, comets proved particularly pernicious for the Western Iberian kingdom, as the decisive defeat of António, Prior of Crato, the Portuguese noble who contested Philip II in the succession struggle, occurred following the appearance of a comet in October 1580 (Pelechá, 1619: 44).

However, comets did not always yield adverse consequences. Viewed from different perspectives, they might be perceived as heralds of exceptional and positive events. Drawing upon the colossal size and distinctive form of the second comet observed in 1618, resembling a vast palm branch, Pedro Mexia predicted great political repercussions for the Spanish Habsburg cause.

> The palm's enduring form, preserved until its culmination, holds significance as celestial signs are never arbitrary. Its extraordinary and admirable nature symbolizes the onset of a period marked by substantial equity, justice, and notable victories. This symbolism, predominantly and unequivocally, alludes to the Turks relinquishing control of the holy house,

[30] For an introduction on early-modern astrological theory of comets, see Schechner Genuth (1997): 51–65.

owing to the historical association of the palm as a symbol of the province of Judea (Mexia, 1619: fol. B3r.).[31]

In this passage, Mexia alludes to the conviction that the objectives of the Spanish monarchy, commonly referred to as the *Monarquía Católica* (the Catholic Monarchy), carried profound significance. The successors to the Catholic Monarchs, Queen Isabella of Castile and King Ferdinand II of Aragon, whose union in 1469 led to the unification of Spain, were believed to possess a transcendent purpose: ruling the world and suppressing all religious dissensions and heresies. The Ottoman Empire, or the Turks as identified by early-modern Iberians, constituted one of Spain's foremost political adversaries, engaging in conflicts in the Mediterranean and Eastern Europe.

The cometary works published in 1618 and the subsequent year resonated with the political and eschatological contexts of the time. As articulated by Pelechá in his *Discurso* on the comet of 1618, "My primary objective in this discourse has been to primarily demonstrate the destruction of the Moors, which God will bring about through His Majesty [the King Philip III]" (Pelechá, 1619: 57). Numerous Iberian astrologers asserted that the impacts of the comets of 1618 would extend, among other regions, to Turkey, the Middle East, and the Maghreb regions under Ottoman rule.[32] Some authors, including Pedro Mexia and Cristóbal de Montalvo, went so far as to predict that the celestial conjunction of 1603 marked the inception of the definitive decline of the Ottoman Empire (Mexia, 1619: fols. 13v-14r. n.n.; Montalvo, 1618: fols. 7r.-7v.). Accordingly, the comets of 1618 were believed to yield positive outcomes for Spain, either due to their anticipated impact on its adversaries (Ferragut, 1618: fol. 8r. n.n.; Pedro, 1618: 7r.) or, more optimistically, because they "indicate an increase and prosperity of our Kingdom, and a prosperous and happy future for our Most Catholic King" (Soto, 1619: fol. 5v.).

However, notwithstanding the conspicuous optimism expressed by some Spanish astrologers, the Spanish Habsburg Empire grappled with a complex and tense political landscape in the late 1610s, ultimately leading to the crisis of 1640. The expulsion of the Moriscos, Spain's Muslim population, in the early seventeenth century had a pronounced economic and social impact, particularly in the Valencian kingdom and other regions such as Aragon. Additionally, Valencia and Catalonia faced a pervasive phenomenon of banditry, posing significant challenges for local and Madrid authorities.

Moreover, the kingdoms of Valencia and Catalonia in the Eastern part of the Peninsula, as well as the kingdom of Portugal in the West, witnessed a gradual deterioration in their political autonomies and liberties. This decline was a consequence of the Castilian hegemony-driven policies initiated by the King's *valido*, Francisco

[31] In the introduction to his treatise, Bocarro Francês strongly criticized the specific correlation drawn between the palm branch form exhibited by the comet and the interpretation suggesting its connection to the decline of the Ottoman Empire. Bocarro Francês (1619): fol. 2v.

[32] These were, for example, the cases of Bargas y Heredia (1618): fol. 10v. n.n., Bautista (1618): fol. 7r. n.n., Ferragut (1618): fol. 10r. n.n., Mexia (1619): fol. 13v. n.n., Montalvo (1618): fols. 7r.-7v., Salaçar (1618): fol. 6r n.n., Valle (1619): fols. 14r., 18v.

Gómez de Sandoval y Rojas, Duke of Lerma, and later expanded by his successor, the Count-Duke of Olivares. This centralistic policy, perceived as favouring Castile over other Iberian kingdoms, reached its pinnacle when, amidst a backdrop of widespread conflict, including hostilities with France, the Duke of Olivares implemented the renowned Union of Arms (*Unión de Armas*). The union aimed to establish an integrated and comprehensive army to safeguard the monarchy's territories. However, for the majority of the Iberian kingdoms, this measure resulted in a substantial augmentation of taxes and the imperative to raise an army for engagements abroad. This fiscal and military burden prompted uprisings in the late 1630s and 1640s, starting with Catalonia and Portugal and subsequently encompassing other regions under Spanish Habsburg rule, such as Andalusia, Aragon, Naples, and Sicily. Portugal successfully secured its political independence through a coup d'état on December 1, 1640, leading to the ascension of the Braganza dynasty to power.

In contrast to England, where official censorship temporarily collapsed during the seventeenth-century Civil War, resulting in a significant proliferation of astrological publications centring on political predictions (Curry, 1989: 19), the Iberian cometary booklets do not explicitly mirror this political and social tension. Iberian Peninsula witnessed a vigilant oversight by the Inquisition over the circulation of ideas, individuals, and books. This vigilance explains the ambiguity and vagueness characterizing Iberian booklets that delved into political forecasts following the appearance of the comets in 1618. These political predictions unequivocally fell within the purview of exercising free will—a practice meticulously monitored by the Inquisitors. It is not coincidental that all the cometary booklets concluded by extolling the power of free will and the sovereignty of the Catholic Church.

Yet, amidst it all, a space emerged for the articulation of political perspectives fuelled by eschatological yearnings and partisan fervour. From France, the Iberians received a pamphlet on the comets of 1618 purportedly penned by the French author Jean Puget de La Serre. Upon its publication in Barcelona, the readers discovered that,

> A powerful and very Western monarch will find himself divested of his kingdom. [...] The kingdom, however, shall not remain bereft of a ruler for long, as a man of humble origin will ascend. Through his industrious efforts, eloquent persuasion (which garners allies and pacifies adversaries), and noteworthy deeds, he shall ascend to the throne as the new sovereign. (Puget de La Serre, 1618: fols. 5v.-6r. n.n).

This sort of prognostication likely exerted a profound impact on Portugal, the westernmost kingdom of the Iberian Peninsula. Writing from Lisbon, Manuel Bocarro Francês substantiated the prognostication that the comet of 1618 portended the "death of a great monarch of the Western parts, indicating significant changes in various kingdoms, particularly in the Iberian Peninsula (Bocarro Francês, 1619: fol. 17r.).[33] A few years later, he expanded on the significance of the comet of 1618 to propose a more ambitious prophecy. In his lengthy poem titled *Foetus astrologici*,

[33] In the early modern period, Iberian authors often employed the terms "Espanha" or "Hespanha," referring to Hispania or the Iberian Peninsula. This usage is exemplified by Bocarro Francês in his treatise.

initially published in Rome in 1626, Bocarro Francês correlated the appearance of the 1618 comet with the conjunction of superior planets in 1603, observed alterations in the eccentricity of the Sun's orbit around its centre, and computations of Saturn's cycles. According to his analysis, these celestial phenomena pointed to the emergence of a comprehensive and powerful monarchy in 1653 that would govern the world, with none other than Portuguese King John IV as the ruler (Carolino, 2004: 490–495). While the prophecy did not come true, Bocarro Francês had already moved to northern Germany by that time, distancing himself from Portugal. Almost three decades earlier, in 1624, he had fled the Iberian Peninsula for two main reasons: religious persecution in Portugal due to his Jewish background and political persecution from Spain as an advocate for the independence of Portugal.

9.8 Concluding Remarks

Like in almost every corner of Europe, the comets that crossed the skies in 1618 were fervently observed in the Iberian Peninsula, particularly the second one. This event sparked a passionate proliferation of booklets that successfully circulated through bookshops, markets, and piazzas across the Iberian kingdoms. During this period, these kingdoms were all under the rule of the Spanish Habsburgs, furthering the exchange of individuals, books, and ideas throughout the Iberian Peninsula. However, despite the ground-breaking controversy that ensued following the observation of the 1572 supernova in the Iberian Peninsula, the comets of 1618 sparked a comparatively more modest astronomical and cosmological debate. With the exception of Manuel Bocarro Francês's heterodox treatise, which perpetuated the cosmological tenets articulated by Jerónimo Muñoz nearly half a century earlier, all authors who published booklets after the observation of the 1618 comets adhered to a traditional Aristotelian cosmology. They enriched this perspective with doctrines derived from Arabic-Ptolemaic astrology.

Their focus was indeed primarily astrological rather than astronomical, as they were chiefly interested in predicting the potential astrological effects of the comets. These prognostications typically encompassed earthquakes, severe weather conditions, resulting infertility, plagues, wars, and subsequent human fatalities. Since it was believed that comets particularly targeted individuals of high standing, some Iberian scholars, fuelled by eschatological beliefs and ideas of political autonomy, seized the opportunity presented by the appearance of the 1618 comets to suggest that significant political changes were imminent in the Iberian Peninsula. In doing so, they encapsulated the social and political sentiments prevalent in substantial segments of Iberian society during a period increasingly perceived by Iberians as one of crisis and redemption.

References

Avelar, L. d. (1619). *Nox attica hoc est dialogus de impressione metheorologica et cometa anni domini 1618*. Nicolau Carvalho.

Bargas y Heredia, V. G. (1618). *Tratado de cometas dividido en dos discursos, En el primero se trata la materia general suya, En el segundo particular de los que se an visto en España en mes de Nobiembre*. Juan Muñoz.

Bautista, J. (1618). *Segunda Pronosticacion y iuyzio de las dos cometas que se han visto desde cinco de Noviembre, con la declaracion de sus efetos, segun Cupernico, Y Tiquimaco, David Origino, y otros graves autores, estudiado y compuesto por..* Chrisostomo Garriz.

Bocarro Francês, M. (1619). *Tratado dos cometas que appareceram em Novembro passado de 1618*. Pedro Craesbeeck.

Brito, M. P. de (1619). *Discurso em os dous phaenominos aereos do anno de mil seiscentos e dezoito*. Pedro Craesbeeck.

Carolino, L. M. (2003). *Ciência, Astrologia e Sociedade. A teoria da influência celeste em Portugal (1593–1755)*. Fundação Calouste Gulbenkian.

Carolino, L. M. (2004). Scienza, Politica ed Escatologia nella formazione dello 'scienzato' nell'Europa del XVII secolo: Il caso di Manuel Bocarro Francês – Jacob Rosales. *Nuncius, 19*(2), 477–506.

Carolino, L. M. (2012). Manuel Bocarro Francês, the comet of 1618, and the impact of Stoic cosmology in Portugal. In M. Á. Granada (Ed.), *Novas y cometas entre 1572 y 1618. Revolución cosmológica y renovación política y religiosa* (pp. 195–224). Universitat de Barcelona.

Carolino, L. M. (2016). Science, patronage and academies in early seventeenth-century Portugal: The scientific academy of the nobleman and university professor André de Almada. *History of Science, 54*(2), 107–137.

Carolino, L. M., & Leitão, H. (2006). Natural philosophy and mathematics in Portuguese universities, 1550-1650. In M. Feingold & V. N. Brotons (Eds.), *Universities and science in the early modern period* (pp. 153–168). Springer.

Casiano, J. (1618). *Breve discurso a cerca del cometa visto en el mez de Nouiembre deste año de 1618 y sus significaciones*. Pedro Craesbeeck.

Cedillo Díaz, J. (2019). *Ydea astronomica de la fabrica del mundo y movimiento de los cuerpos celestiales: traducción de Nicolás Copérnico 'De revolutionibus' I-III* (M. Á. Granada & F. G. Crespo, Eds.). Edicions de la Universitat de Barcelona.

Curry, P. (1989). *Prophecy and power. Astrology in early modern England*. Polity Press.

Dias, J. J. A. (1996). *Craesbeeck. Uma Dinastia de Impressores em Portugal. Elementos para o seu estudo*. Associação Portuguesa de Livreiros Alfarrabistas.

Feingold, M. (1984). *The mathematicians' apprenticeship. Science, universities and society in England, 1560–1640*. Cambridge University Press.

Ferragut, B. (1618). *Coniectura de los efectos significados por los cometas, que aparecieron el mes de nouiembre, en el horizonte de la ciudad de Huesca, año 1618*. Pedro Bluson na imprensa da viuda de Juan Perez de Valdivielso.

Grant, E. (1987). Medieval and renaissance scholastic conceptions of the influence of the celestial region on the terrestrial. *Journal of Medieval and Renaissance Studies, 17*, 1–23.

Guzman, F. de (1609). *Pronostico y lunario de los tiempos del año de mil seyscientos y diez. El mas copioso y curioso que han salido a luz. Compuesto por el Don Francisco de Guzman, natural de la ciudad de Valencia, a cuyo meridiano va calculado*. Pedro Craesbeeck.

Kronk, G. W. (1999). *Cometography. A catalog of comets, Vol. 1: Anciente-1799*. Cambridge University Press.

Lanuza Navarro, T. (2005). *Astrología, ciencia y sociedad en la España de los Austrias*. PhD., University of Valencia.

Lanuza Navarro, T., & Navarro Brotons, V. (2013). Prophecy and politics in Spain: Celestial novelties and the science of stars, 1572-1630. In D. Tessicini & P. J. Boner (Eds.), *Celestial novelties on the eve of the scientific revolution, 1540–1630* (pp. 33–56). Leo S. Olschki Editore.

Leitão, H., & Carvalho, J. M. (2020). O calendário gregoriano: globalizar a marcação do tempo. In C. Fiolhais, J. E. Franco, & J. P. Paiva (Eds.), *História Global de Portugal* (pp. 409–414). Temas e Debates.

Lourosa, M. G. G. (1666). *Polymathia Exemplar. Doctrina de discursos varios offerecido ao Conde de Castel-Melhor. Cometographia Meteorologica do prodigioso, e diuturno Cometa, que appareceo em Novembro do Anno de 1664*. António Craesbeeck de Mello.

Luciano, A. (1618). *Discurso mathematico sobre los dos cometas que se han parecido, en las quales se declaran los efectos que muestran por los signos y señales, a donde se vieron primero el vno desde doze de nouiembre deste presente año de 1618, y el otro desde 24 de dicho mes.* en la impresion de Felipe Mey, junto a S. Esteuan.

Mexia, P. (1619). *Discurso sobre los dos cometas que se vieron por el mes de nouiembre del año passado de 1618*. Pedro Craesbeeck.

Montalvo, C. de (1618). *Discurso a los dos cometas que han parecido en Granada en el mes de Nouiembre de mil y seyscientos y diez y ocho años: regulado con la coniuncion maxima de Saturno y Iupiter*. por Bartolome de Lorençana, en la calle del Pan.

Najera, A. de (1619). *Discursos astrologicos sobre o cometa que apereceo em 25 de Nouembro de 618*. Pedro Craesbeeck.

Navarro Brotons, V. (2012). Las novedades celestes en España entre 1572 y 1618. In M. Á. Granada (Ed.), *Novas y cometas entre 1572 y 1618. Revolución cosmológica y renovación política y religiosa* (pp. 15–41). Universitat de Barcelona.

Navarro Brotons, V. (2014). *Disciplinas, Saberes y Prácticas. Filosofia Natural, Matemáticas y Astronomía en la Sociedad Española de la Época Moderna*. Universitat de València.

Navarro Brotons, V. (2019). *Jerónimo Muñoz. Matemáticas, cosmologia y humanismo em la época del Renacimento*. Publicaciones de la Universitat de València.

North, J. D. (1986). Celestial influence—The major premiss of astrology. In P. P. Zambelli (Ed.), *Astrologi hallucinati'. Stars and the end of the world in Luther's time* (pp. 45–100). Walter de Gruyter.

Pedro, M. (1618). *Ivyzio y presagio natvral, de los cometas que han aparecido por el mes de noviembre, en el Orizonte de la Ciudad de Çaragoça, en este presente año de 1618*. por Pedro Gel. A los Señales.

Pelechá, O. (1619). *Discurso de la naturaleza, causas y efectos de los cometas: primero en general, después en particular de los que se han visto en el año 1618, conforme la opinión vulgar de los filósofos y astrólogos y verdadera de San Juan Damasceno y otros*. Pedro Patricio Mey.

Puget de La Serre, J. (1618). *Breue y curiosa relacion, del discurso ha hecho Moussur [sic] Iuan de Puget, ayuda de camara del rey christianissimo, y su secretario del Consejo de guerra, sobre los cometas que han aparecido este año de 1618*. Geronymo Margarit.

Pumfrey, S., & Dawbarn, F. (2004). Science and patronage in England: 1570-1625: A preliminary study. *History of Science, 42*, 137–188.

Rosselló Botey, V. (2000). *Tradició i canvi científic en l'astronomia espanyola del segle XVII*. Editorial Biblioteca Nueva/Universitat de València.

Salaçar, J. de (1618). *Pronostico de las cometas dos cometas que se na aparecido este año de mil y seyscientos y diez y ocho: y en Granada en el mes de Noviembr.* por Bartolome de Lorençana, en la calle del Pan.

Schechner Genuth, S. (1997). *Comets, popular culture and the birth of modern cosmology*. Princeton University Press.

Seargent, D. (2009). *The greatest comets in history. Broom stars and celestial scimitars*. Springer.

Soto, J. de (1619). *Iuyzio y pronóstico del cometa que se manifesto a 4 de Noviembre deste ano de 1618*. Emprenta de Gabriel Graells, a los quatro cantones del Call.

Valle, B. del (1619). *Explicacion y pronostico de los dos cometas*. por Francisco Heylan y Pedro de la Cuesta.

van Nouhuys, T. (1998). *The age of two-faced Janus. The comets of 1577 and 1618 and the decline of the Aristotelian world view in the Netherlands*. Brill.

Luís Miguel Carolino is an Associate Professor at the University Institute of Lisbon (ISCTE-IUL), Portugal, where he is also a researcher affiliated with CIES. He was a Visiting Professor at the University of California, Berkeley (2019) and the Federal University of Rio de Janeiro (2007). Before joining the History Department at ISCTE-IUL, he held postdoc and research positions at the Institute and Museum of the History of Science (IMSS), Florence, Italy; the Museum of Astronomy and Related Sciences, Rio de Janeiro, Brazil; and the University of Lisbon, Portugal. His main research interests concern the history of science, history of learned institutions, and early modern natural philosophy, cosmology, and astronomy, areas in which he has published extensively. Carolino is a member of the International Commission of the History of Universities.

Part III
Portents, Omens and Astrology

Chapter 10
Writing Past Futures: Celestial Omens and the Historiography of the Thirty Years' War

Andreas Bähr

10.1 Celestial Omens and Historiography

When Johann Baptist Cysat, a mathematician and astronomer at the Jesuit College in Ingolstadt, looked into the night sky in the early hours of December 1, 1618, he was presented with an astounding sight: without the aid of telescope, he discerned a large, bright comet between the constellations of Ursa Major and Virgo (Cysat, 1619: A 4r, 1 f.). Little could dampen Cysat's fascination for this object, save perhaps the realization that he was not the first person in the Holy Roman Empire to do so: The mathematician Benjamin Ursinus had already "attended" this "unanticipated spectacle" in Joachimsthal on 28 November (Ursinus, 1619: A 1r). And Johannes Kepler, with whom Cysat had little in common (not just cosmologically but also confessionally), had beaten him to it by 2 days. As Kepler took a look through his telescope in the early morning of November 29—which he had been doing for months to study two smaller comets—he saw a third, "yellow/ somewhat reddish" *Cometa clarissimus*, whose trail remained visible in Europe until late January.[1]

As there were now few restraints placed on the zeal of astronomers, both Kepler and Cysat published the results of their observations in the following year. While Cysat sought to confirm the Tychonian model of the universe, Kepler's aim was to

[1] Kepler, 1963: 182–186. The quotation is from Kepler, 1993: 187.

This essay is the English version of Bähr, 2024. Translated from the German by Dominic Bonfiglio.

A. Bähr (✉)
European University Viadrina Frankfurt (Oder), Frankfurt (Oder), Germany
e-mail: baehr@europa-uni.de

C. J. Cunningham (ed.), *Cosmic Events*, Historical & Cultural Astronomy,
https://doi.org/10.1007/978-3-031-86950-1_10

Fig. 10.1 The winter comet above Heidelberg. Copperplate engraving by Matthäus Merian the Elder. In Johann Philipp Abelinus, *Theatrum Europaeum*, vol. 1. Frankfurt am Main: Merian, 1635, 119. Herzog August Bibliothek Wolfenbüttel, PURL: http://diglib.hab.de/periodica/70-a-hist-2f/start.htm?image=00165, CC BY-SA

complete his comet study of 1607 (Kepler, 1941: 57–76)—which he did so, interestingly, with data gathered by Cysat.[2]

The comet of 1618 also garnered significant attention from the broader populace. As no technical aids were needed to discern the comet in the night sky, numerous chronicles and autobiographies reported on this extraordinary event, including Merian's *Theatrum Europaeum* (Abelinus, 1635: 116–119) (Fig. 10.1). Yet the comet did not universally arouse a sense of excited fascination. For Volkmar Happe, a court counselor in Schwarzburg-Sondershausen, the appearance of the comet was "horrific" because it indicated that "war, turmoil, bloodshed, pestilence, and a time of burdens and unspeakable misfortune" would soon occur. After all, everyone knew: "No terrible comet is seen that does not great misfortune mean." (Happe, 2008: part I, 24v) In accounting for historical experience, Happe forgets the Greek etymology of the word comet and invents a Latin one. The "hair star" (from Greek

[2] Cf. Gindhart, 2006: 263. Although Kepler advocated heliocentrism, contrary to Tycho Brahe he assumed that comets—which he did not consider celestial bodies—did not move in circular orbits around the sun, but rather traveled in straight lines (Kepler, 1963: 144–156). Cysat, for his part, though in agreement with Brahe in principal, did not want to eliminate this possibility (Cysat, 1619: 58–60).

kométes, meaning "long-haired") becomes an "ally and companion of fatal things": a *comes et socius rerum fatalium*, as the Bohemian prophet Wilhelm Neuheuser put it (1619: 17). Happe, for his part, was translating a verse by the late antique poet Claudian, which had already become proverbial at the beginning of the seventeenth century (and was also quoted by the physician David Herlitz, for example): *et numquam caelo spectatum impune cometen* (Claudianus, 1985: 248, v. 243; Herlitz, 1619: F 2r). (Fig. 10.2) But Happe did not translate it literally. A closer rendering reads: "Never has a comet been seen in the sky with impunity."

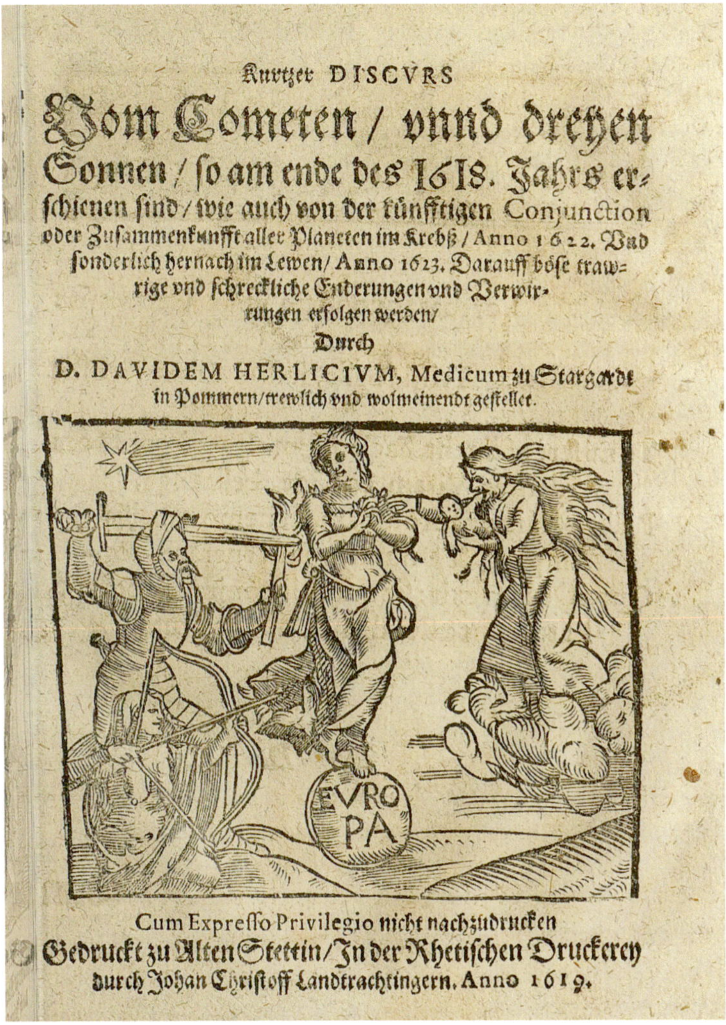
Kurtzer DISCVRS
Vom Cometen / vnnd dreyen Sonnen / so am ende des 1618. Jahrs erschienen sind / wie auch von der künfftigen Conjunction oder Zusammenkunfft aller Planeten im Krebß / Anno 1622. Vnd sonderlich hernach im Lewen / Anno 1623. Darauff böse trawrige vnd schreckliche Enderungen vnd Verwirrungen erfolgen werden /
Durch
D. DAVIDEM HERLICIVM, Medicum zu Stargardt in Pommern / trewlich vnd wolmeinendt gestellet.

Cum Expresso Privilegio nicht nachzudrucken
Gedruckt zu Alten Stettin / In der Rhetischen Druckerey durch Johan Christoff Landtrachtingern. Anno 1619.

Fig. 10.2 War, plague, and famine: The winter comet prophesizes the destruction of Europe. Frontispiece of David Herlitz, *Kurtzer Discvrs Vom Cometen/ vnnd dreyen Sonnen/ so am ende des 1618 Jahrs erschienen sind.* Stettin: Landtrachtinger, 1619. Herzog August Bibliothek Wolfenbüttel, PURL: http://diglib.hab.de/drucke/42-7-astron-18s/start.htm?image=00001, CC BY-SA

Happe, a devout Lutheran, could be sure his readers knew who was imposing this punishment, and why: By dispatching this comet, the righteous Lord in heaven was admonishing mankind for his sins; this was not only clearly expressed in the Revelation of John (8, 10), but also in the apocalyptic verses in Matthew (24, 29), Mark (13, 24), and Luke (21, 25 f.).

However, the scriptures do not specify the punishment due for sinful behavior in individual cases. Only history could reveal this. When Happe completed his 1800-page chronicle, he had already gained sufficient assurance regarding historical events. By 1642 at the latest—because this is the year the text ends—Happe had obtained clarity. In retrospect, it was clear that the comet had predicted decades of war, with all its attendant consequences, including plague, famine, and the overthrow of the secular and spiritual order. There are two remarkable details worth noting.

First: In Happe's eyes (as in the eyes of so many others), the Thirty Years' War did not begin with the Defenestration of Prague, but with the comet of 1618. To be more precise: It was only after seeing the "horrible" comet that he realized that the Prague Defenestration could not be regarded as a regional episode, but rather as the beginning of a major political and military conflict. Happe mentions the events in Prague as the prelude to the "bloody Bohemian War […] which continued for many years, almost enveloped the entire world, and devastated all countries." A note in Latin by a subsequent scribe emphasizes that "this day was the beginning of the cruelest war" and brought "destruction and ruin to the whole of Bohemia and Germany."[3] But it was not these events by themselves that provided Happe with this clarity. Rather, it was the winter comet of 1618.[4]

Second, Happe's understanding of history, which was heavily influenced by theology, shaped his historical writing. As he notes at the outset: "The terrible unrest, wars, murders, inflation, pestilence, change, and assumptions of power that took place after these comets; the persecution of the Protestant religion, its extermination in many places, and the reintroduction of the Papal abomination, can be seen from the following events"—that is, from the events that took place in the German territories up to 1642, especially in Thuringia, but also in other parts of the Holy Roman Empire.[5] Happe's chronicle of the Thirty Years' War was inspired by the winter comet of 1618—which, in his retrospective interpretation, had caused the war—and as the hostilities erupted, Happe began to chronicle them. In this way, the comet turned Happe, a steward and court counselor, into a war chronicler.

In the following, I seek to show that this connection was not arbitrary or coincidental, but rather logical. I am concerned not only with omens and their foreshadowing of future events, but also with the chronologies they brought into being. Texts such as Happe's *Chronicon Thuringiae* collected observations and other evidence

[3] Happe, 2008: part I, 23v: "Dies hic jnitium belli cruentissimi, ruina et perditio totius Bohemiæ et Germaniæ". My translation. On the writer of the marginal note, see below, fn. 6.

[4] See also Bähr, 2017: 13–15.

[5] Happe, 2008: part I, 24v. On the scope of Happe's perceptions and observations, cf. Medick, 2007.

that would enable the exegesis of the miraculous events in the sky. From this perspective, the most important task of such a historiography was to clarify what the omens meant in terms of man's salvation. To demonstrate that this was the underlying concern, I will focus on the *Zeytregister*, a chronicle that was maintained by the farmer and shoemaker Hans Heberle, who witnessed the comet passing over Ulm. Heberle additionally observed the aurora borealis of 1630, as did the Cistercian monk Sebastian Bürster. Bürster, who lived on Lake Constance at the Salem Monastery, reports on these astronomical events in his *Collectanea*.

While it is not possible to reconstruct exactly *when* these three chroniclers wrote down what they saw, given the nature of their accounts, one thing appears clear: Happe, Heberle, and Bürster initially took regular notes of what they observed. Only later did they distill their notes into the manuscripts that have survived to this day. Bürster, for his part, explicitly states that his account was written based on his notes. In the case of the other two authors, this method of production also appears extremely likely.[6] These chroniclers began their records because they suspected a divine message was being imparted by the events in the sky, and they revised these records once their suspicions had become certainties. In this way, Happe, Heberle, and Bürster described their experience of the Thirty Years' War in order to verify a possible prophecy: that is, to disclose a "past future" (Koselleck, 1976).

Contrary to the impression given at the outset, these chroniclers were in no way fundamentally at odds with the astronomers. Cysat also considered the winter comet to have an ominous significance. He was merely more reticent than the others, preferring instead to let the future decide the truth of the matter (Cysat, 1619: 79 f.). Kepler, by contrast, did not display the same reserve (although his work appeared around the same time, at the beginning of 1619). While Kepler refrained from providing an exegetical account of the comet's significance, the basic thrust of his interpretation accords with that of Happe, even if the underlying reasoning was divergent and stemmed from natural philosophy. For Kepler, as well, the comet heralded plague, famine, and war. In Kepler's view, the comet created a zone of infection, a *locus infectus* that had a devastating effect on the air, climate, and the emotions, or *affectus*, of man (Kepler, 1963: 231–233, 251–262). The theologians, by contrast, were influenced by Aristotle, and still generally assumed that comets emitted a poisonous vapor—a vapor upon which the comet fed. In a Christian twist on Aristotelian doctrine, this vapor was the noxious fumes of sin.[7]

[6] On Bürster, see below, Sect. 10.3. The surviving manuscript of Happe's chronicle is a copy from the first third of the eighteenth century. It includes some marginal notes (like those quoted above) which presumably originate from Christian Friedrich Ruhe, the archivist of the comital lines of the House of Schwarzburg, and are implicitly ascribed to Happe as the author. Cf. Medick, 2007: 173 f. On Heberle, see below, Sect. 10.2. Furthermore, cf. Laux, 2002; Merzhäuser, 2002. For a critical discussion of Laux's and Merzhäuser's positions, cf. Bähr, 2017: 87, fn. 61.

[7] For further details, see Bähr, 2017: 64–66. It was not until the late seventeenth century that natural philosophers interpreted the apocalyptic significance of the comet in terms of their possible collision with the earth. Cf. Schechner Genuth, 1997: parts 3 and 4; Briese, 1998: 179–190; Splinter, 2009.

The proof of this fact was provided by a form of historiography that cannot be understood without giving consideration to the omens that animated its production, and the ways in which the authors retroactively ascribed meaning to divine signs by translating the future into stories—that is, into 'histories.' This was a form of historiography that understood both temporal and divine events as representative of God's will and of providence.[8] The comet of 1618 has been repeatedly addressed by scholars, albeit at varying levels of detail.[9] This paper represents the first analysis of the comet's relevance for historiographical practices at the time of the Thirty Years' War—practices that inform much of what we know about the war today.[10]

Contrary to what the literature grounded in secularization theory continues to suggest,[11] this historiography does not contend that future-oriented action is an impossibility (that is, before the temporalization of time and the emergence of historical-philosophical thinking). Rather, this historiography itself represented such a possibility for action. At the intersection of certainty regarding the providential significance of an omen and uncertainty regarding how it should be properly interpreted, the activity of accounting for the events of the war transformed uncertainty into certainty and offered a tool for coping with its suffering.[12]

10.2 The Comet of 1618 over Ulm

The appearance of the comet of 1618 was immediately assumed to have divine significance. Indeed, in regard to the sermons planned for the second Sunday of Advent, the Lutheran pericope order required preachers to speak about the apocalyptic verses of Luke's Gospel that begin with the words: "There will be signs in the sun and moon and stars, and on the earth dismay among the nations, bewildered by the roaring of the sea and the surging of the waves." (Dieterich, 1619: 7) The second Sunday of Advent took place approximately 1 week after the first sighting of the comet in the Holy Roman Empire (according to the old calendar).[13] That could only mean one thing: With the appearance of the comet, the heavens were explaining that

[8] On early modern concepts of history and the practices of writing history, cf. Koselleck, 1975, 1976: 17–66; Burke, 2011; Seifert, 1976; Pomata & Siraisi, 2005; Grafton, 2007; Rau & Studt, 2010; Pohlig, 2007, 2016; Friedrich & Schilling, 2019. Cf. also below, fn. 19.

[9] Cf. in particular Gindhart, 2006; Krusenstjern, 1999; Jerratsch, 2020: chapter 7. From the literature on the Thirty Years' War cf., furthermore, Medick & Krusenstjern, 2001: 30 f.; Medick, 2010: 33 f., and Medick, 2018: 26, 34–36, 53–57; Mortimer, 2002: 19, 76, 137, 180 f.; Duchhardt, 2017: 207 f.; Meid, 2017: 15 f.; Schmidt, 2018: 14–18, 167–173, 689 f.; van de Löcht, 2019; Böning, 2019: 49, 54; Schilling, 2019: 69; Seelbach, 2019b: 301. From the research literature on prodigies, cf. Lehmann, 2007; Mauelshagen, 1998: 121–124; Hille, 2009: 369–371; Barnes, 2016: 286–289.

[10] For an extended version of the following argument, see Bähr, 2017.

[11] Cf. Hölscher, 2016, and also Landwehr, 2014.

[12] From a narratological perspective, cf. e.g. Seelbach, 2019a: 2–10.

[13] According to the old-style calendar, the second Sunday of Advent was on December 6, according to the Gregorian calendar, it was on December 9. Cf. Gindhart, 2006: 27.

the comet and pericope order were both subject to a common plan for the universe. At that moment, the contemporary relevance of the Advent sermon was clear. Throughout the empire, pulpit speakers were now somewhat pressed for time, as they had to quickly consider how to relate the Bible text to current political and military events—namely, to the events in Bohemia.

Conrad Dieterich, an energetic superintendent in Ulm, was also faced with this task. In the many religious sermons and tracts he published over the course of his life, he not only addressed comets,[14] but also dreams, the significance of church bells, and the plague (Dieterich, 1625a, b, 1636). His reflections on the winter comet can be found in his Advent sermon, from which the above biblical quotation is taken. The sermon was published, together with a flood of other relevant sermons, at the beginning of 1619—that is, before the astronomical treatises.[15]

Dieterich was sure that the comet was not to be interpreted as the Star of Bethlehem, even though the Christmas season was approaching. Origen and Thomas Aquinas considered it possible that this celestial phenomenon from the Bible was a comet; this belief was also shared by the painters Giotto di Bondone and Hans Baldung Grien.[16] At the beginning of the seventeenth century, however, this interpretation was generally rejected. At that time, there was a nearly universal conviction that comets heralded disaster, and not salvation; a survey of history made this fact clear. As Heinrich Leuchter, a superintendent in Darmstadt, emphasized on Epiphany Day 1619, the Star of Bethlehem was a *stella lucis*, a "star of light," and not a "star of the cross" (*stella crucis*) (Leuchter, 1619: 8 f.). The winter comet, on the other hand, did not mark Jesus's birth, but rather his death. Alternatively, from the perspective of salvation history, the Star of Bethlehem referred to the first coming of the Lord and the winter comet to the second—that is, to Christ's return before the Last Judgment and the end of the world. This was the view advanced by the Prague pastor Helwig Garth (1619: E 2r).

Against this backdrop, Dieterich quickly realized what was necessary: He had to call for repentance and conversion; and in doing so, he had to fight against two equally problematic attitudes in his eyes: the fearless indifference of unbelievers who did not recognize the danger announced by the comet and the millenarian fear of an eschatological battle fomented by false prophets—a fear that, in Dieterich's opinion, could only lead to violence (Dieterich, 1619: 15, 32–34, 44–52). In their claim to exclusive divine knowledge, these prophets believed that they could not only foresee war, but were also entitled to call on authorities to wage war, to execute the divine will. Johannes Faulhaber, for example, was sharply rebuked by Dieterich in the Ulm comet dispute.[17]

[14] Dieterich, 1619; on the events in Bohemia, see p. 42.

[15] See Christiani, 1619; Burckhardt, 1618; Garth, 1619; Herrenschmidt, 1619; Lungwitz, 1619; Weber, 1618; Samson-Himmelstjerna, 1618. Cf. Gindhart, 2006: section 2.2; Zimmermann, 1999.

[16] Cf. Schechner Genuth, 1997: 38 f., as well as Giotto's *Natività e Adorazione dei Magi* and Baldung's *Die Geburt Christi*.

[17] For details and further references, see Bähr, 2017: 51–71. On millenarianism/chiliasm, see e.g. Penman, 2019, Landwehr, 2014: 41–89; Schmidt-Biggemann, 2007: part 2. On Lutheran lay prophets, see Beyer, 2017; Theibault, 1994; Sabean, 1984: chapter 2.

Thanks to its publication, Dieterich's comet sermon spread rapidly, and not just among the educated classes. The shoemaker and smallhold farmer Hans Heberle of Neenstetten was among the many who came in contact with the sermon. As Heberle was unable to attend service in nearby Ulm, he regularly purchased printed sermons, particularly those of Conrad Dieterich.[18]

In his Advent sermon, Dieterich not only exhorted people to repent but also provided an overview of philosophical and theological explanations for comet phenomena being discussed at the time (Dieterich, 1619: 9–19). He based his warnings on an extensive catalog of examples, which he believed proved the ominous significance of comets.[19] Yet he also did something else: In the printed version of his sermon, he cited Herodotus to exhort his readers to continue gathering historical evidence. In practical terms, this meant observing and noting for oneself whether the current comet would entail the horrors known from history.[20]

Yet Heberle was not just an avid reader of sermons. He was also a prodigious writer. In the evening, after completing his daily chores, he wrote his *Zeytregister*, a chronicle that extended to 350 manuscript pages. In this chronicle, Heberle recorded what happened in the Ulm region during the war (and well after the war) for his God-fearing descendants. In addition to reporting on epidemics and droughts, he describes the constant passage of soldiers of all types, and his more than thirty escapes to the city of Ulm.

We know the "reason and occasion" for Heberle's writing of this chronicle because he named them explicitly. He did so in order to counter the suspicion that he was wasting his time on matters inappropriate to his profession (Zillhardt, 1975: 86). He hardly could have found a better justification. What Heberle, the lowly shoemaker, saw in the sky at the end of 1618 left a lasting impression. The comet was "horrific and wonderful" to look at, so Heberle knew that it would "mean and bring something great" (86 f.). The comet struck fear in Herberle, yet this fear was accompanied by the astonishment and awe that pervades confrontation with the divine (as Dieterich would have wanted). How was one to respond? According to Heberle, the comet "moved me to start writing" (86). His 'emotion,' the 'movement' of his soul, went straight to his hand, and so he immediately reached for pen and ink. He began to write his "report" on what happened after the comet appeared—that is, he began with the exegesis of this divine sign. In this way, the best protection

[18] This can be inferred from Heberle's *Zeytregister*: Zillhardt, 1975: 87 (fol. 2v).

[19] Dieterich, 1619: 23–31, on Ulm: 40 f. On the historical epistemology of examples and case histories, see Krämer, 2014: chapter 3; Ruchatz et al., 2007: part 1; Pomata & Siraisi, 2005; Haug & Wachinger, 1991; Seifert, 1976. This epistemology complied with the logic of collecting; see Eco, 2009: chapters 10–12. See also above, fn. 8. Catalogs of examples found in Dieterich's work could also be used for a diametrically opposed purpose, i.e. to prove the positive significance of comets: Grick, 1619. Cf. Gindhart, 2006: 153–163. However, Grick's satirical perspective represents an exception.

[20] Dieterich, 1619: 21. Dieterich refers to the *Histories* of Herodotus, who depicts the Egyptian practice of portent interpretation (Herodotus, 2022: II, 82.2).

against the misfortune heralded by the comet was to decipher its meaning chronologically (86 f.).

Of course, this presupposed one thing: that disaster was not inevitable. "In anno 1618," it says in the *Zeytregister*, "a great comet appeared in the form of a great and terrible rod, which threatens us violently from and through God, because of our sinful lives, and we have deserved and still deserve this rod daily." (Zillhardt, 1975: 93 f.). The decisive word here is "threaten." In this connection, Conrad Dieterich wrote (1619: 16): "God does not punish anyone unless he has first reminded him, frightened him, and announced the disaster or danger." In other words, the comet warned of war but did not prophesize it.[21] That meant that the war still could have been prevented at the end of 1618. Yet how to avert such disaster? As mentioned, the proper way forward was to repent and convert. Another option was to record written reminders of the need for repentance. This was a means of observing signs sent from the heavens in a God-fearing way. In this way, simply describing the violence that was occurring could diffuse any motivation to commit it. Hans Heberle took this to heart.

Heberle was unable to prevent the war; that was clear in retrospect. But perhaps he could help to prevent similar things from happening again. The message furnished by the *Zeytregister* is therefore: You have ignored the signs in the past; now at least pay attention to them in the future. In this regard, Heberle asserts that he understood the message being communicated by the divine. Indeed, for Hans Heberle and so many others, the Thirty Years' War did not begin with the Defenestration of Prague (which he does not mention anywhere in the *Zeytregister*), but rather with the winter comet, which prompted him to record the events of the war, as an act of heeding a divine warning. For this lowly shoemaker, a belief in omens, in *prodigia*, went hand in hand with a need for caution in rendering religious interpretation. This caution resulted in a form of historiography that not only provided evidence for the presumed meaning of the sign but also evidence for the historiographer's fear of God. It proved that the sign in the sky had been a divine warning—a warning that only retrospectively became a prophecy. It was an omen that became true because people had closed their eyes to it.

10.3 The Aurora Borealis of 1630 over Salem Monastery

While it was predominantly Protestants who were intently preoccupied with celestial signs and who saw the end of days approaching with the outbreak of the war, many contemporary Catholics also showed such proclivities—including Sebastian Bürster, a monk at the Cistercian Abbey of Salem near Überlingen on Lake Constance.

[21] Cf. also Seelbach, 2019b: 301; Schilling, 2019: 69.

Late in the evening of 4 February 1630, Bürster witnessed an incredible occurrence. Above his head, as he reports, "the sky began to turn yellow and red, quite fiery [...] as if there were a terrible great heat." And in addition, the clouds moved to and fro and in confusion, "as if two large armies were attacking each other [...] and a great bloodbath was transpiring." Bürster felt as if he were standing in the middle of the gruesome events: "One could see all the weapons and arms with which they attacked each other; spears, sticks, picks, indeed the smoke of muskets and canons shooting; everything was tossing and turning, running to and fro, as if a real battle were taking place." (Weech, 1875: 12).

Bürster was certain this event was not a natural weather phenomenon; it was not exclusively due to vapors, as Aristotle believed. Rather, he had been confronted by a "prognosticon" with ominous "foreboding" (Weech, 1875: 12).

Yet Bürster was not alone in this regard. Numerous observers, especially Lutherans, described and interpreted the aurora borealis of 1630 in a similar way, including Hans Heberle and the Tübingen astronomer Wilhelm Schickard. As proof, he referred to historical experience—and recalled the comet of 1618, which had "unfortunately taught us all too well that the celestial signs presented were not false threats!"[22]

Schickard did not consider the "miracle" of the aurora borealis to be supernatural, because it did not violate the laws of nature. Yet he did believe it was *praeter naturam* in the sense of out of the ordinary.[23] We do not know if Bürster would have agreed with this scholastic distinction. But one thing is certain: It didn't occur to him to think the event was merely natural.

Bürster observed the aurora borealis from Salem Monastery. In the abbey, he worked as the "bread administrator" and "alms distributor" (Weech, 1875: XV). Since that memorable February evening, however, he has also been active as a chronicler. Like Heberle and Happe, Bürster was concerned with clarifying the meaning of this ominous sign. To do so, he would need to record what happened in Salem and its surroundings in the coming years: "[H]ow and why [God] will punish us for our great and many sins, especially in Southern Germany, over which this splendor has fallen, you, favorable reader, will unfortunately hear enough and too much in subsequent years." (Weech, 1875: 12 f.).

Specifically, this meant that from 1632 onwards, the imperial abbey was repeatedly plundered and extorted by passing troops of different affiliations. In the long term, only those who sought their salvation in flight and exile could save themselves, and so many did so that the abbot felt compelled to dissolve the convent in 1641. Two years later, God gradually allowed his anger to subside and sent the monastery "help from above" (Weech, 1875: 177): The French, who had since entered the war, together with Bernhard von Weimar's soldiers, evacuated their camp outside Salem, such that the priests and lay brothers were ordered to return. But this improvement in circumstances was not long-lasting. For in 1647, the year

[22] Schickard, 1630: 5 f., 17 f. On Heberle, see below, fn. 25.

[23] Schickard, 1630: 17. On the concept of the "preternatural," cf. Daston & Park, 1998: passim.

in which the chronicle ends, Bürster only knows of four clergymen in the abbey, including himself (Weech, 1875: 262).

The act of interpreting the divine sign represented a justification for Bürster's writing. Yet Bürster was less concerned with justifying his writing practice than with the fact that his chronicle was incomplete. If Bürster had wanted to describe, as he says, "the many errant knaves" who "passed by the monastery," "all their wicked antics," he "could not have found or obtained enough time, pen and ink or paper."[24] But none of this could stop the monk from taking up his pen. Bürster notes that, after all, what would future generations say if they wanted to remember the once famous and prosperous abbey, now impoverished and almost completely abandoned, and if they found nothing in writing? (Weech, 1875: 2 f.).

Bürster, who was one of the very few who remained in the monastery until the end of the war, transformed the disorder of violence into the order of *memoria*. He remembered the disturbance, drew the necessary lesson from it, and thus made his contribution to healing the wound. When he began to write in the evenings, he did not present himself as "lazy," but thought he could call himself "hard-working." (Weech, 1875: 2 f.). And this was fortunate, for he only ceased writing when he believed the disaster would soon come to an end; accordingly, his record grew ever longer. In 1643, he put his previous notes into a fair copy and then continued writing the chronicle (Weech, 1875: 3, 13). The result was nearly four hundred handwritten pages.

The conflict described by Bürster was referred to in retrospect by the Kitzingen pastor Bartholomäus Dietwar as the "Swedish war." Dietwar, as well, attributed the outbreak of the war to the "terrible sign that took the form of fire." (Dietwar, 1887: 61 f.). Unlike the winter comet, the *prodigium* that heralded this event was not cryptic. It left little doubt as to what it actually meant; the scope of its interpretation was much more clearly limited. What is more: For authors such as Bürster and Dietwar, what they saw could only be grasped as a natural event in the allegorical sense of a battlefield, as they did not yet have a concept of the phenomenon itself. Here, the interpretation did not *follow*, but went hand in hand with the observation of the phenomenon. The aurora borealis thus provided more concrete and specific expression to the comet of 1618, which had only remotely resembled a bloody sword and a fiery rod of punishment. The aurora borealis 'fleshed out' the omen of the comet, so to speak. The war that it prophesied was already depicted in it, for it was no longer just the future, but—much more so than after the Defenestration of Prague—already in the present and in history.

Naturally, Bürster did not welcome King Gustavus Adolphus's intervention into the war. But even Heberle, who was confessionally on the opposing side, soon realized that what the eagerly awaited Swedes brought was nothing different from what came under imperial Catholics: war and violence. Friends, it would seem, are sometimes a larger liability than enemies. Heberle had interpreted an earlier sighting of

[24] Weech, 1875: 1 f. With this expression, Bürster refers to the topoi of indescribability and ineffability which were prevalent in the seventeenth century. Cf. Bähr, 2013a: section 5.2; Benthien, 2006: chapter IV.

the aurora borealis in August of 1629 as a sign of hope: "a great army of war" that was led by "a mighty horseman with a drawn, shining sword"; this was the "high and praiseworthy king in Sweden" and had "meant that God wanted to save his people from tribulation [...] and protect and redeem them through the horseman with the sword." However, 6 months later, Herberle realized that far from being a sign of salvation, this heavenly host was a messenger of doom.[25]

10.4 1648: Memories of the Comet of 1618

By 1630, the celestial event of 1618 was becoming a distant memory—and a proven prophecy. The comet thus furnished a historical argument for an ominous interpretation of the aurora borealis of 1630. And in this case, as well, the predictive power of heavenly events was to be cruelly verified over the next 18 years. By 1648 at the very latest, it was clear that the aurora borealis represented a warning that had been persistently ignored.

However, something else became clear in 1648. With the signing of the Treaty of Westphalia, many people recalled the comet of 1618, yet not only because the comet had heralded the start of the war (as the Alsatian pewterer Augustin Güntzer states in his *Kleines Biechlin von meinem gantzen Leben*),[26] but also because of how long the comet had been visible. The mayor of Jessnitz, Joachim Rese, noted in his diary that the "broom of fiery sparks" had been visible in the sky for "thirty days"—likewise, "the war in Germany had lasted 30 years, such that every day meant a year, as experience has unfortunately shown." (Paasch, 1998: 12). The symbolic meaning of the comet was thus even greater than initially perceived, for even the total duration of the war had been announced by the comet's passing.

Or should one have made an alternative deduction? Andreas Kothe, a town councilman in Wiedenbrück, selected a telling formulation in his chronicle. The comet was visible for "30 nights," he writes, adding: "It is easy to see why: because the German War in the Holy Roman Empire lasted 30 years." (Flaskamp, 1962: 25 f.). Accordingly, for Kothe, the comet did not prefigure the duration of war. Quite to the contrary: it was the duration of the war that dictated the comet's visibility. In this way, we are not only offered a retrospective interpretation of the sign, but also—at least in part—its retrospective invention.[27]

What is the significance of this? It means that the writing of history as a practice of interpreting omens could have a retrospective effect on past signs, as soon as the exegesis was complete. Indeed, it was the comet of 1618 that led the Thirty Years'

[25] Zillhardt, 1975: 126–128 (fols. 35v–37r), the quotation is on pp. 126 f. (fols. 35v–36r).

[26] Güntzer, 2002: the drawing pasted into the book between fols. 64v and 63[a]r. Furthermore, see 200r.

[27] Further examples are: Anon, 1648; Winckelmann, 1671: 113; Ziegler und Kliphausen, 1695: 242; Crusius, 1730: 89.

War to be named as such.[28] In 1648, after the signing of the Treaty of Westphalia, it was by no means certain that peace was assured. Nevertheless, in 1648, many people recalled the winter comet of 1618. And in doing so, they often expressed their hope and conviction that the violence and hardship had come to an end. To be sure, celebration did not ensue in certain quarters until the Nuremberg Peace Banquet of 1650, when demobilization was finally effected (which technically made the hostilities a "thirty-two years' war"). Nevertheless, the comet of 1618 marked not only the beginning of the Thirty Years' War, but also its end.[29]

Contrary to Georg Schmidt's study, which is heavily influenced by modernization and secularization theory (2018: 689 f.), peace was not achieved thanks to the repression of apocalyptic thinking; it was not the violent experience of the war that 'disempowered' the comet. Rather, it was in fact through the apocalyptic interpretative lens that the Westphalian treaty could be made comprehensible to contemporaries—namely, as an act of God, and not of man.[30] Indeed, in 1648, apocalyptic thinking did not come to an end, but was rather confirmed. To be sure, the Westphalian Treaty forbid an appealing to the Revelation of John as a justification for using military force against a confessional opponent. However, this legitimization practice at no point asserted a right to "tamper with the handiwork" of a God who punishes man with war (Schmidt, 2018: 18, 690)—for this was an argument waged by those who suffered violence, but not by those who perpetrated it. Here, we are not presented with an "inability to consider alternatives" or a "mass formation psychosis" (Schmidt, 2018: 169, 172, 690) that would have all too readily used the supposed inevitability of war as a justification for taking up arms (and which would only have been overcome with the Enlightenment's expulsion of divine powers from politics). As a rule, the apocalyptic exegesis of the comet did not focus on a lack of freedom, but rather aimed at free, future-oriented action that corresponded to God's will. It was marked not by a paralyzing fear, but rather by a right, repentant fear of the Lord. The historical significance of this fact cannot be grasped in modern psychological categories, but only conceptually and analytically, through an Augustinian differentiation. This exegesis called for a "filial" fear of God: not the "servile" fear of divine punishment *(timor servilis)*, but instead the fear of sin that would reap punishment *(timor filialis)*, as a condition for the grace of peace.[31]

---

[28] Konrad Repgen was the first to note that the appellation "Thirty Years' War" was coined immediately after the conclusion of the peace; see Repgen, 1982a, b, 1988: 6 f. and appendix I. However, Repgen explains the emergence of this name with reference to the Defenestration of Prague, and not the winter comet. Benigna von Krusenstjern and Hans Medick have pointed out that the notion of a "Thirty Years' War" can be found not only in printed works, but also in personal narratives: Medick & Krusenstjern, 2001: 30 f.

[29] For details and further references, see Bähr, 2017: 191–194. Cf. also: Anon, 1649. On this work, see van de Löcht, 2019: 103.

[30] Cf. Kaufmann, 1998: chapter IV.

[31] On filial and servile fear of God in Dieterich's work, see Dieterich, 1619: 15, 32–34, 44–52. On these semantics, cf. Bähr, 2013a, b, in particular sect. 3.2.

During the war, some may have gratefully recognized the grace of God in the victories of their own side (including many theologians), but those who endowed their generals with divine attributes (such as in the apotheosis of Gustavus Adolphus), those who not only trusted in God's weapons but thought they could take them into their own hands, had a difficult time in the Lutheran church (Kaufmann, 1998: 24–77). Most comet exegetes, therefore, did not see the sky as an invitation to punish the enemy, but as an indication of their own punishability. In their view, those who believed they could clearly recognize divine will and had a right to execute it, expressed—as strange as the thought may seem—a false fear of divine sanctions, for they sought to avoid the *sanction* and not its cause. For this reason, whoever pointed to the comet of 1618 as a prophetic justification of violence—as in Faulhaber's case—was marginalized by the ecclesiastical and political authorities.

In this way, the achievement of peace did not require the comet to be freed from the supernatural. The Thirty Years' War was not a Hegelian trick of reason; it was not the beginning or catalyst of a happy overcoming of a "static biblical order", as Schmidt has it (2018: 645). Indeed, the conclusion reached at the end of the war was not that the interpreters of comets had been wrong, but rather that they had been right.

For the uncertainty—and not the certainty—surrounding the comet's significance was constitutive to its interpretation as an element in salvation. Only retrospectively, at the moment of peace, could the past become in the eyes of the chroniclers an inevitable event, whose thirty-year duration was predetermined for some from the outset. In 1618 and 1619, by contrast, the events in the sky were generally understood as a divine warning. They set in motion a historiographical practice whose task was to call for repentance and conversion—an exhortation directed at those affected by the war and its violent protagonists. This historiography warned the world to heed the signs and avert the impending danger. It did not aim to fuel the war, but rather to extinguish the conflagration, if possible.

This makes clear that writing about the war was a way of living with it; it represented an attempt to ensure survival and preserve oneself. Celestial events initiated a historiography that not only remembered the war, but also recalled the wondrous omens that threatened its outbreak. This historiography told stories of war as well as stories of celestial phenomena, because earthly and heavenly events explained each other. In collecting the signs, this historiography united past and future under the aspect of eternity—that is, *sub specie aeternitatis*. It filled an exegetical interim moment between the occurrence of an omen and the event which it foretold. In this way, it opened up its own possibilities for action in a world in which man, nature, and God were still intertwined.

References

Abelinus, J. P. (1635). *Theatrvm Europævm, Oder/ Außführliche vnd Warhafftige Beschreibung aller vnd jeder denckwürdiger Geschichten/ so sich hin vnd wider in der Welt/ fürnämlich aber in Europa/ vnd Teutschen Landen/ so wol im Religion- als Prophan-Wesen/ vom Jahr Christi 1617. bis auff das Jahr 1629. [...] begeben vnd zugetragen haben/ etc., depicted by Johann Philipp Abelinus and published by Matthäus Merian*. Merian.

Anon. (1648). *Von dem Dreyssigjährigen Teutschen-Kriege/welcher sich Anno 1618. angefangen/ vnd durch GOttes Gnade Anno 1648. geendiget hat; Eben als wann Gott durch den 1618. erschröcklich leuchtenden Cometen/ welcher dreyssig Tage über Europam gesehen worden/ einen dreyssigjährigen Krieg verkündiget hätte* [...]. No publisher given.

Anon. (1649). *Operation Vnd Würckung/ deß Anno 1618. Erschinen[en] Comet-Sterns/ Das ist: Kurtze Verzeichnuß aller Denckwürdigen Schlachten/ Scharmützel vnd rencontri: So sich von Anno 1618, biß Anno 1648. vnd also jnnerhalb 30. Jahren/ in dem Römischen Reich Teutscher Nation zugetragen haben* [...]. No publisher given.

Bähr, A. (2013a). *Furcht und Furchtlosigkeit. Göttliche Gewalt und Selbstkonstitution im 17. Jahrhundert*. V&R unipress.

Bähr, A. (2013b). Remembering fear: The fear of violence and the violence of fear in seventeenth-century war memories. In E. Kuijpers, J. Pollmann, J. Müller, & J. van der Steen (Eds.), *Memory before modernity: Practices of memory in early modern Europe*. Brill.

Bähr, A. (2017). *Der grausame Komet. Himmelszeichen und Weltgeschehen im Dreißigjährigen Krieg*. Rowohlt.

Bähr, A. (2024). Den Krieg erzählen. Geschichtsschreibung im Zeichen von Prodigien und Wundern. In A. Ackermann, M. Meumann, J. Schmidt-Funke, & S. Westphal (Eds.), *Mitten in Deutschland, mitten im Krieg. Bewältigungspraktiken und Handlungsoptionen im Dreißigjährigen Krieg* (pp. 29–51). De Gruyter Oldenbourg.

Barnes, R. B. (2016). *Astrology and reformation*. Oxford University Press.

Benthien, C. (2006). *Barockes Schweigen. Rhetorik und Performativität des Sprachlosen im 17. Jahrhundert*. Wilhelm Fink.

Beyer, J. (2017). *Lay Prophets in Lutheran Europe (c. 1550–1700)*. Brill.

Böning, H. (2019). Dreißigjähriger Krieg und Öffentlichkeit. Zeitungsberichterstattung als Erste [sic] Rohfassung der Geschichtsschreibung. *Daphnis, 47*, 25–47.

Briese, O. (1998). *Die Macht der Metaphern: Blitz, Erdbeben und Kometen im Gefüge der Aufklärung*. Metzler.

Burckhardt, A. (1618). *Cometen Predigt Als in diesem 1618. Jahre im Novembr: vnd Decembr: ein sehr grosser und erschrecklicher Comet am Himmel/ im Morgen stehend gesehen ward [...]*. Schmidt and Francke.

Burke, P. (2011). Exemplarity and anti-exemplarity in early modern Europe. In A. Lianeri (Ed.), *The Western time of ancient history: Historiographical encounters with the Greek and Roman pasts* (pp. 48–59). Cambridge University Press.

Christiani, C. (1619). *Κομητηστοιχεῖον Hoc est, Synopticus De Cometis Discursus Theologicus [...]*. Betzel.

Claudianus, C. (1985). Bellvm Geticvm. In J. B. Hall (Ed.), *Clavdii Clavdiani Carmina* (pp. 240–262). Teubner.

Crusius, C. (1730). *Der Nieder-Lausitzische Methusalah, d.i. Denck- und Glaubwürdige Lebens-Beschreibung Eines Mannes, welcher zu Drehna unweit Luckau in der Nieder-Lausitz hundert und siebenzehn Jahr alt worden, wie solches in einem Aufferbaulichen Gespräch zwischen dem 147 Jährigen Ertz-Vater Jacob Und den 117 Jahr alt gewordenen Martin Kaschken [...] ans Licht gestellet hat Christoph Crusius*. Höhme.

Cysat, J. B. (1619). *Mathemata Astronomica de loco, motv, magnitvdine, et cavsis cometæ qvi svb finem anni 1618. et initivm anno 1619. in coelo fvlsit; Ex aßiduis legitimisque variorum Phænomenorum obseruationibus deriuata*. Angermaria.

Daston, L., & Park, K. (1998). *Wonders and the order of nature, 1150–1750*. Zone Books.

Dieterich, C. (1619). *Vlmische Cometen Predigte/ Von dem Cometen/ so nechst abgewichenen 1618. Jahrs im Wintermonat sich erstenmahls in Schwaben sehen lassen/ Darinn nachfolgende drey Puncten gehandlet werden/ 1. Was Cometen seyen. 2. Was sie bedeuten. 3. Was vns gegen deren Bedeutung vorzunehmen. Gehalten zu Vlm im Münster den 2. Sonntag des Advents* [...]. Meder.

Dieterich, C. (1625a). *Philosophischer vnd Theologischer Traum Discurß/ Von den Nächtlichen Träumen: Darinnen Bericht geschiehet/ I. Was nächtliche Träume seyn. II. Woher sie insgemein kommen. III. Wie mancherley dieselbige seyn. IV. Was von ihnen zu halten. V. Wie deren allerseyts recht Christlich zugebrauchen* [...]. Unckel.

Dieterich, C. (1625b). *Vlmische Glocken Predigt/ Darinn von erster Erfindung/ rechtem Brauch vnd Mißbrauch der Glocken inn der Kirchen Gottes/ von Anfang deren biß hiehero kurtzlich discurriert vnd gehandlet wirdt [...]*. Meder.

Dieterich, C. (1636). *Ulmische Dancksagungspredig[t] Wegen gnädiger Abwendung der grausamen geschwinden Seuche der Pest [...]*. Meder.

Dietwar, B. (1887). *Leben eines evangelischen Pfarrers im früheren markgräflichen Amte Kitzingen von 1592–1670, von ihm selbst erzählt. Zugleich ein Beitrag zur Geschichte des 30jährigen Krieges in Franken* (V. Wirth, Ed.). Stahel.

Duchhardt, H. (2017). *Der Weg in die Katastrophe des Dreißigjährigen Krieges. Die Krisendekade 1608–1618*. Piper.

Eco, U. (2009). *The infinity of lists: From Homer to Joyce*. MacLehose.

Flaskamp, F. (Ed.). (1962). *Die Chronik des Ratsherrn Andreas Kothe. Eine Quelle zur westfälischen Geschichte im Zeitalter des Dreißigjährigen Krieges*. Flöttmann.

Friedrich, M., & Schilling, J. (Eds.). (2019). *Praktiken frühneuzeitlicher Historiographie*. De Gruyter Oldenbourg.

Garth, H. (1619). *Pragerische Cometen Predigt/ Von den Cometen: Wofür sie anzusehen vnd zu halten: Was sie bedeuten vnd mit sich bringen: Vnd wie man sich dargegen erzeigen vnd verhalten solle: Bey Erscheinung des grossen Langstrahlichten vnd schweiffigen Cometen, im Novemb. vnd Decemb. nechst abgewichenen 1618. Jahrs* [...]. Hoffmann.

Gindhart, M. (2006). *Das Kometenjahr 1618. Antikes und zeitgenössisches Wissen in der frühneuzeitlichen Kometenliteratur des deutschsprachigen Raumes*. Reichert.

Grafton, A. (2007). *What was history? The art of history in Early Modern Europe*. Cambridge University Press.

Grick, F. (1619). *Κομητοδικαιολοπροστασία Oder Cometenbutzer/ Das ist: EIne glaubwürdige Copey Articulierter vnd rechtmässiger Klag/ deß guten/ vnschuldigen Cometen/ welcher im abgeflossenen nächst verwichenen 1618. Jahr erschienen. Wider vnd gegen N. N. wegen viel übel beygelegter Vnwarheit vnd Vnbilligkeit den 25. Januarii jetzt lauffenden Jahrs 1619. dem Gott Apollini Parnasso Durch ermeldtes Cometen wolbestellten Anwald im Rechten Johanne Procopio übergeben*. Zoanettus.

Güntzer, A. (2002). *Kleines Biechlin von meinem gantzen Leben. Die Autobiographie eines Elsässer Kannengießers aus dem 17. Jahrhundert* (F. Brändle & D. Sieber, Eds.). Böhlau.

Happe, V. (2008). Chronicon Thuringiae, ed. by H. Medick, N. Winnige, & A. Bähr. In H. Medick & N. Winnige (Eds.), *Mitteldeutsche Selbstzeugnisse aus der Zeit des Dreißigjährigen Krieges*. Retrieved March 4, 2025, from http://www.mdsz.thulb.uni-jena.de

Haug, W., & Wachinger, B. (Eds.). (1991). *Exempel und Exempelsammlung*. Niemeyer.

Herlitz, D. (1619). *Kurtzer Discvrs Vom Cometen/ vnd dreyen Sonnen/ so am ende des 1618. Jahrs erschienen sind/ wie auch von der künfftigen Conjunction oder Zusammenkunfft aller Planeten im Krebß/ Anno 1622. Vnd sonderlich hernach im Lewen/ Anno 1623. Darauff böse trawrige vnd schreckliche Enderungen vnd Verwirrungen erfolgen werden*. Landtrachtinger.

Herodotus. (2022). *Historiae* (N. G. Wilson, Ed.). Oxford University Press.

Herrenschmidt, J. (1619). *Virga vigilans. Daß ist/ Ein Einfältiger/ doch Christlicher/ vnd Geistlicher Discurs, von jetzigem/ am Himmel stehenden Cometen, darinn Schrifftlich/ vnd Trewhertzig angezeiget/ wofür Teutschland/ bey jetziger vor Augen schwebender Unruh/ selbigen ansehen vnd achten soll* [...]. Meder.

Hille, M. (2009). *Providentia Dei, Reich und Kirche. Weltbild und Stimmungsprofil altgläubiger Chronisten 1517–1618*. Vandenhoeck & Ruprecht.

Hölscher, L. (2016). *Die Entdeckung der Zukunft* (2nd ed.). Wallstein.

Jerratsch, A. (2020). *Der frühneuzeitliche Kometendiskurs im Spiegel deutschsprachiger Flugschriften*. Franz Steiner.

Kaufmann, T. (1998). *Dreißigjähriger Krieg und Westfälischer Friede. Kirchengeschichtliche Studien zur lutherischen Konfessionskultur*. Mohr Siebeck.

Kepler, J. (1941). Außführlicher Bericht/ Von dem newlich im Monat Septembri vnd Octobri diß 1607. Jahrs erschienenen Haarstern/ oder Cometen/ vnd seinen Bedeutungen. Sampt vorgehendem gantz newem vnd seltzamen/ aber wolgegründetem Discurs. Was eigentlich die Cometen seyen/ woher sie kommen/ durch wen ihre Bewegung geregieret werden/ vnd welcher gestallt sie dem menschlichen Geschlecht etwas anzudeuten haben (Halle an der Saale, 1608). In *Gesammelte Werke*, ed. by the Kepler-Kommission, by order of the Bayerische Akademie der Wissenschaften (Vol. 4, pp. 57–76). C. H. Beck.

Kepler, J. (1963). De cometis libelli tres. I. Astronomicvs, Theoremata continens de motu Cometarum, vbi Demonstratio Apparentiarum & Altitudinis Cometarum qui Annis 1607. & 1618. conspecti sunt, noua & παράδοχος. II. Physicvs, continens Physiologiam Cometarum nouam & παράδοχον. III. Astrologicvs, de significationibus Cometarum Annorum 1607. & 1618 (Augsburg, 1619). In *Gesammelte Werke*, ed. by the Kepler-Kommission, by order of the Bayerische Akademie der Wissenschaften (Vol. 8, pp. 129–262). C. H. Beck.

Kepler, J. (1993). Prognosticon, Von aller handt bedraulichen Vorbotten künfftigen Vbelstands/ in Regiments- vnd Kirchensachen/ sonderlich von Cometen vnd Erdbidem/auff das 1618. vnd 1619. Jahr (no place given, 1619). In *Gesammelte Werke*, ed. by the Kepler-Kommission, by order of the Bayerische Akademie der Wissenschaften (Vol. 11, pp. 173–188). C. H. Beck.

Koselleck, R. (1975). Geschichte, Historie. In O. Brunner, W. Conze, & R. Koselleck (Eds.), *Geschichtliche Grundbegriffe. Historisches Lexikon zur politisch-sozialen Sprache in Deutschland* (Vol. 2, pp. 593–717). Klett-Cotta.

Koselleck, R. (1976). Vergangene Zukunft der frühen Neuzeit. In R. Koselleck (Ed.), *Vergangene Zukunft. Zur Semantik geschichtlicher Zeiten* (pp. 17–37). Suhrkamp.

Krämer, F. (2014). *Ein Zentaur in London. Lektüre und Beobachtung in der frühneuzeitlichen Naturforschung*. Didymos.

Krusenstjern, B. v. (1999). Prodigienglaube und Dreißigjähriger Krieg. In H. Lehmann & A.-C. Trepp (Eds.), *Im Zeichen der Krise. Religiosität im Europa des 17. Jahrhunderts* (pp. 53–78). Vandenhoeck & Ruprecht.

Landwehr, A. (2014). *Geburt der Gegenwart. Eine Geschichte der Zeit im 17. Jahrhundert*. S. Fischer.

Laux, S. (2002). "Etwas gross" aufschreiben. Quellenkritische Anmerkungen zum "Zeytregister" des Ulmer Chronisten Hans Heberle (1597–1677). *zeitenblicke, 1(2)*. Retrieved March 4, 2025, from http://www.zeitenblicke.historicum.net/2002/02/laux/index.html

Lehmann, H. (2007). Die Kometenflugschriften des 17. Jahrhunderts als historische Quelle [1985]. In H. Lehmann (Ed.), *Transformationen der Religion in der Neuzeit. Beispiele aus der Geschichte des Protestantismus* (pp. 21–39). Vandenhoeck & Ruprecht.

Leuchter, H. (1619). *Cometa, Oder Ein Predigt von Cometen: Darinn erzehlet wirt/ 1. Was Cometen seyen/ vnd woher sie sich vervrsachen. 2. Was sie bedeuten/ vnd auff was schwere Eventus vnd Händel sie anzeig thun: Vnd dann zum 3. wie man sich Christlich/ wol vnd heylsam/ bey erscheinung derselbigen/ verhalten soll* […]. Hofmann.

Lungwitz, M. (1619). *CoMeta VerVs nobIs Iræ DIVInæ Propheta. Christliche Warnungspredigt/ Aus dem Evangelio am andern Sontage des Advents/ Luc. 21. Und es werden Zeichen geschehen: Von dem grossen/ Erschrecklichen/ Fewerbrennendem Comet Stern/ Welcher im Novembri dieses 1618. Jahres erschienen/ und noch am Himmel gesehen wird* […]. Rehefeld and Gross.

Mauelshagen, F. (1998). Illustrierte Kometenflugblätter in wahrnehmungsgeschichtlicher Perspektive. In W. Harms & M. Schilling (Eds.), *Das illustrierte Flugblatt in der Kultur der Frühen Neuzeit. Wolfenbütteler Arbeitsgespräch 1997* (pp. 101–136). Peter Lang.

Medick, H. (2007). Sondershausen als "Schindershausen". Selbstverortungen und Wahrnehmungshorizonte der Gewalt in Volkmar Happes *Chronicon Thuringiae* aus der Zeit des Dreißigjährigen Krieges. In A. Bähr, P. Burschel, & G. Jancke (Eds.), *Räume des Selbst. Selbstzeugnisforschung transkulturell*. Böhlau.

Medick, H. (2010). The Thirty Years' war as experience and memory: Contemporary perceptions of a macro-historical event. In L. Tatlock (Ed.), *Enduring loss in early modern Germany: Cross disciplinary perspectives* (pp. 25–49). Brill.

Medick, H. (2018). *Der Dreißigjährige Krieg. Zeugnisse vom Leben mit Gewalt*. Wallstein.

Medick, H., & Krusenstjern, B. v. (2001). Einleitung: Die Nähe und Ferne des Dreißigjährigen Krieges. In B. v. Krusenstjern & H. Medick (Eds.), *Zwischen Alltag und Katastrophe. Der Dreißigjährige Krieg aus der Nähe* (2nd ed., pp. 13–36). Vandenhoeck & Ruprecht.

Meid, V. (2017). *Der Dreißigjährige Krieg in der deutschen Barockliteratur*. Reclam.

Merzhäuser, A. (2002). Das 'illiterate' Ich als Historiograph der Katastrophe: Zur Konstruktion von Geschichte in Hans Heberles 'Zeytregister' (1618–1672). *zeitenblicke, 1(2)*. Retrieved March 4, 2025, from http://www.zeitenblicke.historicum.net/2002/02/merzhaeuser/index.html

Mortimer, G. (2002). *Eyewitness accounts of the Thirty Years War 1618–48*. Palgrave.

Neuheuser, W. E. (1619). *Consideratio et Enarratio Brevis de Nova Stella seu Cometa. Das ist/ Sehr wichtige Betrachtung doch in mögeligster kürtze Erklährung vber den Newen Stern oder Cometen, welcher vberall in Deutschen Landen vnd anders wo mehr/ observiert: Von mir aber vom 21. Novembris biß auff den 18. Decembris dieses 1618. Jahrs/ mehrmals ist gesehen worden* [...]. Ehafft.

Paasch, E. W. (Ed.). (1998). Tagebuch des Bürgermeisters Joachim Rese aus den Jahren 1618–1637. *Familienforschung heute; Mitteilungen der Arbeitsgemeinschaft Genealogie Magdeburg, 12*, 8–42.

Penman, L. T. I. (2019). *Hope and Heresy: The problem of chiliasm in Lutheran confessional culture, 1570–1630*. Springer.

Pohlig, M. (2007). *Zwischen Gelehrsamkeit und konfessioneller Identitätsstiftung. Lutherische Kirchen- und Universalgeschichtsschreibung 1546–1617*. Mohr Siebeck.

Pohlig, M. (2016). Was ist Heilsgeschichte? Formen und Funktionen eines Deutungsmusters in Spätmittelalter und Reformation. In L. Grenzmann, B. Hasebrink, & F. Rexroth (Eds.), *Geschichtsentwürfe und Identitätsbildung am Übergang zur Neuzeit, Vol. 1: Paradigms of personal identity* (pp. 54–77). De Gruyter Akademie Forschung.

Pomata, G., & Siraisi, N. G. (Eds.). (2005). *Historia: Empiricism and erudition in early modern Europe*. The MIT Press.

Rau, S., & Studt, B. (Eds.). (2010). *Geschichte schreiben. Ein Quellen- und Studienhandbuch zur Historiografie (ca. 1350–1750)*. Akademie.

Repgen, K. (1982a). Seit wann gibt es den Begriff 'Dreißigjähriger Krieg'? In H. Dollinger (Ed.), *Weltpolitik, Europagedanke, Regionalismus. Festschrift für Heinz Gollwitzer zum 65. Geburtstag* (pp. 59–70). Aschendorff.

Repgen, K. (1982b). Noch einmal zum Begriff 'Dreißigjähriger Krieg'. *Zeitschrift für Historische Forschung, 9*, 347–352.

Repgen, K. (1988). Über die Geschichtsschreibung des Dreißigjährigen Krieges: Begriff und Konzeption. In K. Repgen (Ed.), *Krieg und Politik 1618–1648. Europäische Probleme und Perspektiven* (pp. 1–84). Oldenbourg.

Ruchatz, J., Willer, S., & Pethes, N. (Eds.). (2007). *Das Beispiel. Epistemologie des Exemplarischen*. Kulturverlag Kadmos.

Sabean, D. W. (1984). *Power in the blood: Popular culture and village discourse in Early modern Germany*. Cambridge University Press.

Samson-Himmelstjerna, H. v. (1618). *Cometen Predigt, Das ist Christliche Vnterweisung, Wie man den Cometen (welcher sich Newlich am Himmel hat sehen lassen) soll betrachten: Auch christliche Vermahnung zu wahrer Buss vnd Bekehrung zu Gott* [...]. No publisher given.

Schechner Genuth, S. (1997). *Comets, popular culture, and the birth of modern cosmology*. Princeton University Press.

Schickard, W. (1630). *Beschreibung des Wunder Zaichens/ Welches Montags den 25. Januarij/ eben am Loß Tag S. Pauli Bekehrung/ dieses eingehenden 1630. Jahrs/ Abends von 7. biß zu 10. Vhr Vormittnachts/ am haitern Himmel/ gegen Nord gesehen worden* [...]. Werlin.

Schilling, M. (2019). Vergangene Zukünfte. Prophezeiungen aus den Anfangsjahren des Dreißigjährigen Krieges. *Daphnis, 47*, 68–84.

Schmidt, G. (2018). *Die Reiter der Apokalypse. Geschichte des Dreißigjährigen Krieges*. C. H. Beck.

Schmidt-Biggemann, W. (2007). *Apokalypse und Philologie. Wissensgeschichten und Weltentwürfe in der Frühen Neuzeit* (A. Hallacker & B. Bayer, Eds.). V&R unipress.

Seelbach, S. (2019a). Der Dreißigjährige Krieg: Ereignis und Narration. Einführung. *Daphnis, 47*, 2–22.

Seelbach, U. (2019b). Gottes Kalkül und (k)ein Ereignis. Der Tod Herzog Ulrichs von Holstein und der Untergang der Stadt Schweidnitz (1633). *Daphnis, 47*, 285–312.

Seifert, A. (1976). *Cognitio historica. Die Geschichte als Namengeberin der frühneuzeitlichen Empirie*. Duncker & Humblot.

Splinter, S. (2009). Vom Unglücksboten zur natürlichen Himmelserscheinung. In C. Meinel (Ed.), *Grenzgänger zwischen Himmel und Erde. Kometen in der Frühen Neuzeit* (pp. 113–120). Universitätsverlag Regensburg.

Theibault, J. (1994). Jeremiah in the village: Prophecy, preaching, pamphlets, and penance in the Thirty Years' War. *Central European History, 27*, 441–460.

Ursinus, B. (1619). *Außführlicher Bericht/ Von den Cometen/ welcher im Jahr 1618. im Novembr. erscheinen [sic]/ vnd fast biß zu Ende deß Decembris ist gesehen worden*. Guth.

van de Löcht, J. (2019). "Krieg/ groß Sterben/ vnd Hungers pein/| Bringt mit sich der Cometen schein". Publizistische Folgen der Kometen von 1607 und 1618/19 im Vergleich. *Daphnis, 47*, 85–112.

Weber, J. (1618). *Cometen Predigt: Darin von Christi Geburt her/ biß auff den jetzt Schwebenden/ die vornembsten Cometen/ und was darauff erfolget/ kürtzlich erzehlet werden/ mit angehefften/ trewhertziger Vermahnung/ zu rechtschaffener Buß und Besserung des Lebens [...]*. Birckner.

Weech, F. V. (Ed.). (1875). *Sebastian Bürster's Beschreibung des schwedischen Krieges 1630–1647*. Hirzel.

Winckelmann, J. J. (1671). *Oldenburgische Friedens- und der benachbarten Oerter Kriegs-Handlungen [...]*. Zimmer.

Ziegler und Kliphausen, H. A. v. (1695). *Täglicher Schau-Platz der Zeit/ Auff welchem sich Ein iedweder Tag durch das gantze Jahr mit seinen merckwürdigsten Begebenheiten/ so sich vom Anfange der Welt/ biß auff diese ietzige Zeiten/ an demselben zugetragen/ vorstellig machet. [...]*. Gleditsch.

Zillhardt, G. (1975). *Der Dreißigjährige Krieg in zeitgenössischer Darstellung. Hans Heberles "Zeytregister" (1618–1672). Aufzeichnungen aus dem Ulmer Territorium. Ein Beitrag zu Geschichtsschreibung und Geschichtsverständnis der Unterschichten*. Kohlhammer.

Zimmermann, C. (1999). "Wie man den Cometen [...] soll betrachten". Zwei Predigten des Jahres 1618 aus Riga und Magdeburg im Kontext der frühneuzeitlichen Kometenliteratur. In W. Kühlmann & W.-D. Müller-Jahncke (Eds.), *Iliaster. Literatur und Naturkunde in der Frühen Neuzeit* (pp. 323–344). Manutius.

Andreas Bähr is Professor of Modern European Cultural History at the European University Viadrina Frankfurt (Oder). His publications include the books *Der grausame Komet. Himmelszeichen und Weltgeschehen im Dreißigjährigen Krieg [The Cruel Comet: Celestial Signs and World Affairs in the Thirty Years' War]* (Reinbek b. Hamburg, 2017), *Furcht und Furchtlosigkeit. Göttliche Gewalt und Selbstkonstitution im 17. Jahrhundert [Fear and Fearlessness: Divine Violence and Self-Constitution in the Seventeenth Century]* (Göttingen, 2013), and *Der Richter im Ich. Die Semantik der Selbsttötung in der Aufklärung [Judge within the Self: The Semantics of Self-Killing in the Enlightenment]* (Göttingen, 2002). His research interests comprise the history of religious thought, historical epistemology, the making of the modern self, and the history and theory of historiography.

Chapter 11
The Comets of 1618 and Astrological Prognostication in Seventeenth-Century England

Nicholas Campion

11.1 Introduction

In the seventeenth century, comets were thought to belong to a class of unpredictable new stars, along with meteors and what are now known as supernovas (Baily et al., 1990: 91–99), and so did not form part of the regular canon of predictable astronomical phenomena, such as planetary conjunctions.

The three comets which appeared in the Autumn of 1618 are notable in the study of astronomy and culture for two reasons. The first is the debate they stimulated on their physical nature, partly thanks to the advent of new technology in the form of the telescope (Drake & O'Malley, 2016; Yeomans, 1991: 51–68). The second is the attention paid to them in astrological texts. In this chapter I will explore two English texts concerning the third and most prominent of the three comets of 1618. The first text, by John Bainbridge, was published in 1619, within weeks of the comet's disappearance, and has been studied before (Bainbridge, 1975 [1619]). The second, by William Lilly, was published twenty-four years later, in 1644, and has never before been studied (Lilly, 1644). Yet the comets of 1618 had a huge impact on popular culture. As Roberta Olson and Jay Pasachoff (1998: 20) wrote,

> None of the six comets that followed the spectacular comet of 1618, struck terror in the hearts of Europeans until the Comet of 1665 lit up the sky… In 1665, the plague descended on the Continent, and England with a fury that had not been seen for centuries. Soon thereafter, the city of London burned for four days in the great conflagration of 1666.

In this chapter, I will review the texts available to Lilly and Bainbridge, and contextualise their work in relation to other cometary commentaries. Our understanding of seventeenth-century astrology is complicated by the fact that there was no clearly

N. Campion (✉)
University of Wales Trinity Saint David, Lampeter, UK
e-mail: n.campion@uwtsd.ac.uk

C. J. Cunningham (ed.), *Cosmic Events*, Historical & Cultural Astronomy,
https://doi.org/10.1007/978-3-031-86950-1_11

defined and universally accepted and unambiguous distinction between it and astronomy in the sense clearly understood from the eighteenth century onwards. Claudius Ptolemy, whose *Tetrabiblos*, composed around 120 CE, remained a standard textbook (Ptolemy, 1940: I.1, 3); it defined astronomy as being divided into two forms, one which dealt with the positions of the heavenly bodies, which he detailed in the *Almagest*, and the other with their effects, which he delineated in the *Tetrabiblos*. The latter is now referred to as astrology, although the word did not appear in Ptolemy's text. Another fundamental work for seventeenth-century students of what we now call astrology was Guido Bonatti's thirteenth-century volume, the *Liber Astronomiae*, in which, following Ptolemy, the practical application of astronomy included the highly technical rules for the interpretation of individual, collective and natural destinies (Bonatti, 2007). And as Patrick Curry wrote of the seventeenth century, what we now categorise as astrology could not be easily defined (Curry, quoted in Pumfrey et al., 1991: 277).

> Astrology actually never was a single entity, with a fixed identity. Formally speaking, it was a set of ideas or propositions concerning the relation of cosmic and earthly events. These ideas were themselves rather banal, even empty of experiential meaning; however, they found various different social interpretations and uses. It is these that are crucial to consider, because astrology only acquired its equally various particular meanings and significance in the course of, and in the form of, such uses.

One attempt to provide order in the breadth of claims that are now clustered under the single term astrology is a common distinction between natural and judicial astrology (Geneva, 1995: 8–9). Natural astrology was largely based around natural influences and prioritised weather prediction; it could also include general historical and political analyses. Judicial astrology required the making of a judgement, usually by interpreting the complex technical rules of horoscope interpretation as codified in a horoscope. In practice, the distinction is not absolutely clear cut, for it is not necessary for an astrologer to cast a horoscope to make a judgement; indeed, the whole Ptolemaic prospectus was that horoscope reading, or judicial astrology, was rationalised by the heavenly bodies' natural effects. Ann Geneva herself, in her book *Astrology and the Seventeenth Century Mind* (1995: 75–150), devoted a whole chapter to the interpretation of natural phenomena as portents. Nevertheless the distinction between natural and judicial forms of astrology is a useful one, especially for the understanding of debates around the nature of comets. While the existence of natural effects was universally accepted, the use of horoscopes was subject to much criticism on both evidential grounds (that astrologers made incorrect predictions) and religious principles (that making predictions ran the risk of pre-empting God's sole ability to know the future) (Mahlmann-Bauer, quoted in Omodeo & Wels, 2019: 123–86).

Seventeenth century European astrology was a complex strand of different claims and practices, sometimes distinct but often overlapping—none of the varieties of astrology which existed in the seventeenth-century are mutually exclusive. Astrology in 1618 may be defined as what Robert Peters (1991: 95–96) called as a 'concept cluster': a single term that conflates multiple meanings (Weeratunge, 2000: 251). It is also not dissimilar to Curry's set of ideas or propositions. One example of a concept cluster is nature, in that different cultures have different concepts of what it is to be natural—or might not even have a word equivalent to the English 'nature'—but we still use the one word (Weeratunge, 2000: 252). Similarly, there were and are multiple concepts of what astrology is, whether as the interpretation of divine providence or natural influences, and whether capable of making accurate predications or not (Campion, 2009: ix–x).

Patrick Curry (quoted in Boyd, 1995: Vol 1, 55) has provided us with a simple definition: 'Astrology is the practice of relating the heavenly bodies to lives and events on earth, and the tradition that has thus been generated'. In other words, there is no standard formula. Astrology could be rule-based and logical, as in the variety practiced by Galileo (Campion and Kollerstrom, 2003/2004). Or it could aspire to be completely naturalistic and based on empirical observation, as proposed by Johannes Kepler (1601, in Rossi & Brackenbridge, 1979: 85–163). It could be highly technical and based in complex rules, as set out by Claudius Ptolemy (1940) and practiced by William Lilly (1985 [1647]). It could rely on the harmonisation of the soul with the stars, as proposed by Marsilio Ficino (1989). Harmony was, of course, a central feature of Platonic cosmology. Commenting on the prevalence of Platonic numerology in the early seventeenth-century, and especially at the Stuart Court in London, Vaughn Hart (1994: 136) wrote that, 'Renaissance Neoplatonism, in its various forms, was consistent in spreading the conception that the harmonic structure of the cosmos and man's ability to tune himself to it through art, depended on number'.

> Who has measured the waters in the hollow of his hand?
> and marked off the heavens with a span,
> enclosed the dust of the Earth in a measure.
> and weighed the mountains in scales.
> and the hills in a balance? (Isaiah 40: 12)

Hart (1994: 136) continues, 'Specific numbers in themselves defined qualities, three the Trinity, four the elements, seven the planets, twelve heavenly temples or signs of the zodiac, and so on'. The universe, in other words, was one of potential mathematical perfection. In that comets appeared not to conform to this perfection, by appearing at unpredictable moments, they provided a puzzle.

Further, any distinction we now make between divine providence and natural history is anachronistic as applied to 1618, a time when it was universally believed God speaks through astronomical motions. The Portuguese Manuel Bocarro Francês (c. 1593–c. 1662) was concerned with divine signs rather than judicial astrology, but he also considered the eccentricity of the solar orbit (from a geocentric perspective) and Saturn's motions Carolina & Camenietzki, 2006: 37). Boccaro criticised Aristotle's natural philosophy in favour of the Stoic tradition and played a part in the early modern revival of Stoicism, which may have had a role in the emergence of new cosmologies (Carolino, 2012: 196, 198). The point is that neither astronomy nor astrology existed as discrete categories, and the divine cosmos was a mathematical cosmos, including all that could be measured in God's benevolent creation. As Doris Gruber (2021: 174) wrote, the process of knowledge change is 'multi-layered', paradoxical, sometimes results in dead ends and contradicts narratives of linear scientific progress.

11.2 Cometary Theory

Comets' dual role as heralds of either disaster or potentially positive changes was noted by Olson and Pasachoff (1998: 20), who commented that comets, along with meteors, had been viewed as what they termed 'ambiguously symbolic importance'. The ambiguity they observed was embodied in the distinction between auspicious and inauspicious significance. As they concluded, 'On the one hand, they [comets] ordered auspicious fortune, as well as births of rulers and important persons. On the other, they were omens of disasters (literally, "bad stars"), calamities, and deaths of rulers or important persons' (Olson & Pasachoff, 1998: 21).

As applied to cometary theory, Tabitta van Nouhuys (1998: 395) identifies two varieties. One, the teratological view, consists of 'astrology, or prediction of the heavenly bodies' as 'just one of many possible techniques of practising divination'. The other, the astrological view, prioritised the use of the heavenly bodies in order to interpret the past, manage the present, and predict the future as a distinct practice, often rationalised by natural causes. Complexity and nuance being the essence of the distinction between different approaches to astrology, there is no clear philosophical or technical boundary between the two. The contrast between van Nouhuys' teratological and astrological models is evident in Isidore's *Etymologies,* published in the mid-seventh century CE (2007: III.xxvii). He first of all divided astrology into two, one natural and legitimate and one superstitious and illegitimate. He argued that astrology is

> natural as long as it investigates the causes of the Sun and the moon, all the specific positions of the stars, according to the seasons; but it is a superstitious belief that the astrologers (*mathematicus*) follow when they practise augury by the stars, or when they associate the twelve signs of the zodiac, with specific parts of the soul or body, or when they attempt to predict the nativities and characters of people by the motion of the stars.

For Isidore, superstitious astrology included predictions made from horoscopes. He then listed different kinds of divination, including horoscopic astrology, all of which he considered contrary to Christian practice (Isidore of Seville, 2007: III. xxvii). Van Nouhuys' teratological model of cometary significance is based on all those forms of knowledge-seeking deemed unacceptable by Isidore but, while her astrological model is based in Isidore's. As applied to seventeenth-century cometary theory, the teratological view emphasised comets as divine signs, while the astrological held that they were parts of a chain of natural causes connected to the planets as well as terrestrial events such as weather and earthquakes (Van Nouhuys, 1998: 575).

11.3 Intellectual Contexts

Astrology in early seventeenth-century Europe was coming to the end of a five-hundred year period in which it occupied a distinguished, although not uncontested, position in the European intellectual world. By the end of the century, around 1700, it had almost entirely have lost its credibility, surviving only in popular culture and the production of annual and monthly almanacs. Stillman Drake and C.D. O'Malley (2016: ix) see the debates over the comets of 1618 as crystalising this change in fortunes and, in their view,

> If a scientific revolution took place about 1600—and hardly anyone questions the fact, though few still support the older view that it came unheralded—it can be explained satisfactorily only in terms of some novel element, which intervened about that time. Such an element is found in the emergence of scientific method as a road to truth.
>
> The controversy, which sprang from the comets of 1618, enables us to see in strong contrast the spirit of traditional philosophy and the spirit of modern science—the former with its emphasis upon authority, dogma, and verbal exegisis, the latter, with its stress on observation, skepticism, and mathematical analysis.

Thus, they attribute certain characteristics, such as reliance on authority, to the old astronomy, and others, such as a reliance on observation, to the new. The comets certainly produced vigorous debate as to their physical nature, but how far this played a significant role in accelerating the development of secular, materialist, empirical, observational and experimental science is open to question (see Van Nouhuys, 1998: 15–88).

Luis Carolino and Carlos Camenietzki document the extent of the debate and concerning explanations for and interpretations of comets in the late sixteenth and early seventeenth century, including those of 1618: comets could be signs from God, may be caused by exhalations from the Earth extracted by different planets, or related to planetary conjunctions—or all three. The use of astrology is framed within divine agency, even when the focus of astrological texts is on technical analysis (Carolino & Camenietzki, 2006: 51). Astrologers saw the influence of comets as expressed through three key observations: motion, light and influence (Carolino & Camenietzki, 2006: 41). Duration was also significant: the longer a comet was observed, the longer its influence lasted (Carolino & Camenietzki, 2006: 45). Cometary colour was significant, too. For example, a silvery comet had the nature of the Moon (Carolino & Camenietzki, 2006: 45). There could be seven types of comet, each embodying different a planetary nature (Carolino & Camenietzki, 2006: 44).

A difficulty with the view that the 1618 comets represented an unambiguous break with the past is that two of the key figures in the discussions about the comets' nature were Galileo and Johannes Kepler, both enthusiastic, although by no means uncritical, practicing astrologers (Campion & Kollerstrom, 2003; Greenbaum, 2010). And, for Kepler, there was no point in examining comets' physical nature except as their role in God's creation. As Eric Jorink (2010: 109–80) argued, there was a wider debate on the role of God in nature because there was no concept of a nature without God, and therefore no concept of science without God.

The idea that there was some emergence of scientific method as a road to truth is actually rather complex. In relation to England, for example, Patrick Curry (2024 [1989]) argues that intellectual shifts in the early seventeenth-century were driven more by class and political tensions which were focused on the build-up to, and aftermath of, the trauma of the civil war and republic of 1642–60. He explores the struggles for hegemony between religious divines and natural philosophers, draws out the complex and nuanced positions of each, and makes the concept of some idea of the pure pursuit of truth difficult to sustain. Other readings of intellectual changes at the time include Sarah Schechner Genuth's view (1992: 215–17, 1997), influenced by Curry, that the elites distanced themselves from popular culture, or Tabitta van Nouhuys' (1998: 20–24) identification of what she calls the 'cessation of dialogue', in which attitudes to classical authors changed.

Summarising Schechner and Nouhuys, Dario Tessicini (2013: 59) wrote about the new 'historical insight that astrology and meteorology served as intellectual areas, in which a relevant part of the debate on comets and other celestial novelties took place in the late 16th and 17th centuries'.

11.4 Political Contexts in 1618

In Germany, the comet was different for two reasons, a point stressed by John Theibault (1994: 442). First of these was that 'it arrived at the peak of Europe of apocalyptic speculation that had been building since Luther's death' in 1546 and, second, it coincided with 'a dramatic rise in political tensions within the Holy Roman Empire', tensions that culminated in the Thirty Years War. He points out that more than 100 pamphlets on the comet were produced in Germany, 'most of which… foretold of evils to come' (Theibault, 1994: 441). Of course, the comet of 1618 was not the first astronomical event to produce worried speculation—there had been comets, planetary conjunctions, and new stars in 1585, 1593, 1596, 1604, and 1607, each of which produced its own spate of pamphlets warning of God's anger (Theibault, 1994: 441). Theibault stresses that, because of the Thirty Years War, unlike previous warnings, the 1618 comet was to become associated with the horror of the war: for the first time, the danger traditionally described to such signs actually came to pass on a scale that seemed to match the warning. In terms of what the Thirty Years War meant to the common people, being framed within apocalyptic literature, Theibault (1994: 443) argues that it would have achieved a wide significance through pastoral and confessional works which dealt with immediate experience rather than, say, grand prophecies of the last judgment. He continues that it was not the vicissitudes of battles of the Imperial and Protestant sides that mattered to ordinary people as much as the devastation of the towns and countryside.

Yet 1618 itself came in the middle of a relatively quiet period in English domestic history, following the turbulence of the Reformation (Coward & Gaunt, 2017: 153–62). In the previous century there had been a long period of religious and political tension following Henry VIII's separation of the English church from the authority of Rome in 1531–36. Henry's successor, Edward VI (1547–1553) encouraged a shift to Protestant theology in the new Church of England, a move briefly reversed by his Roman Catholic successor, Mary I (1553–58). Mary's sister and successor, Elizabeth I (1558–1603), inherited a deeply unstable situation in which at least three religious factions competed for power. On the one side were the remaining Roman Catholics, on the other a range of Protestants influenced by Martin Luther and John Calvin, who had varying views of the relationship between

Christianity and the monarchy (Lutherans being generally supportive of monarchical authority, Calvinists often suspicious) but were united in hostility to Rome. In the middle were compromisers who sought a middle way between rival theologies, and whose main goal was to maintain a stable state, reinforced by the religious authority of the monarch as head of the Church of England. The power of the state was prioritised over sectional theological concerns: the supporters of this position in France were known as the politiques (Claussen, 2021). Tensions in English relations with other states were stoked by hostility from Roman Catholic regimes, especially Spain, and conflicts such as the French wars of religion from 1562–98. Elizabeth was highly successful in managing to create a balance between competing forces in England and avoiding conflict abroad, especially after the defeat of the attempted Spanish invasion in 1588.

By a remarkable twist of fate, Elizabeth I's death in 1603 resulted in the king of Scotland, James VI, ascending the throne of England as James I, uniting the two crowns, although the states remained separate. Barry Coward and Peter Gaunt summed the period up: 'Compared to the period, 1603–10, during the second decade of the century, James I's reign seems on the surface less eventful and rather quieter', although they add the caveat that 'this may be a false impression, created by the almost total absence of parliament, and thus of rich parliamentary sources' (Coward & Gaunt, 2017: 162). In other words, the opposition did not have a voice in government, as the king ruled alone from 1610–21. This exacerbated tensions with the Puritans, who constituted a substantial political force and were uncomfortable with the surviving features of Roman Catholic ritual and dogma in the post-Reformation protestant Church of England, of which the king was the head. It also created tensions between the crown and the gentry who had benefited from the break-up of the Roman Catholic church's vast land holdings (around a third of the country) and were represented in Parliament's lower house, the House of Commons. It also made it difficult for the king to raise income from taxes without Parliamentary approval.

The ongoing religious and social tensions set in train by the break with Rome in 1531 eventually erupted in the civil war of 1642, resulting in the execution of the king, Charles I, in 1649, and the republic of 1650–1660. The monarchy was restored in 1660 but tensions were never satisfactorily resolved and the whole period reached a definitive point in the so-called Glorious Revolution of 1688 when the Roman Catholic king, James I, was deposed, and Parliament proclaimed itself supreme. Subsequent Scottish-based Roman Catholic rebellions in 1715 and 1745–6 proved unsuccessful.

The comets of 1618 therefore appeared around eighty years into a period of long-term and recurrent religious conflict and political tension. The most serious political question in 1618 was caused by the protracted negotiations over the 'Spanish Match', James' attempt to resolve long-term hostility between England and Spain by marrying his son and heir, Charles (the future Charles I), to the Spanish Infanta Maria (Turnbull, 2020: 638–60). The negotiations began in 1618 and were widely associated with the third comet, and collapsed in 1624 after massive hostility from Protestant factions. John Bainbridge's pamphlet was published in December 1618 when these negotiations were in their opening stages and it looked as if the stability established by Elizabeth and maintained by James might last. William Lilly's tract of 1644 was published two years after Elizabeth's consensus had collapsed into civil war.

11.5 Astrological Sources on Comets in 1618

In her important study of comets, Sara Schechner Genuth posed questions which are pertinent to all seventeenth century astrology: 'what skills were needed to see the rough waters ahead, and what talents were required to sharpen and decode the terrible vision reflected in what William Green described in 1680/1 as the "Looking Glasses of God's Wrath" (Green, cited in Genuth, 1997: 51). She then listed the ten key factors which could help interpret a comet's significance and meaning: colour, conjunction, zodiacal position, mundane house, direction of tail, site of nucleus, shape, motion, duration, and resemblance to comets of the past (Genuth, 1997: 51).

The available literature on the meaning of comets in 1618 was diverse and rooted in Christian, Jewish and Hellenistic literature. The scriptural concept of prodigies—unexpected events which could be read as signs from God—was universal in Christian Europe. Writing in the 1680s, the English preacher William Green (1680/1681: 22–26) announced,

> when God intends to punish a Nation, Country, or City, he is first pleased to give them notice thereof by Apparitions, or Prodigies; and there has never been any noticeable Apparition or Prodigy Seen in the heavens, but it has been attended in the Sequel with some more than ordinary Troubles.

Anna Jerratsch (2019: 191) points out that, in the religious context of the time, whether comets were signs from God on the one hand, or natural prodigies on the other, they could be seen as miraculous.

> Interpreting the comets as divine symbols is an element of a Christian theology, and a perspective on nature that took different forms according to religious confession in the aftermath of the Reformation. Instead of the Catholic miracles of the Saints, Protestants interpreted highly unusual natural phenomena as miraculous signs. Enthusiasm for prodigies was a largely Protestant phenomenon, particularly in German-speaking regions. However, the boom in literature and prodigies from the mid-sixteenth century was a European-wide event on a comprehensive scale. It occurred both in Latin treatises and vernacular writings, drew sources from ancient philosophy, as well as from contemporary occasional literature, and was equally at home in humanist-influenced elite culture and in popular folk culture.

Comets' malign influence was in early medieval literature set out by Isidore (2007: III.lxxi.16), who wrote. 'A comet (*cometa*) is a star, so named, because it spreads out the "hair" (*coma*) of its light. When this type of star appears it, signifies plague, famine, or war'. Sara Schechner Genuth (1997: 217) summarised the general view of comets prior to, and immediately after, 1618. The view she outlined was fundamentally that of Isidore:

> Until the mid-seventeenth century, comets were interpreted by both the learned elite and common folk as portents of war, murder, civil discord, sickness, poor harvest, tidal waves, and earthquakes. The roots of these beliefs lay in ancient folklore and philosophy, but the young Catholic Church readily appropriated the pagan views. The clergy preached that comets were monsters, prodigies, and miraculous signs of God's wrath.

Amongst the commentators on the 1618 comets, the Lutheran theologian Elias Ehinger's (1573–1653) view (cited in Theibault, 1994: 441) was typical:

> [Comets] generally mean war and the spilling of blood/death and death,
> and one fears that some years hence there will be great political upheavals/ with wretched wars and rebellions among the common man/
> and there will be great persecution. Great lamentation and misery will
> course through the whole world/with war, spilling of blood/robbery/
> murder and arson/ great dearth and pestilence.

However, the Church fathers, from Origen to Aquinas, had enlarged the scope of cometary significance. Genuth (1997: 217) continued:

> according to the expanded interpretation, comets did more than presage the downfall of sinners; they demarcated critical periods in the sacred history of the world and church. It was said, for instance, that comets had augured the birth of Jesus, the inception of Christianity, and the reformation of the church. it was expected that they would blaze overhead as the day of judgement approached.

The literature on the significance of comets in the seventeenth-century is rooted in a series of texts from the classical era, all of which were part of the standard educated reading of 1618. A foundational text was Aristotle's *Meteorologica*, dating to around 340 BCE, which treated cometary effects as weather predictors (Aristotle, 1937: I.vii, 29–30; Heidarzadeh, 2008, chap. 1). For example, 'when comets appear frequently and in considerable numbers, the years are, as we say, notoriously dry and windy', the dryness being a natural consequence of their fiery nature. The logic is clear: comets are visibly made of fire, and so must heat and dry the atmosphere in which they are located. Aristotle's discussion was subsequently expanded in Seneca's *Natural Questions* (2010: Book 7), written around 65 CE, shortly before his death.

A major text on cometary significance available in 1618 was Marcus Manilius' *Astronomica* (1977: xii, I. 817–79), of which the section on comets appears to have been composed before the death of Augustus in 14 CE. Manilius had repeated the naturalistic Aristotelian formula, arguing that comets are an atmospheric phenomenon caused by the igniting of the atmosphere as a result either of heat rising from the Earth or proximity to the Sun, and encouraged by the innate prevalence of fire everywhere in sky, earth and water (as evidenced by lightning, volcanoes and hot springs). It was their fiery nature which underlay their significance for terrestrial and human affairs, disrupting normal affairs as fires do. 'In times of great upheaval', Manilius (1977: I.814–7) wrote, rare ages have seen the sudden glow of flame through the clear air and comets blaze into life and perish'. For Manilius, a comet's visual appearance was generally related to people, and Manlius talks about the head, hair (streaming locks, tresses and beards), shaggy chins and distended punches as well as goats, rectangular beams and rounded columns, although such forms did not possess different meanings. In all cases their significance is dire, including crop failure, plague caused by heat, war and insurrection. Precedents included the slave rebellion of Spartacus in 73–71 BCE, the Civil War in Rome following the assassination of Julius Caesar in 44 BCE, and the massacre of the legions of Varus in the Teutoburg Forest in 9 CE. Yet, Manilius (1977: I.903–5) insists, we do have a choice into how to manage or avert such disaster, writing that 'for the fault of lies within us; we have not sense to trust heaven's message'. The comet's disruptive significance could only be temporary and, in due course, the war following Caesar's assassination came to an end with Octavian's absolute victory. Manilius (1977: I.924–9) concluded optimistically, 'Let fate content itself with this! May wars now cease and, fettered with bonds of adamant, may discord, prisoned fast, be curbed forevermore', and, alluding to Octavian's triumph and Julius Caesar's divinisation, he continued 'unconquered be the father of our father Land; may Rome serve him; and for all that she has given a god to heaven, may she miss him not on earth!'. Comets might portend dire consequences, human agency (in this case, Octavian's military genius) could alter fate, leading to a, perhaps, preferable state of affairs once the trouble has passed.

The most important work on the effects of astronomical phenomena, or astrology, in Medieval and Renaissance Europe, Claudius Ptolemy's second-century encyclopaedia, *Tetrabiblos* (1940: II.13.102, 217), also drew on Aristotele:

> Of occasional phenomena in the upper atmosphere, comets generally, foretell droughts or winds, and the larger the number of parts that are found in their heads and the greater their size, the more severe the winds.

Ptolemy (1940: II.9.90, 193) added certain other details which were to be fundamental to seventeenth-century cometary prognosis, discussing shape un less personal terms than Manilius, and adding planetary significators:

> We must observe, further, for the prediction of general conditions, the comets, which appear either at the time of the eclipse, or at any time, whatever; for instance, the so-called "beams, "trumpets," "jars," and the like, for these naturally produce the effects peculiar to Mars and Mercury– wars, hot weather, disturbed conditions, and the accompaniments of these; and they show, through the parts of the zodiac, in which their heads appear, and through the directions, in which the shapes of their tails point, the regions, upon which the misfortunes impend.

Ptolemy's relationship of comets to Mars and Mercury was entirely consistent with astrological theory, derived both from analogy and the structure of the universe. Mars' reddish tinge indicates that it is fiery in nature, while its sphere is adjacent to the Sun's (Ptolemy, 1940: I.4, 37). Mercury is always close to the Sun and is therefore hot, but its sphere is next to the Moon's and it is therefore moist, the Moon being watery on account of its connection with the tides (Ptolemy, 1940: I.4, 29 and 1.2, 7).

Another major treatment of comets was written around the same time as Seneca's by Pliny the Elder. He was more concerned with comets' portentious nature than was Seneca, and his work is significant because it argued that comets did not possess an essential inauspicious meaning. Instead, their significance changed depending on various factors. They were neither inauspicious nor auspicious, but to be interpreted in accord with various factors which could be supported by either by the evidence of past precedents or a matter of symbol, correspondence and allegory based on their appearance.

1. Shape

   Pliny (1929: II.xxii.89) defined 'Javelin stars' which 'quiver like a dart' as unambiguously 'a very terrible portent'. His example of this was the comet which appeared in the fifth consulship of the Emperor Titus in 76 CE. A comet which resembles a pair of flutes, he concluded (1929: II.xxiii.93), is 'a portent for the art of music'.
2. Position in the sky

   Cometary meanings could also be derived from their position in the sky. For example, that a comet in the western sky is 'usually a terrifying star and not easily expiated' is evidenced by the appearance of such a star during the war between Pompeii and Julius Caesar in 49 and 48 BCE, during the civil disorder in the consulship of Octavian in 43, BCE, and around the time that Nero murdered Britannicus, the heir to the Emperor Claudius, securing the succession for himself, and thereafter shining 'almost continuously during Nero's reign' (Pliny, 1929: II.xxiii.92). Ideally people would have the chance to expiate a comet's ominous significance, or make amends for their bad deeds, but in this case there was little chance.
3. Relation to the fixed stars

   If a comet forms an equilateral triangle or rectangle in relation to certain fixed stars, Pliny wrote, then 'it portends men of genius, and a revival of learning' (Pliny, 1929: II.xxiii.94).
4. Relation to the constellations

   If a comet was in the 'private parts' or genitalia' of any constellation figure, then it would signify immorality, while in the hand of the northern or southern serpent, Ophiucus, then 'it brings poisonings' (Pliny, 1929: II.xxiii.93).

It is not quite clear whether immorality is linked to genitalia in constellation images through an essential link in which they are presumed to share the same nature through a direct link, or whether the connection is analogical, but an Aristotelian universe suggests that analogy only works because such links exist. The most notable example of a comet which had a dual auspicious/inauspicious nature was the one which coincided with the assassination of Julius Caesar in 44 BCE. The appearance of a comet which shone for seven days in the northern sky during Caesar's funeral games, organised by his great-nephew and adopted son, Octavian (the future emperor Augustus) in honour of Venus, was popularly viewed as Julius Caesar's soul on its way to join the immortal gods—the stars. Octavian moved fast to place a statue of Caesar in the temple of Venus with a star above its head (Manilius, 1997: I.799–804), gave the dead dictator his own feast day, and named a month after him—July (Pliny, 1929: II, xxiii.93–94 and xxx.99; Dio Cassius, 1914–1927: xlv, 6.5–7.2; Ramsey et al., 1997). If Caesar was now a god then the implication was obvious: Octavian was the son of a god. Privately, Octavian also interpreted the comet in messianic terms as 'having been born for his own sake' and as containing his own birth within it as well as having 'a health-giving influence over the whole world' (Pliny, 1929: II, xxxiii.94).

The poet Ovid was born in 43 BCE and so would have grown up while there were still clear memories of the comet of 44 BCE. He died in 17/18 CE, five or six years before Pliny was born, and gave an allegorical rendition of the story (Ovid, 2004: lines 839–49). Jupiter was speaking:

> Meanwhile you must rescue his father's soul from his [Julius Caesar's] cut-ridden body
> Make him a comet, that deified Julius's image may always
> gaze on my Capitol Hill from the height of his shrine in the Forum.
> Scarcely had Jupiter ended his speech, when life-giving Venus
> set herself down in the heart of the Senate, though no one could see her,
> And caught the soul of her Caesar up as it passed from his body.
> She did not allow its component atoms to be dispersed
> Into air, but carried it straight as it was to the stars. In the heavens
> During her journey, she felt it glowing and catching fire,
> So she let it escape from out of her bosom and fly right upwards.
> Higher far than the moon it soared, displaying a sweeping
> trail of flame in its way, till it finally took the form,
> of a gleaming star. Now Julius watches his son's achievements.

If we take Aristotle, Seneca, Ovid and Pliny as classical authorities for the theory of comets in 1618, then it is clear that their nature was far from necessarily destructive. Instead, there was a range of meanings from the dire to the profoundly benevolent, and from destruction to renewal. Julius Caesar's death provides a model for all future associations of comets with the death of princes, but the consequence was the inauguration of the divine Augustus's glorious reign. Comets might be disruptive, but disruption can be beneficial.

A standard text on comets as prodigious events, familiar in 1618, was Josephus's *The Jewish War*, written around 75 CE, shortly after Seneca's and Pliny's texts, and which recorded the Jewish revolt against the Romans of 66–70 CE and was cited by Lilly. Josephus's account includes messianic textual oracles and unusual events in the temple, such as disembodied voices, a light shining around the altar, a cow giving birth to a lamb, and a comet which appeared in 66 CE., and which we now know was Halley's comet. As he wrote, 'First a star stood over the City, very like a broadsword, and a comet that remained the whole year' (Josephus, 1981 [1950]: 361). For Josephus the comet was a warning sign, but Lilly (1644: 31) added that the dramatic comment that it was 'terrible':

> (for as *Josephus*, a Captain, no lesse famous in those Wars, than a true Historiographer) reporteth, besides the great and fearful Eclipses preceding, there appeared a terrible Comet in the form of a flaming sword over the City, by the space of a Whole yeer together, as you may read *Lib.* 7.12, *cap.* 12, *de bello Judaica.* And surely, I doubt not, but it pleaseth God, by these heavenly Prodigies, no lesse to participate to, and communicate, and confer with men about his secret determinations, then he was wont of old by dreames and visions.

The study of comets as astronomical phenomena took two directions. One was the attempt to understand their physical nature as discreet objects, while the other was to understand their significance as part of a whole suite of events and phenomena which were considered sources of knowledge and meaning. The question was what could be done with such knowledge. Josephus (1981 [1950]): 361–63) gave two examples of mistaken interpretation, one of the light around the Temple altar, and one of a messianic prophecy, both of which were taken by the Jews as prophecies of success. The key lay in action. He wrote (Josephus, 1981 [1950]): 362),

> Anyone who ponders these things will find that God cares for mankind, and in all possible ways foreshows to His people that means of salvation, and that it is through folly and evils of their own, choosing that they come to destruction.

And yet, the future is preordained: in spite of the Jewish rebels' hopes, their rebellion was crushed, and the temple destroyed. As Josephus (1981 [1950]): 363) concluded,

> But it is not possible for men to escape from fate, even if they see it coming. The Jews interpreted some of the prophecies to suit themselves and laugh the others off, till by the fall of their city and their own destruction, their folly stood revealed.

In Josephus's account, the Jews' folly was to rebel against the Romans while making a square in the Temple in defiance of prophetic warnings. No matter how well the rebels were organised, their altering of the Temple plan constituted an offence against God for which they were duly punished. But there is another paradox which is inherent in the notion of God's plan: the Jewish rebellion was bound to fail, no matter what.

The earliest extant mention of comets in England occurred in the *Ecclesiastical History of the English People*, composed by the Venerable Bede (672/3–735) in around 731. Bede's history was designed to show the progressive triumph of the Christian church in England, and so falls into the salvational category. Bede (1990: V.23, 323) reported popular responses on the comet along with his own observations on political corelations:

> In the year of our Lord 729, two comets appeared around the sun, striking terror into all who saw them. One comet rose early and preceded the sun, while the other followed the setting sun at evening, seeming to portend awful calamity to east and west alike. Or else, since one comet was the precursor of day and the other of night, they indicated that mankind was menaced by evils at both times. They appeared in the month of January, and remained visible for about a fortnight, pointing their fiery torches northward as thought to set the welkin aflame.

He then related the corresponding events; the Saracens launched a terrible incursion into France but were turned back (this was three years before they were decisively defeated by Charles Martel at Tours), while Egbert, the bishop who, in 716, had persuaded the monks of Iona to end the dispute with Rome over the date of Easter and other doctrinal matters, died, as did king Osric of Northumbria. It is actually unclear whether Bede was merely reporting popular beliefs about the momentous import of such celestial omens, or to what extent he shared them; he mentioned the comets of 664 and 678 almost in passing, as if they were of little interest (Bede, 1968: IV.13, 226 and V.24, 334).

Two centuries later, in 837, the emperor Louis the Pious was approaching his sixtieth birthday when he asked a member of his court, known to us only as 'the Astronomer', about the significance of a comet—it was Halley's comet—that had appeared in the evening sky. Instead of being told that it heralded the death of princes (Louis died two years later), the nameless Astronomer issued the emperor with a caution against over interpreting the appearance of celestial phenomena, citing Jeremiah 10.2–3. From the English translation of 1611, just seven years before 1618, this read:

> Thus saith the Lord,
> 'Learn not the way of the nations,
> nor be dismayed at the signs of the heavens
> because the nations are dismayed at them,
> for the customs of the peoples are false'.

In an earlier chronicle, though, the 'Astronomer' had connected a lunar eclipse and an ominous group of comets in the constellation Auriga with the death of Pope Stephen II in 817 (Eastwood, 2007: 151).

That comets were not necessarily regarded omens of disaster is evident from the inclusion of the comet of 1066 in the Bayeux Tapestry, the great visual record of the Norman conquest of England: the image is accompanied by the words 'ISTI MIRANT STELLA'—'they wonder at the star' (Bayeux Museum: panel 32). The comet may have been unfortunate for the king of England, Harold II, who was killed at the battle of Hastings, but equally fortunate for the victorious Norman duke William, who replaced Harold and established his own dynasty. The Julius Caesar–Augustus formula, by which what is unfortunate for one person is fortunate for another (the death of one prince is followed by the accession of another) was repeated in the triumph of Duke William, his success being evident in the fact that he is frequently the first of the list of English monarchs to the present day, just as Augustus began the list of Roman emperors. As John Uri (2021) wrote,

> During its 1066 appearance, Halley's Comet approached within nine million miles of Earth. As it first appeared in April of that year, observers in England saw it as a bad omen, portending great change for the Anglo-Saxon kingdom ruled by King Harold, while Duke William of Normandy believed it was a positive sign from heaven.

In spite of the repetition of calamitous interpretations of cometary significance, there was nothing essential in their presumed nature to signify disaster. For the creators of the Bayeux Tapestry the comet was not a source of fear or a warning of disaster, but a source of wonder. Later, Otto of Freising (c. 1110–1158), discussed the effects of the 1066 comet, adding to the Norman conquest of England its coincidence with the excommunication of the Emperor Henry IV in 1076, the latter a traumatic event for those who believed that the Papacy and Empire were to jointly fulfil a divinely authorised destiny in bringing about the kingdom of God on Earth. Otto (cited in McGinn, 1979: 98–99) compared this event to the shattering of the statue in the book of Daniel, signifying the approach of the Last Times. The comet may have been inauspicious in the near future if it corresponded to the division between Pope and Emperor, but auspicious in the long run if it speeded the return of Jesus Christ. It was, in Olson and Pasachoff's words, ambiguously symbolic. As Sara Schechner Genuth (1997: 217) wrote,

> Religion went hand-in-hand with politics. As blazing swords, comets portended battles over belief, as well as territory. Astrology, moreover, had been a vital part of state craft since the Babylonian period, and during the Roman era cunning individuals marshalled comets as political weapons.

The spectrum of opinions concerning comets as harbingers of change—which might or might not have some fortunate consequence, to warnings of disaster—were familiar in 1618. In the sixteenth and seventeenth centuries, opinions of comets' significance were united in their expectation of disruption, but varied from the relatively calm to the overtly millennial. Tycho Brahe was of the relatively neutral persuasion, claiming that 'such unnatural births in the heavens... Always have had something great to deliver to this lower world' (Christianson, 1979: 137; Genuth, 1997: 51).

The greatest evidence for the potentially auspicious nature of new stars, including comets, was the Star of Bethlehem. John Bainbridge (1975 [1619]) cited the Star as a precedent for comets' auspicious signification in a passage which concludes with a confessional declaration of faith, a statement of salvational history.

> Yet to discend somewhat lower than Universalities: That blessed Starre, which conducted the Magi to Christs poore, but sacred nurcery (of whose incarnation, and happinesse to mankinde, thereby that Starre was an heavenly Harbinger) doth enforce me, often to think that those many new stars and Comets, which had been more this last Century of the world, then, in many ages before, did amongst other things signifie that glorious light of the Gospel, which has lately illumined the whole world.

In the eighteenth century the notion of the gospel bringing light to the world was to segue into the proposition that reason brought enlightenment (Campion, 2015: 42–44). William Lilly (1644: 31) followed suit with his own declaration of faith:

> There are a people that do too much slight the appearance and effects of Comets, as if they signified nothing, but let them consider, That God does not always speak unto men in expresse words and sillables, but many times extra ordinarily, admonishing us by the motions and apparitions of the heavenly bodies(which are the works of his owne fingers) is a sufficiently testified both by that star, which led the *Magi* in the gospel to the knowledge of our Saviour's Nativity, and by the words of Christ himself, presaging the destruction of *Jerusalem*, forewarned the *Jews*, to lift their heads to heaven, to behold the signs that should be in the *Sun*, and in the *Moon*, and in the stars, before that day.

He continued: 'God doth not onely deliver his will unto us by Prophets, Apostles, and Pastors, but sometimes in the Elements composed into several forms and Images' (Lilly, 1644: 31). And regarding the invisibility of the third comet when it was obscured by clouds, he also saw a potential divine message (Lilly, 1644: 31):

> As there was a speech many dayes, and a rumour of the appearance of a Comet, but no exact visibility thereof, so it may well be observed, that the mercy of Almighty God is infinite, that did as it were consult with himself, whether this direfull Comet should not be recalled, and add some more years of Repentance to the world.

11.6 Comets and Confessionalisation

The religious Reformation which disrupted Europe in the century following 1517 was characterised by the political and theological rejection of the political power and dogma of the Roman Catholic Church across much of northern Europe (Louthan

& Zachman, 2004). The process was both bottom-up, often driven by popular revulsion against the abuses perpetrated by the Church, and top-down, driven by rulers who saw an opportunity to curb the oppositional power of Pope and bishops (Haigh, 1993). In either case large numbers of people were required to abandon religious beliefs held by their communities for up to a thousand years, including practices which were central to eternal salvation and avoidance of the risk of eternal damnation. Private conversion was obviously one route to the adoption of new beliefs. But public statements of faith were also required in order to conform to new political realities. The terms now given to such public statements, or confessions of faith (rather than of sin), are confessionalism and confessionalisation (Morrissey, 2014, 2015).

As Mary Morrissey (2015) wrote,

> Confessionalism and conversion are misleadingly straightforward terms. At first glance, they seem to refer simply to the "outward-facing" and the "inward-facing" sides of a change of religion; confessionalism is the outward adaptation of a public religious identity…; adherence to a formulated theological system"; conversion is the inward "change of heart" that finds expression in that outward change. This dichotomy fails to capture the significance of these terms to the current historiography of the Reformation, however, and it is possible to argue that much of the most vital work in Reformation studies is concerned with the questions of identity that coalesce around notions of confessionalism, conversion and conformity.

The term confessionalisation is now applied to other shifts in religious history, as in fourteenth and fifteenth century Islam (Matthiesen, 2023). But our specific context is sixteenth and seventeenth century Europe. Across both the new Protestant territories and the old Roman Catholic ones, a huge amount of official effort, including legal codes, and unofficial enterprise, including prophetic tracts, were devoted to persuading populations to either stay with their existing faith or adopt a new one. Rienk Vermij and Paul Vieth (2022: 290) applied this model to cometary literature, using Bainbridge's work as an example:

> In the sixteenth and seventeenth centuries, the defence of comets, as omens, or its refutation, was not just an abstract, philosophical exercise, a matter of personal belief, a case of following tradition, or of implicit 'worldviews'. Such positions were propagated by powerful political and institutional forces. These forces can best be understood in the context of confessionalization.

Confessionalisation could therefore involve loyalty to political and religious authorities and the renovation of theology, but also, or even in the first place, a social renewal (Vermij & Veith, 2022: 277). The whole adds up to what Vermij and Vieth (2022: 290) call a 'confessional paradigm'. Hence, Bainbridge's condemnation of the moral evils of tobacco smoking were consistent with the confessional paradigm, as was his condemnation of the military weakness which made England vulnerable to Popish invasion. 'Portents', Vermij and Vieth (2022: 278) write, 'were instrumentalised by Church leaders, and other people who worked towards the confessionalisation of society to lend force to their messages, bring their flocks in line, and strengthen the social position of their respective churches'. Vermij and Vieth (2022: 284) conclude that,

> At first sight, the understanding of the comet as a positive sign, as showing divine approval for the English church, appears to contradict the dominant ideas of the time that all comets are signs of divine anger over people's wickedness, and as admonitions to reform. Regardless, in both cases, the comet is seen in confessional terms.

Actually, the supposition that there was a contradiction between auspicious and inauspicious interpretations of comets is an error read into the history of science, for all Biblical eschatological prophecies involve punishment for the wicked, salvation for the faithful, and an enteral kingdom of God once the great crisis has passed, as is evidenced throughout the Book of Revelation. Nevertheless, Vermij and Vieth's point is well made.

It would be wrong to regard confessionalisation as simply a top-down process, but force and persuasion was certainly part of it. The state was deeply involved, to the extent that one can speak of a 'confessional state', a virtual symbiosis of Church and state. Confessionalisation should not be seen as just a renovation of theology, but also, or even in the first place, a social renewal (Vermij & Vieth, 2022: 277).

11.7 Astrology and the Comets of 1618

In the early seventeenth century, as in the early twenty-first century, there was a gap, as well as sometimes an overlap, among mass, popular culture, and elite, intellectual culture. Stillman Drake and C.D. O'Malley (2016: vii) identified a distinction between mass and elite culture when they wrote of the three comets of 1618, 'these events inspired a large number of books and pamphlets, because comets were, at that time generally regarded with superstitious dread, and because even among the learned it was still debated whether they were atmospheric or celestial phenomena'. There is a slight anachronistic tone to Drake and O'Malley's use of the word 'even' in relation to concepts of comets as atmospheric: after all, 'will o' the wisps', lights which appear in marshy areas are a well-known phenomenon so it is reasonable to assume that lights might occur higher up in the atmosphere. However, their point is that there was a range of responses to comets depending on their role either as natural phenomena or as signs of forthcoming events, or both, for the two were by no means distinct. As Darrel Rutkin (2015: 42–54) has argued, astrology could encompass both possibilities, and the warnings it gave of future events could be justified as much but its role as natural philosophy as of divinatory of divine hermeneutics.

The comets of 1618 were significant precisely because of their coincidence with major political events at the heart of Europe. Vladimir Urbánek (2002: 283) points out that while the comet of 1577 is significant in scientific terms because Tycho Brahe's observations contributed substantially to the breakdown of the Aristotelian model of the cosmos, the comet of 1618 deserves more attention because, as he writes, it played 'an important role in the broader context of the intellectual, religious and political history of the period', coinciding with the beginning of the Thirty Years War. The first phase of the war, the revolt in Bohemia, began with the defenestration of Prague in May 1618 and developed into a wider anti-Hapsburg Protestant uprising, which was initially defeated the Battle of the White Mountain in 1620 before the fighting spread and continued until 1648.

John Theibault refers to study of Christian themes in the War, referencing Silvia Tschopp's concept of 'salvational history' in her account of pro and anti-Swedish propaganda in the Thirty Years War (Tschopp, cited in Theibault, 1994: 442).

> The New Year's sermon for 1619 by Conrad Dietrich, the metropolitan of Ulm, which was published as a pamphlet that year, was characteristic of how commentators interpreted the comet in terms of why God sent a warning and what could be done to avert the danger. His topic was "(1) What Comets are, (2) What they mean, (3) what we should do in response to their meaning." His answer was straightforward. God sent the comet as a warning because sin abounded in Germany. Picking up on themes that were a standard part of the preacher's repertoire…Dietrich urged his listeners to 'mend their ways'. The catalogue of sins that Dietrich identified as the cause of God's displeasure was unexceptional. He condemned those who disregarded God's commandments and did not live a moderate and pious life (Tschopp, cited in Theibault, 1994: 444).

The threat was clear, but the remedy equally so: God might not need to follow through on his threat to destroy his people if they heeded the celestial warning and live within the bounds of Lutheran discipline (Tschopp, cited in Theibault, 1994: 442).

David Basilius, at Prague University, followed both the Aristotelian theory of comets as the burning of terrestrial, vapours and exhalations, and Ptolemy and Cardano in interpreting their effects (Urbánek, 2002: 284). For the 1618 comet he anticipated meteorological changes, pestilence, dissension, and the imminent death of a great Prince or monarch: the comet's location in the constellation of Leo (the Lion) indicated difficulties for Bohemia in 1619, but a triumph for the Bohemian lion in 1620 (Urbánek, 2002: 285). Basilius was answered by Andreas Haberweschel, who preferred the eschatological tradition in which the model for all significant or extraordinary astronomical phenomena was the star of Bethlehem, which announced the birth of Christ, and therefore that the new comet was a sign of the second coming (Urbánek, 2002: 287).

Simeon Partlicius argued that the first comet was connected to the deaths of Queen Anne of Bohemia and the emperor Matthias in line with the start of early November (Urbánek, 2002: 290). He took a generally positive gloss on the second comet of November, and the third in the whole series, predicting that after a period of calamities, a new king of Bohemia would bring justice, a new peace and the nation would enjoy improved fortune, gaining strong allies and enhanced reputation. Partlicius also employed a twenty-four stage periodisation of history (along with great conjunctions), the final phase being due to begin in 1621 or 24 with defeat of both the Turks and the Roman antichrist, the conversion of the Jews and heathens to Christianity and an imminent end of the world (Urbánek, 2002: 291). From these examples, then, we can see that comets might have beneficial as well as difficult implications, and that even though they might have terrestrial origins this did not prevent them being part of a wider astrological, and hence celestial, interpretative system. The boundary between the terrestrial and celestial is therefore blurred if the experiential focus is on the observation of the sky.

As Urbánek (2002: 291) writes, Basilius, Partlicius and Haberweschel 'were not especially important in the history of science... my purpose has been to show that they were interesting representatives of broad intellectual currents... Basilius represented traditional, natural philosophy, Haberweschel a mystical current, and Partlicius a late Renaissance tendency to harmonise all learning in such projects as encyclopedias, or universal comprehensive chronologies'. In other words we are taking comets—and astronomy—out of the history of science and into the history of culture.

Referring to all celestial movements, including those of Mars and Venus, as well as comets, Michael Maestlin (1550–1631) wrote (Boner & Granada, 2022),

> Such things on earth, and in the water and air, give the aspects of the stars, certain radiations, by which lower elementary nature is directed in ordered. From this, a shrewd and careful astrologer can predict changes in the air and much more. I strongly doubt whether what may not be achieved in celestial matters, according to astronomical calculation from the motions, may be judged astrologically, with any degree of certainty or probability.

The continuing overlap between what were shortly to be distinguished clearly and definitively as astronomy and astrology, respectively, is evident in Johannes Kepler's work.

Kepler was, of course, Maestlin's protégé. He wrote the *Appendix to the Hyperapistes*, a commentary on the 1618 comets partly in response to Galileo's own commentary and, in 1619, the year after the comets, he published his magisterial work on the harmonic universe, *Harmonice Mundi*, which included a substantial rationale for his own reformed version of astrology (Kepler, in Drake & O'Malley, 2016: 341–55; Galilei, in Drake & O'Malley, 2016: 151–336; Kepler, 1997).

> However, that this determination of the character of the Earth's soul, which I have stated in the second place, was produced by the sensible zodiac, and indeed by the whole sphere of the fixed stars, is also confirmed by the following. At the great coincidence of all periods, it is established that, if some novelty is found in the heaven, whether in accordance with the ordinary course of the heaven, such as the rarer conjunctions of several stars, and as the more remarkable eclipses of the luminous bodies, or outside their ordinary nature, such as comets, or new fixed stars, at the same time, sublunary nature, is perturbed by unusual influences, so that there is an immense and continual violence of rain, beyond that indicated by the aspects, or on the contrary drought, and harshness, and associated earthquakes, and lastly, there is unaccustomed dampness in the air, which brings pestilential catarrhs and other epidemic plagues, particularly on those places from which they are exiled with greater force, or which they are capped by more frequent winds (Kepler, 1997: 369).

For Kepler, comets were novelties which stood outside the ordinary nature of the sky, like novas, but still had the same disruptive natural effects as major planetary conjunctions. He elaborated on his theory in his *De cometis libelli tres,* also published in 1619. In this he argued that all things, people included, possess a *facultas* which can be disturbed by phenomena such as comets (Boner, 2006: 40–45). Given that human emotions can be governed by reason and therefore can resist celestial effects, there is still a disruptive effect.

The Jesuit priest Orazio Grassi (1583–1654) viewed comets as celestial portents but his specific concern was their location, whether above or below the Moon. He then considered whether to add an explicitly astrological view, but said he lacked the necessary skills. Grassi (in Drake & O'Malley, 2016 [1619]: 10) continued, citing the Ptolemaic correspondence between comets and the planets Mercury and Mars:

> If anyone now asks me about the parents of the comet, although this is the duty, not of the astronomer, whose character, I am sustaining, but that of the astrologer, yet I, a man by no means skilled in the physiognomic art, may conjecture in so far as possible from the appearance and colour of the foetus, and I would call it the offspring of Mercury. For thus it bore the eyes, limbs, and countenance; that is, although when the comet was very near to the sun this golden Lucifer, almost outshone it, yet, when it receded far from the sun, it at once grew pale and astonishingly changeable and manifested Mercury in its countenance. Many say that in the beginning, they recognised in it the colour of Mars since it was born in Scorpio, that is, in the principal house of Mars; but an honourable exchange had occurred between Mercury and Mars, and the one then inhabited the house of the other, that is, Mercury was in Scorpio and Mars in Virgo.

When two planets are in each other's house, which in this case means zodiac signs, they are brought into a positive relationship known as 'reception'. In this case Mercury was in Scorpio. ruled by Mars. and Mars was in Virgo, ruled by Mercury. Grassi's message is that the comet is more likely to be benevolent than not. Also significant from an understanding of the astronomy of the time was that the moment of first observation is significant. He continues, leaving the responsibility to comment on such things to astrologers themselves (Grassi, in Drake & O'Malley, 2016 [1619]: 10).

> But here I exaggerate. Let astrologers consider these things more carefully, and as good midwives receive the foetus. They will consider the delineation of its countenance very attentively and, as prophetic Protei, predict for it the various vicissitudes of life, and foretel the final day of its light. I, in so far as I am able, have sought its native land, which, in agreement with all, I declared to have been Scorpio. Although, according to the view of astronomers, this region, like all others, extends from the middle of the earth to the firmament, and, although the comet first looked upon the light of the sun spread through it, yet in addition, it may be inquired whether the comet was produced in the lowest or the highest part of this house.

An astrological spin was given by Willobrord Snell (1580–1626) the discoverer of the law of optical refraction, who linked the comet's malign significance to Saturn's malefic character (Snell, cited in Heilbron, 2021 [1619]: 202):

> What is most important to me is that Saturn was exposed to [the comet's] malign and adverse rays from its first risings, and that Saturn infected it with his evil qualities; consequently, from the glare of this comet I fear strong Saturnine effects. Which evils may Almighty God by his church and our prayers most mercifully avert, I implore on my knees: because only he can do it.

It is unclear why Snell argued for a Saturnine connection, as the planet was retrograde at 5°–3° Gemini while the comet was visible, and so generally invisible at times when the comet could be seen. Moving away from astrological connections, there was a fair amount of literature which sought allegorical meaning, in the style of Ovid, as a means of making political points. One was the English writer Richard Corbett, who, according to James Doelman, lightly dismissed any connections between the comet and the Synod of Dordrecht (which was held in 13 November 1618 to 29 May 1619 to settle major doctrinal questions in the Calvinist Dutch Reformed Church), or to troubles in the French royal family, 'If the comet'[he wrote, 'is going from north to south, and now entering Charles-Wain (Ursa Major), that it might suggest something about Spanish–English negotiations': he concluded: 'brave Charles sit fast and looking into thy Wayne' (Corbett, cited in Doelman, 2007: 31–32).

Allegorical accounts were extended to comment on the king James's death in 1625. Isaac Wake, English ambassador to Savoy and Venice, wrote a verse providing an answer as to why it was not signified astrologically in the actual year (cited in Doelman, 2007: 34).

> *Question.*
> Can a king die, and we no comet see?
> Tell me, astrologers, how can this be.
>
> *Answer*
> Heaven's beacons Byrne, not but to give alarm.
> Unto a state of some ensuing harm.
> The angels, carrying up our blessed king.
> Did with still music, his sweet requiem sing.
> No innovation being to be heard,
> Why should Heaven summon men unto his guard?
> His spirit was redoubled on his son:
> and that was seen on his assumption.

Doelman (2007: 34) commented that, Anne of Tyrol, wife of emperor Matthias, died December 1618, Anne of Denmark, wife of James VI/I died in March 1619; The king himself was also seriously ill in the spring of 1619.

> As with the verses responding to Queen Anne's death, here, the heavens are understood to reflect divine favour of England. Wake's epitaph concludes that there is no astrological response to the death, because the transition has been so smooth: no 'innovation' was heard and James's spirit passed unto his son. Hence, James, his death is unmarked by the heavens, both because of its smoothness and because of his own distain of astrological portents.

William Strode then looked forward to what Doelman (2007: 34) called a 'symbolic astronomical marker' referring to the new king, Charles I, as, 'the newe sun risinge in the east'. Allegory was not purely poetic and enabled writers to comment on contentious issues such as the Spanish Match (Doelman, 2007: 32; see chapter by Cunningham in this book).

There are a number of studies on attitudes toward the comets of 1618 in Portugal which can help shed light on the English situation, including questions of the nature of comets and their application to religious and national agendas. A majority of Portuguese astronomers in 1618 persisted in the belief that comets were meteorological events caused by the burning of terrestrial exhalations. Boccaro, however, followed Girolamo Cardano in arguing that comets are above the Moon, and therefore celestial, an argument he justified through his own measurements from 9 November to 22 December (Carolino, 2012: 204–5). However, this still left unanswered the question of their origin. According to Luis Carolino (2012: 195–225;

Carolino & Camenietzki, 2006: 42), Boccaro followed the Persian astrologer Abu Ma'shar (787–886) in arguing that great conjunctions between Mars, Jupiter and Saturn attracted celestial matter, which then ignited. The process was therefore fundamentally similar to the Aristotelian, except that it took place above the Moon rather than below. In 1618, Boccaro reasoned, the comet was caused by a conjunction between Mars and Saturn and this had been instrumental in producing a comet in the region of Mars, in other words, in Scorpio (see also chapter by Carolino in this book).

However, the analysis of cometary meaning, 'cometary divination' as Carolino and Camenietzki (2006: 43) put it, was not dependent on whether comets were of terrestrial or celestial origin. Rather, comets formed part of a matrix of celestial interpretative connections. It was, for example, significant that the new star of 1604 was in same part of the sky, as observed from the Earth, as the comet of 1580 (Carolino & Camenietzki, 2006: 34). Great conjunctions also formed part of the process. For example, the Jupiter-Saturn conjunction in Leo in 1603 held significance for Portugal's messianic future (Carolino & Camenietzki, 2006: 37). In 1603, astrologers predicted that the expected national transformation would take place in twenty years, by 1623, but the birth of Alfonso VI in 1643 forty years later was seen as the significant moment. The conclusion is that in terms of the search for meaning, all that matters is that an event takes place in the sky. Astronomical events form part of a chain of signification and are interpreted according to their visible appearance and motion, not according to whether they are above or below the Moon, and material or not.

11.8 John Bainbridge: *An Astronomicall Description of the Late Comet*

A substantial allegorical treatment of the comet's significance in England was John Bainbridge's *An Astronomicall Description of the Late Comet*, an examination of the third comet, published first in December 1618 and then, having attracted considerable attention, in January 1619. Bainbridge (1582–1643) was a grammar school master, physician—and a member of the College of Physicians. His work holds considerable retrospective interest for the history of science because he was the first of what came to be known as the Savilian professors of astronomy. As his biography, now on the Royal College of Physicians website, records, 'This introduced him to the notice of Sir Henry Savile, who was then founding the astronomical professorship at Oxford, and who at once appointed Dr Bainbridge to that office' (Munk, n.d.). Savile is notable as one of the team who translated the Bible from Greek into English. Bainbridge remained at Oxford until his death twenty-four years later, and never again published on astronomy.

Bainbridge (1975 [1619]: 4) first observed the third comet on the morning of 18th November, until its last visibility on 16th December. His observations (1975

[1619]: 19) are notable because it was an early use of a telescope, or what he called a 'Telescopium or Trunke-specatacle'. Pages 1 to 17 of his text consisted of discussion of the comet's physical nature and position, while pages 27 to 42 are titled 'MORALL PROGNOSTICKS or Applications of the late Comet or Blazing Star', and concern its meaning. He reiterated (1975 [1619]: 27) that his central concern, then, was with what he called "the morall applications of this new Comet, by divine Providence set on the high Olympian mount, to some of the Herald of wrath, but to others, the joyful Embassador of peace and mercy'.

When Bainbridge's observations began on 18th November, the comet's celestial longitude was 9° 25′ in Scorpio, and its declination—3° 20′, and concluded on 18 December when it was at 7° 45′ in Virgo (having retrograded through Libra) and +61° declination. In 28 days it had passed rapidly though 77° 50′ of celestial longitude (Fig. 11.1). These are the mathematical positions but there was no concept of mathematics distinct from divine truth. As he wrote (1975 [1619]: 13), citing Plato in his support,

> *God is the great Master of Geometrie, having created all things in weight, measure, and number*, as holy writ doth witnesse.

Every movement in the comet's path across the sky therefore contained a potential message about the state of the world. Bainbridge was deeply concerned with the comet's prophetic nature, in the Biblical sense of prophecy, but less in terms of a prediction of the future and more as a warning of what will happen if people disobey God, and promise of the benefits if they do obey. Like the anonymous Astronomer, he took his cue from Jeremiah: 'Present the divine anger, with timely and serious repentance, then dare I say to you with *Jeremiah, feare, not the signs of heaven, at which the heathen are dismaid*' (Bainbridge, 1975 [1619]: 32).

Bainbridge's interpretative use of the zodiac signs, he stated (1975 [1619]: 35), was a matter of 'allegorie'. By this he meant that he was interpreting signs rather than analysing real and essential changes in the relationship between sky and earth. Of the comet itself, he wrote (1975 [1619]: 30), 'What this Comet does in particular signifie is not possible to declare without... divine inspiration'. As such, he carefully distanced himself from what he called 'the rules of vulgar Astrologie, whose precepts I esteem, no better than... Phantasticke dreames', preferring the term 'Hieroglyphicke' for his own method (Bainbridge, 1975 [1619]: 32). He referred (1975 [1619]: 40) to the comet's movement through 'celestial Hieroglyphics', including the constellation Boötes. The comet was above all a signifier of religious reform and Bainbridge enrolled the comet in the Protestant cause, looking back to three previous new stars as a chain of signifiers: the comet in 1558, the new star of 1572, and the comet of 1577, all of which he considered gave support to Protestants. Bainbridge was fully supportive of salvational and confessional history, and he employed the prophetic rhetoric of Jeremiah to condemn the unrighteous (1975 [1619]: 28–9): 'Away, away, prophane, irreligious wretches: it is God, it is God, omnipotent, and omniscient, to the wicked, most fearfull, and terrible, to the repentant, most gracious and merciful, that appears in these celestial signs.'

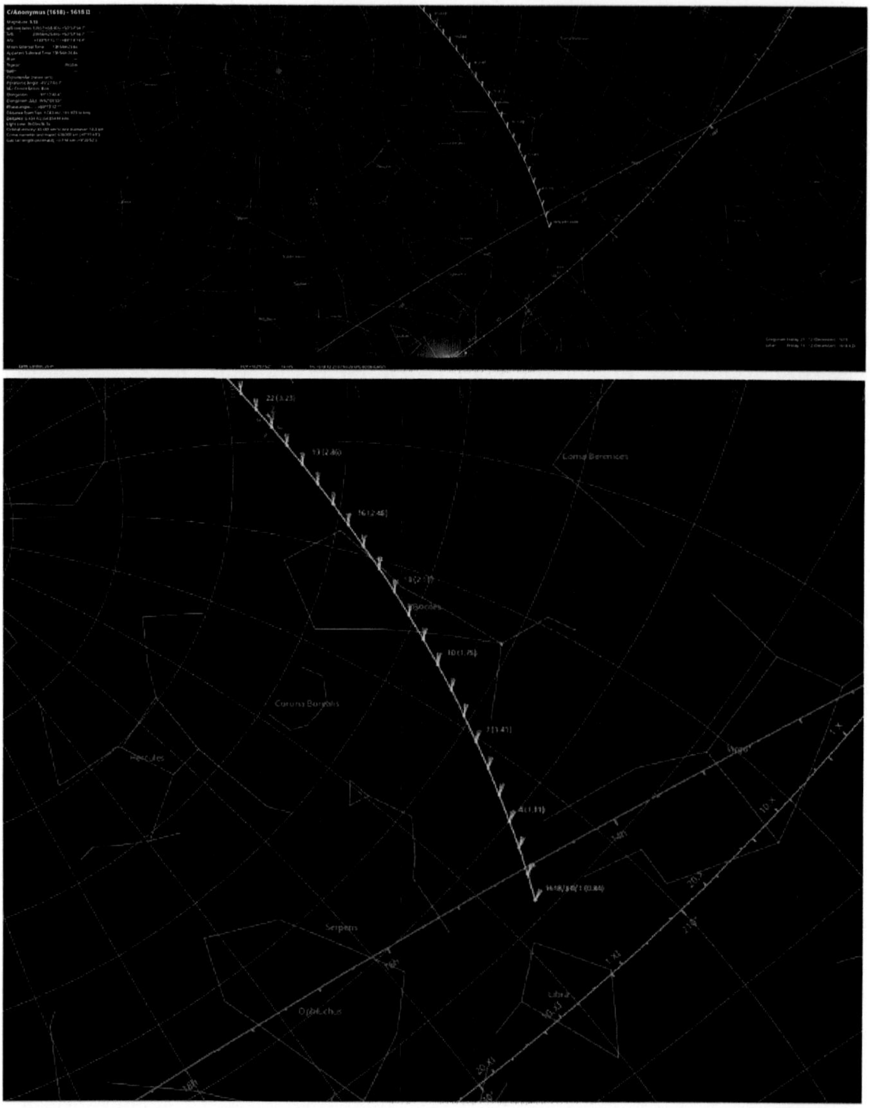

Fig. 11.1 Star map showing the path of the comet. The map features a grid with an orange line (the ecliptic), a blue line (the celestial equator) and constellation outlines, including Corona Borealis, Serpens, and Libra. The comet's trajectory is marked in yellow with daily markers for 4.00 am. The upper image shows a wider field, the lower a narrower focus enlarged. (Images created with Stellarium 23.4)

Protestant evangelism led naturally to a nationalist tone, which was perfectly consistent with England being a Protestant nation, and therefore blessed: 'I dare boldly affirme, that this Comet being followed in his Emblemated motion is to great *Britaine's* Majestie, and Monarchie... An auspicious sign of great honour and happinesse' (Bainbridge, 1975 [1619]: 32–3). He was explicit (1975 [1619]: 38) that his text was a panegyric in praise of the current regime. He concluded (1975 [1619]: 32–3) that,

> Whatsoever evill, this new Comet may presage, the sign be to them that hate us, and the interpretation thereof to our enemies: But whatsoever good, it can promise, the God of Heaven (who there placed it) confirme them to all his royall Majestie and *Great Britaines* Monarchy.

Nationalist themes recur throughout Bainbridge's text. Of the comet's passage through the constellation Arctophylax (Boötes), he wrote (1975 [1619]: 36),

> It is great Britain's royall Court, which divine *Astraea* doth illustrate with her gracious and healthfull rayes, it is this fortunate Iland that is governed with her scales of justice, and enriched with her spike of plenty. The worlds great Monarch has crowned his Majestie with the Imperial Diademe of all the British Isles; and by this his Emassador, doth promise him a long and happie possession thereof, and to his posteritie forever. As for his enemies, he has, and will cloath them with shame, but on him and his shall, his crowne flourish in this life, and an immortall crown of glory in the heavens.

If we relate this passage to Olson and Passachoff's symbolic ambiguity, then the comet actually has an entirely different meaning for the devout Protestant as opposed to the Roman Catholic or Heathen. The former will triumph and the latter be punished, unless they recant and repent. In terms of myth and allegory, it is unclear of there was any reason why Bainbridge related the myth of Boötes, the herdsman to Astraea, the goddess of justice, beyond the need to find any astronomical allusion to make a nationalist and salvational point. This was the England established by Elizabeth I: Bainbridge didn't mention that Elizabeth's propaganda presented her as Astraea, the Greek goddess of justice recast as the Virgin Queen but, then, the imagery was so prevalent that he didn't need to (Yates, 1985 [1973]). Other correspondences between the comet and the recent history of the Reformation were fairly straightforward. Reinforcing his dependence on a salvational confessional history, Bainbridge wrote (1975 [1619]: 42):

> About the preaching of *Luther* were at least five Comets in tenne years, after which followed the happy departure of *Germany*, *England* and many other Northerne parts from the spiritual *Babylon*. This new comet doth give us hope, that the rest of Christendome before long will follow; and so, at length shall be verified the Prophecie of Sybilla upon occasion of these new stars. *Rome shall again become a forlorn and desert village, or sheep-coat.*

In addition, God's word was due to spread to new regions, namely the Indies where the second comet was more visible than in Europe. Like Boccaro, citing the conjunctions of Jupiter and Saturn.

> Besides, these regions of *Europe*, a blessed light has in this age shined to another world, which did long sit in most fearfull darknesse; I mean the East and West Indies. I am verily perswaded that the new Star, which appeared so long from September 1604 to January,

> 1606 in the foot of Serpentarius, having coincidance, with the great coniunction of the three superior planets [Mars, Jupiter and Saturn], and that other so many years in *Cygnus*, doth promise (being with his present Comet conformed) a more clear illustration of those remote regions, with a splendid light of salvation (Bainbridge, 1975 [1619]: 31–2).

The spread of the light of God to confound the darkness of the heathens was a benevolent correspondence of the celestial patterns, and in the apocalyptic mood of the times, Bainbridge's prediction of the likelihood of Christ's second coming was completely standard. Of the current comet's timing, he observed (Bainbridge, 1975 [1619]: 40) the fact that the Synod of Dordrecht was held at the same time:

> I may not forget the coincidence of the present Synode at *Dort*, diune prouidence actually with the rayes of this new Comet dispelling those foggie mists which began to over-shadow the glorious light of Euangelicall veritie.

Bainbridge would have been well aware of the synod as it was attended by an English delegation, as well as delegates of seven other foreign reformed churches. It has been suggested that delegates at the Synod, the general synod of the reformed churches in the Dutch Republic, held between 13 November 1618 and 29 May 1619, did not, unlike Bainbridge, interpret the comet as a sign of divine pleasure (Vermij & Vieth, 2022: 284).

In line with the Ptolemaic formula that comets shared the nature of Mars and Mercury (although without citing Ptolemy), Bainbridge (1975 [1619]: 40) referred to the comet's 'Martial meditations'. The comet's conjunction with Mercury was also notable:

> About the 17 of November, the Comet was in conjunction with *Mercury*, supposed by the ancients, the Messenger, and Interpreter of Heaven, Patrone of Arts, and trafficke: this fortunate meeting was near the Ecliptique of this Northern side; as if Royall *Phebus* had sent him to entertaine this new Embassador, with joyfull, embraces, and honourably to receive him into these arcticke regions: yay, as it were to give him some special charge from the great Monarch of the Starrie Empire. And here (me thinks) the East *Indies* do by these two Legates present Great *Britaine*, with her odiferous and healthfull spices, her precious jewels, and other orientall riches (Bainbridge, 1975 [1619]: 33).

The materialism of the pursuit of wealth is entirely consistent with the proclamations of God's truth, in that any a true Protestant nation should be blessed with riches. The comet's location in Libra was equally beneficial for justice:

> The place of this congresse is most remarkable amidst the balance of *Libra*; which the Comet passing through doth withall, as it were from Heaven proclaim to Prince and people: *discite iusticiam moniti* [literally—'learn the justice of the warning'] (Bainbridge, 1975 [1619]: 33).

After Libra, the comet entered Virgo, where the emphasis on justice was further emphasised, and Astraea was again invoked: any astronomical significator could relate to Astraea in Bainbridge's distinctly flexible form of allegory:

> Which this Starre would blazen forth by displaying his golden lockes over the skirts of *Virgo*, that just Astrea, which, last of all the virtues forsook the Earth, polluted, with so many vices… But this *Syderius nuncios* doth as it were intreat her to return with her flourishing spike, and advise us to give her content, least as in former times our corrupt manners

> make her wearie of the Earth. From *Libra*, the Comet posting over the aethereal planes... at length comes near to *Ariadnes* crown [Corona Borealis], would she take it to him for a princely reward of *Justice*, referuing the Virgin's spike for the people? (Bainbridge, 1975 [1619]: 35).

Bainbridge devoted a page to a description of two different forms of justice, moving into obvious political commentary. Both forms of justice, he wrote (1975 [1619]: 33), are equally valuable, one being 'Geometricall, or distributive' and the other 'arithmeticall or corrective', and both are commended by the comet, this 'heavenly Embassador', especially to inhabitants of the British Isles. The first has enabled resistance to the extremes of Protestantism found in 'Cyclopicall Anarchy... and... Anabaptisticall equalitie', while the second counteracts the decadence exhibited in 'sumptuous, and presumptuous apparel, with strange, and inconstant fashions' (Bainbridge, 1975 [1619]: 33–4). The lesson is that the Comet is a signal of the virtues of moderation, and of a government which rules from the middle, rejecting extremes: on one side were the militant Puritans and on the other foppish aristocrats. The comet had genuine political and social significance but was instrumentalised to make political points with which it had no essential connection. It became a legitimising rather than predictive device.

Bainbridge (1975 [1619]: 36) then set out a manifesto of civil rights and equality before the law, anticipating the United States' declaration of independence by over a century-and-a half.

> The other is corrective Justice, an indifferent and equal administration of right to all... when the Widow and fatherlesse are protected, and the poor man's cause, gratis pleaded; whether scant measure (so abominable to God) is made up; where there is no wicked balance (which he will not justifie), nor secret bagges of deceitful weights.
>
> Wherefore my countrymen, doe justly one with another, and shew mercy; so, shall you be more acceptable to God, then, if you offered a thousand Hecatombs [sacrifices] or ten thousand rivers of oyle.

Bainbridge's commentary on the comet sets out a programme of social and economic justice which anticipated the Christian socialism of nineteenth and twentieth century Britain (Williams, 2016). As a social reformer Bainbridge bemoaned the degeneracy of the current age, of a country grown soft from peace and unprepared for war, and addicted to tobacco, which had been introduced to England in 1586. Bainbridge's text (1975 [1619]: 38–9) is one of the first examples of a moral panic over drug use.

> O, how are wee degenerate from the generous spirits and warlike meditations of our victorious ancestors!... Our over much felicitie has almost drenched us in voluptuousnesse... We feel the breeding evils of Long peace;... I will not ask, where are the bowes and arrowes, wherewith our fathers conquered *France*, and releeved *Spain*: but where are our Muskets? Are they not turned into Tabacco pipes? Where are our English valour and courage? Are they not that with that outlandish weed vanished into smoake?... Who may not from these smoakie parents fear a fumish generation, whose courage may perhaps be soon inflamed, but sooner quenched?

Bainbridge's text is an extraordinary insight into the astronomical-social-political culture of early seventeenth-century England. Astronomy, for Bainbridge, could serve political and social functions and the comet fulfilled this goal as an instrument of God. He found in the messages sent by the stars on God's behalf the promise of a morally upright and better England, one ruled by justice and law, and loyal to God, an embodiment of virtue and a force of good in the world on behalf of divine providence. Bainbridge's text is confessional in that it requires faith in the true Protestant religion for the comet's significance to be benign, but it does not require actual declarations of faith, for these are to be taken as read.

Astronomy therefore came in two parts, the first to establish the positions of the celestial bodies, the second to establish their significance. In this he was following Claudius Ptolemy's formula, mentioned above, in the *Tetrabiblos*, that the first part of astronomy concerns the measurement of celestial positions, the second their effects (Ptolemy, 1940: 1.1). The second part is what later became known as astrology, a designation which Bainbridge deliberately avoids. Or, rather he rejected the term 'vulgar Astrologie' and replaced it with 'celestial Hieroglyphics,' rejecting the endlessly complicated rules of horoscope interpretation with a more impressionistic style, dealing with allegory rather than effects. Yet, being allegorical does not make Bainbridge's text non-astrological, and his search for significance is compatible with Curry's broad definition of astrology. He also adhered to the view that comets were not inherently unfortunate but that their significance depended on human agency, whether they acted in conformity with God's law or against it.

11.9 Reassessing Bainbridge

Previous studies have judged Bainbridge against the standard history of science and find him wanting. Such anachronism removes Bainbridge from the complexities and nuances of his own time. Sara Schechner Genuth's magisterial study of comets (1997: 52) writes of how Bainbridge 'optimistically fancied' that the East Indies would present Britain with aromatic and wholesome spices, precious jewels, and oriental riches' and suggested that his candidacy for he professorship at Oxford 'may have been helped by the unusually upbeat nature of his propagandist tract'. Genuth (1997: 55) added, 'John Bainbridge claimed to deplore the illusory precepts of 'vulgar astrology' but could not resist the temptation to interpret the comet of 1618 according to celestial pictograms of his owned devising'. It all makes Bainbridge sound rather false, but it was utterly consistent to reject the complex technical rules of judicial astrology while maintaining that everything in the sky had potential meaning. This was, after all, the position held by Johannes Kepler (1979 [1601]: 85–163, 1997, 2008).

Vermij and Vieth (2022: 285) followed Genuth's sceptical lead, writing, 'Bainbridge was rather ambiguous on the topic of astrology. Probably, this lack of clarity was intentional'. Actually, he was utterly unambiguous, rejecting the rules of horoscopic astrology while maintaining the completely consistent position that all celestial movements reveal God's plan. Vermij and Vieth (2022: 290) extend the sceptical reading of Bainbridge's motives and suspicion of his integrity into speculation. 'One may doubt', they write, 'whether Bainbridge himself really believed his allegorical reading. To us, the flattery appears so obvious that one wonders how anybody could be taken in. Probably, nobody was'. Yet why not? This was an era in which the literal truth of Biblical miracles went unquestioned, the existence of one's eternal soul was a given, and astrology, magic and alchemy were an integral part of natural philosophy. They (Vermij & Vieth, 2022: 290) continue, 'By the seventeenth century, thinking in analogies and images was no longer a self-evident view of the world or even an authoritative tradition, but rather a conscious and intricate game'. Again, this view is contested by the evidence that the use of art and ritual at the Stuart court was deeply embedded in Neoplatonic cosmology, evidence to which Vermij and Vieth themselves draw attention (Hart, 1994).

As Vermij and Vieth (2022: 288) observe, 'astrological interpretation, in terms of pure allegory, however, uncommon, was not completely unique'. We can point to Marcus Manilius's astrological text, *Astronomica*, written during the reign of Augustus, which was well-known for its poetical and allegorical treatment of the planets, stars and constellations. It was available in 1618 in three editions, the most recent by Joseph Scalliger in 1579 and 1600 (Manilius, 1977: I. 799–806). Bainbridge's use of allegory was actually consistent with a wider cosmology of the time. As Anna Jerratsch (2019: 193) argued:

> The interpretive method, based on the analogy principal corresponds to an emblematic process of interpretation that was not seen as an arbitrary, allegorical presentation of readings, but as the consequence of given factors in nature.

Further, Vermij and Vieth (2022: 290) state, 'although such an allegorical, reading of the skies may have been acceptable in a sermon or edifying work, it was not a part of learned astronomy'. Their evidence is Tycho Brahe's rejection of allegory as nonsense based on uncertain analogies. Yet Tycho was a practicing astrologer, so where does the boundary exist between learned astronomy and learned astrology, and between technical astrology and allegory? Vermij and Vieth follow Genuth in accusing Bainbridge of angling for patronage. They (2022: 279–80) add, "he also repeatedly included political and moral allusions that might be regarded favourably by potential patrons." And they (2022: 279) conclude,

> By all appearances, he used this work to draw attention to himself to further his career. Written in English, it addresses a general audience, but Bainbridge is careful to show his credentials as a scholar. He includes many quotes, some from classical authors, English verse, as well as many Greek terms.

It is not clear how all these claims differentiate Bainbridge's text from so many similar texts. What aspiring scholar did not seek patronage? And what scholar (Bainbridge was Oxford-educated) did not cite Greek or Latin? In addition to which, not only do they offer no evidence for their supposition that Bainbridge did not believe his allegorical readings of the comet, but the whole notion of the word 'belief' in such concepts is misleading, implying as it does a modern binary distinction between belief, which is foolish, and disbelief, which is wise (Campion, 2022: 9–13). Vermij and Vieth's assumptions extend to finding it odd that Bainbridge, who rejected astrology, should be a friend of Christopher Heydon (1561–1623), an astrologer and polymath described by Yenter (2018: 47) as a 'British Rosicrucian Philosopher'. Heydon, incidentally, calculated the position of the 1603 conjunction of Saturn and Jupiter 40 years in advance (Lilly, 1644: 23). In Christopher Heydon's *Astrological Discourse,* written in 1608 but not finally published until 1650 by William Lilly, Heydon (1650; Kepler, 1997) introduced Kepler's justification of astrology according to Harmonic and geometrical proportions into England, even though Kepler's own full treatment, *Harmonice Mundi*, was not published until 1618. Heydon was a member of the landed gentry, a politician and Cambridge graduate who easily crossed between different intellectual circles, and could be a friend of Bainbridge as easily as he was to be an inspiration to William Lilly; then, as now, some people gravitated towards extremes—elsewhere social and philosophical boundaries were porous and distinctions were nuanced.

To write a tract praising the Protestant monarchy of James I marks Bainbridge out not as a career-building sycophant but as an enthusiastic and genuine supporter of the political status quo. As Vermij and Vieth (2022: 290) correctly conclude, Bainbridge 'gives expression to a political and religious programme', seeing the comet as 'a sign of God's approval of the reestablishment of pure Christianity under King James'. Bainbridge was clearly disapproving of decadent lifestyles on the one hand, and ultra-Puritanism on the other. This text stands out as an example of the context within which astronomy was practiced in a world in which all things were created by God, and God was constantly involved in every single detail from the greatest to the smallest.

11.10 William Lilly: *England's Propheticall Merlin, Foretelling to all Nations of Europe until 1663*

William Lilly was born in 1602, the son of prosperous yeoman farmers in the English midlands. He became one of the most prominent astrologers in England in the publication boom which followed the breakdown of censorship in 1642. He had a thriving private practice, and also capitalised on the public thirst for political prediction which was fuelled by the apocalyptic expectations which were an integral part of so much of the new Protestant factions and an atmosphere of uncertainty driven by the outbreak of civil war. As Anne Geneva (1995: 9) writes, he was a genius at practising the… art of judicial astrology'. His text, *England's Propheticall Merlin*, which included an account of the third comet, is retrospective, published after the break down in political order—and censorship—when the Parliamentary opposition took up arms against the king in 1642 (Clegg, 2008). It was therefore written in a very different context to that informed Bainbridge's work. The compromise between different political and the Stuart peace had come to an end. For the first time since the Tudor dynasty was established in 1485 matters of state were being settled by violence.

Lilly's enduring reputation is based on his publication of *Christian Astrology*, notable as the first text-book of astrology to appear in the English language (Parker, 1975; Geneva, 1995; Lilly, 1715). The book contained all the rules for reading horoscopes, including the planets and zodiac signs' psychological, physical, medical, meteorological, geographical and political correspondences. Lilly made his first parliamentary contact in 1643 when he successfully treated the Member of the House of Commons Sir Bulstrode Whitelocke. Grateful to Lilly for saving his life, Whitelocke then became his ally and protector. In the following year, Lilly published his first almanac, tapping into traditions of popular prophecy by naming it after King Arthur's legendary magician; *Merlinus Anglicus Junior*. He sold out, like Bainbridge, very quickly, this time within a week, evidence, if any were needed, of the demand for such work. The Parliamentary licenser (and fellow astrologer) had used his position to censor Lilly's work, but Lilly, never one to accept a slight, complained and was allowed to issue a second edition in the following year, 1645, under the title *Anglicus, Peace or no Peace*. Such was Lilly's reputation that his death was marked by an allegorical poem composed by his patron, Elias Ashmole:

> He must be gone, the Stars had a so decreed;
> As he of them, so they of him, had need.
> This Message 'twas the Blazing Comet brought;
> I saw the pale fac'd Star, and seeing thought.
> (for we could guess, but only LILLY knew)
> It did some glorious Hero's fall foreshew;
> A Hero's fall'n, whose Death, more than a War,
> Or Fire, deserv'd a Comet, th' obsequious Star,
> could be no less than his sad Fate unfold,
> Who had their Risings and their Settings told (Briggs, 1974: 105–6, cited in Geneva, 1995: 97).

And even in 1722, Daniel Defoe (2003 [1722]: 22) recalled the coincidence of the comet of 1665 with the Great Plague of that year:

> The apprehensions of the people were likewise strangely increased by the error of the times; in which, I think, the people, from what principle I cannot imagine, were more addicted to prophecies and astrological conjurations, dreams, and old wives' tales than ever they were before or since. Whether this unhappy temper was originally raised by the follies of some people who got money by it—that is to say, by printing predictions and prognostications—I know not; but certain it is, books frighted them terribly, such as Lilly's Almanack...

Politically, Lilly represented the farming and clerical classes on whose behalf his writings were a political vehicle. His nationalist description (1644: 47) of a virtuous, freedom-loving Protestant nation is reminiscent of Bainbridge:

> There's a Yeomanry of *England*, indures no slavery, it's hard to root it out, they breed their sonns, excellently well, and it's an ill bird defiles his own nest; their sons prove excellent Lawyers, and best know the estate of every County they live in; are valiant and discreet people.

Lilly was, though, in a very difficult situation in 1644, compared to that of Bainbridge in 1618. Political contexts were vital. The peace which Bainbridge celebrated had collapsed and the country was profoundly split between the warring factions of king and Parliament. Lilly (1644: 39) was obliged to follow a careful political path, writing,

> I desire to carry my discourse, and in so equal a balance, as I may not offend, either his majesty, whom I do honour as my Prince; or the Parliament, and whose protection, I now live, and who is just writes, I am bound to maintain by Protestation.

And the correspondence of the comet with attacks on Parliament seriously angered him:

> All our Parliaments have been unluckily, and by the sinister Counsele dissolved, since 1625, what reason can be given, none certainly; only pleased a prevailing faction to cloud His Majesties grace, from his Subjects; rendring him unkind, but themselves worse than *Arctephel*. Were not many of our brave Gentlemen, whom their Countrey sent up for the Publique benefit imprisoned? Did not some die in prison, Gallant, men, of the House of Commons, of the Representative Body of the Kingdom? (Lilly, 1644: 47).

Of the three comets, Lilly (1644: 32) considered the first briefly, did not take the second into account because it was not visible from England, and focused his analysis on the third. His only comment on the second comet is that, as it was visible from the Indies, it must have some significance for Spain and Portugal, the colonial powers, but he had nothing more specific to say. Instead, his focus was on the third comet. The emphasis on visibility is significant. Normally, horoscopic astrology relied on the calculated position of astronomical bodies regardless of whether they were visible. Lilly's cometary astrology therefore harked back to Mesopotamian astral divination of pre-500 BCE, in which, if an omen was not seen, it had no significance.

Lilly (1644: 31, 34) referenced the observations of the Danish astronomer Christen Longomontanus (1562–1647), Tycho Brahe's assistant who, he said, had 'excellently observed' the last comet. He also significantly relied on Bainbridge's measurements and chronology, with which he was clearly very familiar (Lilly, 1644: 34). In England, Bainbridge observed the comet on 18th November and saw it continuously until 16 December. In Denmark, however, due to cloudy skies Longomontanus did not catch sight of it until three days after Bainbridge, on 21 November, as William Lilly (1644: 34) later noted. Therefore, the comet-as-sign of divine wrath was not seen in Denmark, and therefore the message was not received, and perhaps not even dispatched. This delay allowed God, having sent the comet, to withdraw it and, in withdrawing it, withdraw both any warning (of destruction) or instruction (to repent) he chose to send, and physical disturbance the comet might cause.

An obvious, but no less significant, feature of comets, is that they are temporary but move. In this respect, they are of a different order of temporary star to those such as the nova of 1572, which was temporary but fixed (Lilly, 1644: 35). They can also be interpreted in relation to the Jupiter-Saturn cycle, and the 1618 comet had a problematic signification because the 1603 conjunction was difficult, as were subsequent eclipses (Lilly, 1644: 31).

Lilly's astrology operated completely within a Christian context. He wrote how God sent 'his punishing angel', and gave 'direction to the comet', giving both the comet's ominous role and position to God (Lilly, 1644: 37). He (1644: 32) added, for good measure, that, 'It was excellent spoken by the Philosopher Herodotus,

When God is to punish a Nation or City, he first signified his intention by Prodigies'. Any distinction we might try to make between the cosmologies of astrology and astronomy dissolves when celestial movements and their significance are both instruments and omens of God. Astrology was literally divination in the sense that as a reading of messages sent by the Divine. 'The Celestiall bodies', Lilly (1644: 29) wrote, 'whether guided by the immediate finger of God, or whether so appointed by the *Secundreian* Intelligences or Planetary Angels, who love and direct our earthly affairs, as the more secret learning informeth us', give 'permission' for the planets to form significant conjunctions. These included the Jupiter-Saturn conjunction of 1643 which, by his measure, was exact on 15 February (24 February in the modern calendar) at 25° Pisces (Fig. 11.2).

Divine law still operates in the context of demonstrable evidence and an approach which may fairly be called scientific. Referring to Cardano's aphorisms, Lilly (1644: 36) wrote,

> Whether these profound *Aphorismes* have not been exactly verified, even upon the *English* nation in particular, and upon almost every Country in Europe since this Comets appearance, I leave it to every honest Reader to consider: Have an eye to the pressures of *France*, *Germany*, and *Ireland* to consider.

In Lilly's text, comets have a general significance and a particular one. The general significance is disruptive and inclines to violence. He had good authority, citing Guido Bonatti, Girolamo Cardano and Claudius Ptolemy. On Bonatti, he based his

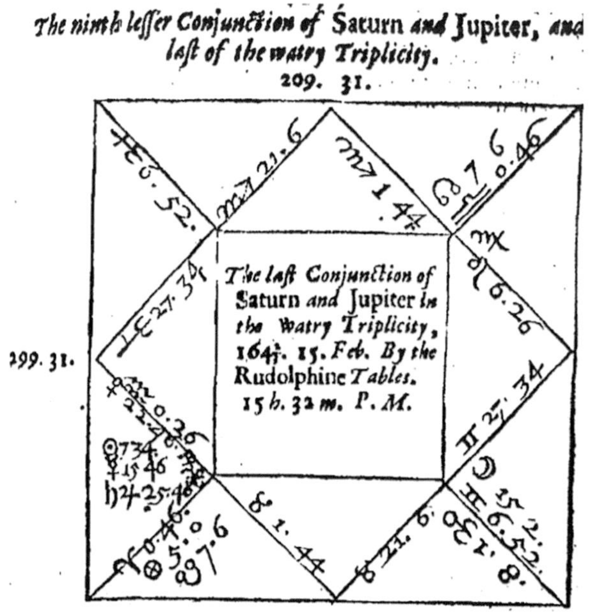

Fig. 11.2 The horoscope for the conjunction between Jupiter and Saturn conjunction in 1643. The square in the middle of the diagram contains the data, time and date. The twelve triangular divisions are the houses and the symbols they contain are the positions of the planets in the zodiac signs. William Lilly, *England's Propheticall Merline*, p. 27

statement that 'All comets signifie Wars, terrours, and strange events in the world' (Lilly, 1644: 32). He reinforced his with Ptolemy's statement that comets are to be observed for general effects which are of the combined character of Mercury and Mars, which means to say that they 'produce effects... *viz* wars, droughts, violent commotions, and other enormities' (Lilly, 1644: 32).

Judgments can be made more specific by reference to the zodiac signs. The first comet 'began and ended in the Signe of *Leo* a Regall signe' (Lilly, 1644: 32). Unlike some of Lilly's other zodiacal relationships, this one did have a clear astrological rationale. This is also significant for two reasons. First, citing the astrologer Rigel, 'Kingdomes are altered by the apparition of Comets in regall signs'—all the more reason to expect threats to kings (Lilly, 1644: 32). Second, Leo rules Bohemia, Hungary and Italy, countries where disturbance might therefore be expected (Lilly, 1644: 32; Lilly, 1985 [1647]: 96). 'Moving Comets', Lilly (1644: 36) wrote, 'shew war from Foraigners'. There may be an unwritten logic here, one which would be consistent with the symbolism, which is that a moving star travels, as do foreigners. Citing Cardan, Lilly (1644: 36) then proceeded to delineate the significance of comets when located in the houses of the horoscope, of which he mentioned only five:

Eighth house: 'a pestilence or the spoil of corn',
Ninth house: 'the hazard of religion and the spoil of the Church', legal disputes, or concerning the legal system itself.
Tenth house: 'death of kings',
Eleventh house: 'the death of many Nobles', legal disputes or concerning the legal system itself,
Twelfth house: 'a pestilence or the spoil of corn' (duplicating the eighth house).

The astrological associations are clear in some cases: the ninth house rules religion, the tenth kings and the eleventh close associates of the King (Lilly, 1985 [1647]: 54–6). The comet added misfortune to the usual house designation. It is not clear, though, why houses eight and twelve should signify the spoil of the corn harvest, unless it's that the comet was last visible in Virgo, which ruled cornfields, and we have an example of symbolic ambiguity (Lilly, 1985 [1647]: 96).

Given that the comet would have appeared in all these houses during the hours of darkness, as well as the seventh, it is not entirely clear on what dates or times such measurements would have been taken. Lilly did provide a horoscope for the comet's first observation from data given as 9:12 pm on 11 November, but actually 9.12 am on 21 November, the date of Longomantanus's first observation (see Fig. 11.3).

To draw on Lilly's previous statements, the comet was in Scorpio, so not a regal sign and not threatening to kings, and in the eleventh house, and so warning of the death of nobles and challenges to the system of law. He did not delineate it, even though there are some interesting features, such as a Moon-Venus conjunction in Capricorn (meaning that the Moon was debilitated) in the unfortunate twelfth house, and a square from Sun in Scorpio, ruled by malefic Mars, to Jupiter in Aquarius, ruled by malefic Saturn, all of which adds up to a profoundly unfortunate picture (Lilly, 1985 [1647]: 58, 62, 66, 97). Why Lilly did not delineate the horoscope is not clear as, even in a publication aimed at a popular market, he could have done so without entering into astrological technicalities. The balance of his text is towards

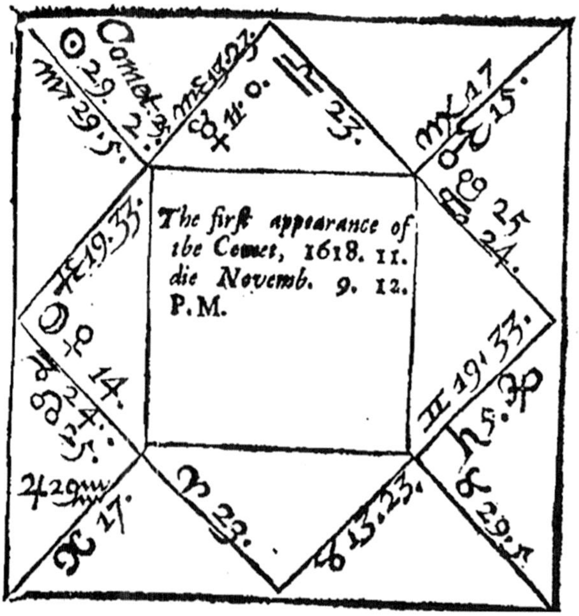

Fig. 11.3 The horoscope for Longornantanus's first observation of the third cornet. The square in the middle of the diagram contains the data, time and date. The twelve triangular divisions are the houses and the symbols they contain are the positions of the planets in the zodiac signs.William Lilly, *England's Propheticall Merline*, p. 33

making political statements and forecasts, justified by the claim that these are signified by God's plan via the comet.

Lilly also gave a horoscope for the Full Moon following Longomontanus's first observation (Fig. 11.3). The data was given as 12:05 pm on 21 November, with the Full Moon at 9° 47′ and 17° Virgo ascendant. A modern recalculation using Solar Fire software shows how accurate Lilly was: the Full Moon was at 9 ° 49′, his calculation being just two minutes out, while the ascendant was 15 ° 40 Virgo. Neither, though, did he delineate this (Fig. 11.4).

In terms of zodiacal longitude, the comet was retrograde. Citing John Dee, Lilly (1644: 38) concluded that, like a retrograde planet, it 'doth seem to break the constant order of Nature'. Therefore, he wrote (1644: 38),

> So that it seems some Princes and Potentates, many Noble men and Ecclesiasticall have ever since the appearance in 1618, been steared in the course of their lives by the influence of this Retrograde Comet. viz. They have laboured hard by a strict hand, and an awing domineering authority to oppresse their Neighbours, Tenants and Subjects, esteeming that the safest for them, whereas in truth, it first undid them in the good opinion of the people, and lastly, when the Comet came to be verticall over the afflicted countries, very necessity, enforce them to rise in Armes for destruction of their Retrograde Lords.

As Olsen and Pasachoff (1998: 1) wrote, the fascination for comets since earliest recorded history has stemmed partly 'from the fact that their appearances seemed to violate what was perceived as the proper, elegant order of the universe'. For Lilly,

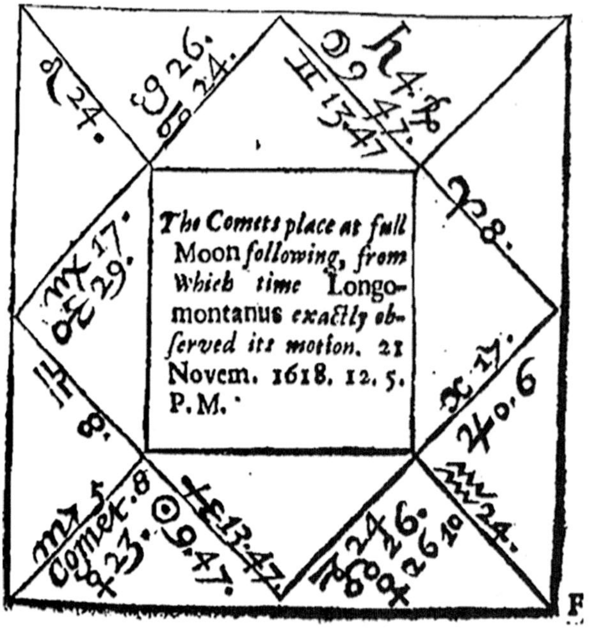

Fig. 11.4 Horoscope for the Full Moon following the third comet's first appearance by Longomontanus. The square in the middle of the diagram contains the data, time and date. The twelve triangular divisions are the houses and the symbols they contain are the positions of the planets in the zodiac signs. William Lilly, *England's Propheticall Merline*, p. 33

the pun on the word 'retrograde' makes the point: just as the Comet was headed in a reverse direction, running opposite to the planets and therefore contrary to nature, so the Princes and Potentates of England moved against nature, against the tide of natural law, in their oppressive laws and practices.

> yet his clemency was such, as that he permitted not the Comet to be Verticall over any part of *England*, until it had been visible, and verticall to many other places, 21 dayes, which, in measure of time, give, according to my limitation, 21 years.
>
> By which his long forbearance, it appears, that God did reflect upon the many intercessions of some *Abrahams, Samuels, Davids*, and *Daniels*, that then lived amongst us: who, with weeping, fasting and mourning, like so many *Hezekiahs*, obtained these 21 years longer, for us to repent it (Lilly, 1644: 37).

Again visibility is significant, while to be vertical—overhead—is significant. He added,

> After the comet, I say, had been visibly apparent, 21 days, it had by the exact observation of Longomontanus, 49° of Northern Declination, and then entered the sign of Virgo, the common ascendant of the *English* Kings, and Monarchy; for many conceive *William* the Conqueror obtained England, presently after a conjunction of Saturn and Jupiter, in Virgo (Lilly, 1644: 37).

He added, 'And we find by experience those kings of England, that have any sign of the earthly Trigon their, ascendant, have been fortunate to the *English*' (Lilly, 1644: 37). Such monarchs included both James I and those of the preceding Tudor dynasty, Henry VIII, Edward VI, and Elizabeth I, all of four of whom who were born with the Moon in Virgo.

As the comet entered Virgo on 11 December, reaching a northerly declination of 49°, 'it came near to be Verticall and perpendicular over England' (Lilly, 1644: 37). Verticality was of profound significance. Citing, John Dee's aphorism 54, Lilly (1644: 37) wrote,

> How much greater the *Axis* of radious beames of any star doth come to perpendicularity upon any elementary superficies; by so much more forcibly, he does imprint his influence upon such a place of his incidence, in a direct manner, by reason of the greater vicinity of the Agent: but in a reflexed way, because such beams bowed or turned back, are near conduplicated to the incidents.

And following up with Dee's aphorism 68: 'Stars that are Verticall have great power, and are those which have as much declination from the *Aequinocyial*, as the elevation of the Pole of that Region' (Lilly, 1644: 37).

Lilly used a suite of established technical methods to interpret the comet's meaning, answering Genuth's rhetorical question concerning what skills were used. The direction of the comet's tail was significant, as Ptolemy had stated.

> The tail of this Comet, all the time of its being, pointed out the West, and North-west parts of the world: if we believe *Charles Piso*, the French physician... What places are principally threatened, and intended to feel the Comet's influence, the parts of the *Zodiack* where first, the matter was collected, and appeared, do manifest: as also the inclinations of the scattered haires, according to the form of the Comet (Lilly, 1644: 44).

He continued.

> *Austria*, certainly, the seat of the *Roman* Empire, was chiefly pointed out by the Comet, to be either the seat of war, or the place from once much mischief should come (Lilly, 1644: 44).

This is a certainly an oblique reference to the Thirty Years War, but it is strange that Lilly did not make more of such an obvious correlation with the comet's appearance, considering its importance for astrologers in Germany. The fixed stars were also important:

> The passage of this Comet through the forms of the fixed Stars is most to be admired, and wanted nothing, but a tongue, and an audible voice to expresse articulately the ends why God did illuminate the same (Lilly, 1644: 44).

Zodiac signs also played an important role, Scorpio, Virgo and Libra being relevant in this case.

> The generation of this Comet in the *Scorpio*, under the Sun-beames, did plainly expresse, and shadow out onto us. Virulent and Scorpion like Counsels and Confederacies of those who in the 19 chap of the Revelation are described: to have the faces of men, but the stings of Scorpions in their tails; I mean, the Jesuits, and Legates of the Pope, out of whose mouth they are sent, and by their Vow of mission are bound to go, whither he shall send them, and execute whatsoever he shall employ them in: Being in this Emblematicall generation of the Comet, described to lurk in the Courts of Kings, and Emperours (as it were under the Sun-beams), thereby more importantly, to solicite and incense them to take any opportunity to sting and extinguish, such as profess the true Protestant Religion in their Dominions. Do

> Kings and Princes harbour such monsters, and will not beleeve it: will they not beleeve there are Bees in the Hive, unlesse they feele their sting first? When they are undone, and their kingdoms, and subjects, destroyed, some belief will bee given to what I write… but The axe is laid to the root, a few more blows do cut down the Tree (Lilly, 1644: 44).

Revelation chapter 19 portrays the devout elders as bowing down before God and the slaying of the armies of Satan. Citing Cardan and Ptolemy, Lilly (1644: 45) linked the comet to the Emperor Matthias's birth horoscope and notable royal deaths.

> It's ever served, those Comets, which move in a Retrograde course, do signify the mutation of Lawes; because that motion is of the first Heaven, and therefore from God, or some supream Rector of our Affairs, (perhaps he means, the Planetary Angel presiding at this time), but Laws are by permission of God: the habit of the Comet to the Sun, shews his beginning to be quick, and sudden, if he be Oriental, &c. and so it was, for *Matthias*, the Emperour survived, not the Comet three Moneths: He had the 16 of *Scorpio*, the Ascendant in his Nativity; he lived when the Comet blazed, but when the Comet cut the 16 of *Scorpio* he sickened, and not long after died: the Empress died a little before, so did Queen *Anne*, in *Anno*, 1619, and yet Comets, have nothing to do with Princes, or signify any thing, as some erroneously hold.

Like Bainbridge, Lilly was concerned with the *derivation* of cometary meaning from the constellations, although they are conflated with the zodiac signs (Fig. 11.5). He knew exactly what constellations the comet was in from Bainbridge's calculations, yet also cast a horoscope for the comet's first appearance showing it in the zodiac signs, in zodiacal longitude (see Fig. 11.3).

> Many shall be the labours of mankind, (when a Comet is in *Scorpio*), fearfull dissension betwixt Princes, tumult, amongst the vulgar, the death of Prime, and Eminent men, scarcity and dearnesse of provision, fear of warres, the anger of Princes, much defection from them by their Subjects, not without civill discord, and intestine slaughter (Lilly, 1644: 45).

> If the Comet pass through *Libra*, Princes of the West part of the World shall die, fearful, murders occultly carried, and mens states consumed they know not how (Lilly, 1644: 45).

In this respect,

> We have had had many Princes in the West part of the world, dead since the Comets appearance, many a dear year (Lilly, 1644: 48).

Lilly's interpretations had some of the same allegorical flexibility as Bainbridge's. Both Scorpio and Libra signified death and destruction, whereas in Lilly's own account Libra is normally a peaceful sign, ruled by Venus, so the prime significator is not the zodiac sign but he comet, and even then he was selecting the inauspicious reading as opposed to any auspicious variant (Lilly, 1985 [1647]: 96–7). He then delineated the comet's passage through the zodiac signs, framing his salvational reading within the overarching role of the Protestant church.

> The comets passage through the sign of Virgo (who Enigmatically represents Religion and Justice, under whose title the Church, and spouse of Christ is figured) and over whom the Comet's tail did bow five days, intimates that, for some years, the Church of Christ should be afflicted by the Pope, Emperour, King of *Spain*, and others, Jesuitically, addicted; But when the church has over-passed those afflictions, they shall fear no more persecutions to follow, but enjoy a perpetual peace and tranquility in this world. Further denotes that the industrious Merchant shall make ill voyages, and men shall, prosecute each other with all kind of injustice; good men shall be vexed, and imprisoned; a rumour and fear of warre to no purpose, but great persons made away suddenly. Let us observe Historically of these

Ephimeris Cometæ.

Moneth.	*Dayes of the Moneth.*	*Its dayly Motion.* G.	M.	*Its Motion in Longitude.* G.	M.	Sig.	*Declination.* G.	M.	*Anno Dom.*	
Novemb	21	3	20	9	25	♏	3	20. *Mer.*	1618	
	22	3	17	7	55	♏	0	35. *Mer.*	1619	1
	23	3	14	6	25	♏	2	20. *Bor.*	1620	2
	24	3	11	4	55	♏	5	30. *Bor.*	1621	3
	25	3	9	3	7	♏	9	0	1622	4
	26	3	6	1	20	♏	12	0	1623	5
	27	3	3	29	30	♎	15	20	1624	6
	28	3	1	27	45	♎	18	10	1625	7
	29	3	0	26	10	♎	20	40	1626	8
	30	2	57	24	10	♎	23	15	1627	9
Decemb.	1	2	47	22	0	♎	26	10	1628	10
	2	2	38	20	0	♎	29	10	1629	11
	3	2	24	18	0	♎	32	0	1630	12
	4	2	19	16	20	♎	34	20	1631	13
	5	2	16	14	40	♎	36	30	1632	14
	6	2	14	12	30	♎	38	40	1633	15
	7	2	12	10	20	♎	40	50	1634	16
	8	2	10	7	40	♎	43	0	1635	17
	9	2	8	5	0	♎	45	0	1636	18
	10	2	6	2	20	♎	47	0	1637	19
	11	2	3	29	30	♍	49	0	1638	20
	12	1	59	26	40	♍	51	0	1639	21
	13	1	55	23	40	♍	53	0	1640	22
	14	1	50	20	50	♍	55	0	1641	23
	15	1	43	18	0	♍	56	40	1642	24
	16	1	33	14	40	♍	58	10	1643	25
	17	1	18	11	10	♍	59	35	1644	26
	18	1	8	7	45	♍	61	0	1645	27

Fig. 11.5 Ephemeris for the third comet in zodiacal longitude, showing secondary progressions. From left to right the columns indicate (1) month, (2) date of the month, (3) daily distance travelled by the comet in degrees and minutes, (4) daily distance travelled in relation to the zodiac in degrees and minutes, (5) daily declination, position north of the celestial equator, (6) equivalent year in secondary progression: Nov 22 = 1619, Nov 23 = 1620 etc., (7) secondary progressed year after 1618 by year number: Nov 22 = 1619 = 1, Nov 23 = 1620 = 2, etc. William Lilly, England's *Propheticall Merline*, p. 35

> axiomes of the learned, have not been true and our age, and Common-wealth (Lilly, 1644: 45–6).

Having set out the theory, Lilly (1644: 46) looked at the evidence, all of which involved violence and misfortune, contrary to Bainbridge's optimistic view. England had suffered internationally through defeat in north Africa and France.

> From the time of the Comets appearance, which was 1618, until it became Verticall to *England*, which was 1639, we did not miserably suffer; but all men know, matter has been preparing for these very times ever since: how did the English groan under the pressures of the Monopolists, about 1618, 1619 and 1620. How much treasure did we spend, and how much disgrace was it, that simple employment to *Algiers*, in the latter end of King *James*, his reign, our brave Sea-men, and the brave spirited, Sir *Robert Mansell*, muzzled up with

> a blank commission: To number the Sea-men captivated in those times, or recount, the unvaluable loss of the *London* marchants were *infandum renovare dolorum*; tears, but not words, best, expresse, it; and immediately, or few years after, the Conjunction of *Saturn* and *Jupiter* in 1623, we lost King *James*, whose reign was spent in peace; what a monstrous plague afflicted the city of *London*, that year, 1625, and all, all the most part of the Kingdome, the next year; Now, or near this time, was our unfortunate attempt to relieve *Rochell*, after our owne Ships had undone them, and then we fight in the mud at the Isle of *Rees*, where we lost the best men we had them living, not in fight, but by treachery in the Ouse and Dirt &c...
>
> The destruction of the *Rochellers*, and in them of the Protestant party in *France*, and the innocent blood of that Protestant party, cries, loud to heaven, for vengeance against the Authour, and procurer of their final desstruction: A City, Queen Elizabeth loved; nor the victorious Henry the fourth King of *France* discast it. To this day the Rochellers believe Buckingham was the Authour of their undoing.

Like Bocarro, Lilly saw the comet's chain of signification engaging with the Saturn–Jupiter conjunctions. Like Bainbridge he used the comet to discuss contemporary affairs and geopolitical developments, assuming that those affairs must, a priori, be related to celestial events, rather than starting with the literary canon on the meaning of celestial events, in this case on the character of the zodiac signs.

11.11 Day-for-a-year Predictions

So far, Lilly's cometary astrology is quite standard. However, it is unusual, perhaps unique, for his use of a system outlined by Johannes Kepler, in which units of time correspond to each other, in what he called 'proportion of time' (Lilly, 1644: 41). The system he used, which in modern astrology is known as secondary progression, equates each day with a corresponding year. In geocentric reasoning, one day and one year are both complete revolutions of the Sun around the earth and must therefore contain the same quality of time. For example, in the case of an individual birth, the second day of life corresponds to the second year of life. The positions of the planets on the second day (in effect, only the Moon will have substantially moved), then indicate the kind of events which will happen in the second year of the child's life. The system is obviously of importance for adults, as part of a package of predictions for the 30th year of a person's life, the astrologer would look at the disposition of the planets on the 30th day of their life. The system was first described by the second-century astrologer Vettius Valens (2010: Book IX, 154):

> Be aware of the transits of the stars and their changes of sign at the various chronocratorships, as I have described. It is necessary to calculate as follows: add a number of days to the birth date equivalent to the age (in years) of the native. Then, having first determined the date, whether in the following month or in the birth month itself, cast a horoscope for that day.

However, Lilly did not cite Valens in his extensive bibliography in *Christian Astrology*. But we do know that he was familiar with Kepler's work in general and utilised Kepler's new planetary aspects, as advocated by Christopher Heydon

(1650). In 1610 Kepler (2008: thesis 41, 102) elaborated the method in his *Tertius Interveniens*:

> On the other hand, the doctrine of directions takes on great appeal for me, when I let the world make its revolution (as per Copernicus), then the proper proportion of a day for a year, one for 365, finds itself for a domicile, but, apartment or ship in which we are led around the world. And this [doctrine] is all the more credible, in that, for the directions and nativities of human beings who are the occupants of this ship, this proportion [one day for one year] should rule - as the astrologers teach us.

> Just I have derived my own special opinion from the various opinions of the most prominent astrologers, and also I have compared these authors with each other. In my opinion, with a dissenting attitude, [I have reached the conclusion that] each day after birth means one year, two days means two years, and so on.

We do not know of any previous accounts of this method in English, so the balance of evidence suggests that Lilly must have borrowed it from Kepler. However, Lilly was highly unusual in applying the system to political affairs and the interpretation of comets, especially as we do not know of him applying it to individuals. He included in his table of the comet's daily positions a list of the years to which each day's positions corresponded. Therefore, as the comet was first observed by Bainbridge on 18th November 1618, the following day, the 19th November, corresponded to 1619, the 20th November to 1620, and so on.

1639 (11th December 1618)
Lilly began his day-for-a-year chronology with 1639, the 21st year from 1618, equating to ninth December 1618, the 21st day of the comet's visibility, dated from Bainbridge's observation. This had been a year of rising political tension, as the gentry and the urban artisan classes who would have had a voice in the House of Commons began to put pressure on the king to summon Parliament. Lilly (1644: 38) wrote,

> who is ignorant of our then approaching troubles? And yet, because the comet was not then the vertical fully over *England*, no great mischief then happened: a cloud was falling upon us, but did not; yet, from the time it had 49° of North Declination, it began to exasperate, and stir up matter in the English Nation, and so, as its declination increased, our courages were stirred up.

The warning was reinforced by an eclipse of the Sun on the 22nd of May at 10° of Gemini. The eclipse began with the sun in the ninth house, ruling religion, and concluded with the Sun in the seventh house, passing through the eighth house, ruling death, on the way (Lilly, 1985 [1647]: 54). For this Lilly's reading (1644: 38) was therefore that 'church men began our misery, war, and death shall end it'. The comet formed part of a chain of signification. On the 12th of December 1618, it was at 26° 40' Virgo, within 1° of the conjunction of Saturn and Jupiter at 25° of Pisces in February 1643.

1640 (12th December 1618)
To add 22 years to 1618 results in 1640, when the comet was 'verticall over many places of *England* and in special manner to London itselfe' (Lilly, 1644: 38). In November 1640 the king finally gave way to pressure and called parliament into

session. Lilly provide a horoscope for this moment, set for the minute at which the king began his speech in the House of Lords, 2:04 in the afternoon on 3 November, under a new Moon at 21° Scorpio.

1641 (13th December 1618)

The 23rd day of the observation of the comet was the 13th of December 1618, corresponding to 1641, by which time the comet had reached 53° of northern declination and was vertical over the Midlands of England (Lilly, 1644: 40). He wrote,

> did not a misery steale upon us, and the cursed rebellions of the *Irish* begin 23 October 1641. And did not many of our inland counties, begin to smart: but to see how God concatenated worldly affairs with Coelestiall operations: All January 1641, *Stilo Anglicano*, was full of fears, mistrust, jealousy, and what not (Lilly, 1644: 40).

The rest of the chain of signification also played a part: 'for then *Mars*, by Transit, past the ninth degree of Sagittarius, directly opposite to the place of the *Sun* Eclipsed 1639' (Lilly, 1644: 40). This was not all: Saturn was at 9° Pisces in a ninety angle from both Mars and the eclipse degree. The angles were difficult and the planets equally so in their essential nature.

On the first of January, Lilly recorded, the apprentices marched to Westminster, and on the tenth of January the king withdrew from London to his palace, Hampton Court outside London, and was never to return until his eventual imprisonment, trial and execution in 1648–9. On the 18th of January 'the *Essex* men came in abundance to *London.* Ten bishops were questioned in Parliament, for now both the Comet passeth the ninth in the Radix, and also the Eclipse in 1639, was in great force' (The Radix is the horoscope for the First appearance of the comet) (Lilly, 1644: 41).

1642 (14th December 1618)

The 24th day observation corresponded to the 12th of December 1618, and therefore 'in proportion of time', 1642 (Lilly, 1644: 41). The comet was by now the vertical over northern England, which now suffered and the king was defeated at the Battle of Edgehill. And here Lilly (1644: 41) contemplates the future, asking how long the violence can continue.

> when I hear His Majesties forces have killed so many of the Parliament, that His Majesty hath still the worst, for he has a speciall interest in the life or death of any, that is slain, or otherwise perisheth; and when we are all killed, by these unkind divisions, if it happen His Majestie, or his posterity to survive, Over, whom shall He, or they reign? The Earth is the same, but where, or whither are the men gone?

1643 (15th December 1618)

In Lilly's counting 1643 corresponds to the 15th of December 1618, by when the comet was vertical over Scotland, warning of disruption there.

1644 (16th December 1618)

In Lilly's scheme 1644 corresponded to the 16th of December 1618. Lilly checked the *figura mundi* (or horoscope of the world) for the year, or the Aries ingress—the horoscope for the moment that the Sun entered Aries. At this moment Mars was in Aquarius and the Moon was in the tenth house, so he wrote (1644: 42) that the year 'is probably like to be very full of action, and many battles to be fought in the North and North-west of *England*, and so also South-west and full West from London'. There would also be, he warned the death of a great queen or lady, prince or church leader. He ended 'God bless his Majestie of England. The pope can find no leisure to die', hedging his political position (Lilly, 1644: 42).

1645 (17th December 1618),
For Lilly, 1645 corresponded to the 17th of December 1618, 27 days after the comet was first observed. Lilly had nothing to say about England, but warned of disruption in Russia, and possible aggression from the Turks and Tartars.

1646 (18th December 1646)
In Lilly's count, the 18th of December corresponded to 1646, and again in international affairs, he pinpointed Finland and the Baltic region as the danger zone.

Lilly's application of Kepler's day-for-a-year time theory to the comet is remarkable. He applied it only to the comet but, in theory, any significant event might initiate its own sequence of day-for-a-year equivalences until the complexity of qualitative relationships and meanings becomes unmanageable. But Lilly's time theory was contained within other considerations. Even though his mathematical universe relied on precise calculations, down to the minute both of time and zodiacal longitude, the whole picture was contained within a political loyalty to the English yeomanry, the country's national interests and a religious devotion to the one true Protestant church, all of which share a part in God's providence. He used the latest astrological techniques from Johannes Kepler, but his text, like Bainbridge's, also cited the standard ancient authorities, such as Herodotus and Josephus, with references to Revelation and the Bride of Christ, placing political events within the prophetic tradition. Lilly's text, like Bainbridge's, was thoroughly confessional as the term is defined by Vermij and Vieth. It was a declaration of faith in God's creation, and every movement of the comet pointed to twists and turns in the unfolding of God's plan.

11.12 Conclusion

There have been many studies of comets but as Rienk Vermij and Paul Vieth (2022: 276) observed, 'Treatises and comets have been mostly studied by historians of science or astronomy, so it is understandable that scientific aspects have drawn most attention'. Yet, to be fully understood, the history of science needs to be contextualised, and in the early seventeenth century this means that it needs to be framed within the prevailing view that the universe was created by God and that God's universe is filled with meaning. As Anna Jerratsch (2019: 193) wrote,

> The idea of comets as penitentiary warnings, which Franz Mauelshagen has described as 'the most important paradigm of religious thought' in the early modern age, is a product of the Christianization of comet-based astrological prognostications.

She (2019: 193) added that,

> Comets can virtually be seen as a theological special case in astrology, based on the comets being integrated into a specific form of Christian religiosity, rooted in an extreme consciousness of sin.

This is perhaps an over statement, for all astrology operated in a Christian context: Lilly's own text-book was titled *Christian Astrology*. Yet, in that comets were unpredictable they constituted an additional and expected warning, in contrast to the predictable apocalyptic events associated with Jupiter-Saturn conjunctions which could be calculated into the future. Anne Jerratsch (2019: 194) wrote that,

> The concept of symbolic interpretation of comets was extended in the sense that the comets became divine media of communication with the revelatory character. What lay behind this metaphysically conceptualised communication was the topos of the Book of Nature with the miraculous signs of the mighty elevated to the level of a holy scripture… In this context, the deciphering of the divine message becomes a Christian duty because God reveals himself in Nature and wants to be recognised through Nature.

Olson and Pasachoff (1998: 20) commented that comets, along with meteors, had been viewed as what they termed 'ambiguously symbolic importance'. Such a description may be applied to the dual auspicious/inauspicious nature of Julius Caesar's comet. For both Jerratsch and Pasachoff and Olson, the separation of comets from the rest of astrology is not borne out by the texts. First, the entire tradition of Jupiter-Saturn conjunctional theory is also intertwined with the unfolding of God's plan (Ernst, 1986: 265–80; Zambelli, 1986: 1–43). Second, both Bainbridge and Lilly combine the mundane readings familiar from all astrology concerning world affairs, with religious contexts (Green, 1680/1681, 26–27, cited in Genuth, 1997: 51). In response to Sara Schechner Genuth's list of the ten key factors which could help interpret a comet's significance and meaning, Bainbridge referred to the comet's shape and motion in the first, technical, part of his text, and to its connections (although not physical relationship) to comets of the past, its conjunctions with planets and zodiacal sign position.

Both Bainbridge and Lilly relied on exact astronomical calculations, but whereas Bainbridge devoted half his work to an account of his observations, Lilly was never an observer and instead relied on others' work, in this case Bainbridge's. And both were modernisers, Bainbridge in his use of the telescope, Lilly's in his application of Kepler's day-for-a-year proportionate predictive method. Bainbridge was similar to Galileo, Grassi and the other astronomical observers of 1618, but paid more attention to meaning than they did, with his moral prognostics. Kepler and Galileo, who both had the skill and inclination to interpret the comet according to the rules of horoscopic astrology did not do so, unlike Lilly.

Lilly used the full range of interpretative possibilities set out by Genuth, including colour, conjunction, zodiacal position, mundane house, direction of tail, motion, duration, and resemblance to comets of the past, ignoring only the site and shape of the nucleus. He also interpreted the comet as a chain of signifiers connected to other astrological factors, including birth horoscopes for important personalities, such as the Emperor, Jupiter-Saturn conjunctions, zodiacal patterns in royal dynasties, and eclipses. The comet's physical nature and location above or below the Moon therefore was of no interest. In the application of both astronomy and astrology to questions of natural influence, temporal significance or divine providence, the concern was with lights in the sky, irrespective of their distance from the Earth or regularity of motion or visibility.

All that mattered was that it appeared in the sky. He combined different frames of reference using coordinates for the comet's zodiacal longitude—its projection on to the ecliptic—and it's actual observed location, noting he dates of its verticality as astrologically significant. He also relied on both Bainbridge, who observed the comet on 18 November, for his day-for-a-year analysis, and Longomantanus' first observation, 21 November, for his horoscope of that event. Lilly's astrology was at once a matter of great precision in time (the time of events) and space (measurement of astronomical positions), thereby respecting Bainbridge's geometer God, evidence based in that he referred to past precedents, and somewhat deterministic in tone. His text was framed in the same apocalyptic, triumphalist Protestant English nationalism as Bainbridge's, and he made the same appeal to God as the ultimate arbiter of all things. The End Times were near and England was to gain victory. Bainbridge, along with Lilly, made worldly predictions, and both framed their forecasts in a sense of destiny. Both berated the failures of their contemporaries, although Bainbridge took a more moral tone, condemning the weakness of the current generation (including its dependence on tobacco).

The major difference between Bainbridge and Lilly is found in the latter's use of horoscopic astrology. Bainbridge found meaning in celestial signs, but not in the complex interpretative structures of the horoscope. He acknowledged the existence of causality in nature, but considered that this did not extend to the relationship between celestial bodies and human behaviour. Writing of the relative importance of comets as opposed to meteors, he wrote, 'the signification [of comets] also does as much transcend their [meteors'] effects, being not so much a cause (as they may be) of elementary alterations, as a celestiall sign of greater consequents' (Bainbridge, 1975 [1619]: 27–8). Yet Lilly's own primary rationale for astrology was as a set of signs to be read in the Book of Nature, rather than of causes. In the front matter to his 1647 work, *Christian Astrology*, he wrote,

> The Author's God, Composer and the Setter
> Of all his works, and there in every letter.
> Heaven is his Book; the stars both great & final
> Are letters Nonperill and Capitall (Lilly, 1985 [1647]: 'Upon the WORKE').

Bainbridge (1975 [1619]: 8) spat contempt at the casters of horoscopes (even though he did not use the term):

> There be not yet lacking some, who erect Chancels to the blind Goddesse Chance, sacrificing to that abominable Idoll, not without impious contumelie of the omnipotent God & vile contempt of his power and providence in the fabrique and regiment of the world.

In spite of Bainbridge and Lilly's different technical approaches, there is a significant connection in that they both take a hermeneutic approach to the sky, using it to interpret current affairs. From this perspective, the debate is about whether comets are meteorological effects which are of a different order to other astronomical phenomena. They are lights in the sky and therefore carry a meaning to be read along with all other celestial objects. Bainbridge and Lilly followed different paths concerning the use of horoscopic astrology but they both operated within a wider concept cluster (as Peters put it), or set of ideas or propositions (as Curry wrote), which read meaning and significance in the sky, practicing an astronomy which was designed to reveal God's creation and provided insights into politics, morality and national destiny.

Lilly's publication was titled England's *Propheticall Merlin*. This tells you that his main mission was the presentation of geopolitical futures and England's national history in a confessional, salvational framework using celestial events as allegory. Technical horoscopic astrology was secondary to this goal. His use of standard authorities in astrology was balanced by his use of standard authorities in prophetic narratives. In this sense, although his work is ostensibly astrological, there is little or no attempt to justify his political forecasts astrologically. Astrology instead forms a support to his political, salvational and confessional narratives. In terms of Tabitta van Nouhuys' identification of two varieties of astrological prognostication, one 'teratological' and divinatory, the other properly astrological, Lilly crossed both. Scripture, natural philosophy and horoscopic astrology combined in one mode of political-cosmological inquiry.

Beyond the differences between Bainbridge's and Lilly's approaches to interpreting the comet lies a shared concern with the divine gaze. God watched everything and knew everything. Christian astronomy was God's creation and a means to communicate with humanity exactly, as Francesca Rochberg (2004) showed, the Babylonian deities had done through their 'Heavenly Writing'. And consequent action was aways required. In confessional cometary prophecy the required action was repentance, in order to fulfil the promise of salvational history, and for both Bainbridge in a time of peace and Lilly in a time of war, the required action was support of the Protestant cause. Neither Bainbridge nor Lilly subscribed to the disaster narratives favoured by Protestant confessionalisation. Bainbridge preferred to take an optimistic line, while Lilly's list of likely conflicts was derived more from standard astrological literature than from the Book of Revelation and prophecies of the End Times. Bainbridge lived at a time of peace and stability in England and celebrated the triumph of the divinely ordained Protestant monarchy as confirmed by the comet. Lilly lived in 1642 when the order of Bainbridge's England was breaking down as the rival forces, which Elizabeth I and James I had held together, were divided into armed camps. That Bainbridge's reading was optimistic and Lilly's equally pessimistic is partly a feature of the variable reading of cometary signification, but also of the time in which they lived. Lilly's text, written almost twenty-four years after Bainbridge, was one of the last full analyses of cometary significance using the methodologies of horoscopic astrology. In relation to Portugal, Luis Carolino and Carlos Camenietzki (2006: 34) wrote, 'From 1577 to the end of the seventeenth century, the belief and the political influence of comets was replaced by an explanation which took only physical features into account', the inference being that they were stripped of their meaning. Yet physicality did not necessarily deny meaning. The early seventeenth century, after all, was a period when the entire natural world was a manifestation of God's creation, and the notion of a distinction between natural and supernatural was unknown. Lilly's text uses precise astrological data in the form of horoscopes not to make precise predictive statements (his comments on the comet's political significations are so general as to apply to anything at any time) but to provide a structure for his confessionalism, his purpose being to outline a salvational history of England at a time of profound crisis, the terrible civil war of the 1640s, a penitentiary warning to his countrymen to obey God's law. His work was closer to the sermons of Elias Ehinger and Conrad Dietrich, than to a reasoned astrological analysis.

Almost forty years after Lilly's *Propheticall Merline*, Isaac Newton contributed to a major revision of the general view of comets, maintaining their role as divine signs, but demonstrating their role as astronomical phenomena free from any associations with horoscopic astrology (Schaffer, 1987: 219–43). As Edmond Halley (cited in Olson & Pasachoff, 1998: 1) wrote in his dedicatory ode to Newton's *Principia*,

> Now we know.
> The sharply veering ways of comets, once.
> A source of dread, no longer, do we quail
> Beneath the appearance of bearded stars.

Acknowledgments I am grateful to Georg Zotti, Stellarium co-developer, for his assistance with Stellarium and Deborah Houlding for advice on William Lilly's use of day-for-a-year forecasting.

Bibliography

Aristotle. (1937). *Meteorologica*. English translation: Lee, H.D.P. Harvard University Press.

Baily, M. E., Clube, S. V. M., & Napier, W. M. (1990). *The origin of comets*. Pergamum Press.

Bainbridge, J. (1975). Published in London 1619: *An Astronomicall description of the late comet*. Walter J. Johnson. https://archive.org/details/bwb_KR-519-598/page/n5/mode/2up

Bayeux Museum. (n.d.). Explore the Bayeux Tapestry Online, Panel 32. https://www.bayeuxmuseum.com/en/the-bayeux-tapestry/discover-the-bayeux-tapestry/explore-online/

Bede. (1968). *A history of the English church and people*. Penguin.

Bede. (1990). *Ecclesiastical history of the English people*. English translation: Farmer, D.H. Penguin.

Bonatti, G. (2007). *Liber Astronomiae* (2 vols). English translation: Dykes, Benjamin. The Cazimi Press.

Boner, P. J. (2006). Kepler on the origins of comets: Applying earthly knowledge to celestial events. *Nuncius, 21*(1), 31–47. https://brill.com/view/journals/nun/21/1/article-p31_2.xml?rskey=fzwxzy&result=

Boner, P. J., & Granada, M. A. (2022). *Michael Maestlin's manuscript treatise on the comet of 1618: An edition and translation of Manuscript WLB Stuttgart, Cod. Math. 4 15b, Nr. 8*. Brill.

Briggs, K. M. (Ed.) (1974). *The last of the astrologers: Mr William Lilly's history of his life and times from the year 1602 to 1681*.

Campion, N. (2009). *A history of western astrology* (The Medieval and Modern Worlds) (Vol. 2). Bloomsbury.

Campion, N. (2015). *The new age in the modern west: Counter-culture, utopia and prophecy from the late eighteenth century to the present day*. Bloomsbury.

Campion, N. (2022). Belief. In G. D. Chryssides & A. R. Whitehead (Eds.), *Contested concepts in the study of religion* (pp. 9–13). Routledge.

Campion, N., & Kollerstrom, N. (2004/2003). Galileo's astrology. Cinnabar Books/*Culture and Cosmos 7*(1).

Carolino, L. M. (2012). Manuel Boccaro Francês, the Comet of 1618, and the impact of stoic cosmology in Portugal. In M. A. Granada (Ed.), *Novas y cometas entre 1572 y 1618: Revolucion cosmólogicay Renovación polítíca y religiosa* (pp. 195–224). Universidad de Barcaleina.

Carolino, L. M., & Camenietzki, C. Z. (2006). Tokens of the future: Comets, astrology and politics in early modern Portugal. *Cronos, 2*, 34–57.

Christianson, J. R. (1979). Tycho Brahe's German treatise on the comet of 1577. *Isis, 70*(1), 110–140.

Claussen, E. (2021). *Politics and 'Politiques' in sixteenth-century France: A conceptual history*. Cambridge University Press.

Clegg, C. S. (2008). *Press censorship in Caroline England*. Cambridge University Press.

Corbett, R. (2007). Poems, 65. Cited in James Doelman, The comet of 1618 and the British royal family. *Notes and Queries, 54*(1), 30–35.

Coward, B., & Gaunt, P. (2017). *The Stuart age: England 1603–1714* (5th ed.). Routledge.

Curry, P. (1989). *Prophecy and power: Astrology in early modern England*. Polity Press.

Curry, P. (1991). Astrology in early modern England: The making of a vulgar discipline. In S. Pumfrey, P. L. Rossi, & M. Slawinski (Eds.), *Science, culture and popular belief in renaissance Europe* (pp. 274–291). Manchester University Press.

Curry, P. (1999). Astrology. In K. Boyd (Ed.), *The encyclopaedia of historians and historical writing*, 2 Vols (Vol. 1, pp. 55–57). Fitzroy Dearborn.

Curry, P. (2024/1989). Prophecy and power*: Astrology in early modern England*. Sophia Centre Press.

Defoe, D. (2003/1722). *A journal of the plague year*. Penguin.

Dio Cassius. (1914–1927). *Roman history* (9 vols). English translation: Cary, E. Harvard University Press.

Doelman, J. (2007). The comet of 1618 and the British royal family. *Notes and Queries, 54*(1), 30–35. https://doi.org/10.1093/notesj/gjm009

Drake, S., & O'Malley, C. D. (trans.). (2016). *The controversy on the comets of 1618: Galileo Galilei, Horatio Grassi, Mario Guiducci, Johann Kepler*. University of Pennsylvania Press.

Eastwood, B. S. (2007). *Ordering the heavens: Roman astronomy and cosmology in the Carolingian renaissance*. Brill.

Ernst, G. (1986). From the watery Trigon to the fiery Trigon: Celestial signs, prophecies and history. In P. Zambelli (Ed.), *'Astrologi hallucinati': Stars and the end of the world in Luther's time* (pp. 265–280). Walter de Gruyter.

Ficino, M. (1989). Three books on life. In C. C. Kaske & J. R. Clark (Eds.), *Center for medieval and early renaissance studies*. State University of New York at Binghamton.

Galilei, G. (2016). *The Assayer*. English translation: Drake, S. In *The controversy on the comets of 1618: Galileo Galilei, Horatio Grassi, Mario Guiducci, Johann Kepler* (S. Drake & C.D. O'Malley, Trans) (pp. 151–336). University of Pennsylvania Press.

Geneva, A. (1995). *Astrology and the seventeenth century mind: William Lilly and the language of the star*. Manchester University Press.

Genuth, S. S. (1992). Comets and their malevolence. Review of the origin of comets, by M. E. Bailey, S. V. M. Clube, and W. M. Napier. *Journal for the History of Astronomy, 23*, 215–217.

Genuth, S. S. (1997). *Comets, popular culture and the birth of modern cosmology*. Princeton University Press.

Grassi, H. (2016/1619). On the three comets of 1618', Rome. In *The controversy on the comets of 1618: Galileo Galilei, Horatio Grassi, Mario Guiducci, Johann Kepler* (S. Drake & C.D. O'Malley, Trans) (pp. 3–19). University of Pennsylvania Press.

Green, W. (1680/1681). *Memento's to the world; Or, an historical collection of divers wonderful comets and prodigious signs in heaven.*

Green, W. (1997 [1680/1681]). *Memento's to the world; Or, an historical collection of divers wonderful comets and prodigious signs in heaven*. In Sara Schechner Genuth, *Comets, popular culture and the birth of modern cosmology* (p. 51) Princeton University Press.

Greenbaum, D. G. (Ed.). (2010). Kepler's astrology. *Culture and Cosmos, 12* (1–2).

Gruber, D. (2021). Knowledge in transition early modern comet apparitions and the printing press in the Holy Roman Empire. In T. Wallnig, M. Romberg, & J. Lahner (Eds.), *Kirche und Klöster zwischen Aufklärung und administrativen Reformen* (pp. 163–178). https://www.vr-elibrary.de/doi/10.7767/9783205214069

Haigh, C. (1993). *English reformations: Religion, politics and society under the Tudors*. Oxford University Press.

Hart, V. (1994). *Art and magic in the courts of the Stuarts*. Routledge.

Heidarzadeh, T. (2008). *A history of physical theories of comets from Aristotle to Whipple*. Springer.

Heilbron, J. (2021). *The ghost of Galileo: In a forgotten painting from the English Civil War*. Oxford University Press.

Heydon, C. (1650). *An astrological discourse with mathematical demonstrations, proving the powerful and harmonical influence of the planets and fixed stars upon elementary bodies, in Justification of the validity of astrology together with an astrological judgment upon the great conjunction of Saturn & Jupiter 1603.*

Isidore of Seville. (2007). *The etymologies*. English translation: Barney, S. A., Lewis, W. J., Beach, J. A., & Berghof, O. Cambridge University Press.

Jerratsch, A. (2019). Celestial phenomena in early modernity: The integrated image of comets. In P. D. Omodeo & V. Wels (Eds.), *Natural knowledge and Aristotelianism at early modern Protestant Universities* (pp. 187–208). Harrassowitz Verlag.

Jorink, E. (2010). *Reading the book of nature in the Dutch Golden Age, 1575–1715*. Brill.

Josephus. (1981/1950). *The Jewish War* (Revised edn). English translation: Williamson, G.A. Penguin.

Kepler, J. (1979/1601). *On the more certain fundamentals of astrology*, Prague. English translation: Rossi, Mary with notes by J. Bruce Brackenbridge. *Proceedings of the American Philosophical Society, 123*(2), 85–163.

Kepler, J. (1997). *The harmony of the world*. English translation: Aiton, E. J., Duncan, A. M., & Field, J. V. American Philosophical Society.

Kepler, J. (2008). *Kepler's astrology. The baby, the bath water and the third man in the middle: The first complete English translation of Tertius Interveniens and other astrological writings*. English translation: Negus, K. Earth Heart Publications.

Kepler, J. (2016). Appendix to the Hyperapistes. In *The controversy on the comets of 1618: Galileo Galilei, Horatio Grassi, Mario Guiducci, Johann Kepler* (S. Drake & C.D. O'Malley, Trans) (pp. 341–355). University of Pennsylvania Press.

Lilly, W. (1644). *England's Propheticall Merlin, foretelling to all nations of Europe until 1663.*

Lilly, W. (1715). *William Lilly's history of his life and times from the year 1602 to 1681*. http://www.gutenberg.org/dirs/1/5/8/3/15835/15835.txt.

Lilly, W. (1985/1647). *Christian astrology* (Facsimile edn). Regulus Publishing.

Louthan, H., & Zachman, R. (Eds.). (2004). *Conciliation and confession: The struggle for Unity in the age of reform, 1415–1648*. Notre Dame University Press.

Mahlmann-Bauer, B. (2019). Attacks on judicial astrology, religious dissent, and the rise of skepticism. In P. D. Omodeo & V. Wels (Eds.), *Natural knowledge and Aristotelianism at early modern Protestant universities* (pp. 123–186). Harrassowitz Verlag.

Manilius, M. (1977). *Astronomica*. English translation: Goold, G. P. Harvard University Press.

Matthiesen, T. (2023). *The Caliph and the Imam: The making of Sunnism and Shiism*. Oxford University Press.

Morrissey, M. (2014). Confessionalism. In *OED Online*. Oxford University Press. https://www.oed.com/search/dictionary/?scope=Entries&q=confessionalism

Morrissey, M. (2015). Confessionalism and conversion in the reformation. *Oxford Handbook Topics in Literature*. Oxford Academic. https://doi.org/10.1093/oxfordhb/9780199935338.013.73

Munk, W. (n.d.). *John Bainbridge*. The Royal College of Physicians. https://history.rcplondon.ac.uk/inspiring-physicians/john-bainbridge

Olson, R. J. M., & Pasachoff, J. M. (1998). *Fire in the sky: Comets and meteors, the decisive centuries in British art and science. Cambridge University Press.*

Otto of Freising. (1979). The two cities 6. 35–36. Cited in Bernard McGinn, *Visions of the end: Apocalyptic traditions in the middle ages* (pp. 98–99). Columbia University Press.

Ovid. (2004). *Metamorphoses*. English translation: Raeburn, D. Penguin.

Parker, D. (1975). *Familiar to all: William Lilly and astrology in the seventeenth century*. Jonathan Cape.

Peters, R. H. (1991). *A critique for ecology*. Cambridge University Press.

Pliny. (1929). *Natural history* (Vol. 1, Books I-II). English translation: Rackham, H. Harvard University Press.

Ptolemy, C. (1940). *Tetrabiblos*. English translation: Robbins, F.E. Harvard University Press.

Ramsey, J. T., Licht, A., & Lewis, A. (1997). *Caesar's comet: The comet of 44 BC and Caesar's funeral games*. Scholars' Press.

Rochberg, F. (2004). *The heavenly writing: Divination and horoscopy, and astronomy in Mesopotamian culture*. Cambridge University Press.

Rutkin, D. H. (2015). Understanding the history of astrology (and magic) accurately: Methodological reflections in terminology and anachronism. *Philosophical Readings, VII(1)*, 42–54.

Schaffer, S. (1987). Newton's comets and the transformation of astrology. In P. Curry (Ed.), *Astrology, science and society* (pp. 219–243). Boydell Press.

Seneca, L. A. (2010). *Natural questions*. English translation: Hine, H. M. University of Chicago Press.

Snell, W. (2021/1619). *Descriptio cometae qui anno 1618 mense novembri primum effulsit* (p. 67). (Leyden, Elzviriana, 1619). Cited in John Heilbron, *The Ghost of Galileo*: *In a forgotten painting from the English Civil War* (p. 202). Oxford University Press.

Tessicini, D. (2013). The comet of 1577 in Italy: Astrological prognostications and cometary theory at the end of the sixteenth century. In D. Tessicini & J. B. Patrick (Eds.), *Celestial novelties on the eve of the scientific revolution 1540 to 1630* (pp. 57–84). Olschki Editore.

Theibault, J. (1994). Jeremiah in the village: Prophecy, preaching, pamphlets, and penance in the thirty years' war. *Central European History, 27*(4), 441–460.

Tschopp, S. (1994). Cited in John Theibault, Jeremiah in the village: Prophecy, preaching, pamphlets, and penance in the thirty years' war. *Central European History, 27*(4), 441–460.

Turnbull, E. C. (2020). Defending the Spanish match in early Stuart England. *The Bulletin of the Institute of Historical Research, 93*(262), 638–660.

Urbánek, V. (2002). The comet of 1618: Eschatological expectations and political prognostications during the bohemian revolt. In J. R. Christianson, A. Hadravova, P. Hadrava, & M. Šolc (Eds.), *Tycho Brahe and Prague: Crossroads of European science* (pp. 282–291). Verlag Harri Deutsch.

Uri, J. (2021). 955 Years ago: Halley's comet and the Battle of Hastings. *NASA*. https://www.nasa.gov/history/955-years-ago-halleys-comet-and-the-battle-of-hastings/

Valens, V. (2010). *Anthologies*. English translation: Riley, M. http://www.csus.edu/indiv/r/rileymt/vettius%20valens%20entire.pdf

van Nouhuys, T. (1998). *The age of two-faced Janus: The comets of 1577 and 1618 and the decline of the Aristotelian world view in the Netherlands*. Brill.

Vermij, R., & Vieth, P. (2022). Confessionalization and comets. John Bainbridge on the comet of 1618. *Annals of Science, 79*(3), 275–291.

Weeratunge, N. (2000). Nature, harmony, and the *Kaliyugaya*: Global/local discourses on the human-environment relationship. *Current Anthropology, 41*(2), 249–268.

Williams, A. (2016). *Christian socialism as a political ideology*. Doctoral thesis, University of Liverpool.

Yates, F. (1985/1973). *Astraea: The Imperial theme in the sixteenth century*. Routledge and Kegan Paul.

Yenter, T. (2018). Harmony in Spinoza and his critics. In B. Lord (Ed.), *Spinoza's philosophy of ratio* (pp. 46–60). Edinburgh University Press.

Yeomans, D. K. (1991). *Comets: A chronological history of observation, science, myth and folklore*. John Wiley and Sons.

Zambelli, P. (1986). Introduction: Astrologers' theory of history. In P. Zambelli (Ed.), *'Astrologi hallucinati': Stars and the end of the world in Luther's time* (pp. 1–43). Walter de Gruyter.

Nicholas Campion is an Associate Professor in Cosmology and Culture, a Principal Lecturer in the Institute of Education and Humanities, and the Director of the Sophia Centre for the Study of Cosmology in Culture at the University of Wales Trinity Saint David. He has played a leading role in defining the scope, nature, and character of the emerging discipline of cultural astronomy over the last 25 years. In 2022, he was awarded the Carlos Jaschek Award for outstanding contributions to cultural astronomy by the European Society for the Study of Astronomy in Culture. His books include the two-volume *History of Western Astrology* (London: Bloomsbury 2008/9) and *Astrology and Cosmology in the World's Religions* (New York: New York University Press, 2012). He is the editor of *Culture and Cosmos*, the journal on the history of cultural astronomy and astrology. His current projects include the six-volume *Cultural History of the Universe* (Bloomsbury, forthcoming), for which he is the General Editor.

A. A Digital Library on the Controversy on the Comets, 1618–1626[1]

Stefano Casati, Adele Pocci, and Giancarlo Truffa

A.1. Introduction

The three comets that appeared between August 1618 and January of the following year gave rise to heated disputes regarding the nature and origin of these suggestive celestial phenomena, in the context of a debate that involved some of the major natural philosophers, the mathematicians, and astronomers of the time. Moreover, the physical constitution of cometary events, their location, and their motion were topics that had been at the center of scientific discussion for a long time, but still remained without a definitive answer.

A.2. Orazio Grassi's Pamphlet

Galileo was one of the great protagonists of the debate, in which he was called to participate in 1619 by various solicitations. In particular, the publication, in February of that year in Rome, of a pamphlet, *De tribus cometis anni MDCXVIII...*, which appeared anonymous, but easily attributable to Orazio Grassi, professor of

[1]This is an English translation of a paper originally published in Italian as Casati, Stefano, Adele Pocci, Giancarlo Truffa. "Una biblioteca digitale dedicata alla disputa sulle comete, 1618–1626", *Galilæana* XXI, 1 (2024): 273–280; https://doi.org/10.57617/gal-33

S. Casati · A. Pocci
Museo Galileo, Florence, Italy
e-mail: s.casati@museogalileo.it; a.pocci@museogalileo.it

G. Truffa
SISFA Società Italiana degli Storici della Fisica e dell'Astronomia, Padua, Italy

C. J. Cunningham (ed.), *Cosmic Events*, Historical & Cultural Astronomy,
https://doi.org/10.1007/978-3-031-86950-1

Fig. A.1 Andreas Cellarius (1596–1665), *Atlas coelestis seu Harmonia Macrocosmica*, Amsterdam, 1660 (Florence, National Central Library, Magl. 5._.81, pl. 7). Visualization of the geo-heliocentric system created by Tycho Brahe. The Earth is motionless in a central position. The Moon and the Sun rotate around it, while the other planets move around the Sun

mathematics at the Collegio Romano in Rome, the main school of the Jesuits, contributed to motivating his intervention. In the work, the Jesuit essentially supported Tycho Brahe's geo-heliocentric cosmological model, a system that kept the Earth immobile at the center of the universe and placed the Sun, orbiting around it, at the center of the orbits of the planets (Fig. A.1). This theory agreed with the new observational data and was also considered acceptable by the Church because it was compatible with the dictates of the Holy Scripture.

The publication of Grassi's pamphlet triggered an intense dispute between the Jesuit and the Pisan scientist. In June 1619, Galileo refuted his theses in the *Discourse on the Comets*, a text almost entirely written by him, but published under the name of his student Mario Guiducci. Grassi responded in October of the same year by publishing *Libra astronomica* under the pseudonym of Lotario Sarsi. Guiducci responded to this text in 1620 with the *Letter to the m.r.p. Tarquino Galluzzi of the Society of Jesus*, while Galileo, urged by the Linceans, developed a detailed point-by-point refutation of Grassi's work, concluded in 1621, but published only 2 years later. *The Assayer*—this is the title assigned to the work—was published in Rome on 21 October 1623, under the aegis of the Lincei who decided it should be dedicated to Maffeo Barberini, who had just ascended to the papal throne with the name of Urban VIII (Fig. A.2). The work, a meticulous critical commentary on Grassi's *Libra*, is notable for its stringent arguments, methodological findings, and the admirable quality of the

Fig. A.2 *Il saggiatore: nel quale con bilancia esquisita e giusta si ponderano le cose contenute nella Libra astronomica e filosofica di Lotario Sarsi ... scritto in forma di lettera all'ill.mo ... Virginio Cesarini ... dal sig. Galileo Galilei ...* [*The assayer: in whom the things contained in the astronomical and philosophical Libra of Lotario Sarsi are weighed with an exquisite and fair balance ... written in the form of a letter to the most illustrious... Virginio Cesarini ... by Mr. Galileo Galilei ...*]. (in Rome: after Giacomo Mascardi, 1623) (Florence, Museo Galileo, MED 1298)

prose, constituting, in all respects, one of the fundamental texts of modern science.

At the end of the year 1626, Grassi's response, always hidden under the pseudonym of Lotario Sarsi, the *Ratio ponderum librae et simbellae*, came to light. Galileo, now busy drafting the *Dialogue Concerning the Two Chief Systems*, did not deem it appropriate to reply.

A.3. Cometary Illustrations

The extraordinary appearance of comets in 1618 provoked the publication of many other writings about their motion in the sky, their nature, and their astrological influences. Volker Fritz Brüning, in his *Bibliographie der Kometenliteratur*, lists over 200 titles. Some of these texts were cited directly by the protagonists of the dispute, others we can only assume were known.

Many treatises were enriched with illustrations, which together constitute an interesting iconographic apparatus, which has so far been little investigated. In conjunction with the spread of the printed book, a true flowering of astronomical iconography had already occurred during the sixteenth century, in its double role as a teaching aid and a scientific resource aimed directly at deepening astronomical knowledge (Fig. A.3).

The representation of comets, therefore, varies depending on the purpose of the image: graphs and maps for the identification, description, and measurement of the cometary path; simple drawings of objects with star-shaped bodies and tails (in accordance with the classic representation of the phenomenon); natural panoramas or views of cities dominated by the celestial event, often with groups of people intent on observing it; and other varied representations of a phenomenon that surprised, terrified, and in each case aroused intense curiosity and a lively desire to understand (Fig. A.4).

Cometary images, however, were not limited to printed texts alone, but were also delivered on celestial globes, coins, medals, and even to the rete of an astrolabe (Fig. A.5).

The Digital Library: *The dispute over the comets, 1618–1626* (https://bibdigtematiche.museogalileo.it/Comete/; Fig. A.6), created with the aim of documenting the important debate just mentioned, was published on the occasion of the exhibition *Celestial Splendors, the observation of the universe from Galileo to the telescope* (Florence, former dormitory of Santa Maria Novella, 16 December 2023–17 March 2024).

The digital library, designed and created by Stefano Casati, Adele Pocci, Giancarlo Truffa, with the collaboration of numerous scholars, allows to consult the resources through various search and navigation methods. The *Esplora* [*Explore*] function offers access to multimedia contents, organized in three thematic environments, through a hierarchical index ordered alphabetically by author or title. The texts section includes around 20 treatises directly cited or presumably known by the protagonists of the dispute and provides descriptive and biographical information on the works and authors; the *Iconografia* [*Iconography*] resource contains approximately 80 digital images indexed and searchable by subject; while, the section entitled *La disputa* [*The controversy*] presents a historical-thematic path with precise references to the main sources of the debate.

The navigation mode *Percorsi tematici* [*Thematic Paths*] offers further investigation possibilities, such as semantic searches through keywords and queries by type

Fig. A.3 Casparus Hersbach, *Eigentliche verzeichnus dieses im jahr 1618 erscheinenden cometen, welcher lauss oder bewegung, duirch die zeichen des 8 himmels, alhie vor augen gestellet wirdt*. [*Actual description of the comet which appeared in the year 1618, the location or movement of which is presented here by the signs of the eighth heaven*] (Cologne 1618). (London, © The Trustees of the British Museum)

of material, language, and format. Conceptual maps allow to reconstruct the relationships between the different entities in which the collections are structured and, through the expansion of the "nodes", offer the possibility of consulting the digital resources starting directly from the graphic representation.

Fig. A.4 Anonymous, *De verschijning van een komeet in november 1618* [*The appearance of a comet in November 1618*]. (Image in the public domain)

Fig. A.5 Representation of the comet on the rete of an astrolabe. Unsigned, possibly made in Germany, dated “Anno Domini 1620”, Chicago, Adler Planetarium, M-34

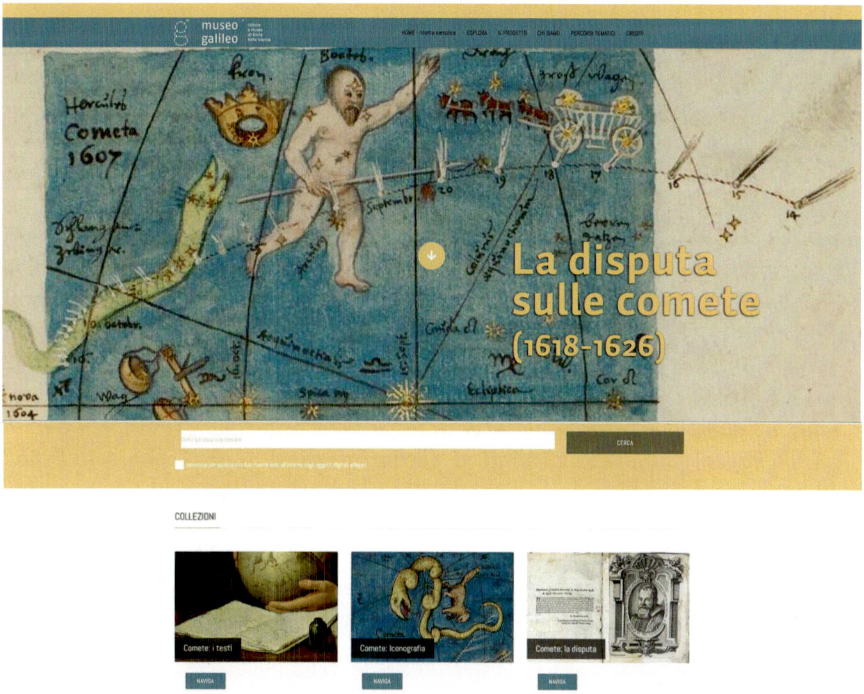

Fig. A.6 Home page of the thematic digital library

The *digital library* presented here, in addition to organizing the contents into structured paths, also constitutes an "open" research environment, susceptible to constant and progressive updates, entrusted to the direct contribution of scholars. It is also able to integrate composite resources into a single space, usable starting from a single access by virtue of the syntactic and semantic interoperability of the data. The *digital library* thus becomes a "node", a platform aggregating and dispensing heterogeneous knowledge, but easily and almost simultaneously available.

Primary Sources

Brüning, V. F. (2000). *Bibliographie der Kometenliteratur* [Bibliography of the literature on comets]. A. Hiersemann.

Galilei, G. (1623). *Il saggiatore: nel quale con bilancia esquisita e giusta si ponderano le cose contenute nella Libra astronomica e filosofica di Lotario Sarsi ...* [*The assayer: in whom the things contained in the astronomical and philosophical Libra of Lotario Sarsi are weighed with an exquisite and fair balance...*]. appresso Giacomo Mascardi.

Galileo, G. (1632). *Dialogo di Galileo Galilei ... doue ne i congressi di quattro giornate si discorre sopra i due massimi sistemi del mondo tolemaico e copernicano proponendo indeterminatamente le ragioni filosofiche e naturali tanto per l'vna, quanto per l'altra parte* [*Dialogue of Galileo Galilei .. where in the meetings of four days we talk concerning the two chief world systems, Ptolemaic and Copernican, indeterminately propounding philosophical & natural reasons, as well for the one part as the other*]. per Gio. Batista Landini.

Grassi, O. (1619a). *De tribus cometis anni MDCXVIII disputatio astronomica publice habita in collegio Romano Societatis Iesu ab uno ex patribus ejusdem Societatis* [*On the Three Comets of the Year 1618 An Astronomical Disputation Presented Publicly in the Collegio Romano of the Society of Jesus by one of the Fathers of that same Society*]. ex Typ. Iacobi Mascardi.

Grassi, O. (1619b). *Libra astronomica ac philosophica qua Galilaei Galilaei opiniones de cometis a Mario Guiducio in Florentina academia expositae atque in lucem nuper editae examinantur a Lothario Sarsio* [*The Astronomical and Philosophical Balance on which the Opinions of Galileo Galilei regarding Comets, presented in the Florentine Academy by Mario Guiducci and recently published, are weighed by Lothario Sarsi*]. ex Typographia Marci Naccarini.

Grassi, O. (1626). *Ratio ponderum librae et simbellae* [*A Reckoning of Weights for the Balance and the Small Scale*]. sumptibus Sebastiani Cramoisy.

Guiducci, M. (1619). *Discorso delle comete* [*Discourse on the comets*]. nella stamperia di Pietro Cecconcelli.

Guiducci, M. (1620). *Lettera al m.r.p. Tarquino Galluzzi della Compagnia di Gesù di Mario Guiducci: nella quale si giustifica dalle imputazioni dategli da Lotario Sarsi ...* [*Letter to the very Reverend Father Tarquinio Galluzzi of the Society of Jesus by Mario Guiducci in which he answers to the imputations made against him by Lothario Sarsi ...*]. nella Stamperia di Zanobi Pignoni.

GPSR Compliance

The European Union's (EU) General Product Safety Regulation (GPSR) is a set of rules that requires consumer products to be safe and our obligations to ensure this.

If you have any concerns about our products, you can contact us on ProductSafety@springernature.com

In case Publisher is established outside the EU, the EU authorized representative is:

Springer Nature Customer Service Center GmbH
Europaplatz 3
69115 Heidelberg, Germany

Batch number: 10756376

Printed by Printforce, the Netherlands